Lectures on Electricity

Comprising Galvanism, Magnetism,
Electro-Magnetism, Magento- and Thermo-Electricity

HENRY MINCHIN NOAD

CAMBRIDGE
UNIVERSITY PRESS

CAMBRIDGE UNIVERSITY PRESS

Cambridge, New York, Melbourne, Madrid, Cape Town,
Singapore, São Paolo, Delhi, Mexico City

Published in the United States of America by Cambridge University Press, New York

www.cambridge.org
Information on this title: www.cambridge.org/9781108052160

© in this compilation Cambridge University Press 2012

This edition first published 1844
This digitally printed version 2012

ISBN 978-1-108-05216-0 Paperback

CAMBRIDGE LIBRARY COLLECTION

Books of enduring scholarly value

Technology

The focus of this series is engineering, broadly construed. It covers technological innovation from a range of periods and cultures, but centres on the
technological achievements of the industrial era in the West, particularly
in the nineteenth century, as understood by their contemporaries. Infra-
structure is one major focus, covering the building of railways and canals,
bridges and tunnels, land drainage, the laying of submarine cables,
and the construction of docks and lighthouses. Other key topics include
developments in industrial and manufacturing fields such as mining
technology, the production of iron and steel, the use of steam power,
and chemical processes such as photography and textile dyes.

Lectures on Electricity

During the early nineteenth-century craze for conducting kite experiments in
lightning, deaths were not unheard of. Electrical physicists, meanwhile, were
often shocked badly enough to collapse in the course of their work. However,
the perils of electricity did not deter its proponents. Published in 1844, this
enlarged collection of lectures by Henry Minchin Noad (1815–77) had proven
immensely popular in earlier incarnations, eventually running to four editions
and recognised as an invaluable textbook for electricians and telegraph
engineers until the turn of the century. An electrical practitioner himself,
Noad includes illustrated explanations of some of the most significant ideas in
the field, and describes many of his own experiments, from his version of the
lightning kite to a battery constructed with fifty jars and a thousand feet
of wire. His work remains relevant to students in the history of science.

Cambridge University Press has long been a pioneer in the reissuing of out-of-print titles from its own backlist, producing digital reprints of books that are still sought after by scholars and students but could not be reprinted economically using traditional technology. The Cambridge Library Collection extends this activity to a wider range of books which are still of importance to researchers and professionals, either for the source material they contain, or as landmarks in the history of their academic discipline.

Drawing from the world-renowned collections in the Cambridge University Library and other partner libraries, and guided by the advice of experts in each subject area, Cambridge University Press is using state-of-the-art scanning machines in its own Printing House to capture the content of each book selected for inclusion. The files are processed to give a consistently clear, crisp image, and the books finished to the high quality standard for which the Press is recognised around the world. The latest print-on-demand technology ensures that the books will remain available indefinitely, and that orders for single or multiple copies can quickly be supplied.

The Cambridge Library Collection brings back to life books of enduring scholarly value (including out-of-copyright works originally issued by other publishers) across a wide range of disciplines in the humanities and social sciences and in science and technology.

NOADS LECTURES ON ELECTRICITY

London: Published by George Knight and Sons, Foster Lane.

LECTURES

ON

ELECTRICITY,

COMPRISING

GALVANISM, MAGNETISM, ELECTRO-MAGNETISM, MAGNETO- AND THERMO-ELECTRICITY.

BY

HENRY M. NOAD,

AUTHOR OF "LECTURES ON CHEMISTRY," ETC.

A NEW AND GREATLY ENLARGED EDITION.

Illustrated by nearly Three Hundred Wood Cuts.

LONDON:

GEORGE KNIGHT AND SONS, FOSTER LANE.

1844.

London :
Printed by STEWART and MURRAY,
Old Bailey.

TO

ANDREW CROSSE, ESQ.

OF BROOMFIELD, NEAR TAUNTON, SOMERSETSHIRE,

TO WHOSE INDEFATIGABLE INDUSTRY FOR A LONG PERIOD

OF YEARS,

ELECTRICAL SCIENCE

IS INDEBTED FOR SO RICH AN ACCUMULATION OF VALUABLE FACTS:

THE INTERESTING RESULTS OF WHOSE

ELECTRO-CHEMICAL RESEARCHES

HAVE TAUGHT US THE VALUE OF PATIENT ENQUIRY:

AND WHOSE

LIBERAL, OPEN, AND COMMUNICATIVE SPIRIT

IS NOT LESS REMARKABLE THAN HIS ENTHUSIASTIC LOVE OF SCIENCE,

These Lectures

ARE INSCRIBED,

IN TESTIMONY OF THE SINCERE ESTEEM OF

THE AUTHOR.

PREFACE.

A NEW EDITION of my Lectures on Electricity having been called for, I have devoted a considerable portion of my leisure to the preparation of a publication which should, in some degree, present a popular view of the present state of this most interesting science. So rapid, however, has been the progress of Electricity during the last few years, and so great has been the accumulation of valuable facts, that on looking over the former edition, with a view to the compilation of the present, I found that the general plan of the work was far too limited to enable me to do anything like justice either to the science or to the reader; I have, therefore, so completely revised it, that the present volume may be considered a new work, bearing but little resemblance to the last, except perhaps in its popular character. The quantity of matter has been considerably increased; but although the present series of lectures is more than twice as comprehensive as the former, no one can be more alive than myself to the fact, that it is far from giving a *complete* view of the present state of electrical science. Becquerel's Treatise on Electricity extends through eight volumes, and can any one say that even *he* has exhausted the subject? From so rich and ample a store, the selection of a series of topics, which should embrace those points of electrical science which might most please the general reader has been by no means an easy task, nor do I pretend to have been completely successful: some will, I fear, complain that I have omitted many facts that ought to have been recorded; but few, I believe, will say that I have inserted anything *devoid of interest*. Had it seemed advisable to have added a second volume I should have performed my task with greater ease to myself, and perhaps with more satisfaction to some of my readers. But such an extension of the work would have injured its usefulness, by necessarily increasing the price; and it would have destroyed the character with which it has been my wish that the work should go forth, viz. : that of giving a *succinct* account of some of the most important facts relating to electrical science. For the convenience of reference I have followed the plan which I am glad to see is becoming more general with writers on scientific subjects, that of arranging the lectures in *paragraphs;* experience has taught me to appreciate the value of these helps, and I trust they will ere long be *universally* adopted.

In order to keep pace, in some degree, with the progress of the science, I have introduced in an Appendix, a condensed account of some of the most important electrical papers that have appeared during the progress of this work through the press, though too late to occupy their proper places in the preceding pages; amongst these I wish especially to allude to a paper on the " Gaseous Voltaic Battery," by Professor Grove ; also to an extract from a valuable essay on "Thunder-Storms," by William Snow Harris, Esq., to whom I take this opportunity of offering my thanks for many valuable suggestions, and for the obliging manner in which he has on several occasions favoured me with his assistance.

To the last number of the Electrical Magazine (a new and valuable periodical, conducted by Mr. Charles V. Walker,) I am indebted for the description of the extraordinary electrical machine which occupies a conspicuous place in the Frontispiece, a drawing of which the Directors of the Polytechnic kindly permitted to be taken from the machine now exhibiting at the Institution.

Those who are in possession of the former edition will perceive that the number of woodcuts in the present volume is nearly *trebled,* which circumstance, together with many unavoidable delays during the progress of the work through the press will, I trust, be received as an excuse for its somewhat tardy appearance. It is, however, at last finished; and I submit it to the public, with the hope that it may receive at their hands, the same favour, which was enjoyed by its predecessor.

Shawford, 1843.

CONTENTS.

~~~~~~~~

The following is a list of the Works to which the Author has been principally indebted in compiling these Lectures.

Faraday's Experimental Researches in Electricity.
Becquerel's Traité de l'Electricité et du Magnétisme.
Daniell's Introduction to the Study of Chemical Philosophy.
Treatises on Electricity, &c., in the Library of Useful Knowledge.
Lardner's Manual of Electricity.
The London, Edinburgh, and Dublin Philosophical Magazine.
Taylor's Scientific Memoirs.
The Philosophical Transactions.
The Encyclopædia Britannica.
Sturgeon's Annals of Electricity, &c.
Golding Bird's Elements of Natural Philosophy.
Proceedings of the London Electrical Society.
Roget's Treatise on Electro-Magnetism.
The Mechanic's Magazine.
Brewster's Treatise on Magnetism.
Cumming's Manual of Electro-dynamics.
Harris's Essay on Thunderstorms.
Singer's Electricity.
Cuthbertson's Electricity.
Graham's Elements of Chemistry.
Kane's Elements of Chemistry.
Turner's Elements of Chemistry.
Brande's Manual of Chemistry.
Brande's Dictionary of Science, Literature, and Art.

The Apparatus and Materials described in these Lectures may be obtained of the Publishers, George Knight and Sons, Foster Lane.

LECTURES

ON

ELECTRICITY, GALVANISM, &c.

~~~~~~~~~~~

LECTURE I.

ELECTRICITY.

Historical Sketch—Primary phenomena— Electric light — Conductors and non-conductors—The terms Vitreous and Resinous, and Positive and Negative—One kind of Electricity cannot be produced without the other — Electroscopes — Electricity set in motion by every form of mechanical change—Attraction and repulsion — Coulomb's electrometer — Distribution of Electricity — Intensity of Electricity distributed on the surfaces of bodies dependent on their forms — Induction — The gold leaf electroscope— Electrophorus— Faraday's experiments and new views of the nature of inductive action — Application thereof to the explanation of electrical phenomena—Theory of the condenser—Practical demonstration of the mechanism of inductive action — Electrical machines — —Amalgam — Theory of the action of the electrical machine — Experiments illustrating the principle and action of the electrical machine.

(1) THERE is perhaps no branch of experimental philosophy which is received by persons of all ages with greater pleasure than Electricity. The reasons are obvious. It is the science susceptible of the most familiar demonstration, and its phenomena, from the striking and ocular manner in which they are presented, are calculated to arrest the attention and become fixed on the mind more powerfully than those of any other science. To this may be added its connexion with the most sublime and awful of the agencies of nature; its secret and hidden influence in promoting at one time the decomposition of bodies, and at another time their re-formation; at one time, in its current form causing the elements of water to separate, and exhibiting them in the form of gases; and at another time, in its condensed form causing these same gases to re-unite, and become again identified with water; now, in its

B

current form, exhibiting the most wonderful and sometimes terrible effects on the muscles and limbs of dead animals, and now in its condensed form moving with a velocity that is beyond conception through the living body, and communicating a shock through fifty or a thousand persons at the same instant ; now exhibiting its mighty powers in the fearful thunder-storm, and now working slowly and quietly in the development of beautiful crystals. With such varied subjects for contemplation and admiration, it is no wonder if Electricity should be a favourite and fascinating study.

(2) The statement that Electricity is a science susceptible of familiar demonstration, must be understood in reference to its *general laws* only; for, as with other branches of natural philosophy, the investigation of its particular details, the analysis of its laws of induction, distribution, attraction and repulsion, &c., are each matters requiring the resources of mathematics, and fitted for the study of the profound philosopher only. With such subjects we have however obviously little to do in the following lectures, the design of which is to give a popular account of the present state of the sciences on which they are to treat, and to show their connexion with each other.

(3) Common, or *statical* Electricity, with which we shall first be engaged, although occupying so prominent a place in modern science, was quite unknown to the philosophers of olden times, and may be said to date its entrance into physics about the beginning of the eighteenth century. In the writings of Pliny and Theophrastus we meet with some remarks on the attractive power of amber and lyncurium stone (supposed to be the same with the modern tourmaline) for light substances when rubbed, but no attempt to explain this property is made. In the year 1600, Dr. Gilbert published a work on magnetism, in which he mentioned several new facts attributable to electrical agency, and greatly increased the knowledge of substances capable of acquiring electrical properties. This drew the attention of philosophers to the subject, and it was in 1730 that the true foundation of Electricity was laid by Mr. Stephen Grey, a pensioner of the Charter House, who, impelled by enthusiasm, engaged in a course of experimental researches, in which some general principles producing important effects on subsequent investigations were developed.

(4) The most considerable discovery made by Grey was, that all material substances might with reference to electrical phenomena be reduced to two classes, *electrics* and *non-electrics*, the latter acquiring the electric state by contact with the former when excited by friction. He also discovered the insulating property of silk, resin, glass, hair, &c., and the *fact*, though not the *principle*, of induction. He was also on the threshold of the discovery of the existence of the two opposite

Electricities, which honour was however reserved for his more intellectual contemporary, Dufaye.

(5) This sagacious philosopher re-produced in a more definite form the principles of attraction and repulsion, previously announced by Otto Guericke. He excluded the class of substances called electrics, and proved that conductors as well as non-conductors may be excited by friction, provided they are insulated. But his grand discovery was the one above alluded to : that of the two distinct kinds of Electricity, one of which he called *vitreous*, or that of glass, rock-crystal, precious stones, hair of animals, wool, and many other bodies; and the other *resinous*, that of amber, copal, gum-lac, silk-thread, paper, and a vast number of other substances. He showed that bodies having the same kind of Electricity repel each other, but attract bodies charged with Electricity of the other kind ; and he proposed that test of the state of the Electricity of any given substance which has ever since his time been adhered to, viz. to charge a suspended light substance with a known species of Electricity, and then to bring near it the body to be examined. If the suspended substance was repelled, the Electricity of both bodies was the same ; if attracted. it was different.

(6) Shortly after the time of Dufaye, the Germans appear to have turned their attention to Electricity, and in the hands of Boze, Winkler, and Gordon, the electrical machine assumed a form very nearly identical with the cylindrical machines of the present day, and very extraordinary effects are related to have been produced by the improved apparatus.

(7) It was in the year 1746 that those celebrated experiments, which drew for many succeeding years the almost exclusive attention of men of science to the new subject, and which led the way to the introduction of the Leyden phial,—were made by Muschenbroek, Cuneus, and Kleist. Professor Muschenbroek and his associates having observed that electrified bodies exposed to the atmosphere, speedily lost their electric virtue, conceived the idea of surrounding them with an insulating substance, by which they thought that their electric power might be preserved for a longer time. Water contained in a glass bottle was accordingly electrified, but no remarkable results were obtained, till one of the party who was holding the bottle attempted to disengage the wire communicating with the prime conductor of a powerful machine ; the consequence was, that he received a shock, which, though slight compared with such as are now frequently taken for amusement from the Leyden phial, his fright magnified and exaggerated in an amusing manner. In describing the effect produced on himself by taking the shock from a thin glass bowl, Muschenbroek stated in a letter to Réaumer, that " he felt himself struck in his arms, shoulders, and breast, so that he lost his breath, and was two days

before he recovered from the effects of the blow and the terror," adding, "he would not take a second shock for the kingdom of France." M. Allamand, on taking a shock, declared "that he lost the use of his breath for some minutes, and then felt so intense a pain along his right arm, that he feared permanent injury from it." Winkler stated that the first time he underwent the experiment, "he suffered great convulsions through his body; that it put his blood into agitation; that he feared an ardent fever, and was obliged to have recourse to cooling medicines!" The lady of this professor took the shock twice, and was rendered so weak by it, that she could hardly walk. The third time it gave her bleeding at the nose. Such was the alarm with which these early electricians were struck, by a sensation which thousands have since experienced in a much more powerful manner, without the slightest inconvenience. It serves to show how cautious we should be in receiving the first accounts of extraordinary discoveries, where the imagination is likely to be affected.

(8) After the first feelings of astonishment were somewhat abated, the circumstances which influenced the force of the shock were examined. Muschenbroek observed that the success of the experiment was impaired if the glass was wet on the outer surface. Dr. Watson showed that the shock might be transmitted through the bodies of several men touching each other, and that the force of the charge depended on the extent of the external surface of the glass in contact with the hand of the operator. Dr. Bevis proved that tin-foil might be substituted successfully for the hand outside, and for the water inside the jar; he coated panes of glass in this way, and found that they would receive and retain a charge; and lastly, Dr. Watson coated large jars inside and outside with tin-foil, and thus constructed what is now known as the Leyden phial.

(9) In repeating the experiments with the Leyden phial, Mr. Wilson, of Dublin, discovered the *lateral* shock, having observed that a person standing near the circuit through which the shock is transmitted would sustain a shock, if he were only in contact with or even placed very near any part of the circuit. Many experiments were also made in France and in England to determine the distance through which the electric shock could be transmitted, and many attempts to explain the phenomena were made; but it was reserved for the celebrated American philosopher, Franklin, to do this in a satisfactory manner.

(10) It was in the year 1747 that, in consequence of a communication from Mr. Peter Collinson, a Fellow of the Royal Society of London, to the Literary Society of Philadelphia, Franklin first directed his attention to Electricity; and from that period till 1754 his experiments and observations were embodied in a series of letters, which were

afterwards collected and published. " Nothing," says Priestley, " was ever written upon the subject of Electricity, which was more generally read and admired in all parts of Europe, than these letters. It is not easy to say whether we are most pleased with the simplicity and perspicuity with which they are written, the modesty with which the author proposes every hypothesis of his own, or the noble frankness with which he relates his mistakes when they were corrected by subsequent experiments." The opinion adopted by Franklin with respect to the nature of Electricity differed from that previously submitted by Dufaye. His hypothesis was as follows : — All bodies in their natural state are charged with a certain quantity of Electricity, in each body this quantity being of definite amount. This quantity of Electricity is maintained in equilibrium upon the body by an attraction which the particles of the body have for it, and does not therefore exert any attraction for other bodies. But a body may be invested with more or less Electricity than satisfies its attraction. If it possesses more, it is ready to give up the surplus to any body which has less, or to share it with any body in its natural state ; if it have less, it is ready to take from any body in its natural state a part of its Electricity, so that each will have less than their natural amount. A body having more than its natural quantity is electrified *positively* or *plus*, and one which has less is electrified *negatively* or *minus*. One electric fluid only is thus supposed to exist, and all electrical phenomena are referable either to its accumulation in bodies in quantities more than their natural share, or to its being withdrawn from them, so as to leave them *minus* their proper portion. Electrical excess then represents the vitreous, and electrical deficiency the resinous Electricities of Dufaye : and hence the terms *positive* and *negative*, for *vitreous* and *resinous*. The application of this theory to the explanation of the Leyden phial will appear in its proper place.

(11) Besides this theory, we are indebted to Franklin for the discovery of the identity of lightning and Electricity, for the invention of paratonnerres, and for the discovery of induction, which latter principle was immediately taken up and pursued through its consequences by Wilke and Œpinus, and soon led to the invention of an instrument, which in the hands of Volta became the *condenser* now so useful in electroscopical investigation.

(12) Franklin's hypothesis was investigated mathematically by Œpinus and Mr. Cavendish between the years 1759 and 1771. About the same time the electrophorus was constructed by Volta; Watson and Canton fused metals by Electricity, and Beccaria decomposed water, although at the time he had no idea he had done so, supposing it to be a simple elementary substance.

(13) In the year 1785 the foundation of *electro-statics* was laid by Coulomb, a most profound philosopher, who reduced Electricity, the most subtle of all physical agents, to the rigorous sway of mathematics, and caused it to become a branch of mathematical physics. By means of his torsion electrical balance, he made three valuable additions to the science; establishing,—1st, That electrical forces, viz., attraction and repulsion, vary *inversely as the square of their distances*, following, it will be observed, the same law as gravitation ;— 2nd, That excited bodies, when insulated, gradually lose their Electricity from two causes ; from the surrounding atmosphere being never free from conducting particles, and from the incapacity of the best insulators to retain the whole quantity of Electricity with which any body may be charged, there being no substance known altogether impervious to Electricity ; Coulomb determined the effect of both these causes ;—3rd, That when Electricity is accumulated in any body, the whole of it is deposited on the surface, and none penetrates to the interior. A thin hollow sphere may contain precisely as much Electricity as a solid of the same size. Hence accumulation is not a consequence of attraction for mass of matter, but, on the contrary, is solely due to its repulsive action. These observations of Coulomb on the distribution of the electric fluid on the surfaces of conductors illustrated satisfactorily the doctrine of points which formed so prominent a part of Franklin's researches.

(14) The identity of Electricity and lightning being proved, it was not long before an explanation of the phenomena of the aurora borealis was offered by *Eberhart*, of Halle, and *Frisi*, of Pisa. They argued that the aurora is nothing more than electrical discharges transmitted through parts of the upper regions of the atmosphere, so refined as to produce that peculiar luminous appearance which they exhibit : — a view which was adopted by most of the electricians of the day, and which recent researches countenance. Much attention was also paid to atmospheric Electricity, and in 1780 it was proved that the Electricity with which the atmosphere is, in its ordinary condition, always charged, and which in its effects assumes all the terrific forms of the tempest and the hurricane, is to be traced to the process of natural evaporation. The famous experiments illustrative of this, were made by Lavoisier, Laplace, and Volta.

(15) With this brief sketch of the history of Statical Electricity, we may proceed to a popular investigation of the phenomena as they are at present understood.

(16) *Primary phenomena.* — For illustrating the primary phenomena of Electricity, we can employ no materials either simpler or better than those used by Stephen Grey in 1730.

1^{0}. If a thick glass tube, previously made dry and warm, be briskly rubbed for a few seconds with a piece of silk or woollen cloth, and held near a pith-ball, suspended by a long silken thread, the ball will be attracted, and after adhering for a short time, it will be repelled to a considerable distance, nor will it again be attracted until it has touched some body connected with the earth, and thus given up the Electricity which it had acquired from the tube, or until, by remaining undisturbed for some time, it has lost it by dissipation into the atmosphere.

2^0. If a stick of excited resin be brought near the pith-ball while under the influence of the Electricity from the glass, it will *attract* it powerfully, but soon repel it, when the *glass* will again attract it, and the ball may thus be kept for some time vibrating between the two substances.

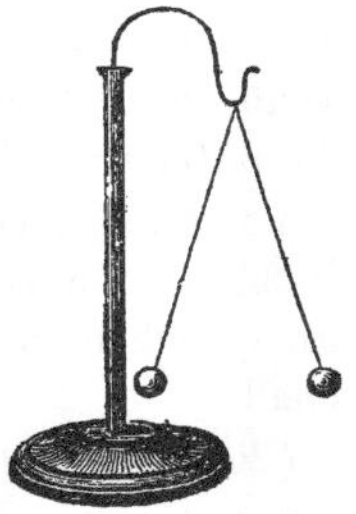

Fig. 1.

3^0. If two pith balls be suspended by two silk threads, and excited either by the glass or by the resin, on removing the exciting material, they will no longer fall into the vertical position, but repel each other in the manner shown in Fig. 1., the balls acquiring a property relative to each other, similar to that which the glass and single ball exhibited after contact in the preceding experiment.

4^0. If the pith-balls be suspended by thin metallic wires or threads made of hemp soaked in salt, they cannot be excited *permanently*, and the moment the glass or resin is removed, they return to their original condition.

(17) From these simple experiments we learn several important electrical facts : —

1^0. That vitreous substances, as glass, become electrical by being rubbed.

2^0. That in this state they attract light substances.

3^0. That having once attracted, they afterwards repel them.

4^0. That resinous substances, such as sealing-wax, are also capable of receiving electrical excitation by being rubbed.

5^0. That *they* also attract and repel light bodies.

6^0. That though excited resin and excited glass agree in their property of attracting light matter, the property called forth by friction in each is very different, for one attracts what the other repels, and *vice versâ*.

7^0. That bodies excited by similar Electricity, exhibit a disposition to *repel* each other.

8^0. That in order that they shall retain for any length of time the Electricity communicated to them, they must be *insulated* from the earth.

9⁰. That silk is a substance which possesses this power of insulating.

10⁰. That metallic bodies and thread soaked in salt and water, do *not* possess this power.

(18) But certain other effects attend the excitation of glass or resin: if either be briskly rubbed while dry and warm, in a darkened room, light will be perceived ; a slight crackling noise will be heard ; and if the hand be held near, a sensation similar to that which is felt when we touch a cobweb, will be experienced.

(19) The difference which in the previous experiments we perceived between bodies such as silk and glass, and bodies such as cotton, thread, and metal, arises from the former conducting Electricity very badly, while the latter offer a ready passage to the same. On this account bodies have been divided into two great groups—*conductors* and *nonconductors ;* the former being in general identical with *analectrics,* and the latter with *ideo-electrics.* But it must be observed that the line of demarcation between these two great classes is by no means strictly definable ; as a large number of substances exist which conduct Electricity when present in large quantities, but insulate it when in small.

(20) Among conducting substances may be classed all metals, water, steam, all animal and vegetable substances containing water, &c. ; while glass and other vitrifications, gems, resins, sulphur, metallic oxides, organic substances perfectly free from water, and ice, are all more or less perfect non-conductors, or ideo-electrics. A substance placed upon a non-conductor, as on a stool with glass legs, is said to be *insulated* from the earth. Atmospheric air must, it is clear, be ranked among non-conducting bodies, for if it gave a free passage to Electricity, the electrical effects excited on the surface of any body surrounded with it would quickly disappear, and no permanent charge could be communicated ; but this is contrary to experience. Water, on the other hand, whether in the liquid or vaporous form, is a conductor, though of an order very inferior to that of the metals ; and as water is always present in the atmosphere, it affects, in a very important manner, all electrical experiments. Hence, one of the reasons why these are made with more facility, and the desired effects produced with more certainty and success in cold and dry weather, the atmosphere then holding but little aqueous vapour suspended in it. Another injurious tendency of the watery vapour in the atmosphere is that which it has to become deposited on the surfaces of bodies, thereby destroying their insulating power ;—hence the non-conducting power of the glass or resinous supports of electrical apparatus can only be preserved by constantly rubbing or wiping them with a dry warm cloth, or silk handkerchief :

all supports of delicate pieces of apparatus should be coated with a thin layer of gum lac, dissolved in spirits of wine, on which aqueous vapour is not deposited nearly so readily as on glass.

(21) *Opposite Electricities.*—We have seen that excited resin and excited glass, though they both attract light substances, exhibit each a *different kind of force.* Hence, the name of resinous Electricity as applied to the former, and of vitreous as applied to the latter. These terms are, however, very objectionable, implying, as they do, that when vitreous bodies are excited, they are always electrified with one species of Electricity, and that when resinous bodies are excited, they are always electrified with the other. But this is by no means the case ; for examples :

1⁰. When a glass rod is rubbed with a woollen cloth, it repels a pith-ball which it had once attracted : but if the cloth be presented it will be found to attract the excited ball. We hence conclude, that as the glass was *vitreously* electrified, the woollen cloth must be *resinously* electrified.

2⁰. When a stick of sealing-wax is rubbed with a woollen cloth, it repels a pith ball which it has once attracted ; but if the cloth be presented it will be found to attract the excited ball. Hence, by a similar reasoning, we are led to the inference that the cloth is *vitreously* electrified.

3⁰. When a piece of polished glass is rubbed first with a woollen cloth, and then with the fur of a cat, and examined after each excitation by a pith ball, it is found in the first case *vitreous*, and in the second, *resinous*. A woollen cloth and a piece of glass may thus be made to exhibit both kinds of Electricity ; the terms vitreous and resinous do not therefore convey to the mind a proper impression of the nature of the two forces.

(22) The terms *positive* and *negative*, though they take their origin in a theory of Electricity which is not now recognised as compatible with observed phenomena, are less objectionable, and have accordingly partially superseded the other terms. *Positive* Electricity, then, is that which is produced upon polished glass when rubbed with a woollen cloth ; and *negative* Electricity is that which is produced upon a stick of sealing-wax when rubbed. *One kind of Electricity cannot be produced without the other ; and of two substances which, by mutual friction, excite Electricity, one is invariably positive, and the other negative, after the friction.*

(23) A pith ball, suspended by a silken thread is, in ordinary cases, sufficient to detect the presence and species of Electricity on any body. It must first be charged by an excited glass rod, and the body to be examined brought near it : if it attract the ball, then its Electricity

is *negative ;* if it repel it, it is *positive ;* if it have no effect on the
ball it is not electrified, or at least not sufficiently so to produce a force
strong enough to overcome the rigidity of the silken string. A more delicate test—the gold leaf electroscope, must then be applied ; but we are
not yet prepared for the rationale of this elegant instrument.

(24) An immense number of experiments have been made with a
view to the discovery of the physical circumstances which determine
the species of Electricity which different substances acquire ; but,
hitherto, this inquiry has not been attended with very satisfactory
results. Of the substances exhibited in the table below, any one, when
rubbed against the body immediately beneath it, becomes positively
electrified, and therefore the other negatively.

1. Fur of the cat.	6. Paper.
2. Polished glass.	7. Silk.
3. Woollen cloth.	8. Gum lac.
4. Feathers.	9. Rough glass.
5. Wood.	

This list was derived from direct experiment : it exhibits facts which
we have no means whatever of explaining ; that for instance of a piece
of polished glass being invariably positive when rubbed against a piece
that is rough, and *vice versâ.*

(25) If two persons stand on two stools with glass legs, and one
strike the other two or three times with a well-dried cat's-fur, he that
strikes will have his body charged *positively*, and he that is struck will
be electrified *negatively*. A spark may, in fact, be obtained from the
face of either, by a person in contact with earth. There is no substance
so easily excited as the fur of a cat ; and most persons are aware of the
fact, that if in dry weather the hand be passed briskly over the back
of a living cat, the hairs will frequently bristle, and be attracted by the
hand, and sometimes a crackling noise will be heard, and a spark
obtained. These effects are occasionally observed with the human
hair, which, when clean, dry, and free from grease, is electrified with
great facility by friction, and this is especially the case with fair hair,
which is in general fine and pliable. Even in damp weather, if a person
stand on an insulating stool, and connect himself with a condenser
connected with a gold leaf Electroscope, (Fig. 9) and any one standing
on the floor draw a comb rapidly through his hair, on drawing back
the uninsulated plate of the condenser, the gold leaves of the Electroscope will diverge with *positive* Electricity : if the person using the
comb, stand on the stool, and connect himself with the condenser, as
he combs, the gold leaves will open with *negative* Electricity. In dry
weather the condenser is not required for this experiment.

(26) But it is not by friction alone that Electricity is set free ; the

natural Electricity of a substance is disturbed by almost every form of mechanical change to which it can be submitted ; mere pressure is quite sufficient for the purpose.   If two pieces of common window glass be pressed firmly together, and in this state brought near a gold leaf electrometer, no disturbance of the leaves will ensue ; but if they be suddenly separated, and one piece brought near the electrometer, (being held by a handle of sealing-wax) the presence of free Electricity will be demonstrated—one piece proving to be positive, and the other negative.

If sulphur be poured whilst melted into a conical glass, and furnished with an insulating handle, or a piece of glass or silk, it will, when cold, indicate no free Electricity, but on removing the cone of sulphur from the glass, and presenting it to the electroscope, it will be found to be negatively excited, the glass itself being positive.

Some minerals, as the tourmaline, become electrical by being heated ; and if a blast of air be directed by a bellows against a plate of glass, it will cause it to assume an electrical condition.

(27) *Attraction and Repulsion.*—It was mentioned (13) that the law of electrical attraction and repulsion was determined by Coulomb, with the aid of his Torsion Electrometer. This exquisite contrivance is shown in *Fig. 2.* a, b, is a thread of silk, or spun glass, from which a needle of shell lac, c, is suspended ; it is attached to the screw d, by which it can be twisted round its axis.   The needle carries a gilt ball of pith, or a disc of paper at one extremity, which is balanced by a counterpoise on the other ; e is a metallic wire passing through the glass shade, and terminated by a metallic ball at each end.   The ball of the needle and the interior brass ball of the wire are brought into contact by turning the screw d, and the index then points to 0 on the scale which is marked upon the circumference of the glass.

FIG. 2.

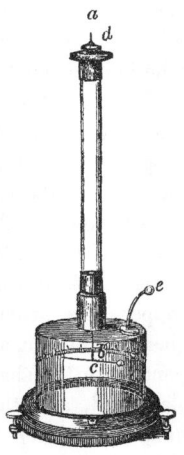

(28) To use this instrument to detect the presence of free Electricity, the rod e is removed, and the ball brought into contact with the substance whose Electricity is to be examined ; the ball acquires some of the free electric fluid, and on being placed in the glass cage, it communicates some of its Electricity to the ball terminating the horizontal needle c : the two, being similarly electrified, repel each other, and as e is fixed, c necessarily moves and describes a certain angle, which it retains until it loses its Electricity : to measure the quantity of Electricity they acquired, the screw to which the silk or glass thread is attached is turned round until by the torsion or twisting of

the thread the ball of c is compelled to come into contact with that of e. The number of degrees described by the index fixed to the revolving screw gives an approximation to the proportion of Electricity acquired by the ball of e during its contact with the electrified body. The delicacy of this instrument is such, that a force amounting to less than the 20,000,000th part of a grain can be rendered actually observable.

(29) By a series of carefully conducted experiments with this instrument, Coulomb succeeded in establishing the universality of the law, " *that bodies electrified by similar Electricities repel each other with a force that diminishes in the same proportion as the square of the distances between them is increased,*" and " *that the mutual attraction or repulsion of two electrified bodies is directly proportional to the quantity of Electricity on the other, and diversely proportional to the square of the distance between them.*"

(30) *Distribution of Electricity.* — When a substance becomes charged with Electricity, it is extremely probable that the fluid is confined to its surface, or, at any rate, that it does not penetrate into the mass to any extent. A ball formed of any material will be equally electrified whether it be solid or hollow, and if it be hollow, the charge which it receives from any source of Electricity will be the same whether the shell of matter of which it is formed be thick or thin.

To demonstrate practically the distribution of Electricity on the surface of a conductor, the following apparatus was contrived by BIOT :—
A sphere of conducting matter a, is insulated by a silk thread, and two thin hollow covers $b\ b$, made of gilt paper or tin, thin paper or copper, are provided with glass handles $c\ c$, and corres-

FIG. 3.

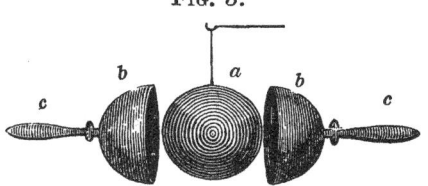

pond with the shape and magnitude of the conductor. The sphere a is electrified, and the covers are then applied, being held by the glass handles. After withdrawing them from a, they are found to be charged with the same kind of Electricity as was communicated to a, which will be found to have lost the *whole of its charge*, proving that it resided on the surface only.

(31) But although Electricity may be considered as confined to the surfaces of bodies, its intensity is not on every part the same. On a sphere, of course the symmetry of the figure renders the uniform distribution of Electricity upon it inevitable; but if it be an oblong spheroid, the intensity becomes very great at the poles, but feeble at the equator. A still more rapid augmentation of Electricity at the extremities takes place in bodies of a cylindric or prismatic form, and the more

so as their length bears a greater proportion to their breadth. Coulomb insulated a circular cylinder two inches in diameter and thirty inches in length, of which the ends were hemispherical; and on comparing the quantities of Electricity collected at the centre and at points near the extremities, he obtained the following results :—At two inches from the extremity the Electricity was to that at the centre as $1\frac{1}{4}$ to 1. At one inch from the extremity it was as $1\frac{4}{5}$ to 1, and at the extremity it was as $2\frac{3}{10}$ to 1. From the observations of the same experimentalist, it appears that the depth of the electric fluid on a conductor always increases in a rapid proportion in approaching the edges, and that the effect is still more augmented at corners, which may be regarded as two edges combined; the effect is still farther increased if any part of a conductor have the form of a point. Now, the pressure of the air is probably the only force which retains the electric fluid on a conductor, and it is evident that if at the edges, corners, or angular points of a conductor, the electric depth be so much increased that the force of the electric fluid shall exceed the restraining pressure of the atmosphere, the Electricity must escape. Accordingly, it is found practically impossible to accumulate any quantity of Electricity on a conductor furnished with points, as will be proved hereafter.

(32) *Induction.*—Amongst the earliest manifestations of the phenomena of Electricity, effects were rendered apparent which proved that contact between two bodies was not absolutely requisite to cause them to assume the electrical state; but, on the contrary, it was found that the force or agency operates at definite distances, producing distinct mechanical effects. Thus an electrified body, or an excited rod of glass or sealing-wax, when brought near to bits of paper, feathers, or other light substances, causes them to move towards it, and if presented to a small suspended unelectrified ball, it draws it aside from the vertical position.

(33) Did it happen that these attractions and repulsions took place between electrified bodies only, the natural inference would be, that the forces are exerted between the electric fluid on one body and the electric fluid on the other; but since an electrified body exhibits an attraction for bodies not electrified, it would seem that the fluid diffused over an electrified body exerts an attraction on the *matter* of bodies not electrified. But it has been satisfactorily ascertained that the Electricity diffused on a body is retained there not by any proper attraction existing between the electric fluid and the matter of the body, but merely by the pressure of the surrounding atmosphere.

Supposing a pith-ball to be insulated by a filament of silk, and electrified, we know by experience that if another similar but unelectrified pith ball be brought near, an attractive force will be exhibited; but if it be true that there exist no attraction between the Electricity diffused

on the pith ball and the matter of the pith, how can it be imagined
that there should exist any attraction between it and the other pith
ball ? But yet the attractive effects are certain : how, then, are they
to be explained ?

(34) Let us first examine the condition of fixed bodies : — Let
$d\ a\ c$ Fig. 4, be a conducting body,

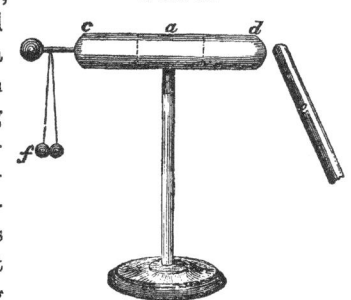

FIG. 4.

such as a cylinder of brass, supported
on a glass stand, and furnished with a
pith ball electroscope, and let e be an
excited glass tube. On approaching
this tube within about six inches distant from d, the pith balls will instantly separate, indicating the presence of free Electricity. Now, in this
case the electric e has not been brought
sufficiently near to the conducting
body to communicate to it a portion of Electricity, and the moment
that it is removed to a considerable distance the balls fall together, and
appear unelectrified ; on approaching e to d the balls again diverge, and
so on. The fact is, this is a case of what is termed *induction*, the
positive Electricity of e decomposes the neutral and latent combination
in $d\ a\ c$, attracting the negative towards d, and repelling the positive
towards e, and the balls consequently diverge, being positively electrified.
On removing e the force which separated the two Electricities in $d\ a\ c$
is removed, the separated elements re-unite, neutrality is restored, and
the pith balls fall together. The Electricity of e *induces* a change in
the electric state of $d\ c$.

In Fig. 5, suppose $s\ s'$ to
be two metallic insulated
spheres, and $a\ a'$ an insulated metallic conductor ;
suppose s to be strongly
charged with positive, and
s' with negative Electricity,
and placed in the position
represented in the figure.
If $a\ a'$ be examined by means

FIG. 5.

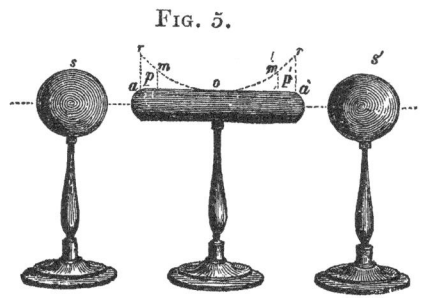

of an electrometer, it will be found that the only part which
is free from Electricity is the centre o, that half of the conductor
extending from o to a is electrified negatively, and that half extending
from o to a' is electrified positively. The intensities of the opposite
electricities at the extremities will be found to be equal, and at any
points equally distant from the centre, as $p\ p'$, the depths of the

electric fluid will be equal, and the electric state of each half may be correctly represented by the ordinates $p\,m$, $p'\,m'$ of two branches of a curve which are precisely similar and equal.

In Figs. 6, 7, and 8, suppose A A' to be a conductor, and the curves of the circles R R' those branches, the ordinates of which represent the densities of the Electricity induced upon it by the spheres $s\,s'$, (Fig. 5); by gradually removing these in an equal manner, the curves will become less and less concave, and the ordinates correctly represent the diminished density. But if the spheres be made to approach the conductor, the accumulation of Electricity

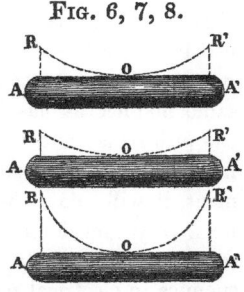

FIG. 6, 7, 8.

towards the extremities will be increased, and the curve representing the electrical densities will take the form in Fig. 8.

(35) These results strongly confirm the idea of the existence of two electric fluids uniformly distributed in equal proportions over a body in its natural state, *and the conductor comports itself exactly as it theoretically should do when charged with equal quantities of the contrary Electricities.**

(36) We saw that when an excited tube of glass was approached to one extremity of an insulated conductor, it caused the pith balls suspended from the other end to diverge. Now, on examining the conductor, it is found that the end nearest the positively excited electric has become *negative*, and the opposite end *positive*, while an intermediate zone is neutral and unelectrified. We found also that when the excited glass was removed, all signs of Electricity in the conductor vanished. Let us pursue this experiment a little further. Whilst the conductor is under the influence of the excited glass, touch it with the finger, the pith balls will collapse because the positive Electricity running off by the finger, escapes to the earth and is lost. The negative Electricity cannot escape in the same manner, because it is firmly held at the opposite end by the attractive influence of the excited glass. Now remove the finger, leaving the conductor insulated, and then remove the glass tube, the pith balls will open, *and will remain so;* but the divergence will be occasioned not as before, by *positive*, but by *negative* Electricity. In fact, the whole conductor is left with a permanent negative charge, the reason of which is simply explained. By touching the conductor with the finger, we remove the positive Electricity, and by taking away the excited glass tube, we remove the influence which retains the negative at one extremity, and it accordingly expands over the whole conductor.

* Lardner.

(37) The practical demonstration of these
facts is best exhibited with the gold leaf electroscope, shown with the condenser attached,
Fig. 9.   Previous to the application of electrified bodies to this delicate contrivance as a test,
we give it a permanent charge, exactly in the
same manner as has just been described with
the pith balls; and as great attention is paid to
insulation, in the construction of the instrument, it will retain the charge thus given to it
for a very considerable time, care being taken
to have it perfectly free from those great
enemies to electrical retention, *damp* and *dust.*

FIG. 9.

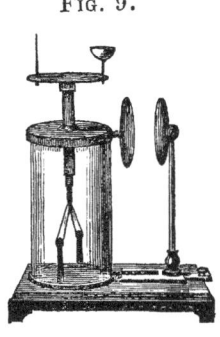

To ascertain the electrical state of any body, we bring it near to the
cap or plate of the electroscope; if on doing this the gold leaves increase
their divergence, we know that the Electricity of the body under examination is *negative;* but if the divergence diminishes, we are certain that
the electrical state is *positive.*

(38) A very instructive and useful instrument, depending on inductive action, is the electrophorus, Fig. 10.   It consists of three parts—
a cake of resinous matter, composed of equal parts of shell lac, Venice
turpentine, and common resin melted at a gentle heat; a conducting
plate or *sole*, which is a circular metallic plate with a rim about a
quarter of an inch deep round the edge, into which the composition is
poured, and a cover which is of metal, provided with a glass handle.

FIG. 10.

To use it, the resinous plate is excited by holding it in the hand in a
slanting direction, and striking it briskly several times with a piece of

dry, warm fur or flannel ; the cover is then laid on, and on removing it by its insulating handle, it is found to have acquired a feeble charge of *negative* Electricity by the contact. Let the metallic plate be replaced, and *uninsulated* by touching it with the finger, and on again lifting it by its handle, it will be found to give a strong spark of *positive* Electricity. The process may be repeated an unlimited number of times without any fresh excitation of the plate being required, and indeed after being once excited, a spark may be obtained from it during many weeks, since the resin acts solely by its inductive influence on the combined Electricities actually present in the plate.

(39) It will not be difficult at once to comprehend this. When the metallic plate is placed on the excited resin, it cannot be considered to be *in actual contact with it*, on account of the inequalities on the surface of the latter. It is therefore in a condition analagous to that of a conductor, under the influence of an electrified surface, its lower surface becoming *positive*, and its upper surface *negative*, by induction. When it is removed from the resin, the separated Electricities re-unite ; but when the plate is uninsulated, while in contact with the resin, the repelled negative Electricity escapes into the earth, and the plate becomes positively charged. It is thus rendered clear that the Electricity of the moveable plate is derived not in the way of *charge* from the resin, but is the result of the process of induction.

The figure represents Mr. John Phillips's modification of the Electrophorus, the object of which is to avoid the trouble and tediousness of establishing a communication between the insulated cover and the earth, by means of the finger, when electrical accumulation, or sparks in rapid succession, is the object. Three methods are proposed : the first consists in raising from the metallic basis above the edge of the resin, a brass ball and wire, to which the edge of the cover, or a brass ball upon it, may be applied ; this method is stated to act very well, especially with small covers, which can with ease and certainty be directed to any particular point of the sole. The second is to fix a narrow slip of tin-foil *b*, quite across the surface of the resinous plate, and unite it at each end with the metallic basis. This construction answers perfectly and instantaneously, and is very convenient with large circles, the covers of which, though uneven, will then be sure to touch some conducting point. The third method is to perforate the resinous plate quite through to the metallic basis at the centre, and any other points, and at all those points to insert brass wires *c, c, c*, with their tops level with the resin. The latter of these methods is preferred, and Mr. Phillips describes an instrument constructed on this principle, with a cast-iron basis 20·5 inches in diameter, resinous surface 19·75 inches, and cover 16·25 inches, which yields loud and flashing sparks two inches long, and speedily charges considerable jars. The cover can be easily charged from fifty to one hundred times in a minute by merely setting it down and lifting it up, as fast as the operator chooses, or as the hand can work. In charging a jar or plate, one knob of the connecting rod is placed near the insulated surface of the jar, and the other some inches above the cover, which is alternately lifted up and set down, and the jar is thus very quickly charged.

(40) A very useful modification of the Electrophorus of Volta is made by coating a thin pane of glass on one side with tin-foil to within about two inches of the edge, placing it with the coated side on the table : the other side is to be excited by friction by a piece of silk covered with amalgam (52), then carefully lifting the glass by one corner, place it on a badly conducting surface, as a smooth table, or the cover of a book with the *uncoated side downwards*. Touch the tin-foil with the finger, then carefully elevate the plate with one corner, and a vivid spark will dart from the coating to any conducting body near it : replace the plate, touch it, again elevate it, and a second spark will be produced. By this means an electric Leyden jar may speedily be charged. This modification of the Electrophorus, or *Electrolasmus*, as it is called by its inventor, Dr. Golding Bird, is a very useful instrument in the chemical laboratory.

(41) It was by an apparatus constructed on the principles of the Electrophorus, that Dr. Faraday succeeded in demonstrating that *induction is essentially a physical action, occurring between contiguous particles never taking place at a distance without polarizing the molecules of the intervening dielectric.*

(42) When an excited glass tube is brought near an insulated conductor in which the electric equilibrium is shown to be disturbed by the divergence of pith balls, we are not to suppose that the disturbance is occasioned by an action at a distance; for it has been shown by Faraday that the intervening dielectric *air* has its particles arranged in a manner analogous to those of the conductor, by the inducing influence of the glass tube. The theory of induction depending upon an action between contiguous molecules, is supported by the fact which would otherwise be totally inexplicable, that a slender rod of glass or resin, when excited by friction, and placed in contact with an insulated sphere of metal, is capable of decomposing the Electricity of the latter by induction most completely, even at the point of the ball equi-distant from the rod, and consequently, incapable of being connected with it in a right line : so that it must either be concluded that induction is exerted in *curved lines*, or propagated through the intervention of contiguous particles. Now, as no radiant simple force can act in curved lines, excepting under the coercing influence of a second force, we are almost compelled

to adopt the view of induction acting through the medium of contiguous particles.

(43) The apparatus employed by Faraday is shown in Figs. 11 and 12. It consists of a shell-lac Electrophorus, on the top of which is placed a brass ball; the charge on the surface of which is examined by the carrier ball of Coulomb's Electrometer. It was always FIG. 11. found to be positive. When contact was made at the under part of the ball, as at (d) Fig. 11, the measured degree of force was 512°, when in a line with its equator, as at (c), 270°, and when at the top of the ball, as at (b), 130°. Now, the two first charges are of such a nature as might be expected from an inductive action in straight lines; but the last is clearly an action of *induction in a curved line*, for during no part of the process could the carrier ball be connected in a straight line with any part of the inducing shell-lac. Indeed, when the carrier ball was placed by Faraday not in contact with the inducteous body at all, as at (e), it was found to be charged to a higher degree than when it had been in contact; and at (a) it was effected in the highest degree, having a result above 1000°.

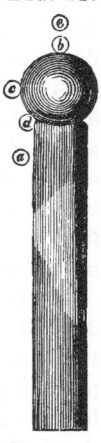

FIG. 12.

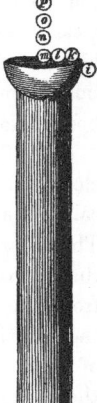

(44) When a disc, or hemisphere of metal was employed, as in Fig. 12, no charge could be given to the carrier when placed on its centre; but when placed considerably above the same spot, a charge was obtained, and this even when a *thin film* of *gold-leaf* was employed; at (i) the force was 112°, at (k) 108°, at (l) 65°, at (m) 35°; the inductive force gradually diminishing to this point. But on raising the carrier to (n), the charge increased to 87°; and on raising it still higher, to (o), it still further increased to 105°. At a higher point still (p), the charge decreased to 98°, and continued to diminish for more elevated positions.

(45) On reflecting on these beautiful experimental results, it seems impossible to resist the conclusion that induction is not through the metal, but through the air, in curved lines, and that it is an action of the contiguous particles of the insulating body thrown into a state of polarity and tension, and capable of communicating their forces in all directions.

(46) We must, in consequence of these decisive experiments, therefore, take a new view of the electric force, and instead of considering the

electric fluid to be confined to the surfaces of the bodies by the mechanical pressure of the non-conducting air (31), we must consider the force originating or appearing at a certain place to be propagated to, and sustained at a distance through the intervention of the contiguous particles of the air, each of which becomes *polarized*, as in the case of insulating conducting masses, and appears in the inducteous body as a force of the same kind, exactly equal in amount, but opposite in its direction and tendencies.

(47) There is hardly any electric phenomena in which inductive action does not come into play. When light substances (16, 32) are attracted by excited glass or wax, it is in consequence of the disturbance of their natural electric states ; in the one case the positive fluid being repelled and the negative attracted, and in the other the negative fluid being repelled and the positive attracted, each being brought about by induction. The following experiment illustrates this development of Electricity by induction in an interesting manner. Support a pane of dry and warm window-glass about an inch from the table by means of blocks of wood, or two books, and place beneath it several pieces of paper or pith balls. Excite the upper surface by friction, with a silk handkerchief, the Electricity of the glass becomes decomposed, its negative fluid adhering to the silk, and its positive to the upper surface of the glass plate. This, by induction, acts on the lower surface of the glass, repelling its positive Electricity, and attracting its negative. The lower surface of the glass thus becoming virtually electrified by induction through its substance, attracts and repels alternately the light bodies placed beneath it, in a similar manner as the excited tube.

(48) The state of a body when under the influence of a distant electric is called *induced Electricity*. The originally active body is called the *inductric*, and that under its influence, the *inducteous* body. The polar state may be excited in a long series of insulated conductors, the intensity however of the forces decreasing rapidly as the distance from the originally charged body increases. Throughout the system, the positive end of one conductor will be opposed to the negative end of another, and the intensity of all will rise by connecting the last with the ground.

In Fig. 13, suppose *a* to represent an excited glass tube, and *b c, d e*, two insulated metallic cylinders ; the feather attached to the end *b*, will be attracted by the glass tube, showing it to be negatively charged ; but the second conductor will also prove to be electrified, for the feather attached to *d* and *c* will mutually attract each other, and this will continue as long as the excited glass tube remains in the vicinity, but the moment it is removed, all signs of Electricity in the two conductors will immediately disappear.

Fig. 13.

(49) We are now prepared to understand the use of the condenser attached to the gold leaf electroscope, Fig. 9, (37). Induction can probably take place through any distance; but according as the extent of the medium through which it is exerted is lessened, it takes place more easily.

The charge of Electricity communicated to an insulated brass plate attached to a gold leaf electroscope is sustained in consequence of induction through the air towards distant surrounding objects. When a second insulated plate is opposed to the first, induction is almost wholly directed through the air to it, being the nearest body, and as it is itself in a polar state, the leaves of the electroscope will collapse, indicating an apparent diminution in the charge. A still further collapse of the leaves will take place on establishing a communication between the second plate and the earth, because the Electricity of the same kind with that of the inductive body, is virtually annihilated by diffusion over the earth. But the charge of the electroscope is only disguised, it has not really sustained any loss, for on removing the second plate, the leaves will re-open to their former extent.

Now, a higher charge of Electricity may be communicated to the electroscope while it is under the influence of the second plate *not insulated*, because the charge will be sustained both by the plate and by surrounding objects, and the gold leaves will open to the same amount as in the absence of the second plate. But when this is again removed, the accumulation which has taken place will be indicated by an expansion of the leaves far beyond their original amount, the whole being now thrown upon the surrounding conductors.—(Daniell). In this manner small quantities of Electricity may be accumulated and rendered apparent.

(50) In the sixty-eighth volume of the Annales de Chimie et de Physique, M. Peclet has published the following description of a new double electrical condenser, by which the slightest trace of electrical tension can be multiplied till it becomes appreciable and capable of measurement by the electroscope.

Three discs of glass are coated throughout with gold leaf. The first, which we will call A, is in connexion with a common gold-leaf electro-scope, and has a coating of varnish on its upper surface; the second, B, is placed on the former, and is likewise coated with varnish. A little brass pin gilt, is adapted to a point of its periphery, and in the centre it is provided with a glass handle, as in the upper plate of the ordinary condenser. On the top of this second disc is laid the third, C, a hole being bored in its centre to allow of the passage of the handle of B. This third disc has a handle likewise, consisting of a glass tube, wide enough to let that of B pass through, but not so long as to come up even with it.

To employ this instrument for ascertaining the electro-motory power excited by a metal in contact with gold, the disc C is to be touched therewith, B being in connexion with the earth : upon which C is raised, and A connected with the earth. This operation is repeated several times, and upon each repetition A and B evidently receive a fresh accession of charge. And when, at last by means of the handle on B, both the discs B and C are together removed, the leaves of the electroscope diverge, and that the more, the greater the number of separate contacts has been.

To give an idea of the condensing power of this instrument, the fol-lowing results are mentioned :—

When the upper disc was touched with an iron wire twice, thrice, four, five, and six times, the divergence of the gold leaves amounted to $9\frac{2}{3}°$, $20°$, $25°$, $31°$, $41°$, and $88°$. When the experiment was made with a platinum wire, freed from all extraneous substances upon its surface by exposure to a red heat, and held in the hand after it had been washed in distilled water, a single contact indicated only a feeble divergence ; but after three contacts it rose to $15°$, and after twenty it amounted to $53°$. This experimental demonstration of the existence of an electro-motory force between platinum and gold, and which was hitherto wanting, has been also obtained by M. Peclet with an ordinary condenser, the sensitiveness of which was carried to the utmost limits.

It appears from M. Peclet's experiments with the double, as well as the single condenser, that all metals are positive with regard to gold, and that their relative order in this respect is as follows : zinc, lead, tin, bismuth, antimony, iron, silver, and platinum. Bismuth, antimony, and iron, behave so like each other, that their order in the series could be made out in no other way than by a very frequent repetition of the experiment.

(51) The mechanism of inductive action, and the practical demon-stration of the fact, that it is from molecule to molecule of any sub-stance, gaseous or solid, that the decomposition of the natural Electrici-ties alone can take place, may be beautifully shown by plunging in a

vessel of oil of turpentine,—which is an excellent fluid insulator,—two brass balls, of which one is in connexion with an electrical machine and the other with the ground. On turning the machine, the latter becomes excited by induction. If now a number of short shreds of sewing silk be mixed with the oil of turpentine, the mechanism of the inductive action is shewn by the little bits of silk attaching themselves mutually by their extremities, by which they transmit the Electricity of the machine, by a series of decompositions, to the ball which is connected with the ground. If the excitation be very violent, the attractions and repulsions become too strong to be regularly transmitted, and this induction is accompanied by a powerful current of the particles of the oil from the first ball to the second. The particles immediately in contact with the directly excited ball acquired its state, and being repelled, immediately pass off to that which has obtained by induction the opposite condition, and those become neutralized. Now what here occurs with the oil of turpentine, takes place in ordinary induction with the air ; every molecule of it interposed between the solid body becomes itself subjected to the inductive action, and forms a chain of alternate positive and negative poles, by which the effect may be transmitted to any distance. If the excitation be very great, the neutralization may occur with violence and rapidity, and generate currents as in the oil of turpentine. It is these currents, which being produced by the repulsion of the particles of air from excited points, are rendered sensible in the effect termed the *electrical aura*, and are shown by the experiment of revolving flies.

(52) *Electrical machines.*—There are two kinds of electrical machines in general use,—the cylindrical, and the plate machine. The former is shown in Fig. 14.

It consists of a hollow cylinder of glass, supported on brass bearings, which revolve in upright pieces of wood attached to a rectangular base. A cushion of leather stuffed with horse-hair, and fixed to a pillar of glass, furnished with a screw to regulate the degree of pressure on the cylinder. A cylinder of metal or wood covered with tin-foil, mounted on a glass stand, and terminated on one side by a series of points to draw the Electricity from the glass,

FIG. 14.

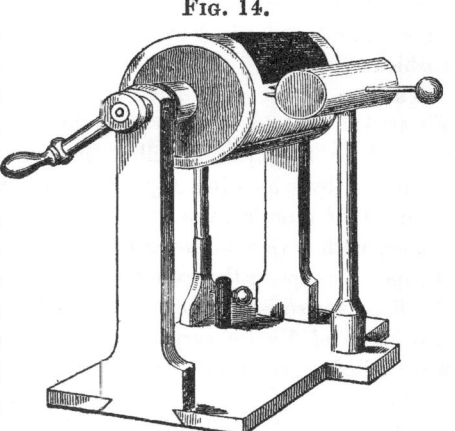

and on the other side by a brass ball. A flap of oiled silk is attached to the rubber to prevent the dissipation of the Electricity from the surface of the cylinder before it reaches the points. On turning the cylinder, the friction of the cushion occasions the evolution of Electricity, but the production is not sufficiently rapid or abundant without the aid of a more effective exciter, which experience has shown to be a metallic substance. The surface of the leather cushion is therefore smeared by certain amalgams of metals, which thus become the real rubber. The amalgam employed by Canton, consisted of two parts of mercury, and one of tin, with the addition of a little chalk. Singer proposed a compound of two parts by weight of zinc, and one of tin, with which in a fluid state six parts by weight of mercury are mixed, and the whole shaken in an iron, or thick wooden box, until it cools. It is then reduced to a fine powder in a mortar, and mixed with lard in sufficient quantity to reduce it to the consistency of paste. This preparation should be spread cleanly over the surface of the cushion up to the line formed by the junction of the silk flap with the cushion ; but care should be taken that the amalgam should not be extended to the silk flap. It is necessary occasionally to wipe the cushion, flap, and cylinder, to cleanse them from the dust which the Electricity evolved upon the cylinder always attracts in a greater or less quantity. It is found that from this cause, a very rapid accumulation of dirt takes place on the cylinder, which appears in black spots and lines upon its surface. As this obstructs the action of the machine, it should be constantly removed, which may be done by applying to the cylinder, as it revolves, a rag wetted with spirits of wine. The production of Electricity is greatly promoted by applying, with the hand to the cylinder, a piece of soft leather, five or six inches square, covered with amalgam. This is, in fact, equivalent to giving a temporary enlargement to the cushion.

The use of the oiled silk flap is to prevent the dissipation of the Electricity evolved on the glass by contact with the air; it is thus retained on the cylinder till it encounters the points of the prime conductor, by which it is rapidly drawn off. It is usual to cover with a varnish of gum lac, those parts of the glass beyond the ends of the rubber, with a view of preventing the escape of the Electricity through the metallic caps at the extremities of the cylinder, and the inside of the flap is also sometimes coated with a resinous cement consisting of four parts of Venice turpentine, one part of resin, and one of bees' wax, boiled together for about two hours in an earthen pipkin over a slow fire.

(53) When the cylindrical machine is arranged for the development of either positive or negative Electricity, the conductor is placed with

its length parallel to the cylinder, and the points project from its side, as in the machine shown in the figure. The *negative* conductor supports the rubber, and receives from it the negative Electricity not by induction, as is the case with the positive conductor, but by *communication.* If it be required to accumulate positive Electricity, a chain must be carried from the negative conductor (which of course is insulated) to the ground. If on the other hand, negative Electricity be required, then the conductor must be put in communication with the earth, and the rubber insulated. We shall return to the consideration of this presently. (59.)

Fig. 15.

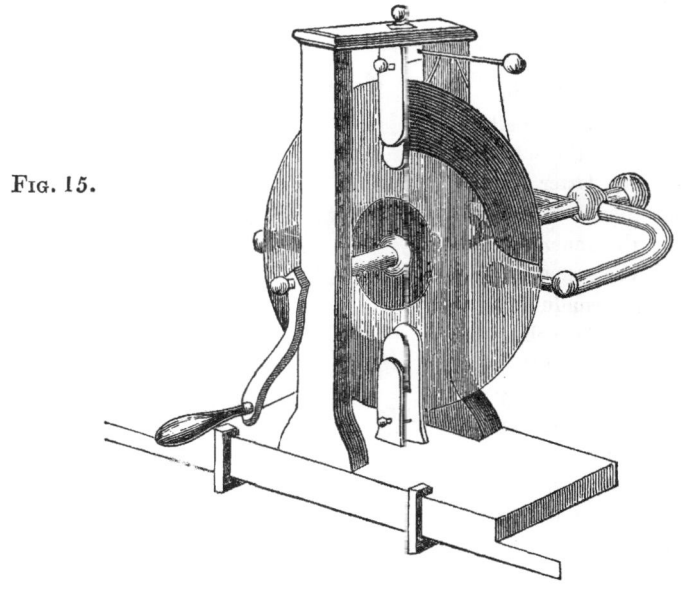

(54) The Plate Electrical Machine is shown in Fig. 15. It consists of a circular plate of thick glass, revolving vertically by means of a winch between two uprights : two pairs of rubbers, formed of slips of elastic wood covered with leather, and furnished with silk flaps, are placed at two equi-distant portions of the plate on which their pressure may be increased or diminished by means of brass screws. The prime conductor consists of hollow brass, supported horizontally from one of the uprights; its arms, where they approach the plate, being furnished with points.

(55) With respect to the merits of these two forms of the electrical machine, it is difficult to decide to which to give the preference. For an equal surface of glass the Plate appears to be the most powerful; it is not, however, so easily arranged for negative Electricity, in conse-

quence of the uninsulated state of the rubbers, though several ingenious methods of obviating this inconvenience have been lately devised.

Fig. 16 represents Dr. Hare's electrical machine, the plate being four feet in diameter, so constructed as to be above the operator, which is very convenient for a lecture-room, being never in the way, and yet always at hand. The prime conductor is supported and insulated by means of wooden posts covered by stout bell glasses, so that the summits of the latter are between those of the posts and the inner surfaces of the caps attached to the conductor. At *c c*, are the collectors. R represents a sliding-rod, which may be drawn out to such an extent as to be brought in contact with any apparatus placed under it upon the table. The machine may be kept in motion by an assistant.

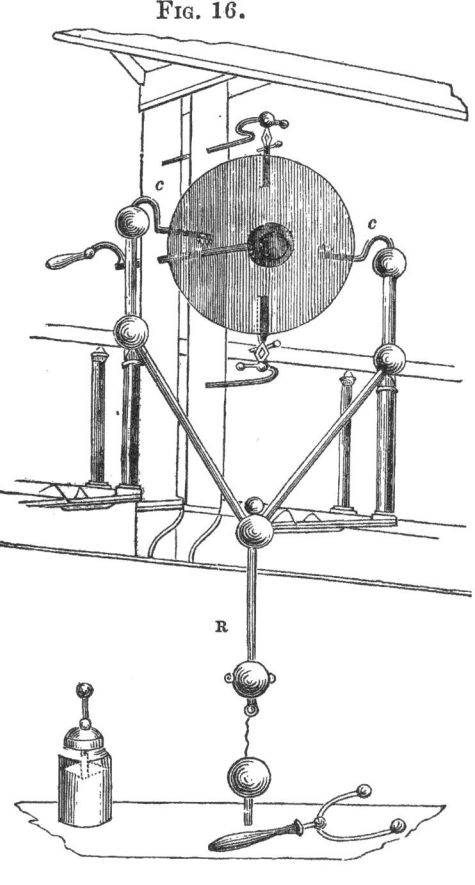

Fig. 16.

(56) One of the best forms of the plate machine is that devised by Mr. C. Woodward, President of the Islington Literary and Scientific Institution. Fig. 17 represents the one in that Institution, presented to the members by the above-mentioned gentleman. The plate, which is two feet in diameter, is fixed in the ordinary manner, between two uprights, to the top and bottom of which are attached the rubbers. The two conductors, A B, insulated on stout glass pillars, are fixed at each end of the mahogany board on which the whole is mounted, and connected together by a brass arm C, which is supported in the centre by a glass pillar E, from these, points project and collect the Electricity from both sides of the plate. This machine possesses the

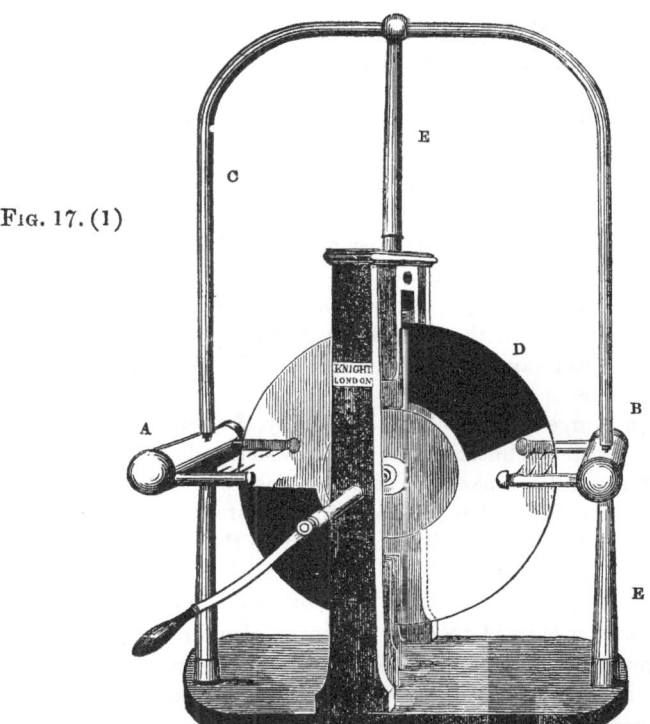

Fig. 17. (1)

following advantages : the insulation is exceedingly good ; it occupies but very little room on the lecture table ; and readily exhibits positive and negative Electricity : for this latter purpose, it is arranged as follows, and the annexed cut will render it perfectly intelligible.   The right-

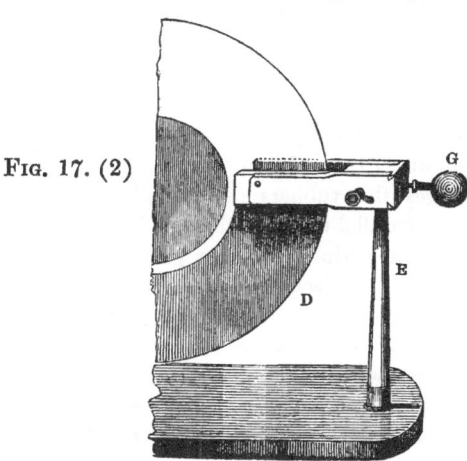

Fig. 17. (2)

hand conductor B, together with the brass arm and support, are removed, and the plate being turned a quarter of a circle, the upper rubber D is brought down on the glass pillar E, and a brass ball G screwed into it. We have now a positive and negative conductor; and although the machine possesses, of course, but half its original power, it is sufficient for all purposes of experiment. Instead of one, this machine is readily mounted with two plates, which work equally well, and it then becomes an exceedingly powerful instrument, occupying scarcely any more space.

Mr. Woodward strongly recommends the covering the glass pillars, and also that part of the plate between the spindle and the rubbers, with sealing-wax varnish, stating that it very much increases the power of the machine.

(57) In Sturgeon's Annals of Electricity, &c. for September, 1841, two very useful modifications of the cylinder and plate machine are described and lithographed. These machines are well worthy of the attention of the Electrician. The principal feature in which the arrangement of Mr. Goodman's cylinders differ from those of the usual construction, consists in their being supplied with *two* rubbers, mounted on glass rods placed parallel to each other on opposite sides of the cylinder, and connected together by means of a brass tube bent twice at right angles. This brass tube rises several inches above the top of the cylinder, so as to be out of the way of the prime conductor, which is so contrived as to answer at the same time for a support for one of the pivots on which the cylinder revolves. Two arms proceed from the upper and lower portion of this upright conductor, passing parallel to, and above and below the cylinder, from which a number of points project to receive the fluid accumulated by the excitation of the rubbers, and brought round by the rotation of the cylinder. To prevent dissipation of the fluid from the extremities of the arms, each is made to terminate in a *lacquered glass ball*. Machines arranged in this manner, are stated by the inventor to possess the desirable qualities of strength and endurance, and for equal surfaces to be twice as powerful as when only one rubber is employed.

Mr. Goodman arranges the rubbers of his plate machine in a similar manner, that is, parallel to each other, and supported by glass pillars on either side of the periphery of the plate, as in the cylinder machine, one end of the axis (of lacquered glass) turns in an insulated conductor, provided with horizontal arms carrying points.

For common purposes, and where extreme cheapness is desirable, the plate may be made of common window-glass, to the centre of which two wood-turned convex caps may be cemented without any perforation of the plate, and the axle is completed by cementing a glass rod to the centre of each cap. The cement recommended by Mr. Good-man is, equal parts of rosin and bees'-wax, made sufficiently thick by the addition of red ochre. The cost of a plate of fifteen inches diameter is about two shillings or half-a-crown.

(58) Scientific men are not agreed as to the *modus agendi* of the amalgam applied to the rubber. It seems pretty clear that the oxidation of the amalgam by the friction employed, is essential to the increased excitation; for the development of Electricity does not appear to be increased when amalgams of difficultly oxidizable metals, such as gold, are employed; and Dr. Wollaston could not succeed in obtaining any signs of free Electricity from a machine worked in an atmosphere of pure carbonic acid. The bisulphuret of tin (aurum musivum) may be employed instead of amalgam; by the friction it probably becomes partially decomposed into bisulphate of tin, as iron pyrites is into sulphate of iron. The chemical influence of friction, indeed, is more energetic than is usually supposed: even siliceous minerals, as mesotype, basalt, and feldspar, become partly decomposed, giving up a portion of their alkali in a free state.

(59) The theory of the action of the electrical machine flows immediately from the principles of induction already illustrated (34): a brief recapitulation may, however, be useful. On turning the handle of the cylinder, or plate, the Electricity naturally present in the rubber becomes decomposed,—its positive adhering to the surface of the glass, and its negative to the rubber: the positive electric portions of the glass coming, during its revolution, opposite to the points on the conductor, act powerfully by induction on the natural Electricities of the conductor, attracting the negative, which being accumulated in a state of tension at the points, darts off towards the cylinder to meet the positive fluid, and thus reconstitute the neutral compound. The consequence of this is, that the conductor is left powerfully positive,—*not*, it must particularly be understood, *by acquiring Electricity from the revolving glass, but by having given up its own negative fluid to the latter*. The rubber is left in a proportionately negative state, and consequently, after revolving the glass for a few minutes, can develop no more free positive Electricity, provided it is insulated: on this account, it is necessary to make it communicate with the earth for the purpose of obtaining a sufficient supply of positive Electricities to neutralize its negative state. In very dry weather, it is necessary to connect the rubber with the moist earth by means of a good conductor; and it is advisable if possible, to establish a metallic connexion with metallic water pipes.

(60) The use of the points attached to the prime conductor is sufficiently obvious; for, suppose a spherical metallic surface to be pre-

sented to the revolving glass, the attraction of the positive Electricity of
the glass would accumulate negative Electricity on the spherical surface,
and this would re-act on the positive Electricity of the glass, and would
have a tendency to collect it in increased quantity at the part nearest
the conductor; but this tendency would be resisted by the non-conducting quality of the glass on which the stratum of free Electricity
would maintain a depth, little, if at all augmented. Under such
circumstances, no Electricity could pass either from the cylinder to the
conductor, or from the conductor to the cylinder, unless the depth of
the electric fluid on the one or the other surface were so great as to
overcome by its force the pressure of the surrounding air. Now, it is
clear that the non-conducting property of the glass conspires with its
cylindrical form to prevent this on the one hand, while the facility to
accumulation offered by the conducting power of the spherical surface
of the conductor is counteracted by the property of that surface, in
virtue of which it favours the uniform distribution of Electricity on the
other. That the conductor may become charged with free Electricity,
either of two effects must be produced. The depth of the Electricity
on the cylinder must be so increased as to overcome the restraining
power of the air, so that it may force its way to the spherical surface
of the conductor: or the depth of the negative fluid on the latter must
be so increased as to surpass the restraining power of the air, so that a
portion of the negative fluid shall pass from it to the cylinder. In
either case the conductor would become positively charged; but by
arming the conductor with points, the perfect freedom of motion of the
electric fluid upon it, enables it to collect at the points with the depth
which the general condition of electric equilibrium requires, and as the
depth greatly exceeds what would give a force equal to that of the
atmosphere, a rapid escape of Electricity takes place.

(61) We will conclude this Lecture with an arrangement of a few
experiments to familiarize the student with the principles and action of
the electrical machine.

Ex. 1. See that the machine is in good working order: the cylinder
or plate being free from dirt and black spots, (52) and perfectly dry;
wipe it well with a piece of warm flannel, and then with an old silk
handkerchief. Take care that the insulating glass stands are clean and
dry, and see that the rubber is uniformly, but not too thickly covered
with amalgam. All these particulars being duly attended to, turn the
handle, and present the knuckle of the other hand to the prime conductor; a vivid spark will dart between them, accompanied by a
sharp snapping sound.*

* It is usual to speak of this spark as the *positive spark,* a term which does not
however convey a correct idea of its nature : for it is not to be regarded as arising

Ex. 2. Continue to turn the cylinder or plate, keeping the knuckle steadily held towards the prime conductor. The sparks will decrease in brilliancy, intensity, and frequency, and after some time no more will be obtained (59). Now establish a good metallic communication between the rubber and the earth, and the sparks will be obtained uninterruptedly, and undiminished in intensity.

Ex. 3. Remove the conductor from its position in front of the glass, and having darkened the room, revolve the cylinder or plate; a series of bright sparks will be observed to pass round the surface of the glass, exhibiting a very beautiful appearance. Let an assistant next take a needle in his hand, and approach its point towards, but at a considerable distance from, the revolving glass. While at the distance of several feet, it will be seen tipped with luminous matter, illustrating in a simple manner the striking influence of points, and their use on the prime conductor (60).

Ex. 4. Remove the ball from the end of the conductor (Fig. 14) disclosing a rounded blunt wire; put the conductor in its place, and turning the machine briskly, attempt to draw sparks from the body of the conductor with the knuckle, you will find that you will obtain very feeble and powerless ones, but you will perceive a beautiful luminous appearance proceeding from the end of the wire, and on holding the hand near it, you will experience a distinct sensation like a gentle stream of wind.

Notice attentively the appearance of the luminous matter at the points at the two opposite ends of the conductor: that on the points immediately opposed to the revolving glass will resemble small stars, and that on the wire at the end of the conductor will resemble a brush or pencil. The appearance of each you will find not unlike Fig. 18. The same luminous appearances will be perceived if a pointed wire be held at a short distance from the conductor and rubber, both being insulated, the *brush* or *pencil* appearing on the wire held towards the rubber, and the star on the wire presented towards the conductor. We shall return to the consideration of this electric light hereafter.

FIG. 18

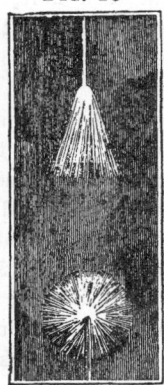

from the mere passage of free Electricity, but as the union of the two electric fluids, and the consequent discharge of the electrified body. According to the principle of induction, the positively electrified prime conductor induces an opposite electric state in any conducting substance approaching it, and when this state has amounted to one of sufficient tension, the negative Electricity rushes towards the positive of the prime conductor constituting the neutral combination. It is this neutraliza-

Ex. 5. Connect the rubber and conductor together by a wire : on revolving the glass no signs of Electricity will be obtained from either : but if the machine be extremely energetic, the wire will appear surrounded with a lambent flame, otherwise the electric fluids will traverse, and the discharge take place invisibly along the wire. But if the conductor be interrupted, vivid sparks will appear at each rupture of continuity, arising from inductive action, and consequently discharge taking place at every one of these spots.

Ex. 6. The last experiment proves that the charges on the conductor and rubber are exactly equal : that they are in opposite electrical states may be proved by suspending from each some light substances, as feathers or pith-balls, which will strongly attract each other when the machine is put in action.

Ex. 7. Place several strips of paper upon a long rod in connexion with the prime conductor, in the centre of a large apartment, they will open out equally, like radii from the centre of a sphere ; but on approximating a conducting body to them in their charged state, they will incline towards it from the concentration of the force upon its nearer surface.

Fig. 19.

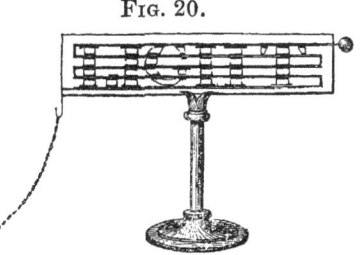

This is illustrated by the ridiculous figure of the *head of hair*, Fig. 19, and is a common electrical experiment. When electrified, the hair stands on end : and each fibre, as if in a state of repulsion from its neighbour, is attracted by, and radiates towards the point which is nearest to it in the oppositely induced state.

Ex. 8. Paste some strips of tin-foil on a plate of glass having portions cut out, so that the space represents letters, as shewn in Fig. 20 : or draw a serpentine line on the glass with varnish, and place on it metallic spangles about one-tenth of an inch apart : or stick the spangles in a spiral direction on a glass tube : in each case, lines of fire, occasioned by sparks passing apparently at the same moment through all the spaces, will be represented on

Fig. 20.

tion, or discharge of the electric state of the conductor, which constitutes the electric spark ; and it is the same with the sparks from an excited glass tube, and from the cover of the electrophorus. All cases of discharge must be preceded by induction.

In order to obtain long sparks from the prime conductor of an electrical machine, the operator must commence by taking short ones, and gradually lengthen them to their maximum.

connecting the first piece of foil with the conductor, and the last with the ground. Fig. 21
represents a little apparatus invented by Mr. Barker
for exhibiting the revolution of a spotted tube. It
is made of a glass tube, blown nice and round at one
end, and open at the other: it should be about ten
inches long, and three quarters of an inch in diameter. A ball or a piece of smooth tin-foil is fixed
at the upper closed end, and the usual spots of tin-foil
carried in a spiral form to the lower open end. A
cap, either of wood or brass, is cemented on the outside of the lower end of the tube, and a strip of foil
placed round it. From this ring four wires project
outwards, having their points bent at right angles.
The tube is then set on an upright wire which
passes upwards into the tube to its top, and this
wire is then set on an insulated stand, and brought
near the prime conductor. It can thus revolve with
great ease.

FIG. 21.

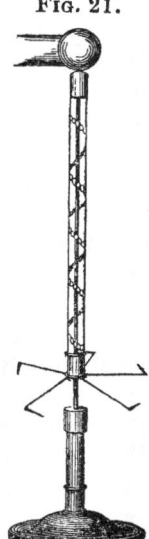

Ex. 9. Provide a stool with glass legs, Fig. 22, and having wiped
it clean and dry, let a person stand upon it, holding in his hand a
chain or wire communicating with the
prime conductor : on setting the machine
in action, sparks of fire may be drawn from
any part of his person; he becomes, indeed, for the time, a part of the conductor,
and is strongly electrified, although without
feeling any alteration in himself. If he hold in his hand a silver spoon
containing some warm spirits of wine, another person may set it on
fire by touching it quickly with his finger.

FIG. 22.

Ex. 10. By employing the little arrangement shown in Figure 23, cold spirits
of wine may be fired. Place it so that the
ball *a* can receive sparks from the prime conductor : pour spirits of wine into the cup *e*,
till the bottom is just covered : place the cup
under the wire *d*, then turn the machine, and
the sparks that are received by *a* will fly from
the wire through the spirits to the cup, and
generally set it on fire.

FIG. 23.

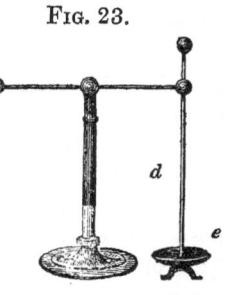

Ex. 11. The phenomena of attraction and repulsion are well
illustrated by the apparatus known as the electric bells, Fig. 24.

They are to be suspended from the prime conductor by means of the hook : the two outer
bells are suspended by brass chains, while the
central, with the two clappers, hang from silken
strings : the middle bell is connected with the
earth by a wire or chain : on turning the cylinder, the two outside bells become positively
electrified, and by induction the central one
becomes negative, a luminous discharge taking
place between them, if the Electricity be in too
high a state of tension. But if the cylinder be
slowly revolved, the little brass clappers will become alternately attracted and repelled by the
outermost and inner bells producing a constant
ringing as long as the machine is worked.

FIG. 24.

  Fig. 25 shows an admirable contrivance for illustrating electrical
attraction and repulsion. Three or four glass balls made as light as

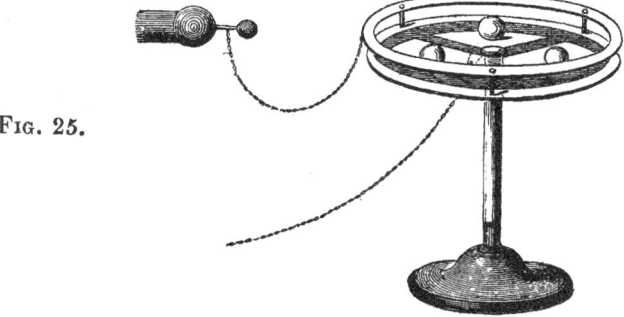

FIG. 25.

possible, are supported on an insulated glass plate, on the under part of
which, strips of tin foil are so pasted as to form a broad circle or border
near the margin, and four radii to that circle ; on the upper part of the
plate is a flat brass ring supported on small glass pillars, so as to have
its inner edge immediately over the exterior edge of the tin foil. The
brass ring being in communication with the prime conductor and the
tin foil with the rubbers of the machine, the ring and foil will be oppositely electrified. The glass balls being attracted by the ring, become
positively electrified in the part which comes in contact with it.
Thus electrified, they will be attracted by the foil, and communicating
the charge, return to the ring to undergo another change. Different
parts undergo in succession these changes, and the various evolutions of
the balls are very striking and curious.

Ex. 12. The current of air which accompanies the discharge of Electricity from points, is pleasingly shown by a variety of toys. Fig. 26 exhibits a little arrangement usually called the electrical planetarium. It is connected with the prime conductor by means of a chain, and when the

FIG. 26.

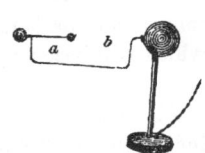

machine is set in action, the current of air discharged from the points at *a* and *b* re-acts on the wires of the apparatus, and it begins to move, the large ball representing the sun, round its axis, the earth round the sun, and the moon round the earth and sun.

Fig. 27 represents a model of a water-mill for grinding corn. A is the wheel, B the cog-wheel on its axis, C the trundle, D the running mill-stone on the top of the axis of the trundle. To set it in motion place it near the prime conductor, in which is inserted a crooked wire terminating in a sharp point. Let this point be directed to the uppermost side of the wheel A. On putting the machine in motion, the current of air, attending the Electricity which issues from the point will turn the wheel, and, consequently, all the other working parts of the mill.

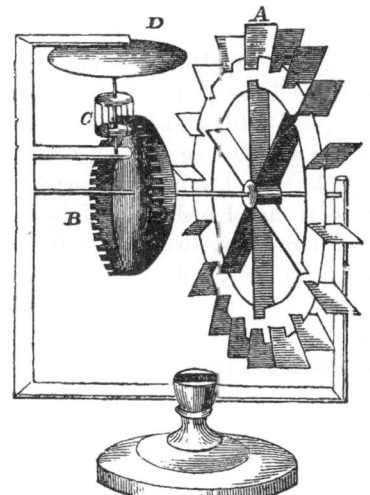

FIG. 27.

Ex. 13. Fill a phial with oil, pass through the cork a copper wire bent near its lower end at right angles, so that its point may press against the inside of the glass, and suspend it by the upper end of the wire from the prime conductor. From the machine the point of the wire in the phial will assume a high state of positive electric tension : bring towards it a brass knob, or the knuckle; induction, and consequent discharge, will take place through the sides of the glass, which will become perforated by a round hole.

Ex. 14. By the following beautiful experiment, the resistance to induction and discharge offered by a dielectric medium, such as atmospheric air, is shown. A glass tube A, Fig. 28, two feet in length, is furnished at either end with a brass ball projecting into its interior, and carefully exhausted of its air by means of a good air-pump : on connecting the end B, with the prime conductor, and the end B' with the

FIG. 28.

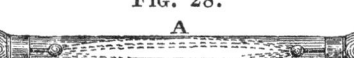

earth, when the machine is turned, B becomes positive, and induces a contrary state on the ball B'; induction taking place with facility, in consequence of the atmospheric pressure being removed, and is followed by a discharge of the two Electricities in the form of a beautiful blue light, filling the whole tube, and closely resembling the aurora borealis (14).

Ex. 15. Attraction and repulsion are amusingly shown by suspending a brass plate, Fig. 29, from the prime conductor, and setting under it a sliding stand, on which is laid a little bran or sand, or little figures made of pith : on turning the machine, the bran or sand is attracted and repelled by the upper plate with such rapidity, that the motion is almost imperceptible, and appears like a white cloud between the plates, and the little figures appear to be animated, dance, and exhibit very singular motions, dependent on inductive action (47).

FIG. 29.

Ex. 16. Fig. 30 represents a small pail with a spout near the bottom, in which is a hole just large enough to let the water out by drops; it is to be filled with water and made fast to the prime conductor : on turning the machine, the water which before descended from the spout in small drops only, will fly from it in a stream, which in the dark appears like a stream of fire ; or a sponge saturated with water may be suspended from the prime conductor, when the same phenomenon will be observed, which is referable to the mutual repulsive property of similarly electrified particles.

FIG. 30.

Ex. 17. Let the tumbler A, Fig. 31, be wiped thoroughly dry, warmed, and the inside charged by holding it in such a direction that a wire proceeding from the prime conductor of a machine in action shall touch it nearly in every part; then invert it over a number of pith-balls; they will be attracted and repelled backwards and forwards, and effect the discharge of the Electricity which induces from the interior towards the plate. They will then remain at rest; but if the Electricity which has been disengaged on the outside towards surrounding objects be removed by a touch of the hand, a fresh portion will be set free on the interior, and the attraction and repulsion of the balls will again take place, and thus for many times successively the action will be renewed until the glass returns to its natural state.

FIG. 31

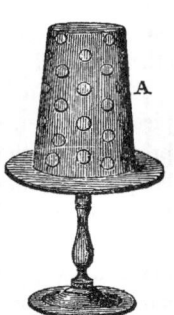

A

FIG. 32.

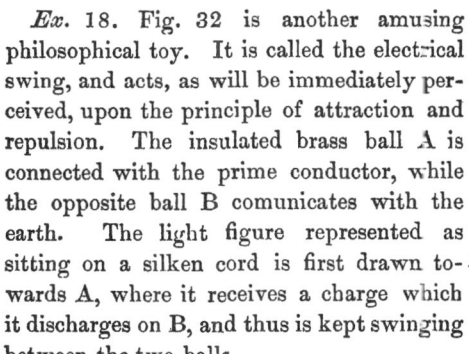

Ex. 18. Fig. 32 is another amusing philosophical toy. It is called the electrical swing, and acts, as will be immediately perceived, upon the principle of attraction and repulsion. The insulated brass ball A is connected with the prime conductor, while the opposite ball B comunicates with the earth. The light figure represented as sitting on a silken cord is first drawn towards A, where it receives a charge which it discharges on B, and thus is kept swinging between the two balls.

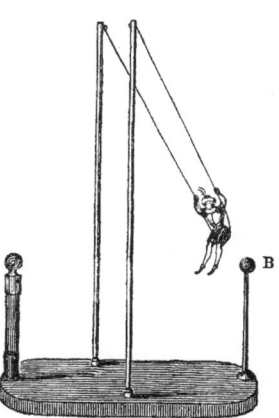

Ex. 19. Fig. 33 represents two hollow brass balls, about three quarters of an inch in diameter, insulated on separate glass pillars, by which they are supported at a distance of about two inches from each other; the upper part of each ball is hollowed into a cup, into which a small piece of phosphorus is to be put. A small candle has its flame situated midway between the balls, one of which is

FIG. 33.

connected with the positive, and the other with the negative conductor of a powerful machine. When the balls are electrified, the flame is

agitated, and, inclining towards the one which is *negative*, soon heats it
sufficiently to set fire to the phosphorus it contains, whilst the positive
ball remains perfectly cold, and its phosphorus unmelted. On reversing
the connexions of the balls with the machine, the phosphorus in the
other ball will now be heated, and will inflame.

Ex. 20. To a wire proceeding from the prime conductor, attach a
piece of sealing-wax ; put the machine in action, no effect will be
produced on the wax : now soften the end by the flame of a spirit
lamp, and while the machine is in action, present a card to the hot
wax, and you will perceive that a considerable quantity of melted wax
will be blown off from the wire, and in the form of fine, soft flexible
filaments will collect on the surface of the card, exhibiting a very
curious appearance. This experiment is interesting, as proving that
the mechanical condition of bodies has an influence on their relation to
Electricity. The sealing wax, when cold, stands high amongst non-
conductors ; but when the physical condition of its atoms is disturbed
by heat, it becomes a conductor.

LECTURE II.

ELECTRICITY—(*continued.*)

(62) HAVING in the preceding lecture endeavoured to give a general account of the leading principles of Electricity, I proceed to enlarge upon these primary facts, and to develop some more of the consequences of induction. Let us first enquire into the nature of the electric discharge, which has been thoroughly investigated by Faraday (Experimental Researches, 13 and 14 Series).

(63) According to this philosopher, both *induction* and *conduction* ought to be considered the same in principle and action—every body appearing to discharge in a greater or less degree, which makes them better or worse conductors—worse or better insulators. He considers the first effect of an excited body upon neighbouring matters, to be the production of a polarized state of their particles, which constitute induction ; and this arises from its action on the particles immediately in contact with it, which again act upon those contiguous to them ; and thus the forces are transferred to a distance. If the particles can maintain this polarized state, then *insulation* is the consequence ; and the higher the polarized condition, the better the insulation ; but if the particles cannot maintain their polarized state, if they possess the power to communicate their forces, then conduction occurs ; and the

tension is lowered, conduction being a distinct act of discharge between neighbouring particles. Thus, as the higher the polarized condition which the particles of the body can assume, the better insulator is that body ; so is a body a better conductor in proportion to the inappetency of its particles to retain a state of polarity.

(64) The discharge which takes place between two conducting surfaces is termed *disruptive* : it is the limit of the influence which the intervening air or dielectric exerts in resisting discharge : all the effects prior to it are inductive, (59) and it consequently measures the conservative power of the dielectric. It occurs not when all the particles have attained to a certain degree of tension ; but when that particle which is most affected has been exalted to the subverting or turning point, all must then give way, since they are linked together, as it were, by the influence of the constraining force, and the breaking down of one particle must, of necessity, cause the whole barrier to be overturned. In every case, the particles amongst and across which the discharge suddenly breaks, are displaced—the path of the spark depending upon the degree of tension acquired by the particles in the line of discharge.

(65) The spark may be considered then, as a discharge, or lowering of the polarized inductive state of many dielectric particles by a particular action of a few of the particles occupying a very small and limited space : all the previously polarized particles returning to their first or normal condition in the inverse order in which they left it, and uniting their powers, meanwhile to produce, or rather to continue the discharge effect in the place where the subversion of force first occurred.

I have given this explanation in the words employed by Dr. Faraday, that no misconception of his meaning may arise. He is of opinion that a peculiar temporary state is assumed by the particles situated where discharge occurs ; that they have all the surrounding forces thrown upon them in succession, and that they are not merely pushed apart ; that the whole terminates by a discharge of the powers by some, as yet, unknown operation, the ultimate effect being exactly as if a metallic wire had been put into the place of the discharging particles.

(66) The electric spark presents different appearances when taken in different elastic media. In air, they have, when obtained with brass balls, a well-known intense light, and blueish colour, with frequently faint or dark parts in their course, when the quantity of Electricity passing is not great. In *nitrogen* they are very beautiful, having the same general appearance as in air, but more colour, of a purple or blueish character ; and Faraday thought that they were remarkably sonorous.

In *oxygen* they are whiter, but not so brilliant as in common air. In *hydrogen* they are of a fine *crimson* colour, and have very little sound in consequence of the physical condition of the gas. In *carbonic acid gas* they have the same general appearance as in air, but are remarkably irregular. Sparks can be obtained under similar circumstances, much longer than in air, the gas showing a singular readiness to pass the discharge. In *muriatic acid gas*, when dry, they are nearly white, and always bright throughout. In *coal gas* they are sometimes green, and sometimes red, and occasionally one part is green and another red. Black parts also occur very suddenly in the line of the spark, i. e. they are not connected by any dull part with bright portions, but the two seem to join directly one with the other.

It is the impression of Faraday, that these varieties of character are due to a direct relation of the electric powers to the particles of the dielectric through which the discharge occurs, and are not the mere results of a casual ignition, or a secondary kind of action of the Electricity upon the particles which it finds in its course, and thrusts aside in its passage. It was remarked by M. Fusinieri, that when a spark takes place between a surface of silver and another of copper, a portion of silver is carried to the copper, and of copper to the silver; and Dr. Priestley observed, that if a metallic chain be laid upon a sheet of paper, or a plate of glass, and a strong discharge sent through it, spots will be produced upon it of the size and colour of each link, parts of which will be found to be fused into the substance of the glass.

(67) *The Electrical brush.* — The phenomenon of the electrical brush has been shown by Professor Wheatstone to consist of successive *intermitting* discharges, although it appears continuous. If an insulated conductor, connected with the positive conductor of an electrical machine, have a metallic rod 0·3 of an inch in diameter, projecting from it outwards from the machine, and terminated by a rounded end or small ball, it will generally give good brushes; or if the machine be not in good action, then many ways of assisting the formation of the brush may be resorted to; thus, the hand, or any large conducting surface may be approached towards the termination to increase the inductive force, (49) or the termination may be smaller, and of badly conducting matter, as wood: or sparks may be taken between the prime conductor and the secondary conductor, to which the termination giving brushes, belongs, or, which gives to the brushes exceedingly fine characters and great magnitude, the air around the termination may be rarefied, more or less either by heat or the air-pump, the former favourable circumstances being also continued. When obtained by a powerful machine, or a ball about 0·7 of an inch in diameter at the end of a long brass rod attached to the positive prime conductor, it has

the general appearance, as to form, represented in
Fig. 34. A short conical bright part or root appears
at the middle part of the ball, projecting directly
from it, which at a little distance from the ball breaks
out suddenly into a wide brush of pale ramifications,
having a quivering motion, and being accompanied at
the same time with a low dull chattering sound.
The general brush is resolvable into a number of
individual brushes, each of which is the result of a
single discharge—each is instantaneous in its existence, and each appeared to Faraday to have the
conical root complete. The sound is due to the
recurrence of the noise of each separate discharge, which, happening at
intervals nearly equal under ordinary circumstances, causes a definite
note to be heard, which, rising in pitch with the increased rapidity and
regularity of the intermitting discharges, gives a ready and accurate
measure of the intervals, and so may be used in any case when the
discharge is heard, even though the appearances may not be seen, to
determine the element of *time*.

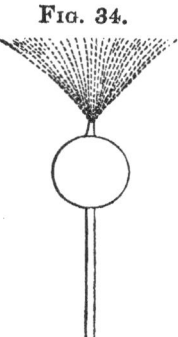

Fig. 34.

(68) The brush is, in reality, a discharge between a bad, or a non-
conductor, and either a conductor or another non-conductor. It is explained by Faraday on the principles of induction, which taking place
between the end of an electrified rod and the walls of a room, across
the dielectric air, polarizes the particles of air; those which are nearest
to the end of the wire being most polarized, and those situated in
sections across the lines of inductive force towards the walls being
least polarized. In consequence of this state, the particle of air at the
end of the wire is at a tension that will immediately terminate in discharge, while in those even only a few inches off, the tension is still
beneath that point. When the discharge takes place, the particle of
air in the immediate vicinity of the rod instantaneously resumes its
polarized state, the wire itself regaining *its* electrical state by induction ; the polarized particle of air exerts a distinct inductive act towards
the farther particles, and thus a progressive discharge from particle
to particle takes place. The difference between the brush discharge
and the spark is, that in the former discharge begins at the root, (67)
and extending itself in succession to all parts of the single brush, continues to go on at the root and the previously-formed parts, until the
whole brush is complete ; then, by the fall in intensity and power at
the conductor, it ceases at once in all parts to be renewed when that
power has risen again to a sufficient degree ; but in the latter, the particles in the line of discharge being, from the circumstances, nearly alike
in their intensity of polarization, suffer discharge so nearly at the same

moment, as to make the time quite insensible to us.   Mr. Wheatstone found that the *brush* generally had a sensible duration, but he could detect no such effect in the spark.

(69) According to Faraday, the brush may be considered as a spark to air ; a diffusion of electric force to matter, not by *conduction*, but by disruptive discharge ; a dilute spark, which, passing to very badly conducting matter, frequently discharges but a small portion of the power stored up in the conductor : for as the air charged re-acts on the conductor, whilst the conductor, by loss of Electricity, sinks in its force, the discharge quickly ceases, until, by the dispersion of the charged air, and the renewal of the excited conditions of the conductor, circumstances have risen up to their first effective condition, again to cause discharge, and again to fall and rise.

(70) By making a small ball positive by a good electrical machine with a large prime conductor, and approaching a large uninsulated discharging ball towards it, very beautiful variations from the spark to the brush may be obtained.   In Fig. 34, the general appearance of a good brush is exhibited ; but if the hand, a ball, or any knobbed conductor be brought near, the extremities of the coruscations turn towards it and each other, and the whole assumes various forms, according to circumstances, as shown in Figs. 35, 36, 37.   The curvature of these ramifications illustrates, in a beautiful manner, the curved

FIG. 35.

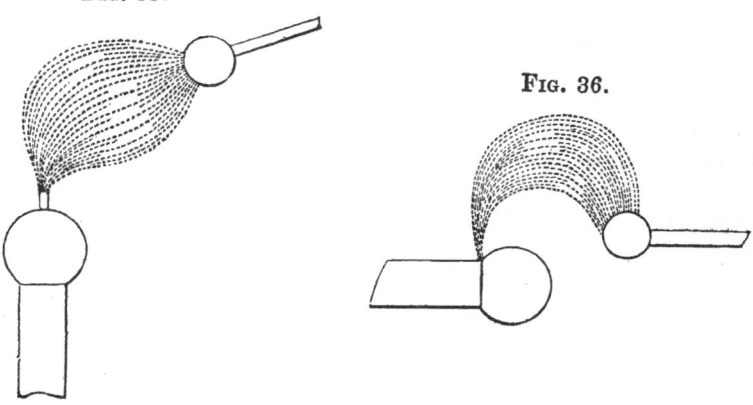

FIG. 36.

FIG. 37.

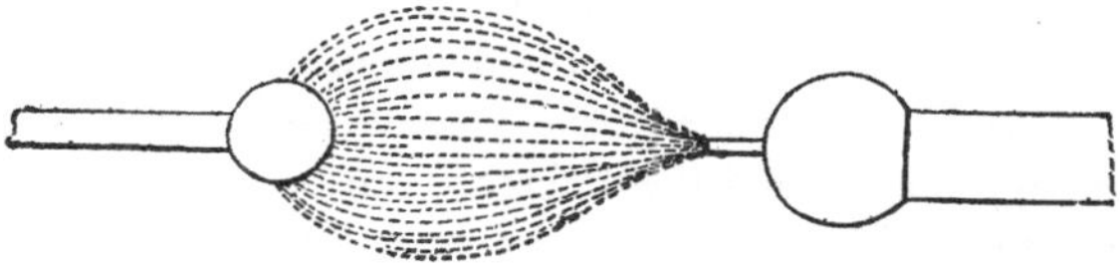

form of the lines of inductive force existing previous to discharge, in
the same manner as iron filings strewed on a sheet of paper placed over
a magnet, represent magnetic curves ; and the phenomena are considered by Faraday as constituting additional and powerful testimony
in favour of induction through dielectrics in curved lines, (42) and of
the lateral relation of these lines by an effect equivalent to a repulsion
producing divergence, or, as in the cases figured, the bulging form.

(71) Discharge in the form of a brush is favoured by rarefaction of
the air, in the same manner, and for the same reason as discharge in
the form of a spark. It may be obtained not only in air and gases,
but also in much denser media. Faraday procured it in oil of turpentine, but it was small, and produced with difficulty. He also found
that, like the spark, the brush has *specific characters* in different gases,
indicating a relation to the particles of these bodies, even in a stronger
degree than the spark. In *nitrogen,* brushes were obtained with far
greater facility than in any other gas ; and when the gas was rarefied,
they were exceedingly fine in form, light and colour; in oxygen, on
the other hand, they were very poor.

(72) The peculiar characters of nitrogen in relation to the electric
discharge must, Faraday observes, have an important influence over
the *form* and even the *occurrence of lightning.* Being that gas which
most readily produces coruscation, and by them extends discharge to
a greater distance than any other gas tried, and is also that which constitutes four-fifths of our atmosphere ; and as in atmospheric electrical
phenomena, one, and sometimes both the inductive forces are resident
on the particles of the air, which, though probably affected as to conducting power by the aqueous particles in it, cannot be considered as a
good conductor ; so the peculiar power possessed by nitrogen to originate
and effect discharge in the form of a brush or of ramifications, has probably an important relation to its electrical service in nature, as it
most seriously affects the character and condition of the discharge
when made.

(73) The characters of the luminous appearances at the ends of
wires charged positively and negatively, are represented in Fig. 18.
Faraday has paid considerable attention to the difference of discharge at
the positive and negative conducting surfaces. According to his observations, the effect varies exceedingly under different circumstances.
It is only with bad conductors, or metallic conductors charged intermittingly, or otherwise controlled by collateral induction, that the
brush and star are to be distinctly distinguished : for if metallic
points project freely into the air, the positive and negative light differ
very little in appearance, and the difference can be observed only upon
close examination. If a metallic wire with a rounded termination in

free air be used to produce the brushy discharge, then the brushes obtained when the wire is charged negatively, are very poor and small by comparison with those produced when the charge is positive : or, if a large metal ball connected with the electrical machine be charged *positively*, and a fine uninsulated point be gradually brought towards it, a star appears on the point when at a considerable distance, which, though it becomes brighter, does not change its form of a star until it is close up to the ball; whereas if the ball be charged negatively, the point at a considerable distance has a star on it as before; but when brought nearer (within about $1\frac{1}{2}$ inch) a brush forms on it, extending to the negative ball : and when still nearer, (at $\frac{1}{8}$ of an inch distance) the brush ceases, and bright sparks pass.

(74) The successive discharges from a rounded metallic rod $0\cdot3$ of an inch in diameter, projecting into air when charged negatively, are very rapid in their recurrence, being seven or eight times more numerous in the same period than those produced when the rod is charged positively to an equal degree; but each brush carries off far less electric force in the former case than in the latter. Faraday also perceived a very important variation of the relative forms and conditions of the positive and negative brush, by varying the dielectric in which they were produced. The difference, indeed, was so great as to point out a specific relation of this form of discharge to the particular gas in which it takes place, and opposing the idea that gases are but obstructions to the discharge acting one like another, and merely in proportion to their pressure. Generally speaking, when two equal small conducting surfaces equally placed in air, are electrified, one positively and the other negatively, that which is negative can discharge to the air *at a tension a little lower* than that required for the positive ball, and when discharge does take place, much more passes at each time from the positive than from the negative surface.

(75) *Glow discharge.*—When a fine point is used to produce disruptive discharge from a positively charged conductor, the brush gives place to a quiet phosphorescent continuous glow, covering the whole of the end of the wire, and extending a small distance into the air. Occasionally this glow takes the place of the brush, when a rounded wire $0\cdot3$ of an inch in diameter is used, and the finer the point the more readily is it produced : thus, *diminution of the charging surface* produces it : increase of power in the machine tends to it, and it is surprisingly favoured by rarefaction of the air. A brass ball $2\frac{1}{2}$ inches in diameter, when made positively inductric (48) in an air-pump receiver, becomes covered with a glow over an area of two inches in diameter, when the pressure is reduced to $4\cdot4$ inches of mercury. By a little adjustment, Faraday succeeded in covering the ball all over with

this light; using a brass ball 1·25 inches in diameter, and making it inducteously positive by an inductric negative point, the phenomenon at high degrees of rarefaction were exceedingly beautiful. The glow came over the positive ball, and gradually increased in brightness, until it was at last very luminous, and it stood up like a low flame, half an inch or more in height. On touching the sides of the glass jar, this lambent flame was affected, assumed a ring form, like a crown on the top of the ball, appeared flexible, and revolved with a comparatively slow motion, i. e. about four or five times in a second.

(76) The glow is always accompanied by a wind proceeding either directly out from the glowing part, or directly towards it. Faraday was unable to analyse it into visible elementary intermitting discharges, nor could he obtain the other evidence of intermitting action—namely, an audible sound. (67) It is difficult to produce it at common pressures with *negative* wires, even on fine points, though in rarefied air the negative glow can easily be obtained.

(77) All the effects tend to show that *glow* is due to a continuous charge or discharge of air; in the former case being accompanied by a current from, and in the latter case by one to, the place of the glow. As the surrounding air comes up to the charged conductor, on attaining that spot at which the tension of the particles is raised to the sufficient degree, it becomes charged, and then moves off by the joint action of the forces to which it is subject, and at the same time that it makes way for other particles to come and be charged in turn, actually helps to form that current by which they are brought into the necessary position. Thus, through the regularity of the forces, a constant and quiet result is produced, and that result is, the charging of successive portions of air, the production of a current and of a continuous glow.

(78) By aiding the formation of a current at its extremity, the brush at the termination of a rod may be made to produce a glow, and on the other hand by affecting the current of air, by sheltering the point from the approach of air, it is not difficult to convert the glow into brushes. The *glow* is assisted by those circumstances which tend to facilitate the charge of the air by the excited conductor, the *brush* by those which tend to resist the charge of the same; and those which favour intermitting discharge in a more exalted degree favour the production of the *spark*. Thus the transition from the one to the other may be established in various ways: by rarefying the air, by removing large conducting surfaces from the neighbourhood of a glowing termination, or by presenting a sharp point towards it, we help to sustain the glow; and by condensing the neighbourhood of a discharging ball, or by presenting the hand gradually towards it, we convert the glow into the brush or spark.

(79) Before proceeding further, it may be useful to give a general summary of the views of Faraday relating to induction. His theory is not intended to offer any thing new as to the nature of the electric force or forces, but only as to their distribution. It undertakes to state *how* the powers are arranged, to trace them in their general relations to the particles of matter, to determine their general laws, and the specific differences which occur under these laws.

(80) The theory assumes :

1^0. That all the *particles*, whether of insulating or conducting matter, are, as wholes, conductors.

2^0. That not being in their normal state *polar*, they can become so by the influence of neighbouring charged particles, the polar state being developed at the instant exactly as in an insulating conducting *mass* consisting of many particles.

3^0. That the particles when polarized are in a forced state and tend to return to their normal or natural condition.

4^0. That being, as wholes, conductors, they can readily be charged either *bodily* or *polarly*.

5^0. That particles which, being contiguous, are also in the line of inductive action, can communicate or transfer their polar forces to one another *more* or *less* readily.

6^0. That those doing so, less readily require the polar forces to be raised to a higher degree before this transference or communication takes place.

7^0. That the *ready* communication of forces between contiguous particles constitutes *conduction*, and the *difficult* communication *insulation;* conductors and insulators being bodies whose particles naturally possess the property of communicating their respective forces, easily or with difficulty ; having these differences just as they have differences of any other natural property.

8^0. That ordinary induction is the effect resulting from the action of matter charged with excited or free Electricity upon insulating matter, tending to produce in it an equal amount of the contrary state.

9^0. That it can do this only by polarizing the particles contiguous to it, which perform this office to the next, and these again to those beyond ; and that thus the action is propagated from the excited body to the next conducting mass, and these render the contrary force evident, in consequence of the effect of communication which supervenes in the conducting mass upon the polarization of the particles of that body.

10^0. That therefore induction can only take place through or across insulators : that induction is insulation, it being the necessary consequence of the state of the particles, and the mode in which the influence of electrical forces is transferred or transmitted through or across such insulating media.

(81) *Accumulation of Electricity.* — *Leyden Phial.* — In the last lecture it was stated (49) that a higher charge may be communicated to the gold leaf electroscope while under the influence of a second plate *not insulated.* To illustrate this property of the second plate, we have only to bring it as close as possible without touching, to the *inductric* plate, and communicate a charge to the latter; then on removing the second plate, the accumulation which has been effected will be indicated by an expansion of the gold leaves considerably beyond the original amount. This divergence of the gold leaves is to be considered as occasioned by the attraction in opposite directions of the oppositely electrified inducteous bodies.

(82) When an excited glass tube is brought near to the cap of the electroscope, the second plate (connected with the earth) being close to it, the gold leaves do not open nearly so much as if the second plate were not there, because induction taking place through the intervening plate of air to the nearest body, viz. the inducteous or second plate, the Electricity of the same kind as that of the cap of the instrument, becomes diffused over the earth; but when the second plate is removed, the leaves diverge much more than if it had not been there, because they have received a higher charge. Now, in this case, the intervening air has received a *higher polar tension* (80), which it will be understood arises entirely from the close proximity of the charged body to a conductor to the earth: the thinner the intervening stratum of air, the higher the degree of polar tension that can be attained, and the rise of force is limited by the *mobility of the particles of the air*, in consequence of which the equilibrium is restored either silently or by a spark.

(83) If instead of a plate or stratum of air, we employ a *solid dielectric*, such as glass, the tension which may be assumed is limited only by its cohesive force. Thus, if we place a plate of glass between two circular pieces of tin, insulated, and connect one plate with the prime conductor of an electrical machine, we shall have an arrangement precisely similar to the condenser, Fig. 9, except that the intervening dielectric will be glass instead of air: on connecting the other plate with the earth to destroy its polar state, and working the machine, the particles of the glass will become powerfully polarized; and if instead of connecting one of the plates with the earth, we touch it from time to time with the knuckle, a series of sparks will be obtained, occasioned by the repulsion of the positive Electricity, naturally present in the tin plate, by induction through the glass from the opposite plate electrified by the machine. After a time these will cease, and on

removing the wire connecting the plate with the prime conductor, it will be charged with *positive*, while the other plate will be charged with *negative* Electricity, both in a high state of tension. If now both plates are connected by means of a curved wire, *discharge* results, attended with a vivid flash, and a loud snap.

(84) The same effects will be produced by coating either side of a pane of glass with tin foil, leaving about $1\frac{1}{2}$ inch all round uncovered, and it is quite clear that the surfaces of dielectrics and conductors may be arranged in different forms without impairing the effects. Glass jars or bottles are found much more convenient in practice than squares of coated glass; and the *quantity* of Electricity which may be accumulated depends upon the extent of the coated surface; its *intensity* on the thinness of the glass.

(85) Glass jars thus arranged are well known by the name of Leyden phials, from their having been first constructed by Muschenbroek and his friends at Leyden (7). In practice it is found impossible to diminish the thickness of the glass beyond a certain extent, as the constrained position of its polarized particles is apt to rise so high as to destroy its cohesive force, and the charge breaks its way through the glass. Fig. 38 represents a Leyden phial of the usual construction, with the dis-

FIG. 38.

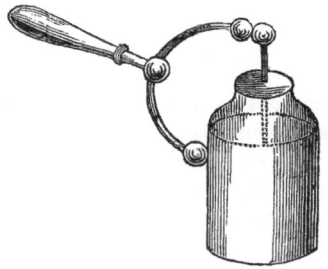

charging rod furnished with a glass handle in the position in which it is placed, in the act of discharging the jar by establishing a metallic communication between the outer and inner metallic coatings. The wire which passes through the varnished mahogany cover of the jar is terminated at one end by a brass ball, and at the other by a chain reaching to the bottom of the jar.

(86) To charge the Leyden phial, its knob should be held about half an inch from the prime conductor, the hand grasping the outer coating. A series of sparks take place between the knob and the conductor, which continue for some time, and then cease. The jar is now *charged*, its inside containing positive, and its outside negative Electricity, their union being prevented by the interposed glass. If the jar be very thin, and the tension of the Electricity considerable, discharge

E

frequently takes place through the glass (85) which thus becomes perforated and useless; or if the metallic coatings extend too near the
mouth of the jar, the discharge is very apt to pass over the uncoated
surface in the form of a bluish lambent brush of flame, constituting a
spontaneous discharge. But if neither of these accidents occur, still the
jar cannot be kept charged long, neutralization taking place more or
less rapidly by the conducting action of the surrounding atmosphere.
It is advisable to varnish the glass above the coating with a solution
of gum lac in alcohol (20) or with the common spirit varnish of
the shops, taking care to warm the jars before and after its application.

(87) In Fig. 38 the Leyden phial is represented as undergoing discharge by an instrument for the purpose; it is not, however, advisable
to discharge large phials by placing one of the balls of the discharging
rod against its side in this manner, there being considerable risk of
breaking them by the explosion, especially if the glass be thin. The
best plan is to place the phial upon a sheet of tin foil considerably
larger than the bottom of the jar, to place the lower ball of the discharging rod upon the metal, and then to bring the other ball quickly
within the striking distance of the knob of the jar; by this method the
Electricity becomes diffused over a larger surface, and is not concentrated to a single point of the glass, the risk of fracture of which is
necessarily diminished in consequence.

(88) When narrow-mouthed jars or bottles, as the common sixteen-
ounce phials of white glass (which from their thinness form excellent
Electric jars) are used, some persons coat them internally with brass
filings instead of tin foil, on account of the difficulty of applying the
latter to their interior; for this purpose some thin glue should be
poured into them, and the bottle turned slowly round until its inner
surface is covered to about three inches from the mouth. Brass filings
are then put in, and the bottle well shaken, so that they may be
diffused equally over its surface; on inverting it, those which are in
excess will fall out, and the bottle will be left tolerably well coated
internally. This method, however, rarely answers well; a better one
is, to melt equal parts of lead and tin, and whilst fused, to add quicksilver enough to keep the whole fluid whilst warm, and in this condition
to pour it into the bottle, turning the latter round and round in various
ways till the whole of the inside is covered with amalgam. A little
bismuth keeps the whole fluid at a lower temperature. This plan
answers very well for coating internally large green glass carboys,
though no experimentalist is advised to go to the trouble of fitting up
these vessels, and they generally prove useless, probably on account of
the imperfection of the dielectric.

FIG. 39.

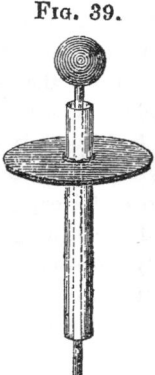

In Fig. 39, a good method of fitting up the Leyden phial is shown: the wire communicating with the interior coating passes through a glass tube extending above and below the cover about six inches. The cover is thus insulated from the inside coating, dust is excluded, and a greater stability is given to the wire. Thus arranged, the jar will retain its charge much longer than on the usual plan. It was contrived by Mr. Barker.

(89) Another arrangement by which the fracture of large jars is, almost with certainty, prevented, is shown in Fig. 40. It was recommended to me by the Rev. F. Lockey. I now always fit up my jars in this manner, and since I have adopted it, I have not broken one, though before, this was a vexatious accident of frequent occurrence. The wire, instead of communicating with the interior coating by means of a metallic chain, screws into the bar of wood a, which is covered with tin-foil, the sides of which press lightly against the inner coating of the jar; two slender pieces of wood, b, c, also covered with tin-foil, are morticed into the bar a, and kept in place by a brass pin at d; the other extremities press against the sides of the jar close to the bottom: wide-mouthed jars should be employed, and if they slope towards the bottom, the firmer can the bar a be fixed:—no covers are required. The advantage of this arrangement will be immediately perceived, there being a metallic communication between the knob and four different points of the inner coating: the force of the discharge is divided into four parts, and not only is the risk of fracture decreased thereby, but *complete discharge* of the jar is ensured. I have fitted up a pretty large electrical battery (96) in this manner, and have usually found the residual charge, after the first explosion, to be very trifling; but those who are conversant with electrical experiments, are well aware that in the usual manner of fitting up batteries, this residual charge is, in large arrangements, very considerable, sufficient to give unpleasant shocks, and frequently giving rise to awkward accidents.

FIG. 40.

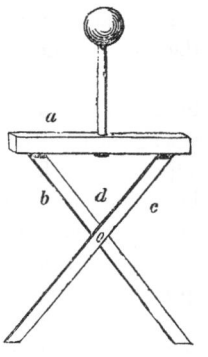

(90) The *quantity* of Electricity accumulated in a jar or battery (96) may be roughly estimated by the number of turns of the machine;

its intensity may be approximately determined by the amount of
repulsion between any two moveable bodies under its influence, or
rather by the amount of their opposite attractions by surrounding
bodies under their inductive influence.

In Fig. 41 is shown the quadrant electrometer, contrived Fig. 41.
by Henley for this purpose. It consists of a graduated semicircle of ivory fixed to a rod of wood d. From the centre of
a descends a light index, terminating in a pith-ball, and readily moveable on a pin. To use it, it is removed from its
stand and screwed upon the jar or battery, the charge of
which it is intended to indicate : as it increases, the pith-ball
moves from its centre of suspension, and measures the intensity upon the graduated semicircle.

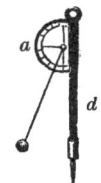

Fig. 42.

(91) A much more accurate instrument, for the same
purpose, is the unit jar of Mr. Harris, Fig. 42. It consists of a small inverted Leyden phial, supported and insulated by a slender glass rod, which is covered with
varnish and fixed on a wooden foot. The wire from
the inner coating of this small jar, is terminated by another wire at right angles to it, and furnished with a
brass ball a, and the outer coating is provided with another
arrangement of wires, terminated by a brass ball c, and so
adjusted as to be moveable by means of a slide or screw,
by which the interval between the balls can be increased
or diminished. From the outer coating, another wire $g\ b$,
proceeds, which is intended to communicate with the
ground by being held in the hand, or otherwise.

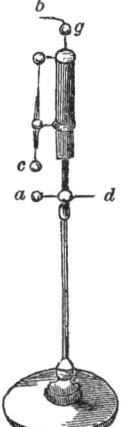

(92) To understand the action of this little instrument, it must be
considered that no charge of any amount can be given to a Leyden
phial, if it be *insulated* :—for, in proportion as the *positive electricity* is
communicated to its interior coating, it is necessary that the same
quantity should be removed from the exterior, which would otherwise

Fig. 43.

counteract the *negative* electricity by which the charge is sustained.
To effect this, a communication is established with the earth, or with

the interior coating of a second jar, the outside coating of which again may communicate with the interior of a third, and thus a series of insulated jars may be charged from each other, as shown in Fig. 43, taking care to withdraw the opposing Electricity of the last.

(93) If a very small jar be thus connected with a large one, the quantity of Electricity which is sufficient to carry the charge of the first up to a high degree of intensity when diffused over the larger surface of the second, will be scarcely appreciable, so that it will require many charges of the former to charge the latter to the same degree. When the unit jar is employed to determine the intensity of the charge of a phial, a communication is established between the prime conductor and the exterior coating by the wire g b, Fig. 42, and the interior coating is connected with the interior coating of the jar by the wire a d. The Electricity evolved by the machine is thus communicated to the large jar from the outer coating of the small one. The discharging rod c, regulating and measuring the distance between its two coatings. The value of the unit measure is determined by the distance between the balls c and a, or the *striking distance* of the spark. The effect of the spark from c to a, is to neutralize the coatings of the unit jar, and to distribute a corresponding quantity of Electricity over the surface of the large jar. On giving a second charge to the unit jar, the large jar receives an increment equal to what it received from the first charge, and the second spark only neutralizes the unit jar as before. Thus, by the number of explosions, or by the number of charges which have passed from the smaller jar, the quantity accumulated in the battery may be very accurately estimated.

(94) When a series of explosions from a Leyden phial is required for any particular purpose, it is useful to have a contrivance by which the discharges can be effected without the interference of the operator. Fig. 44 represents the arrangement of Mr. Lane for this purpose. a is the prime conductor of an electrical machine; b the jar, on the wire communicating with the interior of which is fixed the arm of varnished glass c, on the end of which is cemented the brass knob D; through this ball the wire f d slides, so that the ball d may be brought to any required distance from the knob of the jar

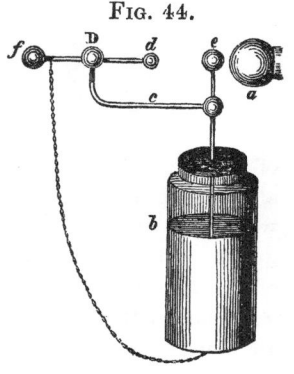

FIG. 44.

e. A simple inspection of the figure will show how this discharging electrometer acts, and how, by increasing or lessening the distance between d and e, the strength of the charge may be regulated.

(95) But a far more useful instrument is the balance electrometer of Cuthbertson, shown in Fig. 45. A B is a wooden stand, about eighteen inches long and six broad, in which are fixed two glass sup ports *d e*, mounted with brass balls; under the ball *d* is a brass hook: the ball *b* is made of two hemispheres, the under one being fixed to the brass mounting, and the upper one turned with a groove to shut upon it, so that it can be taken off at pleasure: it is screwed to a brass tube about four inches long, fitted on to the top of *e*; from its lower end proceeds an arm carrying the piece *f c*, being two hollow balls and a tube, which together makes nearly the same length as that fitted on to *e*: *g h*, is a straight brass wire, with a knife-edged centre in the middle, placed a little below the centre of gravity, and equally balanced with a hollow brass ball at each end, the centre or axis resting upon a proper shaped piece of brass fixed in the inside of the ball *b*; that part of the hemisphere towards *h* is cut open to permit that end

FIG. 45.

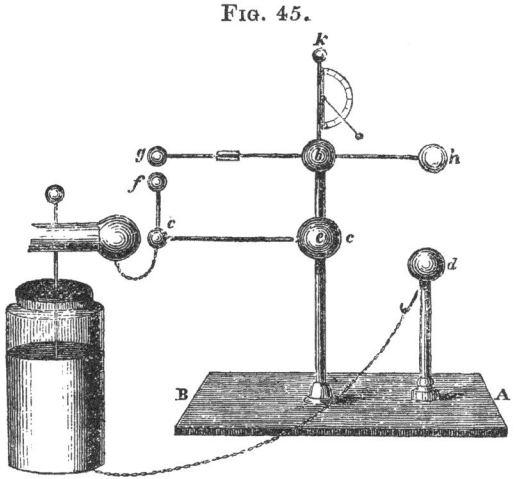

of the balance to descend till it touches *d*, and the upper hemisphere *b* is also cut open: the arm *g* is divided into sixty grains, and furnished with a slider, to be set at the number of grains the experiment requires: *k* is a common Henley's Electrometer screwed upon the top of *b*. The slider is placed loosely on the arm *g*, so that as soon as *g h* is out of the horizontal position it slides forward towards *f*, and the ascending continues with an accelerated motion till *h* strikes *d*.

Now suppose the instrument to be applied to a jar as in the figure; a metallic communication by a wire or chain is established between C and the inside of the jar, *k* is screwed upon *b* with its index pointing towards H, the increase of the charge in the jar is thus shown : suppose the slider to be set at fifteen grains, it will cause *g* to rest upon *f* with a pressure equal to that weight : as the charge increases in the jar the balls *f* and *g* become more and more repulsive of each other ; and when the force of this repulsion is sufficient to raise fifteen grains, the ball *g* rises, the slider moves towards *b*, and the ball *h* coming rapidly into contact with *d* discharges the jar, and as the force of the repulsion depends upon the intensity of the charge, the weight it has to overcome affords a measure of this intensity, and enables the experimenter to regulate the amount.

(96) A very useful piece of apparatus for directing with precision the charge of a jar or battery, is Henley's Universal Discharger, Fig. 46 ; it consists of a wooden stand with a socket fixed in its centre, to which may be occasionally adapted a small table having a piece of ivory, (which is a non-conductor) inlaid on its surface. The table may be raised and kept at the proper height by means of a screw *s*. Two glass pillars P P are cemented into the wooden stand. On the top of each of these pillars is fitted a brass cap having a ring R attached to it, and containing a joint moving both vertically and horizontally, and carrying on its upper part a spring tube admitting a brass rod to slide through it. Each of these rods is terminated at one end either by a ball *a b* screwed on a point, or by a pair of brass forceps, and is furnished at the other extremity with a ring or handle of solid glass. The body through which the charge is intended to be sent, is placed on the table, and the sliding rods, which are moveable in every direction, are then by means of the handles brought in contact with the opposite sides, and one of the brass caps being connected with the outside of the jar or battery, the other may be brought into communication with the inner coatings by means of the common discharging rod, Fig. 38. For some experiments it is more convenient to fix the substance on which the experiment is to be made in a mahogany frame F, consisting of two small boards which can be pressed together by screws, and which may then be substituted for the table. In either of these ways the charge can be directed through any part of the substance, with the greatest accuracy. In Fig. 47, is exhibited an electrical battery of fifteen jars, in the act of being charged from the

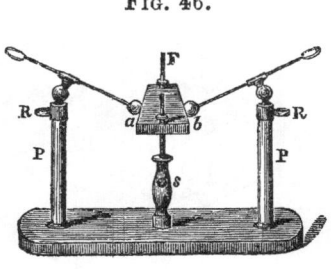

FIG. 46.

machine, with Cuthbertson's balance electrometer, and an arrangement
for striking metallic oxides attached.

FIG. 47.

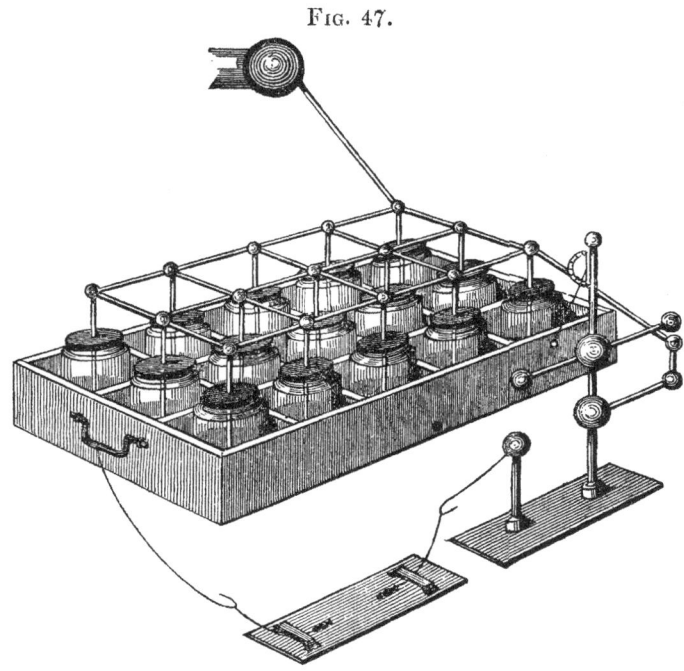

(97) *Experiments with the Leyden phial and battery.*

Ex. 1. Fix to the outside coating of the jar *a*, Fig 48, exposing
about a square foot of coated surface, a curved
wire *b*, terminated by a metallic ball *c* rising to
the same height as the knob of the jar *d*; charge
the jar, and suspend midway between *c* and *d* by
a silken thread, a small ball of cork or elder pith.
The ball will immediately be attracted by *d*, then
repelled to *c*, again attracted, and again repelled,
and this will continue for a considerable time :
when the motion has ceased, apply the discharging rod to the jar, no spark or snap will
result—proving that the phial has been gradually discharged by the pith or cork ball, the
motion of which from *d* to *c* likewise proving
the opposite electrical states (16) of the outer
and inner coatings.

FIG. 48.

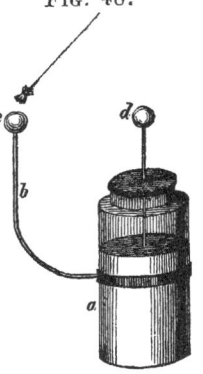

FIG. 49.

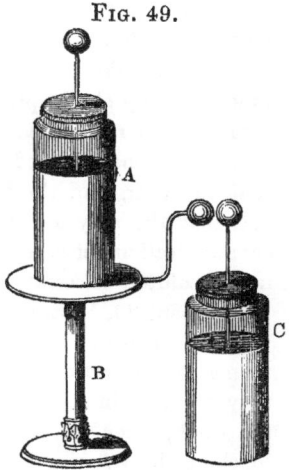

Ex. 2. Place the jar A, Fig. 49, on the insulating stand B, and attempt to charge it from the prime conductor, you will find it impossible (92); now apply the knuckle to the outside coating, and continue to turn the machine : for every spark that enters the jar, one will pass between the outside coating and the knuckle, and on applying the discharging rod, the jar will be found to have received a charge. Instead of the knuckle, the knob of a second *uninsulated* jar C, may be applied, as in Fig. 43, *both* jars will receive a charge.

Ex. 3. Provide a jar, the exterior coating of which is moveable, (it may be made of thin tin plate,) charge this jar in the usual manner (86), and then place it on an insulating stand : touch the knob from time to time with a conducting body; the whole charge will thus ultimately be removed, and the glass will be brought to its natural state : now charge the jar again, remove the outer coating, and replace it on the insulating stand ; in this state it will retain its charge for an indefinite period. The reason of this is, that the wire, by which the charge is communicated to the interior coating, being left attached to it, induction does not take place solely through the glass to the opposite coating, but is partly directed, through the air, to surrounding conductors : this portion is usually called *free charge,* and on removing this, by touching the knob with a conducting body, a corresponding portion of free charge, of the opposite kind, makes its appearance on the outside coating, owing to the induction which is now at liberty to direct itself from that part to surrounding objects. But when the exterior coating is removed, the induction is determined entirely *through the glass,* and the charge on one side is sustained by an exactly equal quantity of the contrary Electricity on the other : all interference with surrounding objects is thus cut off.

Ex. 4. Provide a jar with both coatings moveable (the jar for this purpose must be as wide at the mouth as at the bottom) : let the wire communicating with the interior coating pass through a glass tube, by which it may be removed from the jar without touching the metal : charge the jar in the usual manner, then withdraw the inside coating ; and having set it aside, invert the jar upon some badly conducting body, such as the table-cloth, and remove the exterior coating: then, on applying the discharging rod to the two coatings, no spark or explosion will take place, and they may be taken in the hands without producing any shock, proving them to be quite free from any electrical charge: now replace the coatings on the jar, and complete the circuit with the discharging rod: both spark and explosion will result, proving that the charge of the Leyden jar is dependent on the dielectric glass, and that the only use of the coatings is to furnish a ready means of communication between the charged particles. That the coatings are not absolutely necessary to the charge of a glass jar, is proved by Experiment 29, Lecture 1st.

Ex. 5. Place a charged jar on an insulating stand, and make a communication between the interior coating and the electric bells, Fig. 24: they will remain at rest until the outside of the jar is connected with the earth, when the clappers will be set in active motion: thus, by touching the exterior coating, from time to time, with the finger, the bells may be made to ring at pleasure.

Ex. 6. Place some gunpowder on the ivory slip of the table of the universal discharger, Fig. 42; and having unscrewed the balls *a*, *b*, insert the points of the wires into the powder, about half an inch apart: on passing an explosion from the Leyden phial through the powder, it will be scattered in all directions, but *not* ignited, an effect occasioned, probably, by the enormous velocity (576,000 miles in a second, according to Wheatstone's experiments) with which Electricity travels, not allowing sufficient time to produce the effects of combustion: that this is the reason, is rendered apparent by

Ex. 7. In which some loose gunpowder is placed in the ivory mortar,

FIG. 50.

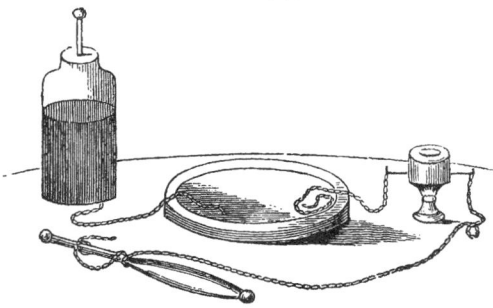

Fig. 50, and the circuit interrupted by ten or twelve inches of water in a porcelain basin: under these circumstances the gunpowder is fired on discharging the jar.

Fig. 51 represents Mr. Sturgeon's apparatus for firing gunpowder.

The powder is placed in the wooden cup A, either dry or made up into a pyramidical form with a little water. The brass ball *b*, which moves on a joint, is brought immediately over it, the chains *c*, *d*, being connected with the outer and inner surfaces of a Leyden jar. The discharge takes place, and the powder is inflamed.

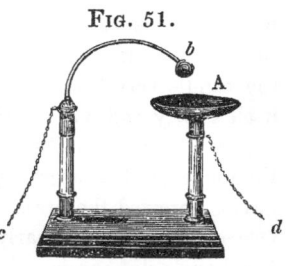

Fig. 51.

Ex. 8. Tie some tow loosely round one of the knobs of the discharging rod, and dip it in powdered resin : place the naked knob in contact with the outside of a charged jar, and bring the other quickly in contact with the ball *a* : discharge will take place, and the resin will burst into a flame.

Ex. 9. Place a thick card or some leaves of a book against the outer coating of a Leyden jar, or between the knobs of the universal discharger : pass the explosion, the discharge will pass through the paper or card, and perforate it, producing a burr or protrusion in both directions, as though the force producing it had acted from the centre of the thickness of the card outwards; a strong and peculiar odour is at the same time developed.

Ex. 10. Drill two holes in the ends of a piece of wood half an inch long and a quarter of an inch thick : insert two wires in the holes, so that the ends within the wood may be rather less than a quarter of an inch distant from each other: pass a strong charge through the wires, and the wood will split with violence. Stones may be split in a similar manner.

Ex. 11. Hang two curved wires, provided with a knob at each end, in a wine-glass nearly full of water, so that the knobs shall be about half an inch asunder: connect *a*, Fig. 52, with the outer coating of a charged jar, and *b*, with the inner coating, by means of the discharging rod; when the explosion takes place, the glass will be broken with great violence.

Fig. 52.

Ex. 12. Remove the press from the universal discharger, and place a lighted candle in the socket : unscrew the balls, and arrange the points of the wires a little above the top of the wick of the candle, and about one inch apart : charge a jar, and having blown out the candle, make the connections between the outer and inner coating : the jar will discharge itself through the smoke of the candle, and relight it.

Ex. 13. Adjust the candle so that the flame shall be exactly on a level with the two points of the discharging wires : set the point of the wire which is to communicate with the interior coating of the jar, at the distance of one inch and a half from the flame, snuff the wick of the candle very low, and complete the circuit, the jar will discharge itself slowly and put out the candle.

Ex. 14. Remove the candle, and screw the table into the socket of the universal discharger: place a lump of sugar on the ivory slip, and having screwed the brass balls on the discharging wires, bring the surface of the sugar to nearly the same height as the centre of the balls. Fix Lane's discharging electrometer, Fig. 44, on the Leyden phial, and interpose the universal discharger between the chain *f* and the outside coating of the jar: darken the room, and turn the electrical machine. When the jar is charged sufficiently high, it will discharge itself over the surface of the sugar, illuminating it, and the light will continue for some time. If five or six eggs be arranged in a straight line, and in contact with each other, they will be rendered luminous by passing a small charge through them.

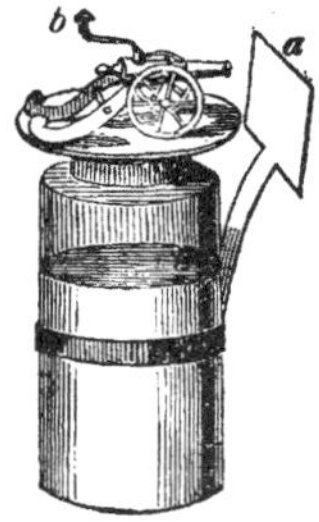

Ex. 15. Place a little model of a brass cannon on a circular brass plate fixed on the top of a Leyden phial instead of the ball, as shown in Fig. 53 : connect the square piece of brass *a* with the exterior coating, and arrange it at the distance of about half an inch from the mouth of the cannon ; bring the knob *b* of the cannon in contact with the prime conductor, and hold a card between the mouth of the cannon and the brass plate *a*, so that it shall not touch either: when the jar has received a sufficient charge, the explosion will pass, and the card will be perforated, as in Experiment 9.

FIG. 53.

Ex. 16. Colour a card with vermillion, unscrew the balls from the universal discharger, and place the points on opposite sides of the card, one about half an inch above the other ; discharge a jar through the card, it will be perforated at the point opposite to the wire connected with the *negative* side of the jar ; a zig-zag black line of reduced mercury will be found extending from the point where the positive wire touches the card to the place of perforation. This curious result arises from the great facility with which positive Electricity passes through air, as compared to negative ; and on repeating the experiment *in vacuo*, the perforation always takes place at a point *intermediate* between the two wires. (Dr. Golding Bird.)

Ex. 17. To the knob of a large jar A, Fig. 54, screw a small metallic stage C, on which place a small jar B, charge the large jar in the usual manner: the small jar, though it will not be charged in the usual acceptation of the term, will nevertheless be in a state of pola-

rization; and on bringing one ball of the discharging rod
in contact with the exterior coating of the large jar, and
the other in contact with the knob of the small jar, a flash
and report will result, arising from the neutralization of a
portion of the negative Electricity of the outside surface of
A, by a corresponding portion of positive Electricity from
the interior of B : *both jars* will now be charged, the inner
surface of A and the outer surface of B being positive, and
the outer surface of A and the inner surface of B negative;
and both jars may be discharged together, by connecting
the inside of B by means of a wire or chain with the outside of A, and bringing one knob of the discharging rod in
contact with this wire or chain, and the other on the

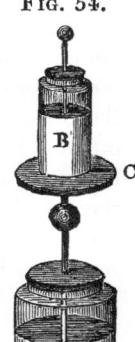

Fig. 54.

stage C, on which the small jar stands. If the large jar A be first discharged in the usual manner, by bringing one knob of the discharging
rod in contact with its outside coating, and the other within striking
distance of the stage C, a second charge will be communicated to it by
the electro-polar influence of the small jar the moment that the discharging rod is removed; and a second small explosion will take place
on applying the discharging rod; after which, both jars will be reduced
nearly to a state of neutrality.

Ex. 18. Fill the bent glass tube, *c d*, Fig. 55, with resin, or sealing-
wax, then introduce two wires, *a b*, through its ends, so that they may
touch the resin and penetrate a little way
into it : let a person hold the tube over a
clear fire by the silk string *e*, so as to melt
the resin and at the same time connect the
wires with the interior and exterior coatings
of a charged jar : while the resin is solid,
the discharge cannot take place through it,

Fig. 55.

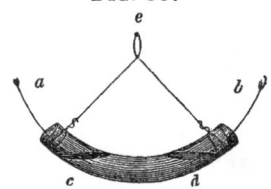

but as it melts it becomes a conductor, (see Exp. 20, Lecture 1st) and
then the discharge passes freely.

Ex. 19. The sudden rarefaction which air undergoes during the
passage of the electric spark through it, is well shown by an apparatus
devised by Mr. Kinnersley, of Philadelphia, and shown in Fig. 56.
It consists of a glass tube ten inches long and two inches in diameter,
closed air-tight at both its ends by two brass caps: a small glass tube,
open at both ends, the lower one bent at a right angle, passes through
the bottom cap, and enters the water contained in the lower portion of
the large tube. Through the middle of each of the brass caps a wire is
introduced, terminating in a brass knob within the tube, and capable of
sliding through the caps, so as to be placed at any distance from each
other. If the two knobs be brought into contact, and a Leyden jar dis-

charged through the wires, the air within the tube FIG. 56.
undergoes no change in volume : but if the knobs
are placed at some distance from each other when the
jar is discharged, a spark passes from one knob to
the other : the consequence is a sudden rarefaction
of the air in the tube, shown by the water instantaneously rising to the top of the small tube, and
then suddenly subsiding ; after which it gradually
sinks to the bottom of the tube, the air slowly
recovering its original volume.

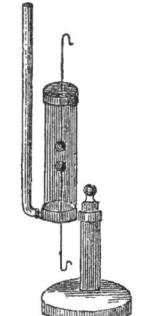

Ex. 20. Fig. 57 represents two small electric jars, coated as usual,
externally, and provided with valves to withdraw the air from them
by means of an air-pump. After the exhaustion, brass balls are screwed
on the necks of the jars over the valves. From the FIG. 57.
brass caps wires proceed a few inches within the
phials, terminating in blunt points. A jar fitted up
in this manner may be charged and discharged like a
common Leyden phial, induction taking place with
great facility through highly rarefied air. When
charged and discharged in a dark room, the extremity
of the wire in the inside becomes beautifully illu--

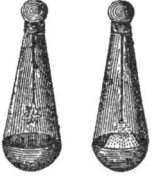

minated with a *star* or *pencil* of rays (as shown in the figure), according as the Electricity happens to be positive or negative. This
experiment is known as the Leyden vacuum.

Ex. 21. One of the most beautiful experiments in Electricity is that
of the "falling star:" it is produced by transmitting a considerable
electrical accumulation through an exhausted receiver. Singer, in his
excellent " Elements of Electricity," recommends a glass tube, five feet
in length and $\frac{5}{8}$ of an inch in diameter, capped with brass at each extremity. When such a tube is exhausted, no ordinary Electricity will
pass through it in any other than a diffused state ; but by employing
the charge of a very large jar, *intensely charged*, a brilliant spark is
obtained through the whole length of the tube. The metallic termination in the tube should be a very small and well polished ball; and if
care be taken to have the brass caps well rounded, and the air within
the tube, not too much attenuated, the experiment will rarely fail. If
the tube be six feet long, it may be four inches in diameter, and a jar
having five square feet of coating should be employed. An assistant
should work the pump, and the operator should occasionally try to pass
the charge down ; when at a certain degree of exhaustion, it does so in
a brilliant line of white light.

Ex. 22. Fig. 58 represents an ingenious contrivance for showing the
explosion of gunpowder by
Electricity. It is generally
made seven or eight inches
long, and nearly the same
height to the top of the roof;
the side, and that half of the
roof next the eye, is omitted in
the figure that the inside may
be more conveniently seen.
The sides, back, and front of
the house are joined to the
bottom by hinges; the roof is
divided into two parts, which

FIG. 58.

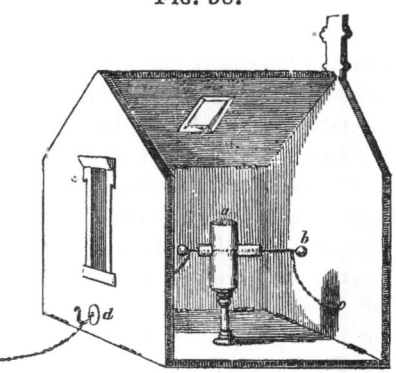

are also fastened by hinges to the sides : the building is kept together
by a ridge fixed half way on one side of the roof, so that when the
building is put together it holds it in its place. Within the house
there is a brass tube $1\frac{1}{2}$ inch long, and $\frac{5}{8}$ of an inch in diameter, screwed
on to a pedestal of wood, which goes through about one-eighth of an
inch, the other end by means of a chain has a communication to the
hook *d ;* at the other side of the tube, a piece of ivory, one inch long,
is screwed, with a small hole for a wire to slide into.

To use this apparatus, fill the brass tube *a* with gunpowder, and ram
the wire *b* a small way into the ivory tube ; then connect the hook *c*
with the bottom of a large jar, interposing a dish of water as in Fig. 50:
charge the jar, and form a communication from the hook *d* to the knob,
discharge will take place, the gunpowder will explode, throwing
asunder the roof, upon which the sides, front, and back will fall down,
without, however, undergoing any damage. The apparatus may be
placed on the ground, or on a table out of doors, communication being
established with the Leyden phial within, by means of insulated wires.

Ex. 23. Fig. 59 exhibits a piece
of apparatus for showing in an
amusing manner the power of the
electric discharge to cause the elements of water, viz. oxygen and
hydrogen, to enter into combination. The metallic wire which
passes through the touch-hole of
the small brass cannon is insulated
from the metal by a hollow tube

FIG. 59.

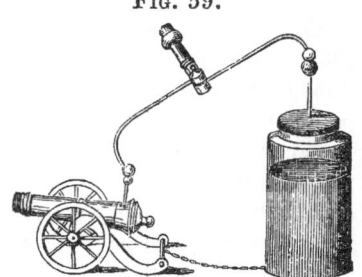

of ivory : this wire reaches nearly but not quite across the bore of the
barrel. The cannon is charged in the following manner :—the mixture

of the oxygen and hydrogen gas being ready in a 4 or 6 oz. stoppered bottle, the cannon is filled with sand, and being held close to the mouth of the bottle, the stopper is removed, and the sand from the cannon entering the gas at the same moment ascending occupies its place. The mouth of the cannon is closed by a cork, which is projected to a considerable distance by the force of the explosion. A single inspection of the figure will show the manner of passing the electric discharge.

Ex. 24. The following experiment is exceedingly beautiful, and highly interesting, as demonstrating the opposite electric states of a charged jar. Make the resinous cake of an electrophorus (38) dry and warm : draw lines on it with the knob of a positively charged jar, and sift over these places a mixture of sulphur and red-lead ; on inclining the plate to allow the excess of the powder to fall off, every line marked by the knob of the jar, will be observed covered with the *sulphur*, whilst the minium will be dispersed. On wiping the plate, and drawing figures with the outside of the jar, the sulphur will be dispersed, and the *minium* collected in a very elegant manner on the lines described by the outside of the jar. The rationale of this experiment is as follows :—the sulphur and red-lead, by the friction to which they have been exposed, are brought into opposite electrical states, the sulphur is rendered negative, and the red-lead positive, so that when the mixture is made to fall on surfaces possessing one or the other Electricity in a free state, the sulphur will be collected on the positive, and the minium on the negative portions of the plate, according to the well-known law of electric attraction. This experiment may be varied by tracing various lines at pleasure on a smooth plate of glass, with the knob of a jar, charged first with positive and then with negative Electricity : on gently dusting the surface with the mixture of sulphur and red-lead, a series of red and yellow outlines will be formed. This experiment was devised by Professor Lichtenberg, of Gottingen. The mechanical effects and calorific phenomena accompanying the discharge of an electric battery, are highly interesting.

Ex. 25. Between the boards of the press of the universal discharger (96): lay a piece of stout plate-glass, and send a powerful charge through it, the glass will not only be broken into fragments, but a portion even reduced to an impalpable powder.

Ex. 26. Lay a fine iron chain, about two feet long, upon a sheet of white paper, and transmit a charge from six or eight square feet of coated surface through it : on removing the chain, its outline will be observed marked upon the paper with a deep stain at each link, indeed, if this charge is sufficiently powerful, the paper is frequently burnt through.

Ex. 27. Place a slip of tin foil, or of gold leaf, between two pieces of paper, allowing the ends to project, and press the whole firmly together between the boards of the press of the universal discharger; transmit the shock of a battery through it, the metals will be completely oxidized ; if gold leaf be the metal employed, the paper will be found stained of a deep purple hue.

Ex. 28. If a piece of paper be laid on the table of the discharger, and a powerful shock directed through it, it will be torn in pieces.

The electrical battery is exhibited in Fig. 47, in the arrangement for fusing metallic wires, and converting them into oxides, and in Fig. 45 a large jar is represented in the experiment of fusing fine iron wire, a wire being substituted in place of the chain at *c.* The best material for this purpose is the finest flattened steel, sold at the watchmakers' tool shops, under the name of watch-pendulum wire. It does not require a large extent of coated surface merely to fuse metallic wires, provided they are sufficiently thin ; but to effect their oxidation, large batteries are necessary. From the experiments of Brooke and Cuthbertson, it has been inferred, that the length of wire melted by the electric discharge, varies as the *square of the quantity of accumulated Electricity* which is sent through it : thus, a combination of two jars, charged to an equal degree, will melt four times the length of wire which one jar will melt.

Fig. 60 represents a useful apparatus for deflagrating metallic wires, devised by Professor Hare. Two brass plates *s, s,* are fixed in a pedestal by a bolt N, about which they have a circular motion. On one of the plates a glass column C is cemented, surmounted by a forceps F ; at the corresponding plate there is a brass rod R, furnished also with a forceps. Between this forceps and that at F the wire through which the electric charge is to be sent is stretched : it may be of various lengths, according to the angle which the plates *s, s,* make with each other. The bottom of the pe-

Fig. 60.

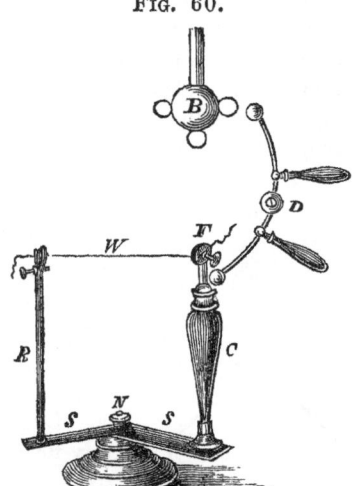

destal is in communication with the exterior coating of a jar or battery which is charged from the prime conductor B, and with which it is allowed to remain in communication. Now, it is obvious that in this case, touching the conductor is equivalent to touching the inner coating of the battery. However, by causing one of the knobs of the discharger D to be in contact with the insulated forceps F, and approximating the other knob to the prime conductor, the charge will pass through the wire W.

The oxides of metals produced by sending powerful electric charges through fine wires, and which may be preserved by stretching them about $\frac{1}{8}$ of an inch above sheets of white paper, as shown in Fig. 44, are exceedingly beautiful: the wires disappear with a brilliant flash, and the paper is found marked as described below (from Singer's Electricity), though no description can convey an adequate idea of the beauty of the impressions.

	Diameter.	Colour of the Oxides on paper.
Gold wire	$\frac{1}{180}$ of an inch	purple and brown.
Silver	$\frac{1}{160}$,,	grey, brown and green.
Platinum	$\frac{1}{180}$,,	grey and light brown.
Copper	$\frac{1}{160}$,,	green, yellow and brown.
Iron	$\frac{1}{180}$,,	light brown.
Tin	$\frac{1}{180}$,,	yellow and grey.
Zinc	$\frac{1}{180}$,,	dark brown.
Lead	$\frac{1}{180}$,,	brown and blue grey.
Brass	$\frac{1}{180}$,,	purple and brown.

Ex. 29. By the following experiment it will be proved that Electricity exerts an agency directly the reverse of the above, viz. that of restoring to the metallic state oxide of tin. If a portion of this oxide be inclosed in a glass tube, and a succession of strong explosions directed through it, the glass will after a time be found stained with metallic tin, and vermillion may be resolved into mercury and sulphur by the charge of a moderate-sized jar.

FIG. 61.

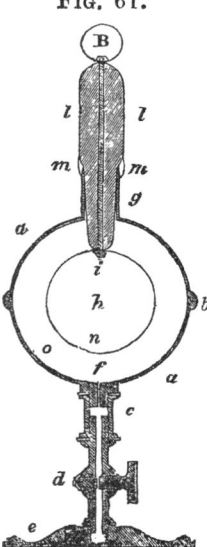

(98) It was with an apparatus constructed on the principles of the Leyden phial, that Faraday succeeded in proving by the most decisive experiment that *induction has a particular relation to the different kinds of matter through which it is exerted.* A section of this ingenious contrivance is shown in Fig. 61. *a, a,* are the two halves of a brass sphere, with an air-tight joint at *b,* like that of the Magdeburgh hemispheres, made perfectly flush and smooth inside, so as to present no irregularity; *c* is a connecting piece, by which the apparatus is joined to a good stop-cock *d,* which is itself attached either to the metallic foot *e,* or to an air-pump. The aperture within the hemisphere at *f* is very small: *g* is a brass collar fitted to the upper hemisphere, through which the shell-lac support of the inner ball and its stem passes; *h* is the inner ball, also of brass: it screws on to the

brass stem *i*, terminating above by a brass ball B; *l, l,* is a mass of shell-lac, moulded carefully on to *i*, and serving both to support and insulate it and its balls *h*, B. The shell-lac stem *l* is fitted into the socket *g* by a little ordinary resinous cement more fusible than shell-lac applied at *m, m*, in such a way as to give sufficient strength and render the apparatus air-tight there, yet leave as much as possible of the lower part of the shell-lac stem untouched as an insulation between the ball *h* and the surrounding sphere *a, a.* The ball *h* has a small aperture at *n*, so that when the apparatus is exhausted of one gas and filled with another, the ball *h* may also itself be exhausted and filled, that no variation of the gas in the interval *o* may occur during the course of an experiment.

(99) The diameter of the inner ball is 2·33 inches, and that of the surrounding sphere 3·57 inches. Hence the width of the intervening space through which the induction is to take place is 0·62 of an inch; and the extent of this place or plate, *i. e.* the surface of a medium sphere, may be taken as 27 square inches, a quantity sufficiently large for the comparison of different substances. Great care was taken in finishing well the inducing surfaces of the ball *h* and sphere *a, a,* and no varnish or lacquer was applied to them, or to any part of the metal of the apparatus.

(100) When the instrument was well adjusted, and the shell-lac perfectly sound, its retentive power was found superior to that of Coulomb's Electrometer, i. e. the proportion of loss of power was less. A simple view of its construction shews that the intervening dielectric or insulating medium may be changed at pleasure with either solids, liquids, or gases; and that it is admirably adapted for investigating the specific inductive capacities of each.

(101) Two of these instruments, precisely similar in every respect, were constructed; and the method of experimenting was (different insulating media being within) to charge one with a Leyden phial, then after dividing the charge with the other, to observe what the ultimate conditions of each were. For a detailed account of the method of manipulating, and the precautions necessary to obtain accurate results, I must refer to the original paper of the author (Experimental Researches, Eleventh Series, 1187 et seq.)

(102) The question to be solved may be stated thus : suppose *a* an electrified plate of metal suspended in the air, and *b* and *c* two exactly similar plates, placed parallel to and on each side of *a* at equal distances and uninsulated; *a* will then induce equally towards *b* and *c*. If in this position of the plates some other dielectric than air, as shell-lac, be introduced between *a* and *c*, will the induction between them remain the same? Will the relation of *c* and *b* to *a* be unaltered notwith-

standing the difference of the dielectrics interposed between them? (Exp.
Resear. 1252.)

(103) The first substance submitted to examination was shell-lac, as
compared with air. For this purpose a thick hemispherical cap of
shell-lac was introduced into the lower hemisphere of one of the inductive apparatus, so as nearly to fill the lower half of the space between
it and the lower ball. The charges were then divided (101), each apparatus being used in turn to receive the first charge before its division by
the other; and as it had previously been ascertained that both the
instruments had equal inductive power when air was in both, it was
concluded that if any difference resulted from the introduction of the
shell-lac, a peculiar action in that substance would be proved, and a
case of specific inductive influence made out.

(104) On making the experiment with all the care and attention
that could be bestowed, an extraordinary and unexpected difference
appeared, and the conclusion was drawn that the specific inductive
capacity of shell-lac as compared with air is as 2 to 1. With glass a
result came out, showing its capacity compared with air to be as 1·76
to 1 ; and with sulphur a result showing its capacity to be as 2·24 to
1. With this latter substance the result was considered by Faraday as
unexceptionable, it being, when fused, perfectly clear, pellucid, and free
from particles of dirt, and being moreover an excellent insulator.

(105) Liquids, such as oil of turpentine and naphtha, were next
tried ; and though no good results could be obtained, on account of
their conducting power, they were nevertheless considered by Faraday
as not inconsistent with the belief that oil of turpentine at least has a
specific inductive capacity greater than air.

(106) Air was then tried, but no alteration of capacity could be detected on comparing together, rare and dense, hot and cold, or damp
and dry : then all the gases were submitted to examination, being compared together in various ways, that no difference might escape detection,
and that the sameness of result might stand in full opposition to the con-
trast of property, composition, and condition, which the gases themselves
presented ; nevertheless not the least difference in their capacity to
favour, or admit electrical induction through them could be perceived.

(107) During the experiments with shell-lac (103), Faraday first
observed the singular phenomenon of the *return charge*. He found,
that, if, after the apparatus had been charged for some time, it was
suddenly and perfectly discharged, even the stem having all Electricity
removed from it, it gradually recovered a charge which in nine or ten
minutes would rise up to 50° or 60°. He charged the apparatus with the
hemispherical cap of shell-lac in it, for about forty-five minutes, to
above 600° with positive Electricity at the balls h and B, Fig. 61

above and within. It was then discharged, opened, the shell-lac taken out, and its state examined by bringing the carrier ball of Coulomb's Electrometer, Fig. 2, near it, uninsulating the ball, insulating it, and then observing what change it had acquired. At first the lac appeared quite free from any charge, but gradually its two surfaces assumed opposite states of Electricity, the concave surface, which had been next the inner and positive ball, assuming a positive state, and the convex surface, which had been in contact with the negative coating, acquiring a negative state; these states gradually increasing in intensity for some time.

(108) Glass, spermaceti, and sulphur, were next tried, all of them exhibited the peculiar state after discharge. Faraday also sought to produce it without induction, and with one electric power, but failed in doing so; a fact in favour of the inseparability of the two electric forces, and an argument in favour of the dependence of induction upon a polarity of the particles of matter.

(109) Faraday was at first inclined to refer these effects to a peculiar masked condition of a certain portion of the forces, but he afterwards traced them to the known principles of electrical action. He took two plates of spermaceti and put them together, so as to form a compound plate, the opposite sides of which were coated with metal. The system was charged, then discharged, insulated, and examined, and found to give no indication to the carrier ball : the plates were then separated, when the metallic linings were found in opposite electrical states. Hence, it is clear that an actual penetration of the charge to some distance within the dielectric, at each of its two surfaces, took place by conduction : so that to use the ordinary phrase, the electric forces sustaining the induction, are not upon the metallic surfaces only, but upon and within the dielectric; also extending to a greater or smaller depth from the metal linings.

(110) The following explanation may be offered :—Let a plate of shell-lac, six inches square, and half an inch thick, or a similar plate of spermaceti, an inch thick, coated on the sides with tin-foil, as in the Leyden phial, be charged in the usual manner, one side positively, and the other negatively. After the lapse of ten minutes, or quarter of an hour, let the plate be discharged and immediately examined; no Electricity will appear on either surface, but in a short time, upon a second examination, they will appear charged in the same way, though not in the same degree as they were at first. Now, it may be supposed, that under the coercing influence of all the forces concerned, a portion of the positive and negative forces have penetrated and taken up a position within the dielectric, and that consequently being nearer to each other, the induction of the forces towards each other will be much greater, and

that, in an external direction, less than when separated by the whole
thickness of the dielectric : when, however, all external induction is
neutralized by the discharge, the forces by which the electric charge was
driven into the dielectric, are at the same time removed, and the penetrated Electricity returns slowly to the exterior metallic coatings, constituting the observed re-charge. According to Faraday, it is the
assumption for a time, of this charged state of the glass, between the
coatings of the Leyden jar, which gives origin to a well-known phenomenon, usually referred to the diffusion of Electricity over the uncoated
portion of the glass, namely the *residual charge* (89). After a large
battery has been charged for some time, and then discharged, it is found
that it will spontaneously recover its charge to a very considerable extent, and by far the largest portion of this is referred to the return of
Electricity in the manner described.

(111) When a large jar or battery is discharged by a metallic wire
held in the hand, without the protection of an insulating handle, a
slight shock is frequently felt in the hand that grasps the wire : and
if a large jar be placed on a table, with its knob in contact with the
prime conductor, and if a chain be stretched upon the table, with one
end nearly touching the outside coating of the jar, by charging the apparatus till it discharges itself voluntarily, a spark is seen to pass between
each link of the chain, which thus becomes illuminated, though it forms
no part of the circuit.

(112) This spark is called the *lateral discharge ;* it is occasioned by
a small excess of free Electricity, which distributes itself over a discharging surface, when a charged system is discharged or neutralized.
It arises from the fact, satisfactorily established by Harris, and acknow-
ledged by Biot, Henry, and others, that the accumulated Electricity is
never exactly balanced between the opposed coatings ; so that there will
always be an excess of positive or negative Electricity *over* the neutralizing quantities themselves, disposed on the coatings of the jar. The
existence of this excess of Electricity, either positive or negative, is
proved by the fact, that if we charge a jar, allow it to remain insulated,
and discharge it gradually, by drawing sparks from the knob, and adding them to the outer coating, we can always take a *finite* spark from
either side alternately, whilst the jar rests on the insulator.

(113) If we place a charged jar upon an insulating stand, and discharge it in the usual manner, with a discharging rod, the excess of free
Electricity exhibits itself in the form of a spark, at the moment of discharge between any body connected with the outer coating, and another
in communication with the earth : the intensity of the spark depends on
the capacity of the jar, being *less* with a large jar, and *greater* with a
small one; the quantity of Electricity discharged being the same (Harris).

After the discharge, the knob, outer coating, and all bodies connected with the jar, are found in the same electrical state, which we may make either positive or negative, by taking a spark either from the knob or coating, previously to discharging the jar.

(114) This small quantity of free Electricity may be obtained even when the jar is connected with the earth, provided we seize it before the conductors have time to carry off the residuary accumulation; it having been proved by Professor Wheatstone, that some portion of time elapses in the passage of Electricity through wires: the effect, however, is greatest when the jar and its appendages are quite insulated.

(115) Although I believe the above to be the true explanation of what is termed the lateral discharge, agreeably to the opinion of Mr. William Snow Harris, it is right that I should mention that different views have been taken of the phenomenon :—Mr. Sturgeon, in his memoir on Lightning Conductors, (see Annals of Electricity, &c. for October, 1839,) states that there are *three* kinds of lateral discharge. The first, he says, takes place at " every interruption of a metallic circuit : by it the air is suddenly displaced, loose bodies are scattered and thrown from the axis of the circuit, and solid inferior conducting bodies are shattered or torn in pieces : by it *waves* of Electricity are produced in the neighbouring atmosphere, and the bodies within it." *The second* kind of lateral discharge, according to Mr. Sturgeon, " is a species of radiation of the electric matter from the surface of good conductors, carrying a *direct* or primitive discharge. It takes place most copiously from the edges of strips of metal, or from the surfaces of ragged or asperous wires." The third kind is " a displacement of the electric fluid, natural to those bodies which are vicinal to a continuous conductor, carrying the primitive discharge."

(116) As I shall have occasion to advert to these views of the lateral discharge in the next Lecture, while on the subject of lightning conductors, I defer for the present their farther consideration.

(117) *Physiological effects of common Electricity :*—The sensation experienced, when the body is made a part of the electrical circuit, is now so universally known, that a description here would be superfluous. The exaggerated accounts of their feelings, on the first transmission of a discharge through the bodies of the first experimentalists, was alluded to in the Introduction (7). It is not, however, easy to explain the cause of the muscular contraction that is experienced. The involuntary action may be produced by the concussion of a material agent (?) passing through the body, by an influence on the nervous system, or by a sudden disturbance of the electric equilibrium : and the dull pain at the joints is probably to be traced to the resistance which the force experiences in passing from one bone to another.

(118) It is stated by Mr. Morgan, that, if a strong shock be passed through the diaphragm, the sudden contraction of the muscles of respiration will act so violently on the air of the lungs, as to occasion a loud and involuntary shout; but that a small charge occasions in the gravest person a violent fit of laughter : persons of great nervous sensibility are affected much more readily than others. A small charge sent through the spine instantly deprives the person for a moment of all muscular power, and he generally falls to the ground. If the charge be very powerful, instant death is occasioned. Mr. Singer states that a charge passed through the head gave him the sensation of a violent and universal blow, which was followed by transient loss of memory and indistinctness of vision. A small charge sent through the head of a bird will so far derange the optic nerve as to produce permanent blindness ; and a coated surface of thirty square inches of glass will exhaust the whole nervous system to such a degree as to cause immediate death. Animals the most tenacious of life are destroyed by energetic shocks passed through the body. Van Marum found that eels are irrecoverably deprived of life when a shock is sent through their whole body : when only a part of the body is included in the circuit, the destruction is confined to that individual part, while the rest retains the power of motion.

The bodies of animals killed by lightning are found to undergo rapid putrefaction ; and it is a remarkable circumstance, that after death the blood does not coagulate.

(119) When the Leyden phial was first discovered, it was imagined that an agent of almost unlimited medical power was raised, and it was applied indiscriminately for the most opposite diseases. The failure consequent on such quackery brought Electricity into disrepute, and for a long time it was discarded almost entirely from our hospitals. It is now again more generally employed, and has been found of great service in many cases, such as palsy, contractions of the limbs, rheumatism, St. Vitus's dance, some kinds of deafness, and impaired vision. It is administered in five different ways :—1st, under the form of a gentle stream or aura, from a pointed piece of wood provided with an insulating handle, and communicating with the prime conductor: by this means it may be directed to parts of great sensibility, as the eye. 2nd, by causing the part to be operated upon to draw sparks from the prime conductor ; or, placing the patient on a stool with glass legs, by drawing sparks from him with a metallic ball. 3rd, by the transmission of shocks, which is the most severe and painful, and which requires great caution. 4th, by galvanism. And 5th, by electro-magnetism and magneto-electricity, both of which latter methods will be described in future Lectures.

(120) There can be no doubt that Electricity is very materially concerned in the economy both of animal and vegetable life, but we possess no precise information on the subject. It is not improbable that it may have something to do with the rise of sap, from the fact that Electricity always increases the velocity of a fluid moving in a capillary tube. On vegetables strong shocks have the same effects as on animals, namely, produce death : a very slight charge is sufficient to kill a balsam. It may further be observed that living vegetables are the most powerful conductors with which we are acquainted. Mr. Weekes found that a coated jar, having 46 inches of metallic surface, was repeatedly discharged by the activity of a vegetable point, in 4 min. 6 sec. ; while the same jar, charged to the same degree, required 11 min. 6 sec. to free it from its electric contents by means of a metallic point: the points in both cases being equi-distant. The same gentleman also found that the gold-leaf electroscope is powerfully affected by a jar at the distance of nearly seven feet, when the cap of the instrument is furnished with a branch of the shrub called *butcher's broom ;* though the same instrument, when mounted with pointed metallic wires, is not perceptibly affected until the charged jar approaches to within two feet of the cap.

If a blade of grass and a needle be held pointing towards the prime conductor of a machine, while the person holding them recedes from the instrument, a small luminous point will appear on the apex of the grass long after it has vanished from the apex of the needle.

(121) The following experiment was made by Pouillet, who drew from it the conclusion that a considerable portion of the Electricity with which the atmosphere is loaded is derived from the gaseous fluids given out by plants during the process of vegetation. But it is right to mention that the experiments of Sir H. Davy, described in the Phil. Trans. for 1826, p. 398, are rather inconsistent with that of the French Electrician.

M. Pouillet arranged in two rows beside each other, on a table varnished with gum-lac, twelve glass capsules, each about eight inches in diameter, coated externally for two inches round the lips with a film of lac varnish : they were filled with vegetable mould, and were made to communicate with each other by metallic wires, which passed from the inside of one to the inside in the next, going over the edges of the capsules. Thus the inside of the twelve capsules and the soil which they contained, formed only a single conducting body. One of these capsules was placed in communication with the upper plate of a condenser by means of a brass wire ; while at the same time the under plate was in communication with the ground.

Things being in this situation, and the weather very dry, a quantity of corn was sown in the soil contained in the capsules, and the effects were watched. The laboratory was carefully shut, and neither fire, nor light, nor any electrified body, was introduced into it. During the two first days, the grains swelled, and the plumulæ issued out about the length of a line, but did not make their appearance above the surface of the earth. But on the third day the blades appeared above the surface, and began to incline to the window, which was not provided with shutters. The condenser was now charged with *positive* Electricity ; consequently, the carbonic acid gas which disengages itself during the germination of the seed is charged with positive Electricity, and is therefore precisely in the same state as the carbonic acid formed by combustion. This experiment was repeated several times with success. But the Electricity cannot be recognised unless the weather is exceedingly dry, or unless the apartment is artificially dried by introducing substances which have the property of absorbing moisture.

These capsules being insulated, and the air being very dry, and the soil so dry that it is an imperfect conductor, it is evident that the Electricity would be retained. Accordingly, when the condenser was brought into a natural state after one observation, and then replaced for experiment, during one second only, it was found to be charged with Electricity.

(122) Mr. Pine, of Maidstone, describes the following experiment,* made to determine the influences of Electricity on germination:—A few grains of mustard seed were sowed in similar soils contained in Leyden jars, electrified positively and negatively. In four days the plants issued from the soils in both jars ; but those in the negative jars were most advanced. Plants under ordinary circumstances did not appear till about two days later. In fourteen days from the time of sowing, the plants in the negative jar had grown to $2\frac{3}{4}$ inches, those in the positive jar to $2\frac{1}{4}$ inches, and those in the ordinary state to $1\frac{1}{8}$ inches. The jars in these experiments were uninsulated for the purpose of allowing a slight current of fluid to pass through the plants. It is difficult to understand the manipulation in these experiments ; and, Mr. Pine having declined to furnish me with any information on the subject, the reader must be satisfied with the above statement of his results in his own words.

The following experiment was made by Mr. Weekes :—In two small flower-pots, filled with rich mould, a few grains of mustard-seed were sown ; both were kept gently watered, but one pot was *insulated* and frequently electrified under circumstances which kept it, as it were, in an electrified atmosphere. The other pot was not interfered with, and the result was, that the vegetation of the electrified seeds appeared

* Proceedings of the Electrical Society, Part III. p. 163.

several days before the others, and continued afterwards to grow with a much greater degree of vigour. The Electricity employed was derived from a galvanic arrangement of thirty pairs of plates, charged with salt water.

(123) It will be proper to mention, however, that the experiments undertaken by myself have failed to show any advantage in favour of electrified soil. They were as follow :—Three small metallic cups were filled with fine vegetable mould, and in each a few seeds of mustard were sown. The soil in each was uniformly moistened with water. Two of the cups were insulated and kept constantly electrified by two batteries of twenty pairs of 5-inch plates, charged with pump-water, the current being made to pass *down* the soil in one cup, and *up* the soil in the other. The third cup was unelectrified. On the third day, the *unelectrified* seeds had germinated, and appeared above the surface of the mould. On the morning of the fourth day, the young plants were fully developed ; but in the seeds in the electrified cups, germination had only just commenced. In this experiment, therefore, Electricity, instead of favouring, appeared actually to retard germination.

Three small garden-pots were then filled with the same soil, and mustard seed being sown in each, they were placed side by side in glass basins containing a little water. Two of them were then kept electrified by the same batteries as before, the wires passing about half-an-inch into the soil at the surface, and through the holes at the bottom of the pots. On the morning of the third day, germination had commenced in all the pots : on the fourth day, the young plants were all out of ground : at the end of a week they were fully developed, but there was not the slightest advantage in favour of those which had been raised under the influence of Electricity.

(124) *Chemical Effects.*—Dr. Priestley first investigated the chemical effects of ordinary Electricity by passing a succession of shocks through a small quantity of water tinged blue by litmus : the liquid in a short time acquired a red tinge, while the air confined in the tube suffered evident diminution : an acid had been formed by the chemical union of the elements of atmospheric air, viz. oxygen and nitrogen. It was Mr. Cavendish, however, that first explained this experiment of Priestley. (See Phil. Trans. for 1784.)

(125) When a succession of discharges are sent through water, a decomposition of that fluid takes place, the elements of which assume the gaseous form. This fact was discovered in 1789, by Messrs. Dieman, Paetz, and Van Troostwyck, associated with Mr. Cuthbertson. For the experiment they employed a glass tube a foot long, and one-eighth of an inch in diameter, through one end of which a gold wire was inserted, projecting about an inch and a-half within the tube ; that

end was then hermetically sealed. Another wire was introduced at the other end of the tube, which was left open, and passed upwards, so that its extremity came to a distance of five-eighths of an inch from the end of the first wire. The tube was then filled with distilled water, from which the air had been extracted by the air-pump, and inverted in a vessel containing mercury. A little common air was let into the top of the tube, in order to prevent its being broken by the discharge. Electrical shocks were then passed between the two ends of the wires through the water in the tube by means of a Leyden jar, which had a square foot of coated surface. This jar was charged by a very powerful double plate machine, which caused it to discharge twenty-five times in fifteen revolutions. At each explosion bubbles of gas rose to the top of the tube ; and when sufficient water had been displaced to lay bare the wires, the next shock kindled the gases and caused their re-union. Thus de-composition and re-composition were effected by the same agent. In the latter case, however, it may be supposed to have acted mechanically, or by the heat evolved in its passage through a badly conducting aëriform fluid.

(126) In 1801, Dr. Wollaston published in the Philosophical Transactions a description of a method of analyzing water by the transmission of sparks instead of shocks. He considered that the decomposition must depend upon duly proportioning the strength of the charge of Electricity to the quantity of water, and that the quantity exposed to its action at the surface of communication, depends on the extent of that surface, he therefore expected that by reducing the surface of communication the decomposition might be effected by smaller machines and with less powerful excitation than were usually considered necessary for the purpose. " Having," he says, " procured a small wire of fine gold, and given to it as fine a point as I could, I inserted it into a capillary glass tube ; and after heating the tube, so as to make it adhere to the point and cover it at every part, I gradually ground it down till with a pocket lens I could discern that the point of gold was disclosed. I coated several wires in this manner, and found that when sparks from a conductor were made to pass through water by means of a point so guarded, a spark passing to the distance of one-eighth of an inch would decompose water, when the point did not exceed $\frac{1}{700}$ of an inch in diameter. With another point, which I estimated at $\frac{1}{1500}$, a succession of sparks one-twentieth of an inch in length afforded a current of small bubbles of air.

(127) In these ingenious experiments, however, true *electro-chemical* decomposition was not effected ; that is, " the law which regulates the transference and final place of the evolved bodies had no influence." The water was decomposed at both poles independently of each other,

and the oxygen and hydrogen gases evolved at the wires are the elements of the water existing the instant before in those places. "That the poles, or rather points, have no mutual decomposing dependence, may be shown," observes Faraday, " by substituting a wire, or the finger, for one of them,—a change which does not at all interfere with the other, though it stops all action at the charged pole. This fact may be observed by turning the machine for some time ; for though bubbles will rise from the point left unaltered in quantity sufficient to cover entirely the wire used for the other communication, if they could be applied to it, yet not a single bubble will appear on that wire."

(128) The following beautiful experiments, made by Faraday (See Exp. Research. series v. 462 et seq.), prove that, so far from electro-chemical decomposition depending upon the simultaneous action of *two metallic poles*, *air* itself may act as a pole, decomposition proceeding therewith as regularly and truly as with metal.

(129) A piece of turmeric paper, not more than 0·4 of an inch in length, and 0·3 of an inch in width, was moistened with sulphate of soda, and placed upon the edge of a glass plate opposite to and about two inches from a point connected with a discharging train arranged by connecting metallically a sufficiently thick wire with the metallic gas-pipes of the house, with those of the public gas-works of London, and with the metallic water-pipes of London. A piece of tin-foil resting upon the same glass-plate was connected with the machine and also with the turmeric paper by the decomposing wire a, (Fig. 62.) The

FIG. 62.

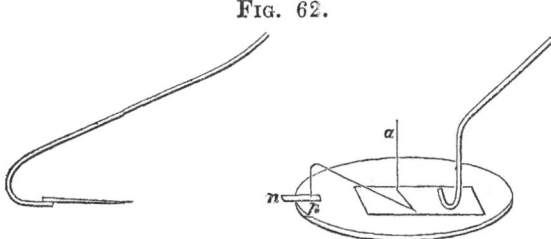

machine was then worked, the positive Electricity passing into the turmeric paper at the point p, and out at the extremity n. After forty or fifty turns of the machine (a plate fifty inches in diameter), the extremity n was examined, and the two points or angles found deeply coloured by the presence of free alkali.

(130) A similar piece of litmus paper dipped in a solution of sulphate of soda (Fig. 63) was now supported upon the end of the discharging train a, and its extremity brought opposite to a point p, connected with the conductor of the machine. After working the machine for a short time, acid was developed at both corners towards the point, *i. e.* at both

FIG. 63.

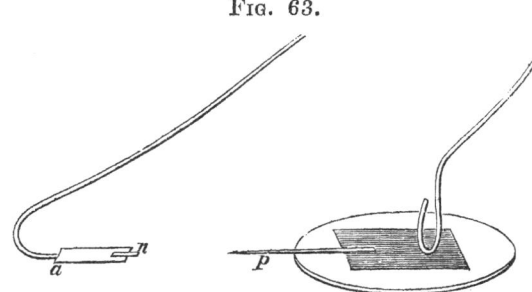

corners receiving the Electricity from the air. Then a long piece of turmeric paper, large at one end and pointed at the other, was moistened in the saline solution and immediately connected with the conductor of the machine, so that its pointed extremity was opposite a point upon the discharging train. When the machine was worked, alkali was evolved at that point; and even when the discharging train was removed, and the Electricity left to be diffused and carried off altogether by the air, still alkali was evolved where the Electricity left the turmeric paper.

(131) Arrangements were then made in which no metallic communication with the decomposing matter was allowed, but *both poles* formed of air only. Pieces of turmeric and litmus paper, *a* and *b*, (Fig. 64)

FIG. 64.

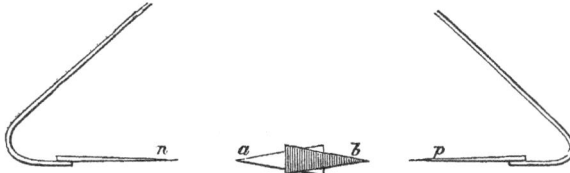

moistened with solution of sulphate of soda, were supported on wax between the points, connected with the conductor of the machine and the discharging train, as shown in the figure; the interval between the respective points was about half-an-inch. On working the machine, evidence of decomposition soon appeared, the points *b* and *a* being reddened from the evolution of acid and alkali.

Lastly, four compound conductors of litmus and turmeric paper were arranged as shown in Fig. 65, being supported on glass rods; and on

FIG. 65.

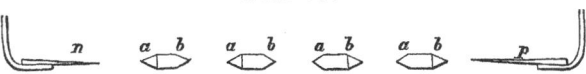

working the machine carefully, so as to avoid sparks and brushes, evidence of decomposition was obtained in each.

(132) Notwithstanding, then, the absence of metallic poles, we have here cases of electro-chemical decomposition precisely similar to those effected under the influence of the voltaic battery; and we appear to have direct proof also that the power which causes the separation of the elements is exerted not at the poles, but at the parts of the body which is suffering decomposition.

(133) The arrangement shown in Fig. 66 was employed by Faraday for effecting electro-chemical decomposition by common Electricity. On a glass plate, raised above a piece of white paper, two small slips of tin-foil, *a*, *b*, were placed : one was connected by the insulated wire *c* with an electrical machine, and the other by the wire *g* with a discharging train, or with the negative conductor. Two pieces of fine platinum wire, bent as in Fig. 67,

FIG. 66.

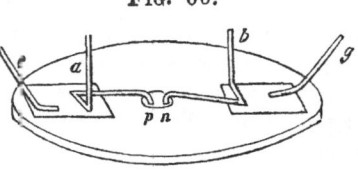

were provided, and so arranged that the part *d*, *f*, was nearly upright, while the whole rested on the three bearing points, *p*, *e*, *f*. The points *p*, *n*, thus became the decomposing poles. They were placed on a piece of filtering paper wetted with the solution to be experimented upon. When litmus paper, moistened

FIG. 67.

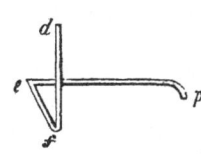

in solution of common salt or sulphate of soda, was employed, it was quickly reddened at *p*; a similar piece, moistened in muriatic acid, was very soon bleached at the same point, but no effects of a similar kind took place at *n*. A piece of turmeric paper, moistened in solution of sulphate of soda, was reddened at *n* by two or three turns of the machine; and in twenty or thirty turns, plenty of *alkali* was there evolved. On turning the paper round, so that the spot came under *p*, and then working the machine, the alkali soon disappeared, the place became yellow, and a brown alkaline spot appeared in the new part under *n*. When pieces of litmus paper and turmeric paper, both wetted with solution of sulphate of soda, were combined, and put upon the glass, so that *p* was on the litmus, and *n* on the turmeric, a very few turns of the machine sufficed to show the evolution of acid at the former and alkali at the latter, exactly in the manner effected by a volta-electric current. (See Exp. Researches, third series, 309 et seq.)

(134) In these experiments the direct passage of sparks must be carefully avoided. If sparks be passed over moistened litmus paper, it is reddened; and if over paper moistened with solution of iodide of potassium, iodine is evolved. But these effects must carefully be distinguished from those due to electro-chemical powers, or true *electrolytic*

action, and must be carefully avoided when the latter are sought for. The effect just mentioned is occasioned by the formation of nitric acid by the chemical union of the oxygen and the nitrogen of the air : the acid so formed, though very small in quantity, is in a high state of concentration, and therefore reddens the litmus paper, and decomposes the iodide.

(135) It does not appear that Faraday was more successful than Wollaston in effecting a true electro-polar decomposition of water. He says (329), "there is reason to believe that when electro-chemical decomposition takes place, the quantity of matter decomposed is not proportionate to the intensity, but to the quantity of Electricity passed; but in Wollaston's experiment this is not the case. If with a constant pair of points the Electricity be passed from the machine in sparks, a certain proportion of gas is evolved ; but if the sparks be rendered shorter, less gas is evolved ; and if no sparks be passed, there is scarcely a sensible portion of gases set free. On substituting solution of sulphate of soda for water, scarcely a sensible quantity of gas could be procured, even with powerful sparks, and almost none with the mere current ; yet the quantity of Electricity in a given time was the same in all these cases." " I believe at present that common Electricity can decompose water in a manner analogous to that of the voltaic pile. But when I consider the *true effect* only was obtained, the quantity of gas given off was so small, that I could not ascertain whether it was, as it ought to be, oxygen at one wire and hydrogen at the other. On substituting solution of sulphate of soda for pure water, these minute streams were still observed ; but the quantities were so small, that on working the machine for half-an-hour, I could not obtain at either pole a bubble of gas larger than a small grain of sand ; and if the chemical power be in direct proportion to the absolute quantity of Electricity which passes, this ought to be the case." In paragraph 359 he says, " It is doubtful whether any common electrical machine has yet been able to supply Electricity sufficient in a reasonable time to cause true electro-chemical decomposition of water."

(136) Mr. Goodman, of Salford, near Manchester who has published a very ingenious essay on the " Modifications of the Electric Fluid," has succeeded in decomposing water *by current alone*, and with *un-guarded poles :* he informs me that the experiment is most readily performed, by exposing $\frac{1}{16}$ of an inch of the ends of two fine platina wires, immersed in distilled water. The extremities thus exposed, are after a few turns of the machine, (which must be a powerful one) covered over with gas bubbles, producing a *frosted* appearance, and at all times in double quantity on the *negative* or *hydrogen* pole. The gas may speed-

ily, and sometimes from the outset also, be seen to ascend, especially with a small convex lens.

(137) Mr. Goodman's Essay is very interesting, and well deserving of attentive perusal; it is published in the sixth volume of Sturgeon's Annals of Electricity. The author describes a method of polarizing frictional Electricity by arranging a series of circular glass plates on an insulating axis, and applying to each two metallic moveable coatings. The first coating is placed by means of a wire, in connection with the positive conductor of an electrical machine, its outer surface and coating being in communication with the inner surface of the next plate, and so on throughout the series: the last surface being connected with the negative conductor of the machine. The physiological effect of this arrangement is described as being entirely novel, the sensations produced by the electromagnetic machine being exactly imitated; but Mr. Goodman did not succeed in effecting the polar decomposition of water by it, the gas generated being a mixture of oxygen and hydrogen at each pole : but when the Electricity was collected by means of points, in connection with each individual surface, the occurrence of shocks in the current was prevented, and the gas was obtained in a manner perfectly identical with galvanic decomposition, though in quantity so minute, that after nearly two hours turning, only a small bubble of about one eighth of an inch in length, was obtained.

(138) In order to obtain as complete a proof as possible of the identity of the galvanic and ordinary Electricities, Mr. Goodman endeavoured so to unite them, that they should act in concert in the decomposition of water: with this view he arranged two wine-glasses of distilled water, and inserted in each two guarded poles,* one of which was from a couronne des tasses of ten or twelve jars, and the other (the opposite pole) from the electrical machine: thus there was a positive pole from the machine, and a negative pole from the battery in one glass, and a negative pole from the machine, and a positive pole from the battery in the other glass. The experiment succeeded, and gas was generated by each of the four poles. It was afterwards found that a similar effect might take place with the frictional fluid alone.

(139) In order to obviate that objection to " guarded points," before alluded to (127), viz., that they have no " mutual decomposing depen-

---

* The guarded poles were thus constructed :—a piece of the finest platinum wire was hammered at its extremity, until it formed a flat plate, whose area was about ten or twelve times the diameter of the wire : then, with a pair of scissors it was cut into as fine a point and filament as possible, placed in a small glass tube, the extremity of which was melted until it adhered to and covered the filament, and afterwards by grinding a point, was exposed sufficiently for a small spark from the machine just to pass,—which point could not be discovered, either with the naked eye or a microscope.

dence," Mr. Goodman determined to attempt decomposition by *un-guarded* ones. Having at hand two hammered platina wires, without any glass or other coating upon them, the point of one was slowly passed beneath the surface of the water, and when about one-eighth of an inch immersed, it became speedily covered with minute bubbles : the decomposition proceeding as usual at the guarded pole during the whole period. The guarded pole was then removed, and the second unguarded one substituted, and on turning the machine for a very short period, both poles were entirely covered with gas, the *negative* in about *two-fold* quantity : thus, decomposition of water was effected perfectly identical with galvanism, from the prime conductor of the machine, alone, and subject to no objections on the ground of the metal poles being covered with non-conducting matter.

(140) Two guarded poles were next immersed in separate glasses of water, one being connected with the outer, and the other with the inner coating of a charged and insulated Leyden phial : the object being to see whether oxygen and hydrogen would be eliminated in *each* vessel, *independently* of the other. Upon attempting to pass shocks, however, no trace of gas could be observed ; but when a communication was established between the glasses, by means of a bent copper wire, decomposition instantly proceeded, visibly at the guarded poles : a piece of copper-wire was then inserted in a glass of distilled water, in which a guarded pole was placed, and after about twenty minutes turning, it became covered with bubbles of gas. Three glasses were then arranged : in the two outer, guarded poles were placed, connected respectively with the positive and negative conductors of the electrical machine, and each was then connected with the middle glass by a bent platina wire, the ends of which dipped about an eighth of an inch below the surface of the water. In five minutes, bubbles of gas made their appearance on the inserted termination of *each* wire, and in twenty minutes, very considerable bubbles were found, as *distinct as in any* galvanic decomposition, and commenced ascending from the surface of the wires.

(141) The currents were next introduced into the glasses, without any guarded pole whatever, copper-wires being substituted : in *two minutes and a half* gas was evident upon every termination : and lastly, two thick unguarded wires were inserted in a *single* wine-glass, one wire proceeding from the positive, and the other from the negative conductor,—in *three minutes* bubbles appeared : in *five minutes*, all parts of the wires below the water assumed a frosted appearance, about double the quantity of bubbles appearing on the negative wire, and in half an hour, the covering, especially of the negative pole, might be seen at the distance of two yards, exhibiting as fair an electro-chemical effect as is ever observed in voltaic Electricity.

(142) Mr. Goodman having kindly furnished me with a pair of the guarded points, employed in his own experiments, I have much pleasure in stating that I have tried them with most satisfactory results. Five turns of a two feet plate-machine, in good action, were sufficient to produce a bubble of gas on the negative point,—twenty turns gave bubbles on both : that on the negative wire being, as nearly as the eye could judge, double the size of that on the positive, and 100 turns sent the gas from the negative point in a shower of minute bubbles, while the bubble on the positive point became as big as the head of a large pin. There is no doubt, that in this experiment, the decomposition was *true electropolar*, although there is no direct proof that such is really the case.

LECTURE III.

ATMOSPHERIC ELECTRICITY.

Franklin's views respecting the identity of Electricity and lightning — Dalibard first obtains sparks from an atmospheric apparatus — Franklin's celebrated kite experiment — Grand result obtained by M. de Romas — Fatal accident to Professor Richmann — Management of electrical kites — Mr. Crosse's atmospheric apparatus — Construction of a thunder-cloud — Splendid experiments — Remarkable electrical fog — Mr. Weekes's atmospheric arrangements and experiments—Simple atmospheric exploring apparatus—Different methods of detecting minute quantities of Electricity in the atmosphere — Sources of atmospheric Electricity — Thunder and lightning — Velocity of Electricity — Wheatstone's experiments—Noise of thunder explained —Back stroke of lightning— Positions of safety during a storm — Lightning conductors and the lightning conductor question fully considered — Case of the ship "Actæon" struck by lightning— Waterspouts and volcanic eruptions by sea — Aurora Borealis.

(143) It was in the year 1749, that the celebrated American philosopher, Franklin, in a letter to Mr. Collinson, stated fully his reasons for considering the cause of Electricity and lightning to be the same physical agent, differing in nothing save the intensity of its action. "When," says he, "a gun-barrel, in electrical experiments, has but little electrical fire in it, you must approach it very near with your knuckle before you can draw a spark : give it more fire and it will give a spark at a greater distance. Two gun-barrels united, and as highly electrified will give a spark at a still greater distance. But if two gun-barrels electrified, will strike at two inches distance, and make a loud snap, to what a great distance may ten thousand acres of electrified cloud strike, and give its fire, and how loud must be that crack?" He next states the analogies which afford presumptive evidence of the identity of lightning and Electricity. The electrical spark is zig-zag, and not straight ; so is lightning. Pointed bodies attract Electricity ; lightning strikes mountains, trees, spires, masts, and chimneys. When different paths are offered to the escape of Electricity, it chooses the best conductor : so does lightning. Electricity fires combustibles : so does lightning. Electricity fuses metals ; so does lightning. Lightning rends bad conductors when it strikes them ; so does Electricity when rendered sufficiently strong. Lightning reverses the poles of a magnet : Electricity

has the same effect. A stroke of lightning, when it does not kill, often produces blindness: Franklin rendered a pigeon blind by a stroke of Electricity intended to kill it. Lightning destroys animal life; the American philosopher killed a turkey and a hen by Electrical shocks.

(144). It was in the June of 1752, that Franklin made his memorable experiment of raising a kite into a thunder-cloud, and of drawing from it sparks with which Leyden jars were charged, and the usual electrical experiments performed. A month earlier, it appears that a French Electrician, M. Dalibard, following the minute and circumstantial directions given by Franklin in his letters to Mr. Collinson, obtained sparks from an apparatus prepared at Marly-la-Ville: and an attempt has lately been made by M. Arago to claim for this philosopher, and Nollet, the honour of having established the identity of lightning and Electricity: it is clear, however, that the just right belongs to Franklin; for although this eminent Electrician was a month later in his capital experiment than Dalibard, it was nevertheless at his suggestion, and on his principles, that the arrangements of the Frenchman were made: and indeed, if the honour of the discovery is to be given to the individual who first obtained sparks from an atmospheric apparatus, it belongs neither to Dalibard, nor to Franklin, but to an old retired soldier and carpenter, named Coiffier, who was employed by Dalibard to assist him in his experiments, and who actually first drew a spark from the apparatus when the curé was absent.

(145). The analogy between the electric spark and lightning was noticed at an early period in electrical science. In 1708, Dr. Wall mentioned a resemblance of Electricity to thunder and lightning. In 1735, Mr. Grey conjectured their identity, and that they differed in *degree* only; and in 1748, the Abbé Nollet re-produced the conjecture of Grey, attended with more substantial reasons. Now, if, as was somewhat whimsically asserted by Arago, " the whole affair of the experiment was useless, the flame on the javelins of the Roman sentinels, of the fifth legion, and the Castor and Pollux so often seen by sailors on their mast tops, being sufficient as experiments," then the credit of the discovery is due to Stephen Grey, who first conjectured the identity between Electricity and lightning, (see the excellent historical sketch of Electricity in the first volume of Dr. Lardner's Manual of Electricity, and No. 130, Cabinet Cyclopædia.)

(146). The following is the account transmitted to us of Franklin's bold experiment:—" He prepared his kite by making a small cross of two light strips of cedar, the arms of sufficient length to extend to the four corners of a large silk handkerchief stretched upon them; to the extremities of the arms of the cross he tied the corners of the handkerchief. This being properly supplied with a tail, loop, and string, could be raised in the air like a common paper kite; and being made of silk, was more capable of bearing rain and wind. To the upright arm of the cross was attached an iron point, the lower end of which was in contact with the string by which the kite was raised, which was a hempen cord. At the lower extremity of this cord, near the observer, a key was fasten-ed : and in order to intercept the Electricity in its descent, and prevent it from reaching the person who held the kite, a silk ribbon was tied to the ring of the key, and continued to the hand by which the kite was held."

(147) " Furnished with this apparatus, on the approach of a storm, he went out upon the commons near Philadelphia, accompanied by his son, to whom alone he communicated his intentions, well knowing the ridicule which would have attended the report of such an attempt should it prove to be unsuccessful. Having raised the kite, he placed himself under a shed, that the ribbon by which it was held might be kept dry, as it would become a conductor of Electricity when wetted by rain, and so fail to afford that protection for which it was provided. A cloud, apparently charged with thunder, soon passed directly over the kite. He observed the hempen cord ; but no bristling of its fibres was apparent, such as was wont to take place when it was electrified. He presented his knuckle to the key, but not the smallest spark was per-ceptible. The agony of his expectation and suspense can be adequately felt by those only who have entered into the spirit of such experimental researches. After the lapse of some time he saw that the fibres of the cord near the key bristled, and stood on end. He presented his knuckle to the key and received a strong bright spark. *It was lightning.* The discovery was complete, and Franklin felt that he was immortal."

" A shower now fell, and wetting the cord of the kite, improved its conducting power. Sparks in rapid succession were drawn from the key ; a Leyden jar was charged by it, and a shock given ; and in fine all the experiments which were wont to be made by Electricity were re-produced, identical in all their concomitant circumstances."

Franklin afterwards raised an insulated metallic rod from one end of his house, and attached to it a chime of bells, which, by ringing, gave notice of the electrical state of the apparatus.

(148) These interesting experiments were eagerly repeated in almost every civilized country, with variable success. In France a grand re-sult was obtained by M. de Romas : he constructed a kite seven feet high, which he raised to the height of 550 feet by a string, having a fine wire interwoven through its whole length. On the 26th of August, 1756, flashes of fire, ten feet long, and an inch in diameter, were given off from the conductor. In the year 1753, a fatal catastrophe from in-cautious experiments upon atmospheric Electricity, occurred to Pro-fessor Richmann, of St. Petersburg ; he had erected an apparatus in the air, making a metallic communication between it and his study, where he provided means for repeating Franklin's experiments : while engaged

in describing to his engraver, Tokolow, the nature of the apparatus, a thunder-clap was heard, louder and more violent than any which had been remembered at St. Petersburg. Richmann stooped towards the electrometer to observe the force of the Electricity, and "as he stood in that posture, a great white and blueish fire appeared between the rod of the electrometer and his head. At the same time a sort of steam or vapour arose, which entirely benumbed the engraver, and made him sink on the ground." Several parts of the apparatus were broken in pieces and scattered about : the doors of the room were torn from their hinges, and the house shaken in every part. The wife of the Professor, alarmed by the shock, ran to the room, and found her husband sitting on a chest, which happened to be behind him when he was struck, and leaning against the wall. He appeared to have been instantly struck dead ; a red spot was found on his forehead, his shoe was burst open, and a part of his waistcoat singed ; Tokolow was at the same time struck senseless. This dreadful accident was occasioned by the neglect on the part of Richmann, to provide an arrangement by which the apparatus, when too strongly electrified, might discharge itself into the earth, a precaution that cannot be too strongly urged upon all who attempt experiments in atmospheric Electricity.

(149) The experiments with the electrical kite are very interesting, but great caution is required in conducting them. When thunder-clouds are about, the string should never be allowed to pass through the hand while raising the kite, even though it have a good connection with the ground : indeed, even under a cloudless sky, during a smart north-east breeze, I have frequently experienced very unpleasant shocks whilst letting out the string. By employing, however, the little apparatus shown in Fig. 68, complete security is insured, and I strongly recommend it to the notice of kite experimenters. A is a square copper box supported on a stout glass pillar (not less than two inches thick), and firmly cemented into a base-board, which is secured to the ground by

Fig. 68.

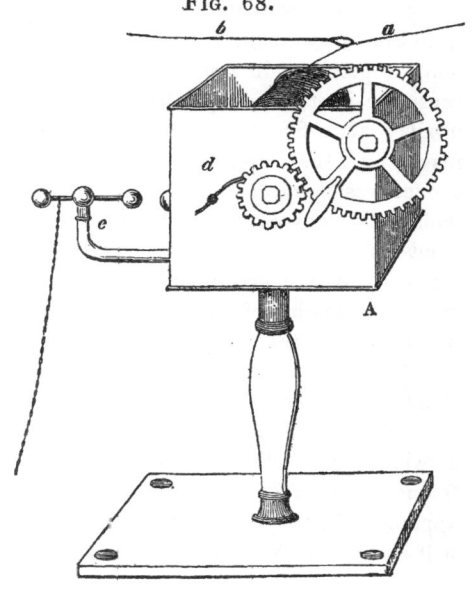

strong iron pegs nine inches long, passing through the holes and driven into the ground. The box contains a reel round which the wired string is wound: it is turned by a glass handle fixed on the multiplying wheel. *d* is a small catch moved by a key furnished with a glass handle, by which the reel may be stopped when required without touching the string. *c* is a Lane's electrometer provided with a screw adjustment, by which the distance between the brass balls may be regulated with the greatest nicety; it is connected with the ground by a chain or wire. The method of using this apparatus will immediately be understood. When the kite, (which may be simply a silk-handkerchief, stretched on a cross of light wood) is raised a sufficient height from the ground for the wind to act upon it, the string need no longer be held in the hand, the kite draws it from the reel, and the experimenter, by means of the catch and key, has it under his complete controul. When a sufficient quantity of string has been taken out, a silk cord, two or three yards long, is thrown over the string, and, by means of a running noose, tightened upon it, and the other end made fast to a post; by this means the strain is taken off the box. On ordinary occasions, that is, when no unusual exhibitions of Electricity are anticipated, I generally set the balls of the discharger about one-fifth of an inch apart, and, instead of connecting the sliding wire with the earth, put it in communication with the interior coating of a Leyden phial. In a few minutes, sparks pass between the balls, and on fine dry days, when the wind is in the north, or north-east, with about half a mile of wired string out, I have frequently had discharges at the rate of one a minute from the jar, through a striking distance of one-fourth of an inch for hours together. In order to test the species of Electricity collected, I cause the jar to discharge through a helix of copper-wire, enclosing a needle: after five or six explosions have passed, the needle becomes magnetized, the direction of the poles indicating the manner in which the jar was charged. If the helix be a *right handed* one, that is, one in which the convolutions take the same direction as that in which the hands of a watch move, then, if the jar be charged *positively*, that extremity of the needle lying in the coil, which is nearest the *negative* or outside coating of the jar, will become a north pole. If the helix be left-handed, the results are exactly the reverse.

(150) It is most interesting and instructive to watch the effect of clouds passing near the kite, their presence being invariably indicated by the increased rapidity of the discharges between the balls: the distance at which the Electricity is communicated is indeed astonishing. I have frequently observed a very marked alteration in the discharges from the approach of a single and small cloud before it could have reached within half a mile of the kite, and I have often astonished by-standers who

have been amusing themselves by drawing small sparks from the string with their knuckles, by watching the opportunity presented by the approach of one of these clouds, and then desiring them to repeat their experiment, and the result has generally been a shock which has taught them to treat the apparatus with far greater respect than before.

(151) One of the first things that the kite experimenter will probably notice, is the peculiar pungency of the spark : we are accustomed to receive sparks an inch long from the prime conductor of an electrical machine, for amusement ; it would not, however, be safe to approach a kite-string from which sparks of such a length might be drawn. The shock from sparks half an inch long are generally very severe, and resembling more the shock from a highly charged Leyden phial, than that from the prime conductor : the length of the spark is not, however, altogether the criterion of the intensity of the Electricity, which depends upon the quantity of string extended, and more still upon the state of the atmosphere.

(152) Mr. Sturgeon's kite experiments appear to have been very extensive. " I have made," says he, " upwards of five hundred electric kite experiments, under almost every circumstance of weather, at various times of the day and night, and in every season of the year ; I have experimented on Shooter's Hill, and on the low lands on the Woolwich and Welling sides of it, and the experiments in the three different places within an hour of each other ; I have done the same on the Chatham lines, and in the valley on the Chatham side of them ; on Norwood Hill, and in the plain at Addiscombe ; also on the top of the monument in London, and on the top of some of the high hills in Westmoreland and in the North Riding of Yorkshire ; and in every case I have found the atmosphere *positive* with regard to the ground. I have floated three kites at the same time at very different altitudes, and have uniformly found the highest to be *positive* to the other two ; and the centre kite *positive* to that which was below it : consequently the lowest was negative to the two above it, but still positive to the ground on which I was standing. I have made more than twenty experiments of this kind, and the results (with the exception of electric tension) were invariably the same, showing most decidedly that the atmosphere in its undisturbed electric state is more abundantly charged than the earth, and, as far as I have been able to explore it, still more abundantly in the upper than in the lower strata."

(153) On the evening of the 14th of June, 1834, Mr. Sturgeon sent up a kite during a thunder storm at Woolwich, and the following is the description he gives of the phenomena observed : — " The wind had abated to such a degree before I arrived in the barrack field, and the rain fell so heavily during the time I was there, that it was with some difficulty that I got the kite afloat, and when up, its greatest

altitude, as I imagine, did not exceed fifty yards. The silken cord also which had been intended for the insulator soon became so completely wet that it was no insulator at all : notwithstanding all these impediments being in the way, I was much gratified with the display of the electric matter issuing from the end of the string to a wire, one end of which was laid on the ground, and the other attached to the silk at about four inches distance from the reel of the kite-string. An uninterrupted play of the fluid was seen over the four inches of wet silken cord, not in sparks, but in a bundle of quivering purple ramifications, producing a noise similar to that produced by springing a watchman's rattle. Very large sparks, however, were frequently seen between the lower end of the wire (which rested on the grass) and the ground ; and several parts of the string towards the kite where the wire was broken were occasionally beautifully illuminated. The noise from the string in the air was like to the hissing of an immense flock of geese, with an occasional rattling or scraping sort of noise.

" Two non-commissioned officers of the Royal Artillery were standing by me the whole of the time. Unaware of the consequence, they would very gladly have approached close to the string ; and it was not until I had convinced them of the danger of touching or even coming near to it at a time when the lightning was playing about us in every direction, that I could dissuade them from gratifying their curiosity too far—probably at the expense of their lives. We anxiously and steadfastly watched what was going on at the end of the string, and the display was beautiful beyond description. The reel was occasionally enveloped in a blaze of purple arborized electrical fire, whose numberless branches ramified over the silken cord and through the air to the blades of grass, which also became luminous on their points and edges over a surface of some yards in circumference. We also saw a complete globe of fire pass over the silken cord between the wire and the reel of the kite-string. The soldiers thought it about the size of a musket-ball. It was exceedingly brilliant, and the only one that we noticed."

(154) Of all the individuals in this country who have distinguished themselves by their researches in atmospheric Electricity, Mr. Crosse, of Broomfield, near Taunton, stands foremost. This gentleman, who, from a modesty which is inseparable from true philosophy, has only within the last few years permitted himself to occupy a portion of public attention, has devoted more than thirty years of his life to a close investigation of the science of Electricity, and his experiments have been carried on, on a scale, and conducted with a degree of skill, which have astonished every one who has had the good fortune to witness them. With a liberality that is truly gratifying, Mr. Crosse has not only permitted me to make a minute inspection of the whole of his

magnificent arrangements, but has favoured me with written details of some of his observations, which, with his permission, I gladly take the present opportunity of presenting to the public.

(155) Mr. Crosse collects the Electricity of the atmosphere by means of wires supported and insulated on poles fixed on some of the tallest of the magnificent trees which ornament his grounds. As far as the eye can reach, these poles may be seen, though at present, in consequence of some extensive damages which happened during a late violent storm, there are not more than 1600 feet of wire insulated. The wires are insulated on the poles by means of funnels, one of which is repre-

sented in Fig. 69. It is made of copper, about four and a half inches in diameter, and eleven inches in length, and into a cavity or socket of about two inches deep formed at the closed end of the funnel is firmly cemented a stout glass rod of sufficient length to reach to the open end of the funnel, where it is mounted, by means of strong cement, with a metallic cap and staple. The latter appendage receives the hook of a very strong wire, which passes through a circular plate of copper placed about four inches from the mouth of the funnel, and terminates in a hook to which one end of the exploring wire is fixed. The object of the metallic disc is to preclude the admission of snow, rain,

Fig. 69.

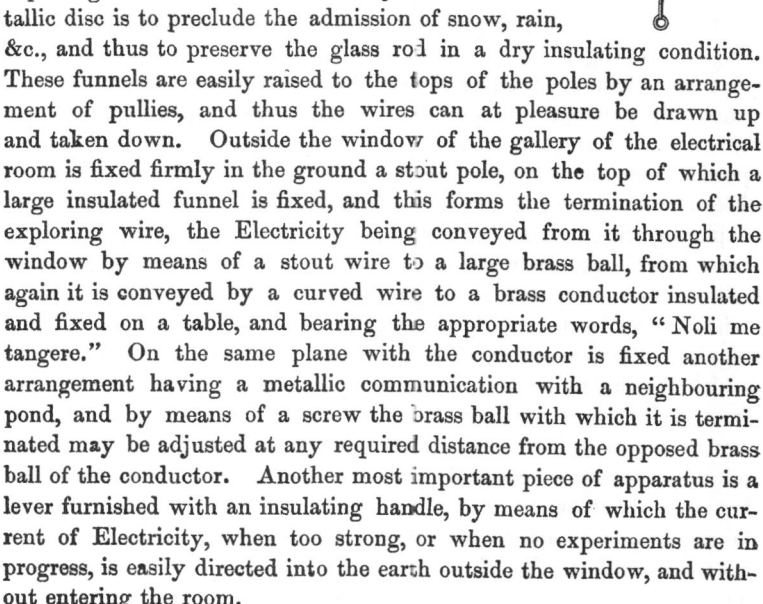

&c., and thus to preserve the glass rod in a dry insulating condition. These funnels are easily raised to the tops of the poles by an arrangement of pullies, and thus the wires can at pleasure be drawn up and taken down. Outside the window of the gallery of the electrical room is fixed firmly in the ground a stout pole, on the top of which a large insulated funnel is fixed, and this forms the termination of the exploring wire, the Electricity being conveyed from it through the window by means of a stout wire to a large brass ball, from which again it is conveyed by a curved wire to a brass conductor insulated and fixed on a table, and bearing the appropriate words, " Noli me tangere." On the same plane with the conductor is fixed another arrangement having a metallic communication with a neighbouring pond, and by means of a screw the brass ball with which it is terminated may be adjusted at any required distance from the opposed brass ball of the conductor. Another most important piece of apparatus is a lever furnished with an insulating handle, by means of which the current of Electricity, when too strong, or when no experiments are in progress, is easily directed into the earth outside the window, and without entering the room.

(156) The electrical battery employed by Mr. Crosse consists of

fifty jars, containing seventy-three square feet of coated surface : to charge it requires two hundred and thirty vigorous turns of the wheel of a twenty-inch cylinder electrical machine : nevertheless, with about one-third of a mile of wire, Mr. Crosse has frequently collected sufficient Electricity to charge and discharge this battery twenty times in a minute, accompanied by reports as loud as those of a cannon. The battery is charged through the medium of a large brass ball, suspended from the ceiling immediately over it, and connected by means of a long wire, with the conductor in the gallery : this ball is raised from, and let down to, the battery by means of a long silk cord, passing over a pully in the ceiling : and thus this extraordinary Electrician, while sitting calmly at his study table, views with philosophic satisfaction the wonderful powers of this fearful agent, over which he possesses entire control, directing it at his will ; and, with a simple motion of his hand, banishing it instantaneously from his presence.

(157) The following account of the construction of a thunder cloud, as examined by means of the exploring wires, has been kindly furnished me by Mr. Crosse :—

" On the approach of a thunder cloud to the insulated atmospheric wire, the conductor attached to it, which is screwed into a table in my electrical room, gives corresponding signs of electrical action. In fair cloudy weather the atmospheric electricity is invariably positive, increasing in intensity at sun-rise and sun-set, and diminishing at mid-day and midnight, varying as the evaporation of the moisture in the air : but when the thunder cloud (which appears to be formed by an unusually powerful evaporation, arising either from a scorching sun succeeding much wet, or vice versâ) draws near, the pith balls suspended from the conductor open wide, with either positive or negative Electricity ; and when the edge of the cloud is perpendicular to the exploring wire, a slow succession of discharges takes place between the brass ball of the conductor and one of equal size, carefully connected with the nearest spot of moist ground. I usually connect a large jar with the conductor, which increases the force of, and in some degree regulates the number of the explosions ; and the two balls between which the discharges pass can be easily regulated, as to their distance from each other, by a screw. After a certain number of explosions, say of negative Electricity which at first may be nine or ten in a minute, a cessation occurs of some seconds or minutes, as the case may be, when about an equal number of explosions of positive Electricity takes place, of similar force to the former, *indicating the passage of two oppositely and equally electrified zones of the cloud :* then follows a second zone of negative Electricity, occasioning several more discharges in a minute than from either of the first pair of zones ; which rate of increase appears to vary according to the size and power of the cloud. Then occurs another cessation, followed by an equally powerful series of discharges of positive Electricity, indicating the passage of a second pair of zones : these, in like manner, are followed by others, fearfully increasing the rapidity of the discharges, when a *regular stream commences*, interrupted only by the change into the opposite electricities. The intensity of each new pair of zones is greater than that of the former, as may be proved by removing the two balls to a greater distance from each other.   When the centre of the cloud is vertical to the wire, the greatest effect consequently takes place, during which the *windows rattle in their frames*, and the bursts of thunder without, and noise within, every now and then accompanied with a crash of accumulated fluid in the wire, striving to get free between the balls, produce the most awful effect, which is not a little increased by the pauses occasioned by the interchange of zones.   Great caution must, of course, be observed during this interval, or the consequences would be fatal. My battery consists of fifty jars, containing seventy-three feet of surface, on *one side* only.   This battery, when fully charged, will perfectly fuse into red-hot balls thirty feet of iron wire, in one length, such wire being $\frac{1}{270}$ of an inch in diameter.   When this battery is connected with three thousand feet of exploring wire, during a thunder storm, it is charged fully and instantaneously, and of course as quickly discharged.   As I am fearful of destroying my jars, I connect the two opposite coatings of the battery with brass balls, one inch in diameter, and placed at such distance from each other as to cause a discharge when the battery receives three-fourths of its charge.   When the middle of a thunder-cloud is over head, a crashing stream of discharges takes place between the balls, the effect of which must be witnessed to be conceived.

" As the cloud passes onward, the opposite portions of the zones, which first affected the wire, come into play, and the effect is weakened with each successive pair till all dies away, and not enough Electricity remains in the atmosphere to affect a gold-leaf electrometer.   I have remarked that the air is remarkably free of Electricity, at least more so than usual, both before and after the passage of one of these clouds. Sometimes, a little previous to a storm, the gold leaves connected with the conductor will, for many hours, open and shut rapidly, as if they were panting, evidently showing a great electrical disturbance.

"It is known to Electricians, that if an insulated plate, composed of a perfect or an imperfect conductor, be electrified, the Electricity communicated will radiate from the centre to the circumference, *increasing* in force as the squares of the distances from the centre ; whereas, in a thunder-cloud the reverse takes place, as its power *diminishes* from the centre to the circumference.   First a nucleus appears to be formed, say of positive Electricity, embracing a large portion of the centre of the cloud, round which is a negative zone of equal power with the former; then follow the other zones in pairs, diminishing in power to the edge of the cloud. *Directly below this cloud,* according to the laws of inductive Electricity must exist, on the surface of the earth, a nucleus of opposite or negative Electricity, with its corresponding zone of positive, and with other zones of electrified surface, corresponding in number to those of the cloud above, although each is oppositely electrified. A discharge of the positive nucleus above into that of the negative below, is commonly that which occurs when a flash of lightning is seen ; or from the positive below to the negative above, as the case may be : and this discharge may take place, according to the laws of Electricity, through any or all of the surrounding zones, *without influencing their respective electricities* otherwise than by weakening their force, by the removal of a portion of the electric fluid from the central nucleus above to that below : every successive flash from the cloud to the earth, or from the earth to the cloud, weakening the charge of the plate of air, of which the cloud and the earth form the two opposite coatings. Much might be said upon this head, of which the above is but a slight sketch."

(158) Magnificent and astounding as a spectacle such as above so forcibly described must be, it was, if possible, exceeded in brilliancy by the electrical phenomena observed some years since by Mr. Crosse, during a dense November fog, of which he has favoured me with the following interesting account. " Many years since, I was sitting in my electrical room, on a dark November day, during a very dense driving fog and rain which had prevailed for many hours, sweeping over the earth, impelled by a south-west wind. The mercury in the barometer was low, and the thermometer indicated a low temperature. I had at this time 1,600 feet of wire insulated, which crossing two small vallies, brought the electric fluid into my room. There were four insulators, and each of them was streaming with wet, from the effects of the driving fog. From about eight o'clock in the morning until four in the afternoon, not the least appearance of Electricity was visible at the atmospheric conductor, even by the most careful application of the condenser and multiplier ; indeed, so effectually did the exploring wire conduct away the Electricity which was communicated to it, that when it was connected by means of a copper wire with the prime conductor of my 18-inch cylinder in high action, and a gold leaf electrometer placed in contact with the connecting wire, not the slightest effect was produced upon the gold leaves. Having given up the trial of further experiments upon it, I took a book, and occupied myself with reading, leaving by chance the receiving ball at upwards of an inch distance from the ball in the atmospheric conductor. About four o'clock in the

afternoon, whilst I was still reading, I suddenly heard a very strong explosion between the two balls, and shortly after many more took place, until they became one interrupted stream of explosions, which died away and re-commenced with the opposite Electricity in equal violence. The stream of fire was too vivid to look at for any length of time, and the effect was most splendid, and continued without intermission, save that occasioned by the interchange of electricities *for upwards of five hours*, and then ceased totally. During the whole day, and a great part of the succeeding night, there was no material change in the barometer, thermometer, hygrometer, or wind ; nor did the driving fog and rain alter in its violence. The wind was not high, but blew steadily from the S. W. Had it not been for my exploring wire, I should not have had the least idea of such an electrical accumulation in the atmosphere : the least contact with the conductor would have *occasioned instant death*,—the stream of fluid far exceeding any thing I ever witnessed, excepting during a thunderstorm. *Had the insulators been dry, what would have been the effect ?* In every acre of fog there was enough of accumulated Electricity to have destroyed every animal within that acre. How can this be accounted for ? How much have we to learn before we can boast of understanding this intricate science ?"

(159) Amongst those individuals in whom the splendid electrical achievements of Mr. Crosse have excited an ardent taste for atmospheric investigations, my friend Mr. Weekes, of Sandwich, must not be allowed to pass unnoticed. This gentleman, at considerable trouble and expense, has erected an exploring wire, about 365 yards horizontally over the town in which he resides, insulating it against the balls from which arise the vane-spindles of the two churches, and conducting the termination to an insulated arrangement in his laboratory. " The scenes enacted by this apparatus," to use my friend's own expressive words, " are occasionally distinguished by a magnificence and interest which nothing short of ocular demonstration can serve to pourtray : nor, perhaps, are the almost hourly varying phenomena of its minor indications less deserving attention from the inquisitive admirer of natural science. When the gathering storm-cloud, pregnant with infuriated lightnings, and momentarily gaining additional sublimity from reverberating peals of deafening thunder, lingers over the line of wire, and deluges the earth with rain, or batters its beautiful foliage with unrelenting showers of hail,—then, tremendous torrents of electric matter, assuming the form of dense sparks, and possessing most astonishing intensity, rush from the terminus of the instrument with loud cracking reports, resembling in general effect the well known running fire occasioned by the vehement discharge of a multiplicity of small fire-arms.

Fluids are rapidly decomposed : metals are brilliantly deflagrated; and large extents of coated surface repeatedly charged and discharged in the space of a few seconds. When these phenomena occur incidental to the hours of darkness, the lightning flash is seen harmlessly to play in various zig-zag and fantastic shapes amidst the several contrivances by means of which its power is subdued; thus augmenting the sublimity of a scene, compared to the correct delineation of which, the efforts of language are but imbecility. Again, relinquishing its claims to the terrific and sublime for features of a more gentle complexion, even the light and feathery aggregations of the summer cloud are found capable of imparting to a pair of delicate gold-leaf pendulums, a test by which the philosopher assigns a character to inaccessible regions of the atmosphere."

(160) In the sixth volume of "Sturgeon's Annals of Electricity," Mr. Weekes has given a most interesting description of some brilliant electrical phenomena observed during a grand hail-storm which occurred at Sandwich on the 9th of May, 1841; but as it would be impossible to do justice to the account without the plate which accompanies it, and which is somewhat too large to be inserted here, I must satisfy myself with referring my readers to the original paper. In the same volume, page 98, will be found another equally interesting communication from Mr. Weekes, in which he describes some electrical phenomena observed by him during a thunder-storm in the autumn of 1840, particularly the alternation of the Electricity from positive to negative, indicating the passage of zones, and verifying Mr. Crosse's idea of the construction of a thunder-cloud. This same paper contains an account of some experiments made by the Author with a view of insulating *ozone*, the name given by M. Schoenbein to that peculiar odorous principle which appears to be developed when ordinary Electricity passes from the points of a conductor to the surrounding air, and which is also disengaged whilst water is decomposing by a voltaic current. It is the opinion of M. Schoenbein that this odorous principle should be classed in that genera of bodies to which chlorine and bromine apparently belong; that it is always disengaged in the air in sufficiently notable quantities during stormy weather; and he suggests the following method of rendering it evident, founded on its property of rendering gold and platinum electro-negative; viz. to place plates of platinum in situations sufficiently elevated, taking care that they communicate with the earth : as soon as these plates become negatively polar, M. Schoenbein thinks that it may be concluded that ozone is developed.

(161) Mr. Weekes found that if a piece of platinum foil three or four inches square, fastened to a wire, and held in the hand of the operator, be brought for a few seconds into the vicinity of the terminal

ball of his aërial apparatus, when a free current in the form of sparks is passing, the plate acquires a negative polarity, which may be immediately shown by the galvanometer, and the experiment repeated with the greatest facility as often as desired. He constructed an apparatus by which dense sparks of Electricity were caused to pass for about fifteen seconds between two metallic plates enclosed in a cylinder of glass; and he states, that on removing the cover of the arrangement the atmosphere of the cylinder had become so strongly impregnated with a powerful pungent phosphoric odour, that he found it exceedingly inconvenient to respire over it ; nor could he find any individual willing to permit its approach towards his nostrils beyond a second or two. Several highly interesting facts connected with this peculiar odorous principle have been collected ; and in a future lecture I shall have occasion to advert to them.

(162) A very simple and inexpensive, though at the same time a highly interesting method of examining the electric state of the atmosphere, is by raising a long pole, twenty-five or thirty feet high, on the top of the highest part of the house, and conveying the insulated wire from the top of this pole through the window into an adjoining apartment.

Some time since I erected an apparatus of this description, about eighty feet from the ground ; and as I have many times experienced great gratification from the observations I have been enabled to make with it, I have thought it worth while to give a wood-cut of the arrangement.

A, Fig. 70, represents the top of a tall fir-pole, thirty-five feet in height, fixed on the chimney of a factory ; b, a painted copper funnel, surmounted with a brass four-pronged fork, from which the copper wire c proceeds, and is conducted to a second insulated funnel d, enclosing a thick brass wire, which, insulated by a stout glass tube, passes through the stone-work of my laboratory window, and terminates in a two-inch brass ball e ; f is a smaller ball connected with a bent wire passing into the gun-barrel g, and capable of being brought to any required distance from the ball e ; the gun-barrel passes through the stone-work of the window, and is metallically connected with the earth by the iron rod h. There is also an arrangement not shown in the figure, consisting of a stout iron rod capped with a brass ball fixed at about two inches distant from the funnel d ; this rod passes into the earth, and thus prevents any accident which might arise from a flash of lightning striking the apparatus. The fir-pole A is terminated by a stout glass rod eighteen inches long, to which the funnel serves as a protection. I generally have a pair of small pith balls suspended from the ball e,

H

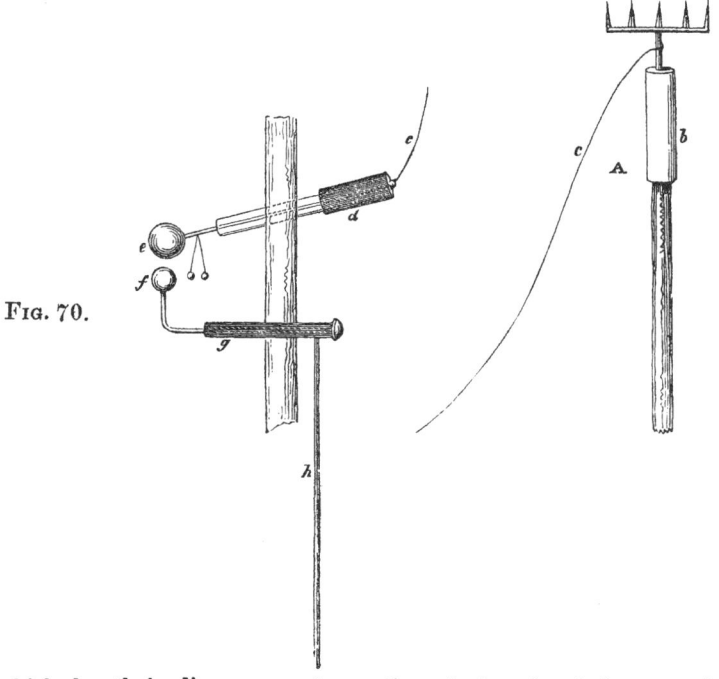

Fig. 70.

which by their divergence give notice of the electrical state of the apparatus.

Although the situation in which this miniature atmospheric exploring apparatus is placed is low, it has occasionally afforded me highly interesting exhibitions of electric phenomena. During hail-storms I have repeatedly charged large Leyden phials from the ball *e*; and every electrical cloud that passes over the factory, indicates its vicinity, either by the passage of small sparks between the balls *e* and *f*, or by the divergence of the pith balls.

(163) There are four Electroscopes, any of which may be employed to ascertain the comparative electrical state of the atmosphere.

1. De Saussure's Electrometer, which consists of two fine wires, each terminated by a small pith ball, and adapted to a small metallic rod fixed in the upper part of a square glass cover, upon one of the faces of which a divided scale is marked in order to measure the angles of deviation of the two balls.

2. Volta's Electrometer, formed of two straws about two inches long, and one-fourth of a line broad, suspended from two small very moveable rings adapted to a metal rod: to measure the deviation of the straws, a telescope with a nonius is employed.

3. Singer's Electrometer, consisting of two slips of gold-leaf suspended from the rod.

4. Bohnenberger's Electroscope, formed of a single strip of gold-leaf suspended from the conducting rod between two dry piles, the negative pole of one, and the positive pole of the other, being uppermost. This arrangement has the advantage of indicating the kind of Electricity communicated to the conductor.

When the quantity of Electricity in the atmosphere is too small to be rendered evident by either of these instruments, I have frequently successfully adopted a method of detection which was furnished me by my friend Andrew Crosse, Esq., and which I find is also adopted and recommended by Mr. Weekes.* A lighted candle, in a metallic candlestick, is insulated, and placed in the open air (if the wind be not too rough) ; one of the plates of Volta's condenser is then applied to the top of the candlestick, and by the usual operation the charge acquired is transferred to the plate of a gold-leaf electroscope : divergence instantly takes place.

(164) It is confirmed by the observations of all electricians, that in serene weather the Electricity of the atmosphere is always positive with regard to the earth, and that it becomes more and more positive in proportion to its elevation above the earth's surface. It has further been ascertained by the observations of De Saussure, Schubler, Arago, and others, that the positive Electricity of the atmosphere is subject to diurnal variations of intensity, there being two maxima and two minima during the twenty-four hours. The first minimum takes place a little before the rising of the sun : as it rises, the intensity at first gradually, and then rapidly increases, and arrives at its first maximum a few hours after. This excess diminishes at first rapidly, and afterwards slowly, and arrives at its minimum some hours before sunset ; it re-ascends when the sun approaches the horizon, and attains its second maximum a few hours after, then diminishes till sun-rise, and proceeds in the order already indicated. The intensity of the free Electricity of the atmosphere has also been found to undergo annual changes, increasing from the month of July to the month of November inclusive, so that the greatest intensity occurs in winter, and least in summer.

In cloudy weather, the free Electricity of the atmosphere is still positive during storms ; and when it rains or snows the Electricity is sometimes positive and sometimes negative, and its intensity is always much more considerable than in serene weather. The electro-

* See Proceedings of the Electrical Society, No. 2, page 89. See also No. 1, page 41, Mr. Weekes's description of his atmospheric arrangements at Sandwich.

scope will, during the continuance of a storm, frequently indicate several changes from positive to negative. (157)

An instrument for investigating the electric state of the atmosphere was constructed by M. Colladon, of Geneva, on the following principle. He found that if the two ends of the wire of a galvanic multiplier, consisting of very numerous coils well insulated from each other, were brought in contact, one with a body positively, and the other with a body negatively charged, a current of Electricity passes through the wire until equilibrium is restored, the energy and direction of this current is indicated by the deviation of the needle from the zero point of the scale. This instrument is applied to the purpose of ascertaining and measuring the atmospheric Electricity by communicating one end of the wire with the earth, and allowing the other to extend into the region of the atmosphere, the electrical state of which is intended to be compared.

(165) The principal source of atmospheric Electricity is undoubtedly evaporation. If a little water be placed in the cup of the gold-leaf electroscope, Fig 9, and a red-hot cinder dropped in, the rapid evaporation will occasion the leaves to diverge with *negative* Electricity; and although M. Pouillet has rendered it probable that the evaporation of perfectly pure water is not attended with any development of Electricity, the impregnation of all the water on the surface of the earth with various saline matters is quite sufficient to destroy any objection which might be raised on that score; the positive Electricity is carried with the vapour into the air, and probably communicates a charge to the minute drops into which it is condensed.

It is stated, in some work on Electricity, that if a small portion of acid be mixed with the water in the metallic cup, the evaporation occasioned by dropping in a hot cinder will cause the leaves to diverge with *positive* Electricity, the vapour being negatively electrified. I do not, however, find this to be the case; for when a few drops of sulphuric acid were mixed with distilled water in a small tin cup, and rapid evaporation effected, the divergence of the gold leaves was still *negative*.

(166) The appearance of the heavens previous to and during a thunder-storm was first diligently studied by Beccaria.* He noticed that a dense cloud was first formed, increasing rapidly in magnitude, and ascending into the higher regions of the atmosphere. The lower end is black, and nearly horizontal; but the upper is finely arched, and well defined. Many of these clouds often seem piled one upon the other, all arched in the same manner; but they keep constantly

* Lardner's Manual of Electricity, &c.

uniting, swelling, and extending their arches. When such clouds rise, the firmament is usually sprinkled over with a great number of separate clouds of odd and bizarre forms, which keep quite motionless. When the thunder-cloud ascends, these are drawn towards it, and as they approach they become more uniform and regular in their shapes, till coming close to the thunder-cloud their limbs stretch mutually towards each other, finally coalesce, and form one uniform mass. But sometimes the thunder-cloud will swell and increase without the addition of these smaller adscititious clouds. Some of the latter appear like white fringes at the skirts of the thunder-cloud, or under the body of it ; but they continually grow darker and darker as they approach it.

(167) When the thunder-cloud, thus augmented, has attained a great magnitude, its lower surface is often ragged, particular parts being detached towards the earth, but still connected with the rest. Sometimes the lower surface swells into large protuberances, tending uniformly towards the earth ; but sometimes one whole side of the cloud will have an inclination to the earth, which the extremity of it will nearly touch. When the observer is under the thunder-cloud after it is grown large and is well formed, it is seen to sink lower and to darken prodigiously, and at the same time a great number of small clouds are observed in rapid motion driven about in irregular directions below it. While these clouds are agitated with the most rapid motions, the rain generally falls in abundance ; and if the agitation be very great, it hails.

While the thunder-cloud is swelling and extending itself over a large tract of country, the lightning is seen to dart from one part of it to another, and often to illuminate its whole mass. When the cloud has acquired a sufficient extent, the lightning strikes between the cloud and the earth in two opposite places, the path of the lightning lying through the whole body of the cloud and its branches. The longer this lightning continues, the rarer does the cloud grow, and the less dark in its appearance, till it breaks in different places, and shews a clear sky. When the thunder-cloud is thus dispersed, those parts which occupy the upper regions of the atmosphere are spread thinly and equally, and those that are beneath are black and thin also, but they vanish gradually, without being driven away by the wind.

(168) The following is the account given by Dr. Thomson :—
" A low dense cloud begins to form in a part of the atmosphere that was previously clear. This cloud increases fast, but only from its upper part, and spreads into an arched form, appearing like a large heap of cotton wool. Its under surface is level, as if it rested on a smooth plane. The wind is hushed, and everything appears preternaturally calm and still.

" Numberless small ragged clouds, like teasled flakes of cotton, soon begin to make their appearance, moving about in various directions, and perpetually changing their irregular surface, appearing to increase by gradual accumulation. As they move about, they approach each other, and appear to stretch out their ragged arms towards each other. They do not often come into contact, but after approaching very near each other, they evidently recede, either in wholes or by bending away their ragged arms.

" During this confused motion, the whole mass of small clouds approaches the great one above it ; and when near it, the clouds of the lower mass frequently coalesce with each other before they coalesce with the upper cloud ; but as frequently the upper cloud coalesces without them. Its lower surface, from being level and smooth, now becomes ragged, and its tatters stretch down towards the others, and long arms are extended toward the ground. The heavens now darken apace, and the whole mass sinks down. Wind rises and frequently shifts in squalls. Small clouds move swiftly in various directions. Lightning darts from cloud to cloud. A spark is sometimes seen co-existent through a vast horizontal extent of a zig-zag shape, and of different brilliancy in different parts. Lightning strikes between the clouds and the earth, frequently in two places at once,—a heavy rain falls,—the cloud is dissipated, or it rises high and becomes light and thin. These electrical discharges obviously dissipate the Electricity :— the cloud condenses into water, and occasions the sudden and heavy rain which always terminates a thunder-storm. The previous motions of the clouds, which act like electrometers, indicate the electrical state of different parts of the atmosphere."

(169) A great difference will be observed in the appearance of the flashes of lightning during a thunder-storm. The scene is sometimes rendered awfully magnificent by their brilliancy, frequency, and extent ; darting sometimes, on broad and well-defined lines, from cloud to cloud, and sometimes shooting towards the earth ; they then become zig-zag and irregular, or appear as a large and rapidly-moving ball of fire—an appearance usually designated by the ignorant a *thunderbolt*, and erroneously supposed to be attended by the fall of a solid body. The report of the thunder is also modified according to the nature of the country, the extent of the air through which it passes, and the position of the observer. Sometimes it sounds like the sudden emptying of a large cart-load of stones ; sometimes like the firing of a volley of musketry : in these cases it usually follows the lightning immediately, and is near at hand : when more distant it rumbles and reverberates, at first with a loud report, gradually dying away and returning at intervals, or roaring like the discharge of heavy artillery. In accounting for

these phenomena, it must be remembered that the passage of Electricity is almost infinitely rapid. A discharge through a circuit of many miles has been experimentally proved to be instantaneous : the motion of light is similarly rapid ;* and hence the flash appears momentary,

* Light is about eight minutes thirteen seconds in passing from the sun to the earth, so that it may be considered as moving at the rate of one hundred and ninety-two thousand miles in a second, performing the tour of the world in about the same time that it requires to wink with our eye-lids, and in much less than a swift runner occupies in taking a single stride. (Herschel.)

The sun is ninety-five millions of miles from the earth, and almost a million times larger : the sun being 882,000 miles in diameter, and the earth 8400 miles. Yet its magnitude, as viewed from the earth, scarcely exceeds that of the moon, which is not more than one-fourth the diameter of our globe, being 2160 miles in diameter. Yet such is the velocity of light, that a flash of it from the sun would be seen in little more than eight minutes after its emission ; whereas the sound evolved at the same time (supposing a medium like air capable of conveying sound between the sun and the earth) would not reach us in less than fourteen years and thirty-seven days ; and a cannon ball, proceeding with its greatest speed, in not less than twenty years.

The velocity of Electricity is so great, that the most rapid motion that can be produced by art, appears to be actual rest when compared with it. A wheel, re-volving with a rapidity sufficient to render its spokes invisible, when illuminated by a flash of lightning, is seen for an instant with all its spokes distinct, as if it were in a state of absolute repose ; because, however rapid the rotation may be, the light has come and already ceased before the wheel has had time to turn through a sensible space. The following beautiful experiment was made by Wheatstone :—A circular piece of pasteboard was divided into three sections, one of which was painted blue, another yellow, and a third red ; on causing the disc to revolve rapidly it appeared white, because a sun-beam consists of a mixture of these colours, and the rapidity of the motion caused the distinction of colours to be lost to the eye: but the instant the pasteboard was illuminated by the electric spark, it seemed to stand still, and each colour was as distinct as if the disc were at rest.

By a beautiful application of this principle, Wheatstone contrived an apparatus by which he has demonstrated that the light of the electric discharges does not last the millionth part of a second of time. His plan was to view the image of a spark reflected from a plane mirror, which, by means of a train of wheels, was kept in rapid rotation on a horizontal axis. The number of revolutions performed by the mirror was ascertained by means of the sound of a siren connected with it, and still more successfully by that of an arm striking against a card, to be 800 in a second. The angular motion of the image being twice as great as that of the mirror, it was easy to compute the interval of time occupied by the light during its appearance in two successive points of its apparent path, when thus viewed, and it was ascertained that the image passed over half a degree (an angle, which being equal to about an inch seen at the distance of ten feet, is easily detected by the eye) in 1,152,000th part of a second. The result of these experiments, as regarded the duration of the spark, was, that it did not occupy even this minute portion of time : but when the electric discharge of a battery was made to pass through a copper wire of half a mile in length, interrupted both in the middle and also at its two extremities ; so as to present three sparks, they each gave a spectrum considerably elongated, and

however great the distance through which it passes : but sound is
infinitely slower in its progress, travelling, in air, with a velocity of
only 1130 feet in a second, or about twelve miles in a minute. Now,
supposing the lightning to pass through a space of some miles, the
explosion will be first heard from the point of the air agitated nearest
the spectator; it will gradually come from the more distant parts of
the course of the Electricity : and, last of all, will be heard from the
remote extremity, and the different degrees of the agitation of the air,
and likewise the difference of the distance, will account for the different
intensities of the sound and the apparent reverberation and charges.

(170) Thunder only takes place when the different strata of air are
in different electrical states : the clouds interposed between these
strata are also electrical, and owe (according to Dr. Thomson) their
vesicular nature to that Electricity. They are also conductors. The
discharges usually take place between two strata of air : very rarely
between the air and the earth ; and sometimes without noise, in which
case the flashes are very bright : but they are single, passing visibly
from one cloud to another, and confined, usually, to a single quarter of
the heavens.   When the discharge is accompanied by thunder, a
number of simultaneous and different-coloured flashes may generally be
observed stretching to an extent of several miles. These seem to be
occasioned by a number of successive discharges from one cloud to
another, the intermediate clouds serving as intermediate conductors, or
stepping-stones as it were for the electric fluid. It is these discharges
which occasion the rattling noise. Though they are all made at the
same time, yet as their distances are different, they only reach the ear
in succession, and thus occasion the lengthened rumbling noise—so
different from the snap which accompanies the discharge of a Leyden
jar.

(171) The snap attending the spark from the prime conductor of an
electrical machine and the awful thunder-crash are undoubtedly similar
phenomena, and produced by the same action. The cause is the
vibration of the air agitated by the passage of the electric discharges
with a greater or less degree of intensity : and two explanations may
be given of the manner in which the vibration is produced. On the
one hand, it may be imagined that the electric fluid opens for itself a

indicating a duration of the spark of the 24,000th part of a second. The sparks at
both extremities of the circuit were perfectly simultaneous, both in their period of
commencement and termination : but that which took place in the middle of the
circuit, though of equal duration with the former, occurred later by at least the
millionth part of a second, indicating a velocity of transmission from the former
point to the latter of nearly 288,000 miles in a second,—a velocity which exceeds
that of light itself.

passage through air, or other matter, in the manner of a projectile, and
that the sound is caused by the rush of the air into the vacuum
produced by the instantaneous passage of the fluid: or, on the other
hand, the vibration may be referred to a decomposition and recomposition
of Electricity in all the media in which it appears. On this hypothesis,
the continued roll is the effect of the comparatively slow propagation of
sound through the air, and it may be thus illustrated.* Suppose a
flash of lightning 11,300 feet in length, or that the spark is instantaneously seen from one end to the other of this line. At the same
instant that the flash is visible the vibration is communicated to the
atmosphere through the whole extent of the line. Now, suppose an
observer placed in the direction of the line of the flash, and at the distance of 1,130 feet from one end: then since sound travels at the rate
of 1,130 feet in a second, *one* second will elapse after the flash has been
seen before any sound will be heard. When the sound begins, the
vibration communicated to the nearest stratum of air has reached his
ear; and since the line of disturbance has been supposed to be 11,300
feet in length, the vibrations of the more distant strata will continue
to reach his ear in succession, during the space of *ten seconds*. Hence,
the length of the flash determines the duration of the sound: and it
follows that the same flash will give rise to a sound of greater or less
duration, according to the position of the observer with respect to its
direction. Thus, in the above instance, suppose a second observer to
be placed under the line, and towards its middle, he would only hear
the sound during half the time it was heard by the first observer; and
if we suppose the line to be circular, and the observer to be placed near
its centre, the sound would arrive from every point at the same instant
in a violent crash.

(172) Although the vibratory motion is communicated to all the
strata of air along the whole length of the flash, they will not all
receive the same impulsion unless they are all at the same temperature,
and in the same hygrometric state, which can rarely happen. Hence,
although proceeding from the nearest point, the first impression of the
sound is not always the most intense.

The latter of these two ways of accounting for the vibration, seems
to accord best with facts; for, in the first place, it has been objected
that if the noise was occasioned by the electric fluid forcing for itself a
passage through the air, a similar sound ought to be produced by a
cannon-ball: and a still stronger objection is that experiments seem to
indicate that the electric fluid is not transferred from point to point like

* See Brand's Dictionary of Science, Literature, and Art. Article, *Thunder*.

a projectile of ponderable matter, but by the vibration of an elastic medium, as sound is conveyed through the atmosphere.

The equilibrium of the clouds is sometimes restored by a single flash of lightning : at other times the accumulation is so immense, and the neighbouring strata of air so strongly charged, that the flashes continue for hours before they terminate in a storm of rain.

(173) A person may be killed by lightning, although the explosion takes place twenty miles off, by what is called the back stroke. Suppose that the two extremities of a cloud highly charged hang down to the earth, they will repel the Electricity from the earth's surface if it be of the same kind as their own, and will attract the other kind : if a discharge should suddenly take place at one end of the cloud, the equilibrium will instantly be restored by a flash at that point of the earth which is under the other. Though this back stroke is often sufficiently powerful to destroy life, it is never so terrible in its effects as the direct shock.

When a building is struck by lightning, the charge is generally determined towards the chimney, owing to its height, and to the conducting power of the carbon deposited in it ; for it has been demonstrated experimentally, that the electric fluid will pass with facility to a considerable distance over a surface of carbon.

(174) The directions to be given as to the best positions of safety during a thunder-storm are few and simple. If out of doors, trees should be avoided ; and if from the rapidity with which the explosion follows the flash it should be evident that the electric clouds are near at hand, a recumbent posture on the ground is the most secure. It is seldom dangerous to take shelter under sheds, carts, or low buildings, or under the arch of a bridge : the distance of twenty or thirty feet from tall trees or houses is rather an eligible situation, for, should a discharge take place, these elevated bodies are most likely to receive it, and less prominent bodies in the neighbourhood are more likely to escape. It is right also to avoid *water*, for it is a good conductor ; and the height of a human being near the stream is not unlikely to determine the direction of a discharge. Within doors we are tolerably safe in the middle of a carpeted room, or when standing on a double hearth rug. The chimney, for reasons above stated, should be avoided : upon the same principle, gilt mouldings, bell-wires, &c. are in danger of being struck. In bed we are tolerably safe—blankets and feathers being bad conductors, and we are, consequently, to a certain extent, insulated. It is injudicious to take refuge in a cellar, because the discharge is often from the earth to a cloud, and buildings frequently sustain the greatest injury in the basement stories.

(175) *Lightning Conductors.*—To Franklin, whose active mind was constantly directed to practical applications of the facts disclosed by science, we are indebted for the suggestion of a method of partially defending buildings from the dreaded effects of lightning. His method was to erect by the side of the building to be protected, a continuous metallic rod in perfect communication with the earth, and experience has fully demonstrated the value of this precaution.

In the choice of a conductor, preference should be given to copper, and it is well to divide the extremity into three or four points: it should penetrate the ground sufficiently deep to be in close contact with a stratum of moist soil, and be carried above the highest point of the building: great care should be taken that every part of the rod *be perfectly continuous,* and that its substance be sufficient to prevent any chance of its being melted; perfect security on this head is arrived at by having a rod three quarters of an inch thick. It has been proved that conductors erected with these precautions will protect a circular space of a radius double their height above the highest point of the building to which they are attached. (Daniell).

(176) The little arrangement Fig. 71, amusingly illustrates the use of a continuous conductor. A board about three quarters of an inch thick, and shaped like the gable end of a house, is fixed perpendicularly upon another board, upon which a glass pillar also is fixed in a hole about eight inches distant from the gable-shaped board. A small hole, about a quarter of an inch deep, and nearly an inch wide, is cut in the gable-shaped board, and this is filled

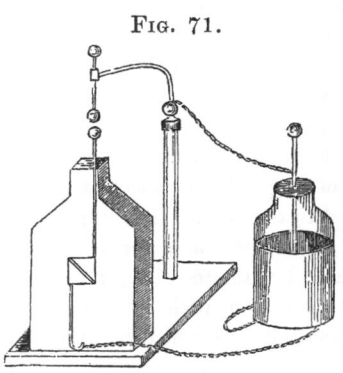

FIG. 71.

with a square piece of wood of nearly the same dimensions. It should be *nearly* of the same dimensions, because it must go so easily into the hole, that it may drop off by the least shaking of the instrument. A brass wire is fastened diagonally to this square piece of wood, and another of the same dimensions, terminated by a brass ball, is fastened on the gable-shaped board, both above and below the hole. From the upper extremity of the glass pillar a crooked wire proceeds, terminated also by a brass ball, and sufficiently long to reach immediately over the ball or the wire of the board. The glass pillar is loosely fixed in the bottom board, so that it may move easily round the axis. It is evident that, with this arrangement, a shock from a Leyden jar may easily be sent over the square hole by connecting the exterior coating with the wire in the gable-shaped board below it, and the interior with the wire on the glass pillar which comes within the striking distance of the wire in the gable-shaped board below it.

Suppose now the square piece of wood to be placed in the hole in such a manner, that the wire attached to it diagonally shall be in *contact* with the wires above and below it, a shock may evidently be transmitted without any disturbance taking place; but if it be put into the hole in an *opposite* direction, so that the shock from the jar shall be obliged to pass over it altogether in the form of a spark in its passage from wire to wire, the concussion it will occasion will throw the square piece of wood to a considerable distance from the apparatus. The square piece of wood may here be supposed to represent a window, and the wire a continuous or broken conductor passing by the side of it, and the violent effects produced by the minute quantity of Electricity accumulated in a Leyden jar may be considered as a humble imitation of the effects of a stroke of lightning. When the passage is uninterrupted, the Electricity passes quietly down, but when impeded it produces the most violent effects.

(177) To exemplify the method of defending ships, a small model may be made, with a glass tube for the mast. Into this tube two wires are to be inserted through its opposite ends, until within half an inch of each other. The tube is then to be filled with water, and the ends stopped. Connect the lower wire with a small metallic thread tied to the stern. The upper wire is to be surmounted by a brass ball. A moveable conductor may be formed of a thin copper-wire, placed parallel with, and rising above the mast: this wire is to be connected at the bottom with the metal thread. If a powerful charge be passed along the mast when the conductor is attached no effect is produced; but if the conductor be removed the mast is shattered to pieces.

(178) Much has lately been written on the subject of lightning conductors; and much also, unfortunately, with a degree of acrimony totally inconsistent with the spirit of true philosophy, which is disgraced by being thus made the handle for the exhibition of petulance and ill-will. The object of every follower of science is, or ought to be, the *investigation of truth,*—but how frequently is this lost sight of! how often is *exaltation of self* the feeling substituted! Yet, after all, how little do we know how much have we to learn; and how ill does it become us to allow our difference of opinion to destroy our charity,— to occupy time, which should be spent in calm investigation, in useless recrimination and abuse!

(179) The grand point of discussion, with respect to lightning conductors, appears to be this: Does a continuous metallic rod, arranged with the precautions above adverted to, afford security to the building to which it is attached; or is there a danger of *lateral explosions* passing from the rod to vicinal conducting bodies? It will not be possible, nor would it be desirable, to enter into a consideration here of all that has been said upon this subject. I shall therefore endeavour to select the chief arguments on both sides, and leave the matter to the judgment of my readers.

(180) It is well known that that eminent electrician, Mr. William Snow Harris, of Plymouth, has invented a system of lightning conductors, which, after undergoing careful examination by Professors Faraday, Wheatstone, and others, has been introduced generally into the British navy. In the Report of the Committee appointed by the House of Commons to investigate the proposed plan,* the conductors are described as being composed of two plates of copper rivetted together, so as to form an elastic and continuous line of metal: the inner plate being $\frac{1}{16}$th, and the outer $\frac{1}{8}$th of an inch in thickness: their breadth varying according to the class of ship, and the description of the spar. These plates are inserted in dovetailed grooves, in the after part of the masts, and extend from the truck to the keelson; a copper plate, of the same dimensions, is led over the cap, and the continuity is preserved at all times by a tumbler on the caps, consisting of a short copper bar, with a hinge at the base, by which it leans against the conductor of the top mast, whether fidded or haused, a stop being placed on the exterior by which the tumbler is prevented from falling backward. Copper plates, of equal dimensions to those on the lower masts, are placed under the keels and steps of the masts, and are thence led along the keelson, in contact with the copper fastenings. In order to insure connexion with the copper sheathing, bolts are driven transversely through the keel, so as to meet those passing down from the keelson. Copper plates are likewise led along the underside of the beams of the lower and orlop decks to the principal copper fastenings, and ultimately terminate in the sheathing; thereby combining all the chief masses of metal in the hull and spars of a ship with the conductors, and affording, by means of its ultimate connexion with the copper sheathing, a vast surface in contact with the water for the dispersion of the Electricity.

(181) The first objector to Mr. Harris's system of conductors was

---

* See also Nautical Magazine for 1837, p. 742, which contains a plate illustrating Mr. Harris's method of fitting his conductors.

Mr. Martyn Roberts, who, in a paper read before the Electrical Society, June 27th, 1837, submits " that it is highly dangerous to conduct such an immense accumulation of electric fluid, as that in a lightning cloud, into the body of the vessel, close, generally speaking, to the powder-magazine, or at all events among many substances that would produce awful effects from its action on them, on account of the " lateral explosion," which he says can easily be proved, by experiment, to take place, even in the transmission of the feeble quantity of Electricity generated by our machines ; and is therefore much to be dreaded, from the enormous quantity of fluid which will be conveyed into the hull of a ship. Mr. Roberts suggests the employment of a metallic rope, to be fixed to a copper point at the highest mast-head, led down the after part of the mast until it arrives at the lower mast-head, and from thence led, as a backstay, to the outside of the ship, and there fastened to her copper sheathing.

(182) In answer to Mr. Roberts's objections, Professor Wheatstone stated to the Committee, " that the lateral explosion is a phenomenon which has been observed and experimented on for the last half century ; that he conceived it has no application to lightning conductors ; and that it was physically impossible that the least accident could occur to a ship, if the known conditions of a good conductor were fulfilled :" and Professor Faraday declared, that " in his opinion Mr. Harris's conductors met every case of division of charge that he could possibly conceive to occur ; " that a lateral discharge could not be obtained from them, provided the continuity were not interrupted : that there was not on record, as far as he could learn, any instance of lateral explosion ; and that he believed a man could receive no injury if he were leaning against a conductor when the Electricity descended : any opinion to the contrary being only assumption.

(183) The next attack on Mr. Harris's system was made by Mr. Sturgeon, who, in a memoir addressed to the British Association for the Promotion of Science, insists on the danger arising from lateral explosions, and brings forward a variety of experiments in support of his position. His definition of three kinds of lateral discharges has been already given (113). The following experiments are adduced as instances of the third kind. Let the Leyden jar (Fig. 72) be discharged through a good conducting rod $c\ c$, standing on the same metallic plate as that on which the jar rests. At the same moment a spark will appear at the opening o, between the metallic *insulated* body B and the conductor $c\ c$. The same will occur if the jar be discharged by the direct application of the discharging rod in the usual manner, and the effect is much increased, by connecting the body B

with the ground.   In Fig. 72, *c c* is taken to represent the conductor
of a ship's mast connected with the iron knee K, by means of the copper

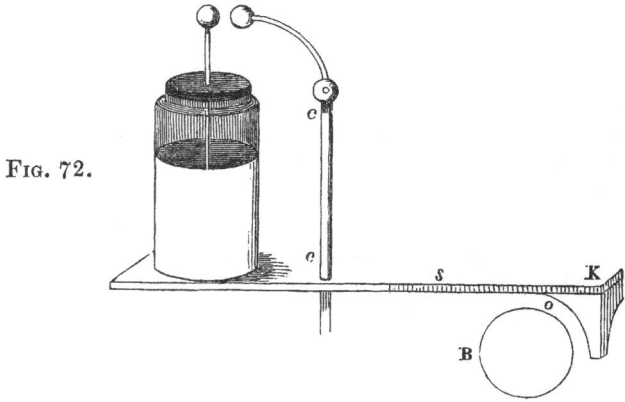

FIG. 72.

strap *s*, as proposed by Mr. Harris.   Then, according to Mr. Sturgeon's
views, a discharge from the jar, through the conductor *c c*, would
imitate a flash of lightning striking a similar conductor in a mast : and
as in the experiment a spark takes place at *o*, between the body B and
the iron knee K, a similar lateral discharge would take place from any
part of the copper strap *s*, or from any metallic appendage or branch of
the conductor; and it is obvious, that men who happened to be near to
any of these conductors, straps, knees, &c., would experience all the
effects of these lateral discharges.   From the results of many experiments, Mr. Sturgeon has satisfied himself that this kind of lateral
discharge will always take place when the vicinal bodies are sufficiently
capacious, and near to the principal conductor which carries the
primitive discharge, or to any of its metallic appendages.

   (184) Another objection brought forward by Mr. Sturgeon to Mr.
Harris's conductors, is, that in consequence of their passing through
the hull of the vessel, a flash of lightning would be exceedingly injurious
to the chronometers on board, by magnetizing every piece of ferruginous
matter which enters into their construction.   " Not only," he observes,
" would the principal conductor in the mast be productive of these
effects, but every strap, knee, and other metallic appendage to that
conductor, would communicate permanent magnetism to those morsels
of steel which form so essential a part to those valuable horological
instruments."

   Instead of inserting copper plates into grooves in the masts, Mr.
Sturgeon proposes to protect the lower masts and rigging by cylindrical
rods of copper, four to each mast, the upper extremities to be attached

to the fore, main, and mizen tops, as distant from the masts as circumstances will allow ; the lower ends to be fixed to the chains on the
outside of the fore and aft shroud of each mast, and continued by broad
and stout straps of copper to the copper sheathing of the vessel.   To
prevent lightning from entering the lower rigging from ahead, he proposes a conductor on each side of the fore-stay, their upper ends to be
united with the conductors of the foremast, and the lower ends with
the sea in the most convenient way.

The topmast and rigging are to be furnished with a similar system of
conductors, and the top-gallant rigging is to be protected by three
strips of copper let into the mast, according to Mr. Harris's plan.

(185) In reply to Mr. Sturgeon's observations relative to the lateral
discharge, Mr. Harris opposes the following experiments : —

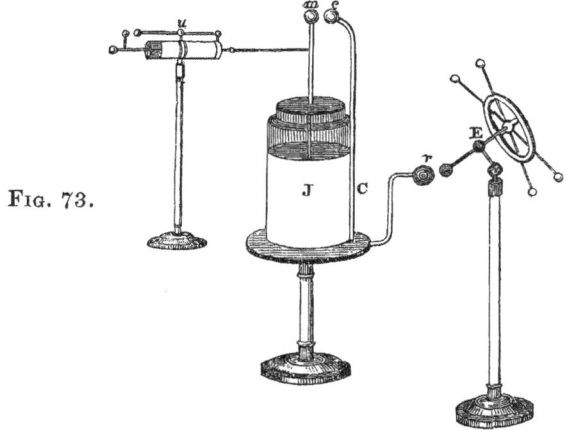

Fig. 73.

Ex. 1. Let the jar J, Fig. 73, be charged *positively*, removed from
the machine and insulated—under this condition discharge it.   When
discharged, let the electrical state of the knob *m*, discharging conductor *e* c, and the outer coating J, be examined ; they will all be
found in the same electrical state, which state will be precisely that
exhibited by the outer coating and knob, whilst charging, and the
small residuary charge will be *plus*.

Ex. 2. Charge the jar as before ; but before discharging it withdraw
the free Electricity from the knob.   The electrical state of the coating
and appendages will now be changed, and the small residuary spark will
be *minus*—thus showing that the Electricity of the spark varies with
the coatings.

Ex. 3. Immediately *after* the discharge, apply a metallic body to
the coating J ; a residuary spark will be thrown off, which spark

obviously cannot be caused by any lateral explosion caused by the discharging rod.

Ex. 4. After this residuary spark has been taken from the outer coating, examine the jar, and it will be found again slightly charged as at first, showing the spark to be merely a residuary accumulation.

Ex. 5. Charge a jar, exposing about two square feet of coating with a given quantity of Electricity, measured by the unit jar *u*, Fig. 73 : let a conducting rod terminating in a ball *r* project from the outer coating, and place near it the electroscope E. Discharge the jar through the rod *c c* as before, and observe the amount of divergence of the electroscope. Double the capacity of the jar, and again accumulate and discharge the same quantity. The divergence of the electroscope will be very considerably decreased : add a second and a third jar to the former, and the effect will be at last scarcely perceptible : connect the jar with the ground, and with a given quantity the spark will vanish altogether.

Fig. 74 represents Mr. Snow Harris' electroscope, which acts on the principle of divergence. A small elliptical ring of metal, *a*, is attached obliquely to a small brass rod, *a b*, by the intervention of a short tube of brass at *a* : the rod *a b* terminates in a brass ball, *b*, and is insulated through the substance of the wood ball, *n*.

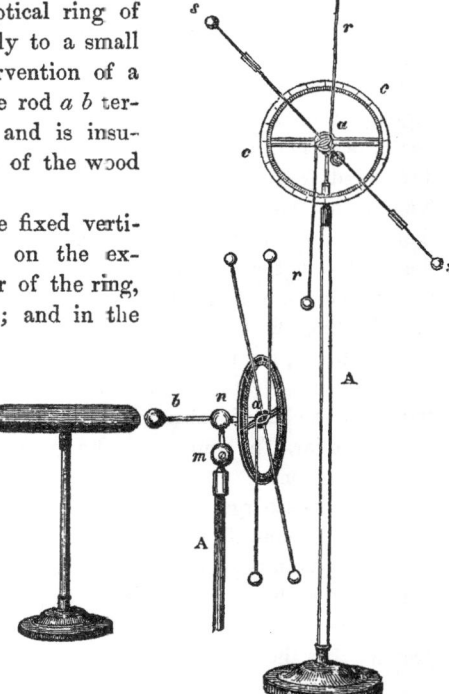

FIG. 74.

Two arms of brass, *r r*, are fixed vertically in opposite directions on the extremities of the long diameter of the ring, and terminate in small balls; and in the direction of the shorter diameter within the ring there is a delicate axis set on extremely fine points : this axis carries, by means of short vertical pins, two light reeds of straw, *s s*, terminating in balls of pith, and constituting a long index, corresponding in length to the fixed arms abovementioned.

The index thus circumstanced is susceptible of an extremely minute force; its tendency to a vertical position is regulated by small sliders of straw, moveable with sufficient friction on either side of the axis. To mark the angular position of the index in any given case, there is a narrow graduated ring of card-board or ivory placed behind it. The graduated circle is supported on a transverse rod of glass by the intervention of wood caps, and is sustained by means of the brass tube, a, in which the glass rod is fixed. The whole is insulated on a long rod of glass, A, by means of wood caps terminating in spherical ends.   In this arrangement, as is evident, the index diverges from the fixed arms whenever an electrical charge is communicated to the ball b, as shown in the lower figure.   The instrument is occasionally placed out of the vertical position at any required angle by means of a joint at m, and all the insulating portions are carefully varnished with a solution of shell-lac in alcohol.

Ex. 6. Accumulate a given quantity as before, and observe the effect of the residuary charge on the electroscope.   Let a double, treble, &c., quantity be accumulated and discharged from a double, treble, &c., extent of surface — that is to say, for a double quantity employ two similar jars and so on; the effect will remain the same.

These two last experiments prove that the spark is of different degrees of force when the Electricity is discharged from a greater or less extent of surface, whilst double, treble, &c., quantities, when discharged from double, treble, &c., surfaces, give the same spark. Now, as no one can doubt but that the effect of a double, &c. quantity should be greater than a single, &c. quantity, it is again evident that the spark is not caused by any lateral explosion from the discharging rod, it being a well established law that the same quantity has the same heating effect on wires, whether discharged from a great surface or a small one, from thick glass or thin : some little allowance being made for the greater number of rods, &c., when the surface is increased by an additional number of jars.   The effect, therefore, depending on the jar, Mr. Sturgeon had a greater chance with a small jar than with a large one.

Ex. 7. Discharge a jar by means of discharging circuits of different dimensions, from a large rod down to a fine wire, which the charge in passing can make red-hot.   Observe the effect on the electroscope in each case : it will be found nearly the same, being rather less where the tension in the discharging wire is very considerable— proving that the tension on the rod is not of any consequence.

Ex. 8. Connect the jar with the ground, and place a small quantity of percussion powder inclosed in thin paper between the discharging conductor *c*, Fig. 73, and a metallic mass placed near it. The powder will not be inflamed even in the case of the discharging conductor becoming red-hot, whereas in passing the slightest spark it inflames directly, which shows that no kind of lateral action arises during the passage of the charge.

(186) The following instructive experiments are also quoted by Mr. Harris, to prove that an electrical explosion will not leave a good conductor constituting an efficient line of action, to fall upon bodies out of that line :—

Lay some small pieces of gold-leaf on a piece of paper, as represented in Fig. 75, pass a dense shock of Electricity (from not less than eight

Fig. 75.   Fig. 76.

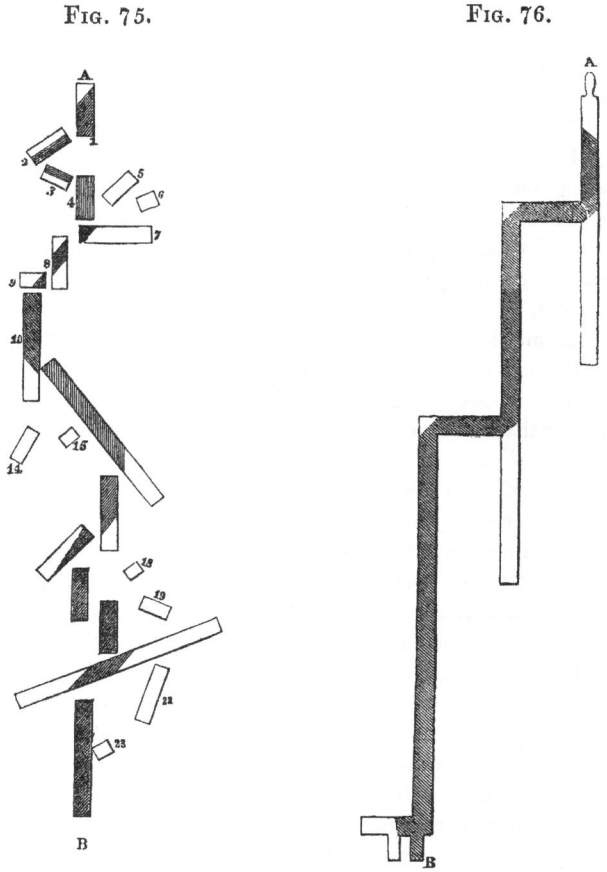

B

square feet of coating) over these, from the commencement at A to the termination at B, so as to destroy the gold: the line which the discharge has taken will be thus shown by the blackened parts, and the result will be as in Fig. 75, which is copied from the actual effects of the electrical discharge, Mr. Harris having, in a very obliging manner, furnished me with several specimens to experiment upon. By the result of the explosion represented in Fig. 76, it is shown that the portions of the mast below the striking parts are out of the line of discharge, and not involved in the result.

(187) In Fig. 75 it is particularly worthy of remark, that not only are the pieces 5, 6, 14, 15, 18, 19, 22, 23, untouched, being from their positions of no use in facilitating the progress of the charge, but even portions of other pieces which have so operated are left. perfect, as in 2, 3, 8, 9, 10, &c.; so little is there any tendency to a lateral discharge even up to the point of dispersion of the metallic circuit in which the charge has proceeded: indeed, as Mr. Harris observes, so completely is the effect confined to the line of least resistance, that percussion powder may be placed with impunity in the interval between the portions 4, 5, and he contends that the separate pieces of gold leaf thus placed may be taken to represent detached conducting masses fortuitously placed along the mast and hull of a ship, and that therefore Mr. Sturgeon's assertion that a conductor on a ship's mast would operate on the magazine is quite unwarranted. Mr. Harris then proceeds to state a few cases of damage to certain ships of the navy, where metallic bodies happened to be so disposed about the rigging and hull as to approximate to that perfect state of defence against the expansive force of the electrical discharge in which a ship would become placed by perfecting the conducting power of the mast, and uniting them into one general continuous system with the metallic masses in the hull and with the sea, and from all these cases it is shown that though a shock of lightning may divide in the absence of any good conducting course, and branch out into a variety of other courses, no damage occurs from a shock *out of its direct path*.

(188) Mr. Harris describes the following experiment, which he considers an important one, as bearing on the theory of lightning conductors, and which he made at Plymouth before the Navy Board. A model of a mast about ten feet in length, was made in parts, and an interrupted line of metal placed in the heart of it. Percussion powder was placed between these interruptions, and on the surface of the model a continuous conductor was fixed, having at the bottom a metallic connection with the interior line of metal. To make the experiment more complete, bands of metallic leaf, &c., were here and there made to surround the mast, together with other metallic

bodies, which could enter into the mast itself, and touch the metallic line within.

An intense electrical accumulation from four jars of five square feet each, highly charged, was now allowed to fall upon the ball surmounting the mast, and connected with the exterior and interior lines of metal ; but the percussion powder was not inflamed as it would have been by the least spark of Electricity, showing that the whole charge passed down the continuous conductor without occasioning any lateral result.

(189) In answer to Mr. Sturgeon's objection, " that a flash of lightning passing down on one of Mr. Harris's conductors, would magnetize every piece of ferruginous matter entering into the construction of the chronometers, and would render them and the compasses useless," Mr. Harris replies by an appeal to facts, and shows that out of a hundred cases all the facts of which are especially known, *only one case* occurs in which the compasses and chronometers became damaged, and in this case the electric discharge invaded the place in which these instruments were, demolished all the bulk heads and fittings of the cabins, and passed directly through them, or near them, in its course to the sea; and he also gives a table containing some remarkable instances in which discharges of lightning traversed lightning conductors in ships of Her Majesty's navy *without* producing any effect whatever on the compasses.

(190) I have now to notice a paper " On the Action of Lightning-Conductors," by Mr. Walker, Secretary to the London Electrical Society, printed in the 6th Number of the Proceedings of that body. In this essay, Mr. Walker endeavours to show—

1st, That the discharge of a Leyden jar *does not* resemble a flash of lightning ; and, therefore, that Leyden jars should not be employed in these experiments.

2nd, That the discharge of a prime conductor *does* in all essential points resemble a flash of lightning ; and *is*, therefore, admissible in these experiments.

3rd, That a wire on which sparks are thrown from the prime conductor represents a lightning-rod.

4th, That sparks will pass from such a wire, and, therefore, from a lightning-rod, to vicinal conducting bodies ; and—

5th, That these sparks may be prevented by connecting the vicinal bodies with the rod itself.

(191) These positions are supported by Mr. Walker in a very ingenious manner, though with what success, each will judge for himself. It is highly pleasing, however, to observe the open, diffident, and liberal manner in which Mr. Walker writes, and to find him declaring, that as

his only object is to arrive at the *truth*, he is willing to withdraw his
faith from the conclusions to which he has arrived, should they be
shown to be based on a false foundation.

(192) In the prosecution of his experiments, Mr. Walker had the
advantage of the finest piece of electrical apparatus probably in the
world, viz. the large machine at the Polytechnic Institution, the plate
of which is 7 ft. 3 in. in diameter, and the spark from the conductor of
which is said to have force enough to fell a man to the ground.

(193) That the discharge of a Leyden phial does not resemble a flash
of lightning, Mr. Walker concludes from the following considerations :
—If C, Fig. 77, be
taken to represent a
plate of glass, and A
and B respectively the
coatings of tin-foil, of
which (for example)
A is positively charged,
the equilibrium is re-
stored by a spark pass-
ing at *c*, between the
balls *a* and *b*.   Now,
if A be taken to re-

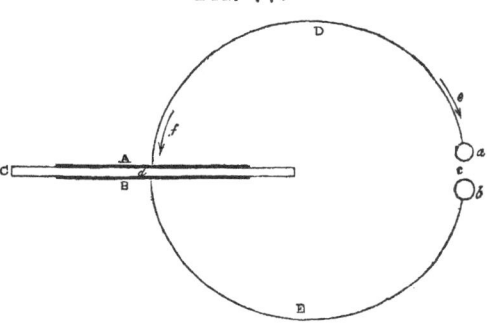

FIG. 77.

present a cloud, B the earth, and C the intervening stratum of air, then
to complete the analogy, the flash ought to take place between the
layers of tin-foil ; and consequently the only case in which the discharge of a Leyden jar resembles a flash of lightning is when the glass
gives way, and the Electricity passes through it.

Again, there is a fundamental difference between the direction of the
respective discharges.   In the Fig. 77, the tendency towards the restoration of the equilibrium is exercised in the direction pointed out by
the arrow *f*, and in that direction *alone*.   But when the discharging
arrangement D is introduced, there is a tendency to discharge in the direction of the arrow *e*; and these forces being in opposite directions, the
length of the striking distance *c* is due to the *difference between these
forces*; and thus Mr. Walker accounts for the fact that the spark from
the Leyden discharge is short and straight, while that from the prime
conductor is, like a flash of lightning, devious and zigzag ; and, as in
the discharge from the prime conductor, there are no counter-forces acting in opposition to each other, it resembles a flash of lightning, both
having a direct path to the earth.

(194) Mr. Walker connected a stout copper wire with the gas-
fittings of the house, being insulated on glass rods at different parts of
the room ; and on drawing sparks from the prime conductor by means

of a brass ball five inches in diameter, attached to the other end of the wire, and held in the hand by means of a glass rod, sparks were drawn at the same time, not only from the gas-fittings of the room in which the experiments were made, but also from the burners in the workshops two stories below.   Hence he draws the conclusion that if the prime conductor represents a cloud, the spark a flash of lightning, and the wire a lightning-conductor, there *is* a danger of a *division of the charge**** among vicinal conducting bodies.

(195) A circular piece of wood between two and three feet in diameter, covered with tin-foil, was then placed on a stool, and connected with the earth in the manner shown in Fig. 78; sparks were taken between the prime conductor and *a,* placed immediately over it, and the arrangement was supposed to verge closely

FIG. 78.

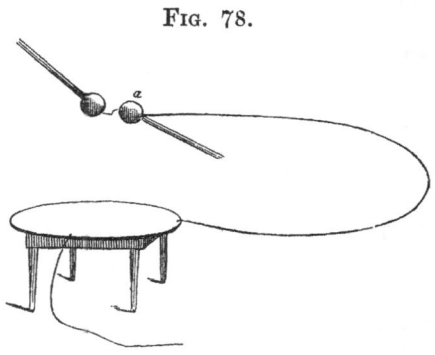

upon the conditions attributed to the action of a lightning cloud, still sparks could always be obtained from the wire whenever a conductor was approached to it.

(196) The following experiment was next made :—On a deal board, about two feet square, were pasted slips of tin-foil, *a, b, c,* Fig. 79.

The upper slip, *a,* was designed to represent the cloud ; the lower one, *b,* the earth; the perpendicular, *c,* the lightning rod; the other slips, bent at right angles towards *x, x, x,* are vicinal conductors in good connexion with the slip *b,* which is itself connected with the earth by the wire B. When sparks were passed from the machine upon *a,* they discharged themselves at

FIG. 79.

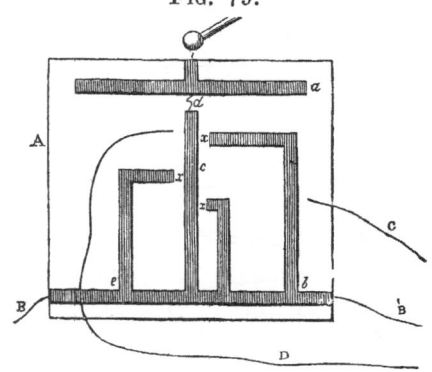

d to the conductor *c,* and passed along it to B ; but under no circumstances would they pass the spaces *x, x, x,* on which was placed per-

---

* It is worthy of remark that Mr. Walker uses the expression "division of charge" in preference to "lateral discharge," a very different affair, the existence of which was the principal topic of discussion between Mr. Harris and Mr. Sturgeon.

cussion powder. The wire B was now removed to the position B';
connecting it with a good discharging train, and the experimenter took
in his hand the wire C, connected with the same pipes, and in the same
direction sparks were passed as before at *d*, and by applying the wire
C to any part of any of the slips of tin-foil, he was enabled to draw off
sparks. But when the wire C was placed in a position similar to that
represented by D, *touching the tin-foil b* at *e*, the sparks ceased to
appear.

(197) Lastly, a long brass rod,
terminating in the 5-inch ball A, Fig.
80, was connected with the prime
conductor; beneath A was a corresponding ball B, mounted on a similar
brass rod; the latter was screwed
into a small brass plate, which was
fixed to the floor of the room. A
stout wire *b* connected the plate with
the discharging train: a smaller brass
rod C, terminating in a small brass
ball, was connected with the same
discharging train by the stout wire *c*.
Here A was taken to represent a
cloud, the small brass plate and the wood-work about it the earth, B a
lightning rod and C a vicinal conductor, the wires *b* and *c* employed for
producing effectual discharges. The machine was put in action,
and a series of long and vivid sparks were passed between A and B.
The experimenter rested the rod C *on the brass plate*, and brought the
ball near the conducting rod, but *no sparks occurred*. But when C

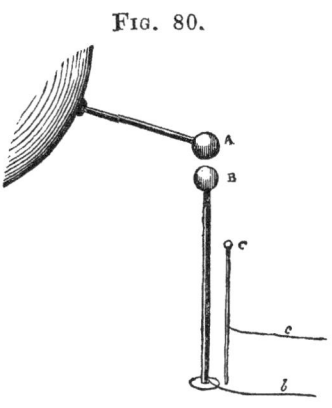

FIG. 80.

was moved into the position represented in
the drawing,—that is, when its lower end
was made to rest on the *floor of the room,
long and bright sparks passed in abundance
from the rod* B; and this, whether the
wire *c* was or was not attached to the
smaller rod.

(198) Mr. Walker attaches particular
value to these two last experiments,
because he thinks they afford him a
means of explaining the following experiment of Mr. Snow Harris :—" Insulate a
circular conducting disc, M, Fig. 81, of
about 4 feet in diameter, and oppose to it a

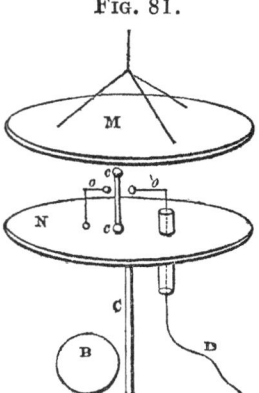

FIG. 81.

similar disc, N, at about 6 inches distance, and connected with the ground. These discs may be of wood, covered with tin-foil. Stand a conducting-rod, *c, c,* terminating in a ball, on the lower disc, and place near it the metallic body *o.* Electrify the upper disc : dense sparks may be caused to fall on the rod, but no effect is observable on *o,* even if percussion powder be placed in the opening between them.

Now, Mr. Walker thinks it clear that the reason that no sparks pass from *c* to *o,* is because they are in *direct* metallic connexion ; for, although in his own experiment, in which sparks were obtained between B and C, (Fig. 80,) on disconnecting C from the brass plate, the two rods were in metallic connexion, (both uniting in one train, some fifteen or twenty feet distant,) yet the resistance of the air between the rods being less than the sum of these resistances, the division of charge takes place ; and he considers, from these illustrations, that it will be evident that we do not imitate nature, in selecting a metallic disc to represent the earth, and experimenting on conductors erected on this disc, because we make such a connexion as never occurs naturally ; and he draws the general conclusion, that the only method of preventing sparks from passing from a lightning-rod, while conveying a flash of lightning to vicinal conducting bodies, is to connect these bodies with the rod itself.

(199) The following is the substance of the reply of Mr. Harris :— In the first place, he submits that Mr. Walker has failed to verify his views, by an appeal to the actual operations of nature, but putting forth certain notions about " *lateral discharges,*" and the action of lightning-rods, he does not give a single instance from observation in nature in which his views are borne out,—*not one case* in which a building or a ship,. having a lightning-rod, presented, when struck with light-ning, the phenomena insisted on.

(200) Secondly, he thinks that Mr. Walker has not given a sufficiently definite explanation of what he means by " lateral explosion." If he means to say, that when a great variety of circuits are open to a passing charge, the charge will divide upon them all, or tend to do so, that is by no means a new fact. If, for instance, A B (Fig. 82) be a lightning-rod, *c, d,* and *e, f,* other rods very near it, each having a common connection with the earth or

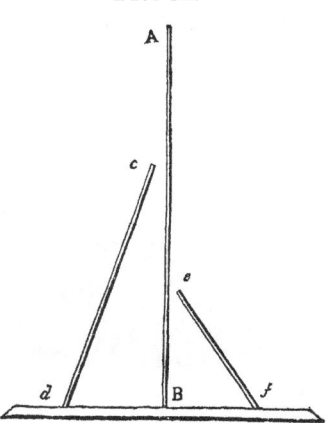

Fig. 82.

sea at B; then supposing the resistance on the points c and e to be either equal to or less than the resistance of the conductor A B, it is clear that the passing charge will divide upon the three rods : hence it is that we should give lightning conductors great capacity, and so place them that there should be no circuit open of which they do not form a part. If, on the other hand, Mr. Walker would infer, that in neutralizing the opposite electrical forces a lightning rod will send off a portion of the passing charge to semi-insulated masses of metal, merely because they are vicinal bodies, as in Fig. 80, then he must protest against there being any such instance in common Electricity, or in the operations of nature.

(201) Thirdly, with respect to the difference of discharge from the machine and the discharge of a jar, he thinks Mr. Walker's views are quite hypothetical and unfounded. The essence of the discharge is the neutralization of two opposite electrical forces, made active and accumulated on a peculiar arrangement of opposed conductors, with an intermediate insulated medium. It is these opposed conductors that are solely and exclusively concerned ; and it can possibly make no difference, whether the discharge be effected through a rod A c B (Fig. 83), in the centre of the planes, or by a rod a, m, b, at the

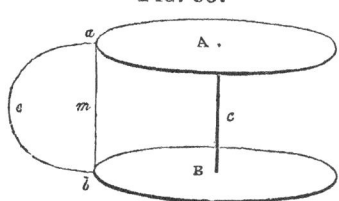

FIG. 83.

edge of the planes, or by a curved rod a, o, b, exterior to the planes. The forces neutralize through the least resisting points ; and it is an electrical impossibility that, in the act of this neutralization, any portion of either of those forces, or any portion of the combination of them, should be left behind, as it were, on the road, upon vicinal bodies, to the exclusion of such portion from the system upon which the whole result depends. In nature, the charged clouds are not always neutralized *directly* through the intermediate air.

FIG. 84.

Thus let S (Fig. 84) be the position of a ship on the plane of the sea B, in respect of a thunder-cloud A. Then if the resistance to the discharge of the system A, C, B, should be less from the edge of the cloud in the direction of the ship's mast at S than at the interval C, there will be a side circuit immediately, A, S, d, B, and flash will pass from the cloud.

(202) Mr. Walker's hypothesis about opposing forces appears to amount to this ; that if two curved wires a, b,—c, d, (Fig. 85,) be placed exactly opposite each other, on the charged surfaces of a coated pane A B, there will be as much tendency to discharge in the points a, c, as in b, d; or, in other words, that the whole coating on one side tends, in every point

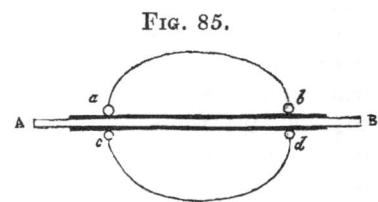

Fig. 85.

equally to discharge upon the whole coating on the other side ; and the supposition of a second coated pane at c, (Fig. 77,) is virtually the case of two coated panes, or two jars connected together at their inner and outer coatings,—in short, the case of an electrical battery. The instant the balls of a discharging-rod are brought to touch the inner and outer coatings of a charged battery, there is no longer any opposing force whatever between the coatings.* The discharging balls are them-selves the coatings to the interval of air at C, which *breaks down actually under the charge ;* and as this circuit in no sense differs from the position of two wires on a single pane, the discharge at c is reduced to Mr. Walker's supposititious case, of breaking through the pane itself in a weak point.†

(203) Fourthly, with respect to the difference between the spark from a machine and that from a charged surface, in any case of taking sparks from the prime conductor, the conditions of the Leyden jar are complied with.   The opposed balls become the hemispherical coatings to the interposed air, and the spark is nothing more than the repeated discharge of the system, and an excess of Electricity *forced over* upon the opposed ball, in order to be *returned to the rubber and negative conductor of the machine.*   But this is not the state of things in nature ;

---

* Even during the approach of the balls, all the action commences to operate in that direction, as may be seen by the falling of the electrometer. The opposing forces tend to neutralize through the circuit, and not through the glass, because that circuit becomes least resisting.

† Mr. Harris's view appears to be this :—The laws of discharge are in all cases the same, the forces neutralizing through lines offering the least resistance ; and consequently if a line of conduction or transit be provided of indefinitely less resistance than any other line, that will in all cases determine the line of discharge without reference to other bodies. This line is a *perfectly* applied lightning-rod. If there be no such line provided, the electrical forces in action will make one for itself, and will actually pick out any set of detached metallic masses which happen to lie as points in such a line, leaving others untouched, as already exempli-fied in Fig. 75.

the discharges take place deliberately, and at intervals : whenever, by the evolution of Electricity from natural causes, the accumulation can break through the insulating interval ; and to imitate this faithfully by the electrical machine, and obtain effects corresponding with cases of buildings or rods struck by lightning, the author should have placed such rods, in the interval, immediately between a similarly charged system, which should be gradually and cautiously brought up to the point of discharging by a slow turning of the machine.   Thus, in

Fig. 86, the lateral discharge, to be examined, should have been by means of conductors placed in the interval *c*, between the hemispherical coatings of the system N, *c*, P, and not outside of the system, as at W and S. It is only the interval at *c* which can be taken to represent the condition of a thunder-storm,

FIG. 86.

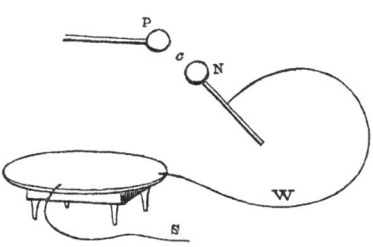

and this only partially, so long as we turn the machine at random.

(204) The length, or striking distance of the spark from the machine is a very casual matter, and greatly depends on the arrangement of the conductors.   If it be taken between very large balls or flat surfaces, the length will be reduced to that of a jar discharging ; and if a jar be discharged over a card covered with vermilion between the points of the universal discharger at $1\frac{1}{2}$ inch apart, then a *black line of a zig-zag form* will be left, and so it is between small discharging balls, only the eye cannot catch the form.

(205) If sparks be taken from a machine between two equally large balls bearing a good proportion to the quantity of Electricity, or force or intensity of the conductor as expressed by the electrometer on a given surface, they will be found as compact and as concentrated as in any discharge of a jar.   If very small balls or points be used, then the jar may be discharged at a very great distance, especially when the balls are placed in an exhausted receiver, in which case the zig-zag Mr. Walker speaks of is very apparent through two feet or more.   Again, in Singer's experiment of the shooting-star, a spark is produced of two or even five feet long, as closely resembling lightning as anything can possibly be imagined.

(206) The distance through which an electrical battery can discharge with a little assistance is considerable.   Thus a slight trace of sulphuric acid will enable a jar to discharge over twelve inches or more ; and a trace of smoke or other light conducting particles in the air will operate in the same way.

(207) Beccaria has shown that a discharge of lightning is greatly facilitated by particles of vapour,—by conditions of partial rarefactions of the atmosphere; that the electrical agency seizes upon any kind of matter or light substances, however imperfect its conducting power, which may be floating in the air, and can assist its progress; we have therefore sufficient explanation of the great length of a flash of lightning. The longest flashes are really produced in this way, very similar to the artificial methods of increasing the distance of discharge of a battery; in both cases, there is a concentrated and dense globular spark produced, intensely luminous, which traversing the intermediate space with very great rapidity, gives the impression of a brilliant and continuous line of light, of momentary duration. On the other hand, when we find that the laws of the discharge are in both cases the same, that in passing amongst a series of detached and semi-insulated metallic masses, the electrical agency seizes only upon those which happen to assist its progress, *selecting* some, and *neglecting* others, *however near;* that before the discharge takes place, there is a sort of feeling of the way, or marking out as it were in advance, by a wonderful species of influence, the course it is about to take; that in both cases the track is concise and definite, and between given planes; that both space and time are, as it were, economized in the greatest possible degree in the neutralization of the charged system; that metallic wires are fused, and the most compact bodies broken or scattered by a powerful expansive force communicated to the air, and a variety of mechanical and chemical effects produced,—it is impossible not to recognize in the Leyden jar all the effects and operations of lightning. Accordingly, continues Mr. Harris, we find that the differences assumed by Mr. Walker to exist in the nature of the two discharges, are unsupported by any good evidence whatever.

(208) Lastly, with respect to the Experiment, Fig. 87. The ball A, and the plate N, and wood-work on which it rests, cannot possibly be taken to represent the cloud and earth so long as the ball B is equal to the ball A, especially as this last is continually discharging, and is very small in comparison to the size of the floor on which N is placed. It is well known as an electrical principle, that supposing a square of glass to have coatings of unequal size or extent, that glass *cannot be charged* except in the points covered by the lesser coating. A cloud only operates in this way upon a corresponding and opposed

FIG. 87.

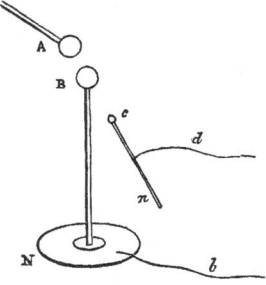

portion of the earth and sea. Evidently, then, all the operation of a charged system is between A and B, and the arrangement wholly different from the conditions of the opposed discs, Fig. 81. If we could fancy ourselves at the time of a thunder-storm, under a spreading conducting mass of equal size to the cloud, stuck on the end of a long metallic pole, then the pole might be in some such condition as that of the rod B N, Fig. 87 ; and any arrangement of wires or balls about it might be affected in a similar way to the wires *c, d*, that is, supposing at the time of the neutralization of the force overhead, some free Electricity pervaded the pole, which would be doubtful.

(209) The reason why sparks are not obtained at *c*, when the rod *c n* touches the plate N, is simply because the resistance in the direction B N is less than in the direction B *c d* ; and the reason why sparks *do* appear when the rod *c, n, does not touch* the plate N, is in consequence of the resistance in the direction B *c d* being less than in the direction B *n*. The action of the wires *b, d*, are directly upon the negative conductor,—simply sources of supply from the prime conductor, which becomes charged with one of the disturbed forces by the action of the glass. The connecting those wires with the gas-fittings, &c. is in no way essential to the experiment, and they might as well have terminated at once in the negative conductor.

(210) From all these considerations, Mr. Harris thinks that sufficient evidence is obtained to show :—

1st. That the discharge of an electrical jar, or battery, does in every way correspond in its nature and effects with a discharge of lightning.

2nd. That any particular arrangement of wires exterior to, and disconnected with the discharge of an electrical accumulation between two opposed conductors, *does not represent the condition of a lightning rod.*

3rd. That in electrical discharges similar to lightning, the course of the discharge upon detached metallic bodies is always according to certain laws of resistance and distance ; and

4thly. That a lightning-rod, so far from sending out sparks to such bodies, *diverts the passing charge from them altogether ;* and that, consequently, the mere connection of the rod with such bodies is quite unnecessary and useless.

(211) I have thus endeavoured to present to my readers a fair and impartial account of the whole question of lightning conductors, the very great importance of which is the only excuse I shall offer for the space I have allotted to it : any remarks from me on the subject would be superfluous. I cannot, however, quit it without observing that, in my own opinion, Mr. Harris has made out his case most fully and completely ; and as far as I can understand the matter, he has demon-

strated most clearly, that there is really no such thing as a " lateral discharge," in the sense in which it has been taken, as respects a lightning-
rod : that the lateral discharge in Electricity, as it is called, is merely
a little excess of free Electricity which distributes itself over the discharging circuit when a charged system is discharged or neutralized (112); but the two forces are never precisely balanced, as has
been shown by Biot, Henry, and others.

(212) I think also it is clear that in Fig. 87 the charging is between
the balls A, B; that it is the air between those balls *alone* that is
charged, and that the ball B absorbs the whole force of that charge,
and, consequently, that the whole is a conventional arrangement of
conducting wires *outside* a small charging system ; and the sparks
which take place are occasioned by the free Electricity pervading the
wires in directions giving the least resistance.

(213) If this be a right view of the case, it is obvious that the wire
B, N, is not in the condition of a lightning-rod ; for suppose A, Fig.
88, to represent a mass of cloud positively electrified, and B an extensive and opposed tract of land ;

FIG. 88.

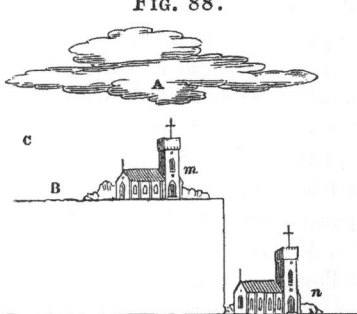

electrified *negatively;* between the
two is the non-conducting stratum of air, C : now, the position of a lightning-conductor or
building is not outside the planes
at *n*, near a wire or rod carrying a current, and in the vicinity of bodies in opposite or
neutral states, but *between* the
planes A B at *m*, and actually
forming a point in one of them, and *sharing its electrical state ;*
and when the discharge breaks through the medium at C, it appears
to me to be virtually the discharging rod of the Leyden experiment ; and I confess I cannot perceive that it can make any difference
whether the building be placed in that circuit between the plates immediately, or whether it operates as a discharging circuit in any other
direction.

(214) With respect to Mr. Harris's ship conductors, one fact is
worth a volume of theorizing ; and I shall conclude the whole subject
with the history of a case which has recently come to my knowledge
from unquestionable authority, and then leave it to the public to judge
whether it does or does not prove satisfactorily, that as far as human contrivance can grapple with an agent so awful as lightning, security the
most perfect is provided by Mr. Harris's system.

(215) H. M. ship " Actæon," lately returned from South America,

was provided with conductors in her masts. She encountered a tremendous tropical hurricane and thunder-storm. The officer who had charge of the watch, and the carpenter, who was close to the mainmast, say that the thunder and lightning were most awful. It was in the middle watch of the night,—pitch dark—the ship rolling fearfully in the squalls, and lightning breaking in all directions around. At last came a fearful crash as if all the main-deck guns had been fired. A momentary gleam of bright light appeared upon the conductor. The ship is reported to have shook from end to end, and the cutlasses stored in a frame round the mast, to have shook and rattled like hail. The carpenter, who was leaning against the pump-handles, says he felt them vibrate. Still no harm was done; no lateral explosion; no ill effects to the men near the mast, and the sails all around the conductors were safe. The people declared that though they had often been in thunder-storms, and had witnessed the destruction of ships' masts by lightning, they had never encountered any thing like this; and, as might be expected, the conductors acquired a great character in the ship. This was not the only time she was struck during the same voyage, without any ill effect.

It is important to notice, that at the time the cloud broke on the mast, there was heard about the line of the crash a furious *whiz*, as if the valve of a steam-engine had been let loose.

(216) Waterspouts, and volcanic eruptions in the sea, are generally attended by thunder and lightning, and may be classed among electrical phenomena. In June 1811, Captain Tilland observed off the island of St. Michael one of these marine volcanoes, of which he has given the following account in the Philosophical Transactions. "Imagine," says he, "an immense body of smoke rising from the sea, the surface of which was marked by the silver rippling of the waves occasioned by the slight and steady breezes incidental to those climates in summer. In a quiescent state, it had the appearance of a circular cloud, revolving on the water like a horizontal wheel, in various and irregular involutions, expanding itself gradually on the lee side, when suddenly a column of the blackest cinders, ashes, and stones, would shoot up in the form of a spire, rapidly succeeded by others, each acquiring greater velocity, and breaking into various branches resembling a group of pines; these again forming themselves into festoons of white feathery smoke. During these bursts, the most vivid flashes of lightning continually issued from the densest part of the volcano, and the columns rolled off in large masses of fleecy clouds, gradually expanding themselves before the wind, in a direction nearly horizontal, and drawing up a quantity of waterspouts, which formed a striking addition to the scene. In less than an hour, a peak was visible, and in three hours

from the time of our arrival, the volcano, then being four hours old, a crater was formed twenty feet high, and from four to five hundred feet in diameter. The eruptions were attended by a noise like the firing of cannon and musketry mixed; as also with shocks of earthquakes, sufficient to throw down a large part of the cliff on which we stood. I afterwards visited the volcanic island : it was eighty yards high, its crater upon the level of the sea was full of boiling water; it was about a *mile* in circumference, and composed of porous cinders, and masses of stone."

(217) The aurora borealis is unquestionably connected with Electricity in some way, though it does not appear at present in what precise manner : its appearance may be imitated with great exactness by artificial Electricity ; for if a tube be partially exhausted of air, and a stream of Electricity sent through it, the same variety of colour and intensity, the same undulating motions, and occasional coruscations, and the same inequality in the luminous appearance, are exhibited as in the aurora ; and when the rarefaction is considerable, (Mr. Singer observes,) various parts of the stream assume that peculiar glowing colour, which occasionally appears in the atmosphere, and which is regarded by the uninformed observer with astonishment and fear.

(218) The aurora borealis is seldom seen in perfection in this country, and of late years has rarely been noticed at all. But Captain Parry, in his second voyage for the discovery of a north-west passage, had abundant opportunities of observing it in the greatest splendour. That highly distinguished philosopher and chemist, Dr. Dalton, has also furnished us with an account of an aurora, noticed by him, on the 15th of October, 1792. I shall take the liberty of inserting an extract from it, and also an abstract of Captain Parry's description of an aurora observed by him on the 11th of December, 1821.

(219) "Attention was first excited," says Mr. Dalton, " by a remarkably red appearance of the clouds to the south, which afforded sufficient light to read by at eight o'clock in the evening, though there was no moon nor light in the north. From half-past nine to ten, there was a large, luminous, horizontal arch to the southward, and several faint concentric arches northward. It was particularly noticed that all the arches seemed exactly bisected by the plane of the magnetic meridian. At half-past ten o'clock streamers appeared, very low in the south-east, running to and fro from west to east; they increased in number, and began to approach the zenith apparently with an accelerated velocity ; when all on a sudden the whole hemisphere was covered with them, and exhibited such an appearance as surpasses all description. The intensity of the light, the prodigious number and volatility of the

K

beams, the grand intermixture of all the prismatic colours in their utmost splendour, variegating the glowing canopy with the most luxuriant and enchanting scenery, afforded an awful, but at the same time, the most pleasing and sublime spectacle in nature. Every one gazed with astonishment, but the uncommon grandeur of the scene only lasted *one minute ;* the variety of colours disappeared, and the beams lost their lateral motion, and were converted into the flashing radiations.

" Notwithstanding the suddenness of the effulgence, at the breaking out of the aurora, there was a remarkable regularity in the manner. Apparently a ball of fire ran along from east to west, with a velocity so great as to be barely distinguishable from one continued train, which kindled up the several rows of beams one after another. These rows were situated before each other with the exactest order, so that the base of each row formed a circle, crossing the magnetic meridian at right angles ; and the several circles rose one above another, so that those near the zenith appeared more distant from each other than those near the horizon, a certain indication that the real distances of the rows were nearly the same. The aurora continued for several hours. There were many meteors (falling stars, as they are commonly called) seen at the same time ; but they appeared to be below, and unconnected with the aurora." *

(220) " The aurora," says Captain Parry, " began to shew itself as soon as it was dark. Innumerable streams of white and yellowish light occupied the heavens to the southward of the zenith, being much brighter in the south-east, from which it often seemed to emanate. Some of these streams were in right lines, others crooked, and waving in all sorts of irregular figures, moving with inconceivable rapidity in various directions. Among them might frequently be observed shorter bundles of rays, which, moving even with greater velocity than the rest, have acquired the name of ' merry dancers.' In a short time the aurora extended itself over the zenith, about half way down to the northern horizon, but no further, as if there were something in that quarter of the heavens that it did not dare to approach. About this time, however, some long streamers shot up from the horizon in the north-west, but soon disappeared. While the light extended over part of the northern heavens, there were a number of rays assuming a circular or radiated form, near the zenith, and appearing to have a common centre near that point, from which they all diverged. The light of which these were composed appeared to have inconceivably rapid motion in itself, though the form it assumed, and the station it occupied in the heavens, underwent little or no change for perhaps a

* Dalton's Meteorological Essays.

minute or more. This effect is a common one with the aurora, and puts one in mind, as far as its motion alone is concerned, of a person holding a long ribbon by one end, and giving it an undulatory motion through its whole length, though its general position remains the same. When the streams or bands were crooked, the convolutions took place indifferently in all directions. The aurora did not continue long to the north of the zenith, but remained as high as that point, for more than an hour. After which, on the moon rising, it became more and more faint, and at half-past eleven was no longer visible.

"The colour of the light was most frequently yellowish white, sometimes greenish, and once or twice a lilac tinge was remarked, when several strata appeared as it were to overlay each other by very rapidly meeting, in which case the light was always increased in intensity. The electrometer was tried several times, and two compasses exposed on the ice during the continuance of this aurora, but neither was perceptibly affected by it. We listened attentively for any noise that might accompany it, but could hear none ; but it was too cold to keep the ears uncovered very long at one time. The intensity of the light was something greater than that of the moon in her quarters. Of its dimming the stars there cannot be a doubt. We remarked it to be in this respect like drawing a gauze veil over the heavens in that part, the veil being most thick when two of the luminous sheets met and overlapped. The phenomenon had all the appearance of being full as near as many of the clouds commonly seen, but there were none of the latter to compare them with at the time."

(221) Although Capt. Parry did not observe any electrical or magnetic disturbances, during the aurora, of which he gives the above description, yet both were noticed by Captain Franklin.* A hissing sound was also heard by Nairne, Cavallo, and others. There can be no doubt, therefore, that the phenomenon is to be ascribed, in a great measure, to the operation of natural Electricity.

* Journey to the Shores of the Polar Sea.

LECTURE IV.

VOLTAIC ELECTRICITY.

(222) THAT remarkable form of Electricity, known by the name of *Galvanism* or *Voltaism*, owes its origin to an accidental circumstance connected with some experiments on animal irritability, which were being carried on by Galvani, a professor of anatomy at Boulogne, in the year 1790. It happened that the wife of the professor, being consumptive, was advised to take as a nutritive article of food, some soup, made of the flesh of *frogs :* several of these animals, recently killed and skinned, were lying on a table in the laboratory, close to an electrical machine, with which a pupil of the professor was making experiments. While the machine was in action, he chanced to touch the bare nerve of the leg of one of the frogs with the blade of a knife that he held in his hand, when, suddenly, the whole limb was thrown into violent convulsions. Galvani was not himself present when this occurred ; but received the account from his wife, and being struck with the singularity of the phenomenon, he lost no time in repeating the experiment, and in investigating the cause : he found that it was only when a spark was drawn from the prime conductor, and when the knife or any other good conductor was in contact with the nerve that the contractions took place ; and pursuing the investigation with unwearied industry, he at length discovered that the effect was independent of the

electrical machine, and might be equally well produced by making a metallic communication between the *outside muscle* and *crural nerve*.

(223) Galvani had previously entertained notions respecting the agency of Electricity, in producing muscular action : these new experiments, therefore, as they seemed to favour his views, had with him more than ordinary interest. He immediately ascribed the convulsive movements in the limb to electrical agency, and explained them by comparing the muscle of an animal to a Leyden phial, charged by the accumulation of Electricity on its surface, while he imagined that the nerve belonging to it performed the function of a wire, communicating with the interior of the phial, which would, of course, be charged *negatively*. In this state of things, if a communication by a good conductor were made between the muscle and nerve, a restoration of the electric equilibrium, and a contraction of the fibres would ensue.

(224) It had been observed many years before this period, that, when a piece of silver is placed upon the tongue, and a piece of zinc or lead under it, a slight sensation and a peculiar saline taste is experienced whenever the ends of the metals are brought into contact, and, that if one metal be placed between the upper lip and the gums, the eyes are affected as by a flash of light, when contact between the metals is established, though no such effect is noticed as long as the metals are kept separate : these previous observations do not, however, at all interfere with the originality to which Galvani has a most undoubted claim, as they excited no attention, and called forth no efforts of the mind. It is curious to notice how frequently the progress of discovery in the sciences is influenced by fortuitous circumstances, and in no case is it more striking than in the present. Had Galvani been as good an electrician as he was anatomist, it is probable that the convulsions of the frog would have occasioned him no surprise; he would immediately have seen that the animal formed part of a system of bodies under *induction*, and he would have considered the movements of the limbs of the frog as evidence of nothing more than a high electroscopic sensibility in its nerves.

(225) To perform the experiment with the frog's legs successfully, the legs of the frog are to be left attached to the spine by the crural nerves alone, and then a copper and a zinc wire, being either twisted or soldered together at one end, the nerves are to be touched with one wire, whilst the other is to be applied to the muscles of the leg. Fig. 89 shows the arrangement. There are several ways of varying this experiment, the following one may be practically applied to a useful purpose. If a piece of copper, as a penny, be laid on a sheet of zinc, and if a common garden snail be put to crawl on the latter, he

FIG. 89.

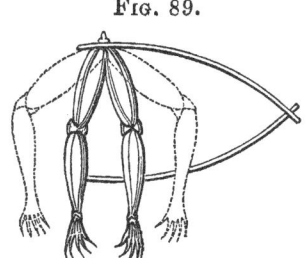

will be observed to shrink in his horns and contract his body whenever he comes into contact with the penny : indeed, after one or two contacts he will be observed to avoid the copper in his journey over the zinc. It occurred to the author last summer, to try whether by constructing a low narrow double wall, consisting of a strip of copper, and another of zinc, soldered together, and fixed in the ground, round small flower-beds, delicate plants might not be protected from the incursions of these destructive reptiles, and he has reason to think that the experiment was successful, and may furnish a valuable hint to gardeners.

(226) The experiments of Galvani excited much attention among the men of science at that period : they were repeated and varied in almost every country in Europe, and ascribed to various causes. Some imagined them the effect of a new and unknown agent : others adopted the views of the discoverer, and recognized them as peculiar modifications of Electricity. The hypothetical agent which passed under the name of the " nervous fluid," now gave way to Electricity, which, for a time, reigned as the *vital principle*, by which " the decrees of the understanding, and the dictates of the will were conveyed from the organs of the brain to the obedient member of the body,"* and this theory for a time so fascinated physiologists, that it was with difficulty that the explanations of *Volta*, viz. that the electric excitement is due to the mutual contact of two dissimilar metals ;—that, by the contact the natural Electricity was decomposed, the positive fluid passing to one metal, and the negative one to the other :—and that the muscle of the frog merely played the part of a conductor, obtained assent.

(227) It is to Professor Volta, of Pavia, that we are indebted for the first galvanic or voltaic instrument, viz. the *voltaic pile ;* it was described by him in the Philosophical Transactions of 1800, and to him therefore, the merit of laying the foundation of this highly interesting branch of science is due. The main difference between common and voltaic Electricity (which are modifications of the same force) will be found as we proceed, to be this : the first produces its effects by a comparatively small quantity of Electricity, insulated, in a high state of *tension*, having remarkable attractive and repulsive energies, and power to force its way through obstructing media : the latter is more intimately associated with other bodies, is in *enormous quantity*, but rarely attains a high state of tension, and exhibits its effects while flowing in a continuous stream along conducting bodies.

* Lardner.

(228) We will first direct our attention to the nature of those arrangements which are the sources of galvanic power : it may, however, be worth while previously to endeavour to obtain a clear notion of the *electrical* meaning of the words *tension* and *intensity*,—terms respecting which some confusion appears to exist in the writings of many electricians.

(229) For the following remarks I am indebted to Mr. Snow Harris; and while I confess myself perfectly satisfied with the distinction he has drawn between the two terms, it is right that I should state that Mr. Goodman, in his Essay (136), has taken a different view of the subject, and that he understands *tension* as referring to a " polarizing power," " transferring force," " capability of passing," " a forcible stretching or expansion from the centre towards the circumference ;" and *intensity* as signifying condensation or concentration from the circumference towards the centre, that of large quantity reduced to a very small compass, &c. (See his Essay.)

(230) According to Mr. Harris's views, *intensity* in common Electricity should be limited to the indications of an electrometer employed to determine by certain known laws of its relations to an accumulated charge,—the quantity accumulated, or any other electrical element required to be known. Thus, by the use of certain instruments, it is found that with a quadruple attractive force there is *twice* the quantity of Electricity accumulated, and so on: the surface remaining the same ; again, with a double extent of surface, the same quantity is accumulated as before, when only one-fourth the force is indicated by the electrometer.* The relations of the indications of the quadrant electrometer, or of any other electrometer, to the quantity accumulated, &c. &c., Mr. Harris considers as coming under the term *intensity ;* for they shew, at the same time, the force of the charge upon surrounding bodies. *Tension,* Mr. Harris applies to the *actual force of a charge* to break down any non-conducting or dielectric medium between two terminating electrified planes. For example, take a coated pane of glass, and charge it in the usual way ; then the absolute force exerted by the charge in the intervening glass — the force exerted by the polarized particles of the glass to get out of their constrained state may be expressed by the term *tension ;* and there would be no contradiction or superfluity of terms to talk of the *intensity of the tension* in this sense.

(231) The sum of the matter appears to be this :—*tension* applies to the particles of the electric agency itself,—to a force, in short, such as Faraday has shewn to exist in the polarized state of particles of matter, to unfetter themselves, as it were ; whereas *intensity* applies to the

* See Mr. Harris's papers in the Transactions of the Royal Society for 1836, Part 2 ; and for 1839, Part 2.

attractive forces between the terminating plates which are the boundaries of the system, as when a plane counterpoised at the end of a beam is caused to descend upon another plane beneath it, by electrical attraction, the weight in the scale pan requisite to balance this force is the intensity between the planes; whereas the *tension* of the charge between them refers to the polarized particles of the dielectric medium, —that is, to the force, whatever it be, by which they endeavour to return to their primitive state. Now, the attraction between the planes may be conceived to be the result of the induction sustained by the particles of the dielectric between them, the force of which may be called intensity; and this may differ from the re-active force in the polarized particles themselves,—that is, the force they exert to return to their primitive state. It may be also that this last force is in proportion to the quantity of disturbance in the particles, or in proportion to the quantity of Electricity developed in the terminating planes or coatings; whilst the intensity, or force of attraction between the coatings, supposing them free to move, might be as the square of the quantity of Electricity.

(232) It is very justly observed by Mr. Harris, that it would be almost as well perhaps if the term "tension" were banished from common Electricity altogether, as being too hypothetical a word for our present knowledge of Electricity, inasmuch as it is essentially applicable to some species of elastic force. Now, we do not know whether Electricity be a force of this kind or not. The term " intensity" is not open to this objection, because it simply expresses the energy or degree of power with which a particular force operates, be that force what it may.

(233) *Galvanic arrangements.*—If we take two equal-sized plates of well-polished copper and zinc, each furnished with a glass handle for insulation, and bring them into contact, holding them by the glass, it will be found by appeal to the condenser (Fig. 9) after their removal, that there is a slight accumulation of positive Electricity on the zinc, and an equal negative charge on the copper: the quantity is exceedingly small, and it requires a good instrument to exhibit it when only one contact is made; but if the operation be repeated six or eight times, taking care to restore the electric equilibrium of the metals by touching them with the finger after each contact, the effect is very distinct on drawing back the uninsulated plate of the condenser. If the same experiment be made in vacuo, or in perfectly dry hydrogen or nitrogen gas, no Electricity whatever can be discovered; hence it has been inferred that the development of the small quantity of Electricity when the plates are brought together in atmospheric air, is due to the slight chemical action of the oxygen of the air on the zinc occasioned by its compression by the copper. This was not the explanation given by the illustrious discoverer of the Voltaic pile, who attributed it to a peculiar electro-motive force,

under which metals by simple contact tend to assume opposite electrical states; and this theory has now some strenuous supporters among the German philosophers, notwithstanding the powerful mass of evidence that has been adduced against it by De la Rive, Fabroni, and particularly by our own illustrious countryman, Faraday. We shall endeavour, in the next Lecture, to present an unbiassed view of both sides of this interesting philosophical question.

(234) Presuming that the Electricity excited by the contact of the copper and zinc plates is traceable to slight chemical action, it is easy to understand that increase of chemical action must give rise to increased augmentation of the electrical force. If we take two plates of different kinds of metal, platinum and zinc, for example, Fig. 90, and 91, and immerse them in pure water, touching each other, a galvanic circle will be formed, the water will be slowly decomposed, its oxygen becoming fixed on the zinc (the oxidable metal), and at the same time a current of Electricity will be transmitted through the liquid to the platina on the surface of which the other element of the water, namely, hydrogen, will make its appearance in the form of minute gas bubbles : the electrical current passes back again into the zinc at the points of its contact with the

FIG. 90.

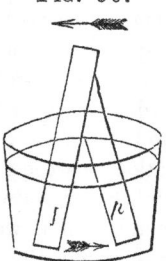

platina, and thus a continual current is kept up : and hence it is called a galvanic circle. The moment the circuit is broken by separating the metals, the current ceases, but is again renewed on making them again touch either in or out of the water, as shewn in the figures.

(235) If we now add a little sulphuric acid to the water, this effect will be much increased, because, in the first place, we make the liquid a better conductor ; and secondly, because the oxide of zinc is removed from the surface of the metal as fast as it is formed, being dissolved by the acid; and thus a new and clean surface is continually exposed. It cannot, however, be too soon impressed, that the great increase in the quantity of Electricity generated is to be attributed solely to the increased facility afforded for the

FIG. 91.

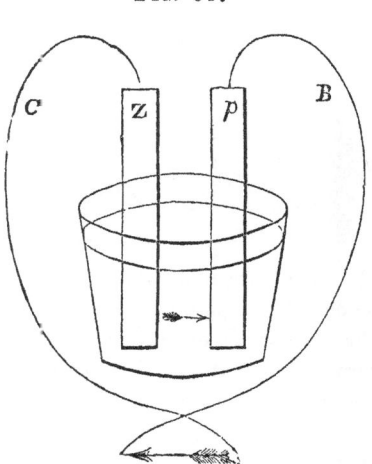

decomposition of water, and has nothing whatever to do with the formation of sulphate of zinc; not chemical action alone, but chemical decomposition being absolutely essential to the development of current Electricity. The force originates with the zinc, passes in the direction of the lower arrow (Fig. 91) to the platina, and thence back through the wires B, C, to the zinc. This is called a simple galvanic circle.

(236) To prove that the wire connecting the platinum and zinc plates is conducting a current of Electricity, we have only to place a nicely-balanced magnetic needle above or below it, and we shall find that the needle will deviate from the magnetic meridian in obedience to laws that will be described hereafter; but, how are we to account for the singular appearance of hydrogen gas on the platinum? If we amalgamate the zinc plate, by immersing it in dilute sulphuric acid, and then rubbing it over with mercury, we shall find that a mixture of one part of sulphuric acid and ten of water, will have no action on it *while alone;* the bright metallic surface will be soon seen covered with bubbles of hydrogen gas, which will adhere to it with considerable force, and thus protect it from further action: but on establishing a metallic communication between the zinc and the platinum, no matter in what manner, or by what circuitous a route, torrents of bubbles will rise from the latter metal, as if it were undergoing violent chemical action, while the zinc (the metal alone undergoing change) is oxidated and dissolved tranquilly and without any visible commotion. It is evident that we cannot explain this singular phenomenon on chemical grounds alone; but we must consider the transference of the hydrogen to take place by the propagation of a decomposition through a chain of particles extending from the zinc to the platinum, as in Fig. 92, in which, for the sake

FIG. 92.

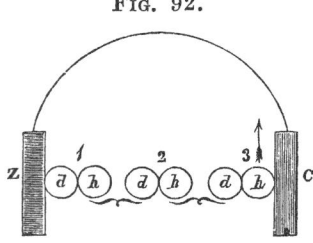

of simplicity, the exciting liquid is supposed to be hydrochloric acid: when the metallic communication is established between the plates, that particle of hydrochloric acid in contact with the zinc, undergoes decomposition, its chlorine combining with the metal, and its hydrogen displacing and combining with the chlorine of the second particle, the hydrogen of which combines with the chlorine of the third, and so on, till the platinum plate is reached, against which the hydrogen of the last particle of decomposed hydrochloric acid is evolved in a gaseous form, because it can find no particle of chlorine to combine with, and because it cannot enter into chemical union with the platinum. These changes and interchanges are precisely similar when dilute sulphuric acid is employed, substituting oxygen and hydrogen (from decomposed water) for chlorine and hydrogen; for, as we have already stated, the formation of sulphate of zinc has nothing to do with the business, it being to the decomposition of water that the effects are to be ascribed.

(237) Now there is nothing in the appearance of the liquid between the plates, which would indicate the transfer of the disunited elements above alluded to ; and the vessel which contains the acid may be divided by a diaphragm of bladder or porous earthenware, and the plates placed on each side of it without interfering much with the general result. The force must be conceived to travel by a species of *convection*, and Mr. Daniell has offered the following illustration, to assist us in forming a first notion :*—

"Where a number of ivory balls are freely suspended in a row, so as just to touch one another, if an impulse be given to one of the extreme ones, by striking it with a hard substance, the force will be communicated from ball to ball without disturbing them, till it reaches the more distant, which will fly off under its full influence. Such analogies are but remote, and must not be strained too far : but thus we may conceive that the force of affinity receives an impulse in a certain direction, which enables the hydrogen of the first particle of water, which undergoes decomposition, to combine momentarily with the oxygen of the next particle in succession : the hydrogen of this again with the oxygen of the next : and so on, till the last particle of hydrogen communicates the impulse to the platinum, and escapes in its own elastic form."

(238) But it is not in the exciting liquid alone, that this remarkable transfer of elements takes place ; the same power is propagated through the wire which connects the platinum and zinc plates together. To prove this, let the wire be divided in the middle, and having attached to each end a long slip of platinum foil, let each be immersed in a glass jar containing hydriodic acid : in a few seconds *iodine* will appear on that slip of foil which is in connection with the platinum plate and hydrogen gas on the other ; so that supposing a decomposing force to have originated in the zinc plate, and circulated through the exciting acid in the jar to the platinum, and onwards through the wires and the hydriodic acid back to the zinc : then the hydrogen of the hydriodic acid followed the same course, and discharged itself against the slip of platinum foil in communication with the zinc.

(239) It does not require two metals to form a galvanic circle, or even two different liquids, if other conditions are attended to. A current is established when a zinc plate is cemented into a box and acted upon on one side by diluted acid, and on the other by solution of common salt : or, by acting on both sides by the same acid, one surface being rough and the other smooth, a communication being of course established between the two cells. Common zinc affords a good illustration of a simple galvanic circle : this metal usually contains about one per

* See his Introduction to Chemical Philosophy, p. 413.

cent. of iron mechanically diffused over its surface. On immersion into diluted sulphuric acid, these small particles of iron and zinc form numerous voltaic circles, transmitting the current through the acid that moistens them, and liberating a large quantity of hydrogen gas.

(240) An important fact, of which a beautiful practical application was attempted by Davy, was early observed :—In proportion as the contact of two metals in an acid or saline solution increases the affinity of one of them, for one element of the solution, it diminishes the liability of the other metal to undergo change. Thus when zinc and copper are united in diluted acid, the zinc is acted upon *more* and the copper *less* than if they were immersed separately. A sheet of copper undergoes rapid corrosion in sea-water, the green oxy-chloride being formed ; but if it be associated with another metal more *electropositive* than itself, such as zinc, it is preserved, and the zinc undergoes a chemical change. Davy found that the quantity of zinc requisite to effect a complete preservation of the copper, was proportionably very small. A small round nail will preserve forty or fifty square inches, wherever it may be placed ; and he found that with several pieces of copper connected by filaments the fortieth of an inch in diameter, the effect was the same. Sheets of copper protected by $\frac{1}{40}$ and $\frac{1}{100}$ part of their surface of zinc, malleable, and cast-iron were exposed during many weeks to the flow of the tide in Portsmouth harbour, their weight, both before and after the experiment, being carefully noted. When the metallic protector was from $\frac{1}{40}$ to $\frac{1}{150}$, there was no corrosion or decay of the copper ; with $\frac{1}{200}$ to $\frac{1}{460}$ there was a loss of weight : but even $\frac{1}{1000}$ part of cast-iron saved a portion of the copper. Davy hoped to apply this principle to the preservation of the copper sheathing of ships ; but unluckily it was found that unless a certain degree of corrosion take place in the copper, its surface becomes foul from the adhesion of sea weeds and shell-fish. The oxy-chloride, formed when the sheathing is unprotected, acts probably as a poison to these plants and animals, and thus preserves the copper free from foreign bodies, by which the sailing of the vessel is materially retarded.

(241) There are many modifications of the simple galvanic circle ; a very useful one is the cylindrical battery, Fig. 93, which consists of a double cylinder of copper closed at the bottom to contain the acid, and a similar but smaller cylinder of zinc, which is kept from touching the sides of the copper, by pieces of cork ; both are furnished with wires terminated by caps to contain mercury for the convenience of making and breaking the circuit. The quantity of Electricity set in motion by these simple

FIG. 93.

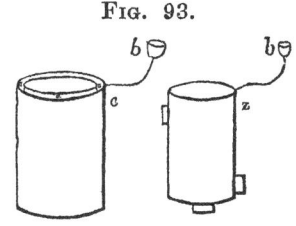

circles, when on a large scale, is very great, though the intensity is very low. No physiological effects are experienced when the body is included in the circuit, nor is water decomposed ; their heating powers are, however, so great, that they were called by Dr. Hare *calorimotors*. An arrangement on a very extensive scale, was made at the Royal Institution, under the direction of Mr. Pepys. A sheet of zinc, and one of copper, were coiled round each other, each being sixty feet long, and two feet wide : they were kept asunder by the intervention of hair ropes, and suspended over a tub of acid, so that by a pulley they could be immersed and removed. About fifty gallons of dilute acid were required to charge this battery, and when it is stated that a piece of platinum wire may be heated to redness by a pair of plates, only four inches long, and two broad, the calorific power of such an arrangement as the above, may be imagined to have been immense. The energy of the simple circle depends on the size of the plates, the intensity of the chemical action on the oxidable metal, the rapidity of its oxidation, and the speedy removal of the oxide.

(242) In order to increase the *intensity* of the electrical current, with a view to the exhibition of its chemical and physiological effects, we increase the *number* of the plates ; an arrangement of this sort is called the *compound* voltaic circle : it has been stated (227) to have been the invention of Volta, and hence is called the voltaic pile. Now, the *quantity* of Electricity obtained from the voltaic pile, is no greater than that from a single pair of plates, it is its *tension* alone that is increased ; an important fact which will be clearly understood when we have given a brief account of the important labours of Faraday.

(243) The original instrument of Volta is shown in Fig. 94. It consists of a series of silver and zinc plates, arranged one above another, with moistened flannel or pasteboard between each pair. A series of thirty or forty alternations of plates, four inches square, will cause the gold leaf electroscope to diverge : the zinc end with positive, and the silver end with negative Electricity ; a shock will also be felt on touching the extreme plates with the finger, when moistened with water. This lat-

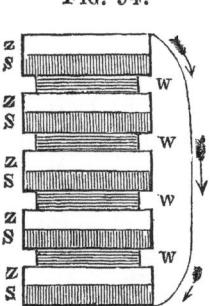

FIG. 94.

ter effect is much increased when the flannel or pasteboard is moistened with *salt* and water ; in this case a small spark will be seen on bringing the extreme wires into contact, and water will be decomposed : from this we learn that the increase of chemical action, by the addition of the salt, materially increases the *quantity* of Electricity set in motion ; but the pile will not in any sensible manner increase the divergence of the gold leaves,—its *intensity* therefore, is not materially augmented.

(244) An electric pile was constructed by De Luc, from which much useful information respecting the direction of the electric current in these cases of excitation may be derived. This instrument consists of a number of alternations of two metals, with paper interposed : the elements may be circular discs of thin paper covered on one side with gold or silver leaf about an inch in diameter, and similar-sized pieces of thin zinc foil, so arranged that the order of succession shall be preserved throughout, viz. zinc, silver, paper, zinc, silver, paper, &c. About five hundred pairs of such discs, enclosed in a perfectly dry glass tube, terminated at each end with a brass cap and screw to press the plates tight together, will produce an active column. The late intelligent electrician, Mr. Singer, constructed a *dry pile* on a much more extensive scale. It consisted of twenty thousand series of silver, zinc, and double discs of writing paper : it was capable of diverging with ball electroscopes, and by connecting one extremity of the series with a fine iron wire, and bringing the end of this near the other extremity, a slight layer of varnish being interposed, *a succession of bright sparks could be produced*, especially when the point of the wire was drawn lightly over the surface. A very thin glass jar containing fifty square inches of coated surface, charged by ten minutes' contact with the column, had power to fuse one inch of platina wire, $\frac{1}{3000}$ of an inch in diameter. It gave a disagreeable shock, felt distinctly in the elbows and shoulders, and by some individuals across the breast. The charge from this jar would perforate thick drawing-paper, but not a card. It did not possess the slightest chemical action, for saline compounds tinged with the most delicate vegetable colours underwent no change, even when exposed for some days to its action.

(245) On examining the electrical state of the electric column, it is found to resemble that of a conductor under induction : in the centre it is *neutral*, but the ends are in opposite electrical states ; and if one extremity be connected with the earth, the Electricity of the opposite end becomes proportionally increased : the zinc extremity is *positive*,

FIG. 95.

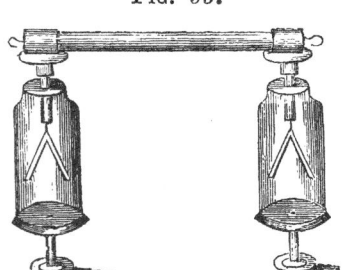

and the silver or gold extremity *negative :* as may be proved by laying the column on the caps of two gold leaf electroscopes in the manner shewn in Fig. 95, the leaves will diverge with opposite Electricities : if a communication be made between the instruments by a metallic wire the divergence of the leaves will cease, but will again be renewed when such com-

munication is broken. It is better to employ, in these experiments, an
electroscope in which the gold leaves are suspended singly as shewn in
Fig. 96, and so arranged as to admit of their

FIG. 96.

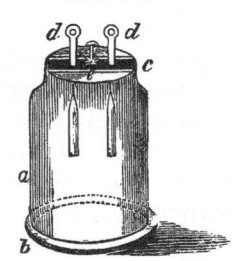

being brought nearer to or carried further from
each other. If in such an instrument the leaves
are adjusted at a proper distance from each
other, and the wire from which one is suspended connected with the zinc end, and the
wire from which the other is suspended connected with the silver end of the column, a
kind of perpetual motion will be kept up between the leaves; for, being oppositely excited,
they will attract each other; and having by contact neutralized each
other, they will separate for a moment, and again attract and separate
as before. If both silver ends, or both zinc ends of two columns are
connected with the two gold leaves a continued repulsion will be kept
up between the leaves, they being then similarly electrified.

(246) A variety of amusing experiments have been devised, dependent upon this curious property of De Luc's column. Thus a small
clapper may be kept constantly vibrating between two bells. This was
the contrivance of Mr. Forster, who constructed a series of fifteen hundred groups, and by its continued action kept up the vibrations of the
pendulum for a very long time. With twelve hundred groups, arranged
by Mr. Singer, a perpetual ringing during fourteen months was kept
up. We are informed by Mr. Singer, that De Luc had a pendulum
which constantly vibrated between two bells for more than two years.
A convenient modification of De Luc's column was contrived by Zamboni, by pasting on one side of a sheet of paper finely laminated zinc,
and covering the other side with finely powdered black oxide of manganese. On cutting discs out of this prepared paper, and piling them
upon each other to the number of 1000, taking care to press them
together, a little pile is obtained, capable of diverging the gold leaves of
the electrometer to the extent of half an inch. Mr. Gassiot describes*
an arrangement which he has constructed, consisting of a series of
10,000 of Zamboni's piles. With this arrangement, he charged a
Leyden battery to a considerable degree of intensity, and obtained
direct sparks of $\frac{3}{50}$ of an inch in length. He ultimately succeeded in
obtaining chemical decomposition of a solution of iodide of potassium,
the iodine appearing at the end composed of the black oxide of
manganese.

(247) Philosophers are divided in opinion respecting the source of
the electric charge of the " column," some supposing it due to the con-

* In a paper read before the Royal Society, Dec. 19, 1839.

tact of the metals, while others trace it to the contact of the zinc with the small portion of moisture which is contained in the paper in its common hygrometric state. It is certain that a degree of moisture is indispensable to the action of the instrument; for the Electricity disappears altogether when the paper discs have lost their humidity by spontaneous evaporation, and the zinc becomes slowly corroded in the course of years; but, on the other hand, no physiological or chemical effects have been obtained from the column, its charge being altogether one of intensity, and after discharge requiring an interval of time for renewal. It is not improbable that the state of the atmosphere is in some way connected with the phenomenon, for the motion of the pendulum is subject to much occasional irregularity. De Luc and Mr. Hausman both observed that the action of the column was increased when the sun shone on it; but they conceived that the effect was not due to the heat of the sun's rays, because it was found that an instrument put together after the parts had been thoroughly dried by the fire had no power whatever, but that it became efficacious after it had been taken to pieces, and its materials had remained exposed all night to the air from which the paper imbibed moisture. Mr. Singer, however, remarks that the power of the column is increased by a moderate heat, as his apparatus vibrated more strongly in Summer than in Winter, and the electrical indications were stronger when there was a fire in the room. Care should be taken not to allow the ends of the column to remain for any length of time in contact with a conducting body; for, after such continued communication, a loss of power will be perceived. When, therefore, the instrument is laid by, it should be insulated; and if it had previously nearly lost its action, it will usually recover it after a rest of a few days.

(248) When a series of some hundred couples of zinc and copper cylinders are arranged voltaically, and charged with common water, a battery is obtained, the Electricity of which is of a high degree of intensity resembling that of the common electrical machine; indeed, by connecting the extremities of such an arrangement with the inner and outer coatings of a Leyden battery, it becomes charged so instantly that almost continuous discharges may be produced. A very extensive series of the water-battery has been constructed by Mr. Crosse, and the phenomena which it exhibits are of a highly interesting character. It consists of 2500 pairs of copper and zinc cylinders, most of which are enclosed in glass jars: they are all well insulated on glass stands, and are ranged on three long tables, well protected from dust and from the light,—a situation which experience has shewn Mr. Crosse to be most favourable for this peculiar form of the voltaic battery.

(249) The following are some of the results obtained from this

battery:—30 pairs afford a slight spark, sufficient to pierce the cuticle of the lip, the hand making the communication being wetted;—130 pairs open the gold leaves of the electrometer about half an inch;—250 pairs cause the gold leaves to strike their sides;—400 pairs give a very perceptible stream of Electricity to the dry hand, making the connection between the poles, the light being very visible;—500 pairs occasion that part of the dry skin which is brought in contact to be slightly cauterized more especially at the *negative* side;—1200 pairs give a *constant small stream* of the fluids between two wires or two pieces of tin-foil, placed $\frac{1}{100}$ of an inch apart, such wires or pieces of foil *not having been previously brought into contact*. This stream, when received by the dry hands, is exceedingly sharp and painful. A pith-ball $\frac{1}{4}$ inch in diameter, suspended by a silk thread, will constantly vibrate between the opposite poles : 1100 pairs will produce this latter effect. If the foot of a gold leaf electrometer be connected with one of the poles, and the hand of another person connected with the other pole be brought over the cap of the instrument, even when held at several inches' distance, the leaves will strike their sides. Again, if the cap of the same electrometer be connected with either pole of the battery of 1100 pairs, the *opposite pole not being connected with the foot of the instrument*, the leaves will continue to strike the sides. This latter is a proof of the great waste occasioned by the imperfect insulation of the cylinders. A much more powerful effect would be produced by a superior insulation :—1600 pairs of cylinders produce the above effects in a much greater degree. In a tolerably well insulated battery every additional ten pairs after the first 100 produce an evidently increased effect ; and after 1000 pairs, the next 100 constitute a much greater addition to their power than one might promiscuously have imagined. With 1600 pairs the stream between two wires *not previously brought into contact* is very distinct ; the light, however, is not great ; the stream is of great intensity, but of small diameter. The method adopted by Mr. Crosse for exhibiting this interesting experiment is this :—he takes a small glass stick, and ties on it with waxed thread very securely, two wires of platina, with the two extreme ends ready to be plunged into two cups of mercury connected with the opposite poles of the battery : the two other ends of the wires are brought to the distance of about $\frac{1}{100}$ of an inch from each other. The moment the connexion is made with the poles of the battery, a small stream of fire takes place at the interval between the wires, which may be kept up for many minutes, nor does it appear inclined to cease. This experiment never fails ; though with a much greater number of plates, each pair not being separately insulated, it would never succeed. The author has had the pleasure of witnessing this most satisfactory experiment as well as all the others here described, with the full power of the whole battery of 2400 pairs.

The light between charcoal points, even with the whole series, is feeble, there is no flame or even approach to it : the conducting power of the water used in the cells being inadequate to transmit a sufficient current to produce great light and heat, even supposing such current to have been excited.   Mr. Crosse has, however, a water battery, consisting of eighty pairs of very large cylinders, which gives very brilliant sparks between two points of charcoal when rubbed together.

(250) When the opposite poles of the 2400 pairs are connected with the inner and outer coatings of the large electrical battery, containing 73 feet of surface, a continual charge is kept up, each discharge being attended with a loud report, heard at a considerable distance.   Each of these discharges will pierce stout letter-paper, and fuse a considerable length of silver leaf, which it deflagrates most brilliantly, attended with loud snappings of light, more than a quarter of an inch in length. Platinum wire is fused at the extremity, and the point of a pen-knife is soon demolished.   Light substances are attracted a distance of some inches and repelled again : the physiological effects would undoubtedly be exceedingly violent.   I have not, however, heard that any person has yet ventured to experience them.

(251) To avoid the trouble of using this large electrical battery, Mr. Crosse constructed one of mica, and having myself made one with some little modification, and on a somewhat larger scale than here described, I can recommend it as being a most useful and instructive piece of electrical apparatus.   It is made in the following manner :—Seventeen plates of thin mica, each five inches by four, are coated on both sides to within half an inch of the edge with tin-foil, and let into a box lined with glass, with a glass plate between each mica plate.   Slips of tin-foil are pasted to each side of each plate of tin-foil, of which all those connected with the lower ones are brought together at the extremities furthest from the plate, and pasted to one end of the interior of the box ; whence, by a tin-foil communication, a connection is made with a brass stem, secured to the outside of the box.   This represents what may be called the outer coating of the battery, and is capped with a ball.   The remaining strips of tin-foil or those connected with the upper surface of each plate, are brought together at the other end of the interior of the box, and turned back upon the tin-foil or upper part of the top plate.   A brass plate, three inches square, is then laid flat upon those combined slips, a cover is fitted on the box with screws, and a glass tube carrying a brass stem, passing through it and the cover, is fixed in the centre of the cover : such stem being cut at the lower part into a screw, which passes through a female screw cut in a cap, cemented to

the lower end of the glass tube within the cover, pressing on the brass plate. The upper part of the stem passes through a cap on the top of the glass tube, and is terminated with a brass ball, and may be termed the inner coating of the battery. By screwing the stem a perfect contact is made between this ball and all the upper surfaces of the mica plates. The two balls are placed on the same level, and a brass wire of $\frac{1}{16}$th inch diameter passes horizontally through the ball of the outer coating, cut into a screw to meet a similar one passing through the opposite ball. These wires are furnished with fine platinum points, and can be brought into contact, or made to recede at pleasure. A micrometer screw may be attached. By means of holes made in the opposite stems, the mica battery may be connected by wires with the opposite poles of a voltaic battery, and the *striking distance* accurately measured between the points.

The whole arrangement will be understood by inspecting Fig. 97. A, is a sectional view of a dry wooden box, lined with glass, containing

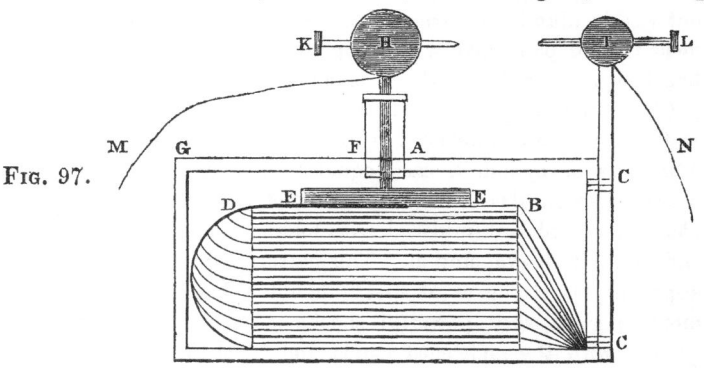

Fig. 97.

the plates of covered mica, a plate of window-glass being interposed between each. B, strips of tin-foil a quarter of an inch wide, each of which has one end pasted to the tin-foil *under* each mica plate, and the other end brought to the bottom of the box, and secured together by paste, and attached by a conducting communication of metal to the rod C C. D, similar strips having one end pasted to the tin-foil *over* each mica plate, and the other ends turned back on the upper part of the upper plate. E, E, a thin brass plate three inches square, placed horizontally on the combined ends of the strip. F, a glass tube, capped at each end, passing through the cover of the box G. Through this tube passes a brass screw, the lower end of which presses on the brass plate E, E, the upper end bearing the brass ball H. I, a brass ball, capping the stem C, C. Both H and I are pierced by the horizontal wires K, L, placed on the same level, cut into screws; and having each a platinum point at one end, and a nut at the other. In each of the upright stems immediately under the balls, is a hole drilled to receive the wires of communication M, N.

(252) The peculiar merits of this apparatus consist in its compactness, and its not being liable to injury from damp. When charged to a certain extent the shock is surprisingly painful, and is equivalent in power to many superficial feet of common coated glass. It is not calculated to be charged to a high intensity, as in such case the thin plates of mica would be pierced. Connected with the water battery, the following results were obtained by Mr. Crosse :—three pairs of cylinders produce light : twenty pairs produce a stream of light : 200 pairs produce a stream of scintillations, by drawing fine iron-wire over the lacquered knob of the mica battery : 300 pairs fire gunpowder : 500 pairs give a smart shock to the dry hands ; 1200 pairs give a shock not easily borne,—felt across the breast and shoulders, and cause a constant stream of light to pass between two wires $\frac{1}{8}$ of an inch apart, in an exhausted glass globe of four inches diameter, that globe being faintly but visibly illuminated over the whole of its interior during the experiment : 1600 pairs give a shock perfectly insupportable, which nearly knocked a person down who received it.

(253) Shortly after the above account of the performances of his water-battery was published by Mr. Crosse, I constructed a series of 500 pairs of cylinders, each equal to a five inch plate ; they were placed in green glass tumblers, insulated with the greatest care, and placed in a cupboard furnished with folding doors, to keep out the dust and to diminish evaporation. This battery, which is still in action, and apparently as strong as at first, (though it has been working eighteen months) gives very powerful shocks when the terminal wires are grasped with the moistened hands, and when the positive wire is held in one hand, and the dry knuckle brought into contact with the binding-screw attached to the negative, a spark is obtained, and a small blister is raised on the cuticle ; a spark is also obtained between the knuckles of two persons touching, respectively, the positive and negative terminations, and bringing their knuckles into contact. This battery has very slight decomposing power : the emission of gas from platinum points in acidulated water, is not so great from the whole series of 500 as from 100, and from 100 not so great as from 40 ; this is evidently occasioned by the great resistance which the current has to encounter from the bad conducting power of the water with which the battery is charged ; a resistance which it cannot overcome, and consequently by far the greater portion of the Electricity generated is checked in its passage, while the small quantity that passes is brought to a high state of intensity. The spark obtained on bringing the ends of the terminal wires into contact, is small, but brilliant, and when the ends are placed

within the flame of a large candle the phenomena are very beautiful, the carbon being deposited in an arborescent form, and with great rapidity on the positive wire : while on the negative wire it is thrown down in much less quantity, though in a more compact form; occasionally, indeed, filaments start from the latter like the quills on the back of a porcupine. I have seen few more beautiful experiments than this,—it was first made, I believe, by Mr. Gassiot; the carbon on the *positive* wire assumes the form of every variety of tree and shrub, some particles starting up into the lengthened form of the poplar, whilst others spread laterally, assuming the appearance of fern : in less than a minute the flame of the candle becomes darkened by the quantity of precipitated solid matter, which, as long as both wires remain in the flame, goes on increasing. Occasionally the carbon on the wires comes into contact,— when a bright spark is seen, and the arborescent appearance for a moment vanishes. When the finely divided carbon on the wires is brought into contact *out* of the flame, the spark is exceedingly brilliant, and four or five times as large as the spark from the clean wires, especially when *hot;* a snap also is heard.

Connected with the mica battery (consisting of twenty plates of mica, each four inches square,) 100 pairs scintillate iron wire, and give a pretty strong shock, the whole series gives a brilliant spark, accompanied by a pretty loud snap, and a powerful shock : it also causes brilliant scintillations of iron-wire, deflagrates gold, silver and copper leaf, and explodes gunpowder ; it also charges a Leyden battery, containing about twelve square feet of glass, sufficiently high to give unpleasant shocks.

By soldering the terminal wires to two copper plates, about two inches square, and fixing them upright on a turned mahogany frame, under a glass shade, perpetual vibration of a pith ball $\frac{1}{4}$ of an inch in diameter, suspended by a filament of silk, is kept up rapidly between the plates, placed $\frac{1}{3}$ of an inch apart. I have had the motion of the ball kept up unceasingly for a fortnight and three weeks together.

(254) By a paper recently published in the journal of the Electrical Society, I see that Mr. Gassiot is occupied in arranging a water battery, which is to consist of 3600 *insulated copper cells*, and that we are shortly to be favoured with a detailed account of its construction. In this paper the author gives a satisfactory explanation of some curious phenomena connected with the polarity of the water battery ; but I must refer my readers to the original paper for information on this interesting subject, having digressed already somewhat freely from the main subject of this lecture. I return therefore, to the historical description of galvanic apparatus.

(255) It is easy to see that many inconveniences must attach to the

pile of Volta, when the plates are numerous : in addition to the trouble
of building it up, it is frequently rendered comparatively inactive by the
moisture pressed out of the lower part by the weight of the upper :
hence, the substitution of troughs and other arrangements. The
most simple of these is the " Couronne des tasses," shown in Fig. 98 :

FIG. 98.

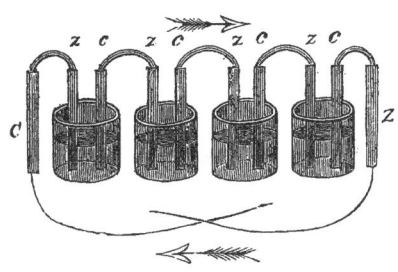

which consists in a row of
small glasses or cups, containing very diluted sul-
phuric acid, in each of which
is placed a small plate of
copper, about two inches
square, and another similar
sized plate of zinc, not touching each other, but so constructed that the zinc of the
first glass may be in metallic
communication with the copper of the second, the zinc of the second
with the copper of the third, and so on throughout the series. By this
arrangement, when glasses are employed, we can see what is going on
in each cell : and if the zinc plates be amalgamated (234), it will be
observed that when the wires are connected, and consequently, when a
current is passing, all the copper surfaces rapidly evolve hydrogen gas,
while the solution of the zinc proceeds quietly ; but, that when the
connection between the extreme plates is broken, the evolution of gas
ceases. Eighteen or twenty pairs of plates will decompose acidulated
water rapidly, and thirty will give a distinct shock to the moistened
hands.

(256) Another arrangement of the plates is shown in Fig. 99, where
they are represented as fixed in pairs into a trough of wood : this

Fig. 99.

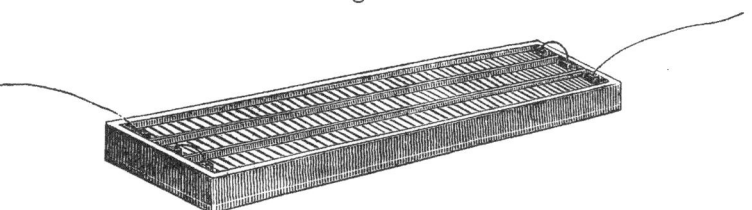

constitutes Cruickshank's battery. It is very convenient when
solution of sulphate of copper is used as the exciting agent,
which, as Dr. Fyfe has shown,[*] increases the electro-chemical
intensity of the electric current, as compared with that evolved
by dilute sulphuric acid in the proportion of seventy-two to

* L. & E. Phil. Mag. vol. xi. page 145.

sixteen. An important modification was that suggested by the late
Dr. Babington, and shown
in Fig. 100.: the plates of
copper and zinc, usually
about four inches square,
are united together in pairs
by soldering at one point
only; the trough in which
they are immersed is made
of earthenware, and divided
into 10 or 12 equal por-
tions. The plates are at-
tached to a strip of wood,
and so arranged that each
pair shall enclose a partition

Fig 100.

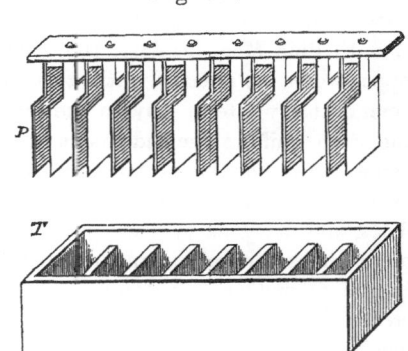

between them : by this means the whole set may be lifted at once into
or out of the cells ; and thus, while the fluid remains in the trough, the
action of the plates may be suspended at pleasure, and when corroded,
easily replaced. The piece of wood to which the plates are attached
should be well dried, and then varnished, in order to render it a non-
conductor of Electricity. When several of these troughs are to be
united together, it is necessary to be cautious in their arrangement, as a
single trough *reversed* will very materially diminish the general effect.
Care must also be taken to insure perfect communication between the
several plates. A battery of two thousand double plates, on this plan,
was constructed several years ago for the Royal Institution; the surface was one hundred and twenty-eight thousand square inches, and
its power immense.

(257) A great improvement in the construction of voltaic batteries
was made by Dr. Wol-
laston, in 1815. It con-
sisted in doubling the
copper-plate, so as to
oppose it to both surfaces
of the zinc, as shown
in Fig. 101. *A*, repre-
sents the bar of wood to
which the plates are
screwed; *B*, *B*, *B* the
zinc plates connected with
the copper plates *C*, *C*,
which are doubled over
the zinc plates. Contact
of the surfaces is pre-

Fig. 101.

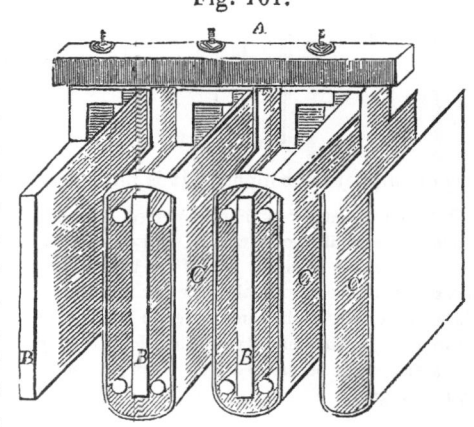

vented by pieces of wood or cork placed between them. Ten or twelve troughs, on this construction, form a very efficient voltaic battery. It appears, from the experiments of Mr. Christopher Binks, detailed in the L. & E. Phil. Mag. for July, 1837, that a still further extension of the copper would be attended with a considerable increase of power. He remarks that whatever may be the care taken to procure two plates of zinc of an uniform size and thickness, and however alike the attendant circumstances may be, no two couples will be found to give the same results in the same time when associated with corresponding copper plates, and acted on by acids in the usual way. While one plate will lose perhaps ten grains; another, apparently similar, will lose five or six grains; and another, fifteen or sixteen in the same time: these differences he finds to be independent of accidental differences in the distances of the plates from one another: zinc plates he also finds to lose less the first time of immersion than during the second and third.

(258) In the London and Edinburgh Phil. Mag. vol. x. p. 244, a battery in which sulphate of copper is the exciting agent, is described by Mr. De la Rue. The author remarks, " That if sulphate of copper be used in charging a battery instead of acid, oxygen is supplied to the zinc by the oxide of copper, no evolution of gas therefore takes place, except the small quantity attributable to local action; and the action is thus rendered continuous, the effect being fully equal to that mo-

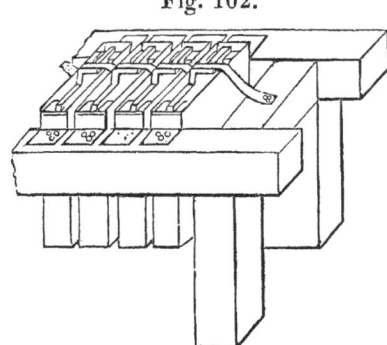

Fig. 102.

mentarily produced by immersion in acids."* Experiments may also be conducted with batteries thus charged, which cannot be exhibited at all with the same arrangements excited by acids, the momentary power being exhausted before the battery can possibly be brought into action. Fig. 102. represents a form of battery suggested by Mr. De la Rue as well adapted to the use of sulphate of copper. It is more economical than any in use at the

* When a battery is first immersed into an acid exciting agent, small wires placed between its poles will be ignited; but this heating power continues for a very short space of time: and if the battery be immersed without connecting the poles, and allowed to remain for a few minutes, and the connection then made with the same length of the same wire, no ignition whatever is produced. This speedy diminution, in the quantity of Electricity rendered available, is considered by Mr. De la Rue, to be occasioned by a large portion being annulled or carried off by the hydrogen gas.

time of its invention; besides possessing the advantage of the zinc plates being easily replaced when worn out. Fig. 103,

Fig. 103.

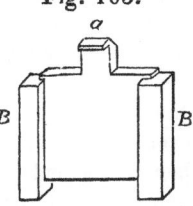

shows the zinc plate, the top part at *a* being tinned prior to the amalgamation of the plate. *B, B* are two slips of wood grooved out to within three-fourths of an inch of the bottom, and in- *B* tended to retain the zinc-plate in its proper position. The copper plates are formed into cells, as represented in Fig. 104. five inches square, and one inch wide. *E, E* are two ears of cop-

Fig. 104.

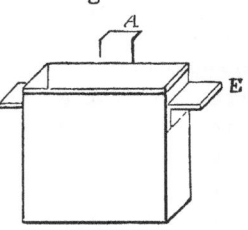

per, by which the cell is suspended in its place. *A*, is a slip of copper, to form a connection, by means of solder, to the zinc plate in the adjoining cell. The cells are *E* painted on the outside to protect them from the action of the acid, and are supported on a long wooden frame, and retained in their place by tacks driven through them, as represented in the figure. Fig. 105, represents a con-

Fig. 105.

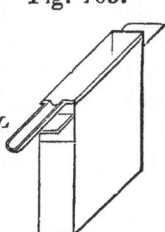

trivance, by which the charge may be renewed while the battery is in action; at the top of each cell may be placed a lip or spout, *L*, a quarter of an inch deep, which must overhang a wooden gutter running the length of the frame. The solution may be renewed by a funnel having a long neck, which must be inserted nearly to the bottom of the cell. When fresh solution is poured in, the spent liquor will run out of the lip into the gutter. Immediately after a series of experiments, the batteries should be emptied, and the plates well cleaned by dashing water between the cells. If this be not immediately attended to, it will be exceedingly difficult to remove the deposit from the cells. A Cruickshank battery is best cleaned by laying it on its side.

(259) The arrangement shown in Fig. 106 is that of Professor Hare of Philadelphia. It combines the advantages of the *compound* trough

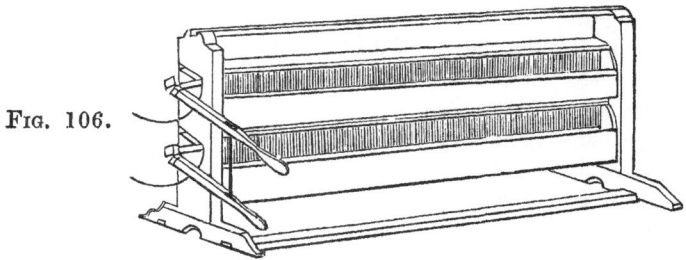

Fig. 106.

and the calorimotor or deflagrator. A voltaic series fixed in a trough is combined with another trough destitute of plates, and of a capacity sufficient to hold all the acid necessary for an ample charge. The trough containing the series is joined to the other lengthwise, edge to edge; so that, when the sides of the one are vertical, those of the other must be horizontal. The advantage of this is, that by a partial revolution of the two troughs, thus united, upon pivots which support them at the ends, any fluid which may be in one trough must flow into the other, and, reversing the movement, must flow back again. The galvanic series being placed in one of the troughs, and the acid in the other, by a movement such as has been described, the plates may all be instantaneously subjected to the acid or removed from it. The pivots are made of iron, coated with brass or copper, as less liable to oxidizement. A metallic communication is made between the coating of the pivots and the galvanic series within. In order to produce a connection between one recipient of this description and another, it is only necessary to allow a pivot of each trough to revolve on one of the two ends of a strap of sheet copper. To connect with the termination of the series the leaden rods, (to which are soldered the vices or spring forceps for holding the substances to be exposed to the deflagrating power,) one end of each is soldered to a piece of sheet copper. The pieces of copper thus soldered to the leaden rods are then to be placed under the pivots, which are, of course, to be connected with the termination of the series; the last-mentioned connection is conveniently made by means of straps of copper severally soldered to the pivots and the poles of the series, and screwed together by a hand-vice. Each pair consists of a copper and zinc plate, soldered together at the upper edge, where the copper is made to embrace the edge of the zinc. The three remaining edges are made to enter a groove in the wood, being secured therein by cement. For each inch in the length of the trough there are three pairs. In the series represented in Fig. 106, there are seven hundred pairs of seven inches by three, and in that shown in Fig. 107, one hundred pairs of fourteen inches by eight. The latter will deflagrate wires too large to be ignited by the former, but is less powerful in producing a jet of flame between two charcoal points, or in giving a shock. Dr. Hare exhibited two of these

Fig. 107.

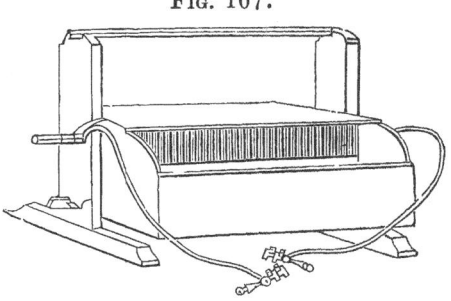

batteries at the meeting of the British Association at Bristol in 1836. Their power was very great in proportion to their size.

(260) A modification of Dr. Hare's battery is described by Mr. J. Young, chemical assistant in the Andersonian University.* It possesses all the advantages of approximation of the plates, and the compactness of Dr. Hare's battery; and an equal effect is produced with *half* the quantity of sheet copper, which arises from both sides of the copper plates being presented to surfaces of zinc.

The sheet copper and sheet zinc to be used are first cut into long ribbons, of the breadth which it is intended to give the plates. Suppose the ribbons two inches broad: both the copper and zinc ribbons are then divided into lengths of five inches, and a portion

Fig. 108.

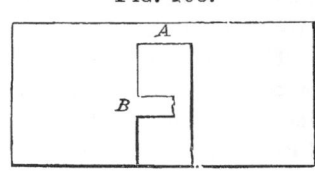

cut out as in Fig. 108. The slip is thus divided into two squares of two inches each, which are connected at *A*, and a piece is left projecting at *B*. The zinc and copper sheets are cut up exactly in the same way. Fig. 109 represents therefore either a single zinc or a single copper plate. The plate is then bent at *a*, and presents the appearance represented in the figure.

Fig. 109.

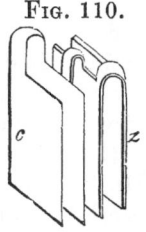

In Fig. 110 we have two plates, one of copper, *c*, and the other of zinc, *z*, which are exactly alike in construction, but placed differently, as shown in the figure. The thin projecting parts, *B*, Fig. 108, are soldered together, and this is the only metallic communication between them that is allowed to exist. Fig. 110, therefore, is only one copper and one zinc plate, or it is one pair of plates. Each pair is made up in the same way. In arranging a number of pairs to form a battery, they are interlaced,

Fig. 110.

so that a copper square comes in between each couple of zinc squares, and a zinc square between each couple of copper squares. It is easy to see how this arrangement can be made, when the plates are in the hand, though it is difficult to describe it. At the positive end of the battery there is a single copper plate, which is soldered at the top to the last double copper plate, as seen in Fig. 111, which figure represents three pairs properly arranged, and also the manner in which they should be fitted up, and kept steadily apart in a wooden frame. This frame consists of two cross bars, E E, E E, in front, and the

Fig. 111.

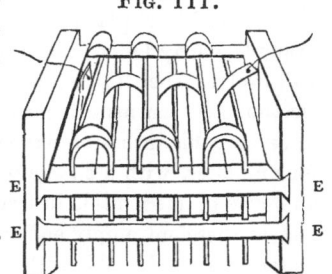

* L. & E. Phil. Mag. vol. x. p. 242.

same behind, dove-tailed into solid ends.   The channels in the cross bars
for the reception of the edges of the plates, are formed by placing four
cross bars together, and sawing a little way into one side of them all,
every eighth of an inch or so in their length, so as to form a set of
parallel grooves.   We have by means of this frame a much greater
security that no metallic contact will occur between contiguous plates,
than when they are separated by wedges of cork, as in Dr. Hare's construction, which may slip out.

The frame and plates are introduced into a trough, which may be of
wood or stoneware, containing the exciting liquor.   Dr. Hare's revolving arrangement of the two connected troughs, may be adopted for this
battery, although the inventor has been led to give a preference in
practice to a single trough to contain the frame.   To the solid ends of
the frame are attached two cords, which are fixed to two pulleys, on
which they are wound up on turning a winch, as represented in Fig.

FIG. 112.

112, by which means the frame and battery can be raised out of the liquid.   If
the axis on which these pulleys are fixed
can be moved a little backwards and forwards on its bearings, it is easy, by means
of a little projecting peg at P, which fits
into a hole in the side of the pulley, to fix
and support the frame in a position above the trough, and out of the
exciting fluid, when that is desirable.   But the form of the trough to
contain the frame and plates may be varied according to the object in
view, or to the purposes to which the battery is to be applied.

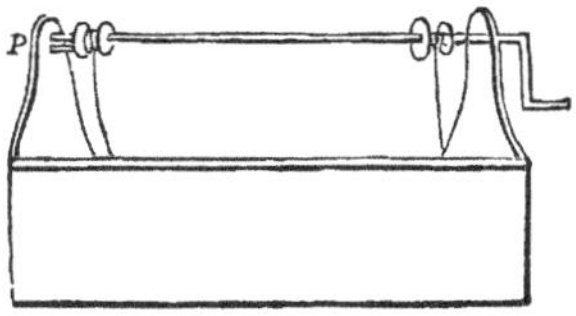

(261) In comparing a battery of the form here described, either with
Dr. Hare's, or with any of the other forms in use, it is to be remembered that the plates or elements of the battery are all of double the
size they appear to be; or that in this construction we have half the
number of pairs, but each of double the dimensions of a pair in any of
the old batteries having the same appearance.

A small battery of this construction, containing twelve pairs of two
inches breadth of plates, (the size which we have taken above as an
example,) may be contained in a trough eight inches in length, and
will evolve, when its terminal wires are soldered to a Faraday's volta-
electrometer, (an instrument to be described in the next Lecture,) six
or seven cubic inches of oxygen and hydrogen gases mixed, in three or
four minutes, with a charge of half an ounce of sulphuric acid and half
an ounce of nitric acid, in twenty-four ounces of water, fluid measure,
and is therefore amply sufficient to demonstrate the decomposition of
water on a considerable scale.

It is proper to use the thickest sheet zinc which can be had, in the construction of the plates, although the thinnest sheet copper will suffice, from its being so well supported. When the zinc plates are worn out, the cross bars may easily be pulled out of the solid ends, and the elements of the battery separated. New zinc plates being soldered to the old copper, the whole may be again quickly re-arranged in the old frames.

(262) But a far more useful voltaic arrangement than any that has yet been described, is that of Professor Daniell : it is called the "*constant battery*," from the regularity and duration of its action, which renders it applicable to the determination of many important questions, which have hitherto been perplexed with the variable conditions of the voltaic current.

In a series of papers in the Philosophical Transactions for 1836, the author describes these batteries, and the circumstances which led to their adoption. It has been remarked in the former part of this lecture, that the evolution of hydrogen gas from the negative metallic surface in the common galvanic battery, greatly interferes with the development of *available* electricity, for, that a considerable portion of the electricity that is actually generated, is probably spent in giving a gaseous form to the hydrogen of the decomposed water. But besides this, Mr. Daniell found that not only were the oxides of copper and zinc reduced by the *nascent* hydrogen, at the moment of its formation, when salts of these metals were purposely dissolved in the fluid of the cells of the battery ; but the oxide of zinc itself formed at the generating plates, *was reduced* at the conducting plates, which became ultimately so incrusted with metallic zinc, as entirely to destroy the circulating force. The variations and progressive decline of the power of the ordinary voltaic battery are thus accounted for, since the transfer of the electro-positive metal must eventually cause two zinc surfaces to become opposed to each other, the use therefore of the nitric acid in the battery charge is to remove the hydrogen by combination. Since, therefore, the hydrogen has a two-fold injurious tendency, its absence altogether becomes a desirable object to effect. In the battery constructed by Mr. Warren De la Rue, this is done by the employment of sulphate of copper as the exciting agent, and in the arrangement of Professor Daniell, the same is accomplished, but under circumstances rather different, as will presently appear.

(263) Fig. 113 represents a section of one of the cells of Daniell's original " sustaining " or " constant" battery ; *a b c d* is a cylinder of copper, six inches high and three and a half inches wide ; it is open at the top *a b*, but closed at the bottom, except a collar *e f*, one inch and a half wide, intended for the reception of a cork, into which a glass syphon tube *g h i j k*, is fitted. On the top *a b*, a copper collar,

Fig. 113.

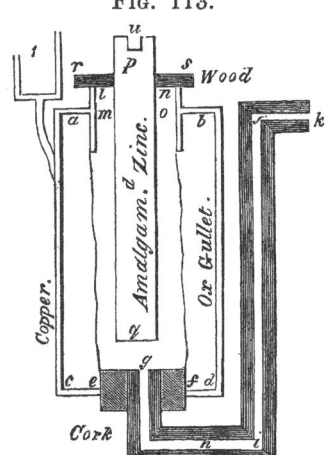

corresponding with the one at bottom, rests by two horizontal arms. Previously to fixing the cork syphon tube in its place, a membranous tube, formed of part of the gullet of an ox, is drawn through the lower collar $e\,f$, and fastened with twine to the upper, $l\ m\ n\ o$, and when tightly fixed by the cork below, forming an internal cavity to the cell communicating to the syphon tube, in such a way, as that when filled with any liquid to the level $m\ o$, any addition causes it to flow out at the aperture k. In this state, for any number of drops allowed to fall into the top of the cavity, an equal number are discharged from the bottom a, at the top of the zinc rod. Various connections of the copper and zinc of the different cells, may be made by means of wires proceeding from one to the other. In the construction of this battery, Mr. Daniell has availed himself of the power of reducing the surface of the generating plates to a minimum. The effective surface of one of the amalgamated zinc rods, being less than ten square inches, whilst the internal surface of the copper cylinder to which it is opposed, is nearly seventy-two inches. His principal objects were to remove out of the circuit the oxide of zinc, which has been proved to be so injurious to the action of the common battery, as fast as the solution is formed, and to absorb the hydrogen evolved upon the copper, without the precipitation of any substance that might deteriorate the latter.

(264) The first is completely effected by the suspension of the zinc rod in the interior membranous cell, into which the fresh acidulated water is allowed slowly to drop, from a funnel suspended over it, and the aperture of which is adjusted for the purpose; whilst the heavier solution of the oxide is withdrawn from the bottom at an equal rate by the syphon tube. When both the exterior and interior cavities of the cell were charged with the same diluted acid, and connection made between the zinc and the copper, by means of a fine platinum wire, $\frac{1}{500}$ of an inch in diameter, he found that the wire became red hot, and that the wet membrane presented no obstruction to the passage of the current.

The second object is obtained by charging the exterior space surrounding the membrane, with a saturated solution of sulphate of copper, instead of diluted acid ; upon completing the circuit the current passed freely through this solution ; no hydrogen made its appearance on the conducting plate, but a beautiful pink coating of pure copper was deposited upon it, and thus perpetually renewed its surface.

When the whole battery was properly arranged and charged in this manner, no evolution of gas took place from the generating or conducting plates, either before or after the connexions were complete ; but when a voltameter was included in the circuit, its action was found to be very energetic. It was also much more steady and permanent than that of the ordinary battery, but still there was a gradual but very slow decline, which Mr. Daniell traced at length to the weakening of the saline solution, by the precipitation of the copper, and consequent decline of its conducting power.

(265) To obviate this defect, some solid sulphate of copper was suspended in muslin bags, which just dipped below the surface of the solution in the cylinders, which gradually dissolving as the precipitation proceeded, kept it in a state of saturation. This expedient fully answered the purpose, and Mr. Daniell found the current perfectly steady for six hours together. This arrangement he has since improved, by placing the salt in a perforated colander of copper, fixed to the copper collar.

Fig. 114 represents a section of this additional arrangement. The colander with its central collar, rests by a small ledge upon the rim of the cylinder. The membrane is drawn through the collar, and turning over its edge is fastened with twine. After this alteration, the effective length of the zinc rods exposed to the action of the acid, was found to be no more than four inches and a quarter.—*Philosophical Transactions for* 1836.

FIG. 114.

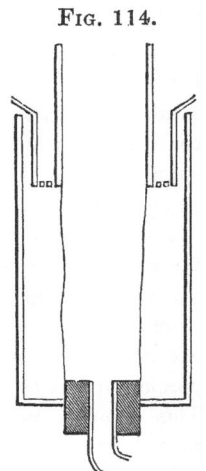

(266) The advantages of this battery over those of the ordinary construction are very great ; it secures a total absence of any wear in the copper; it requires no nitric acid, but the substitution of materials of great cheapness, namely, sulphate of copper, and oil of vitriol; it enables us to get rid of all local action, by the facility it affords of applying amalgamated zinc, and allows the replacement of zinc rods at a very trifling expence; it secures the total absence of any annoying fumes; and, lastly, it produces a perfectly equal and steady current of electricity for many hours together.

With a battery of twenty cells arranged in a single series, twelve cubic inches of mixed oxygen and hydrogen gases may be collected from a voltameter in every five minutes of action, and when they are first connected in pairs, and afterwards, in a series of ten, the quantity amounts to seventeen cubic inches. Eight inches of platinum wire, $\frac{1}{200}$ of an inch in diameter, may be kept permanently red hot by the same arrangement, and the spark between charcoal points is very large and brilliant.

Mr. Daniell has even made it the source of the purest oxygen for laboratory purposes. To this end he constructed an *oxygen cell*, by substituting a plate of platinum for the rod of zinc, enclosing it in the membranous tube, which is closed at the upper end by a glass tube, bent in a proper form to deliver the disengaged gas, under a receiver. In this arrangement the hydrogen is absorbed as before, by the oxide of copper, but the oxygen to the amount of eighty cubic inches per hour, is given off from the platinum.

(267) Fig. 115 represents a single cell of the constant FIG. 115. battery, a cylindrical vessel of porous earth being substi-
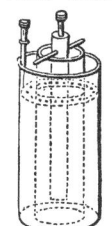
tuted for the bladder diaphragm, which proved very inconvenient on account of its becoming rapidly corroded, and pierced by the sharp edges of the crystals of metallic copper, deposited on the copper plate. These porous jars were, it seems, first employed by Mr. Dancer,* of Liverpool, and they are now composed of the thinnest unglazed biscuit ware, a most excellent material. The battery, shown in Fig. 115, consists of a cylinder of copper, containing a tube of biscuit ware, which has a solid rod of zinc supported in its centre; the cylinder is furnished with a perforated shelf, upon which a supply of crystals of sulphate of copper is placed, so that the battery being once charged, will maintain an equal action for many hours.

Fig. 116 represents a set of six of the above batteries, and Fig. 117 a set of ten large ones, the copper cylinders being eighteeen or twenty-one inches high, with zinc rods, and porous earthen tubes in proportion. This forms a powerful voltaic arrangement, evolving eight or ten cubic inches of oxygen and hydrogen gases in the voltameter per

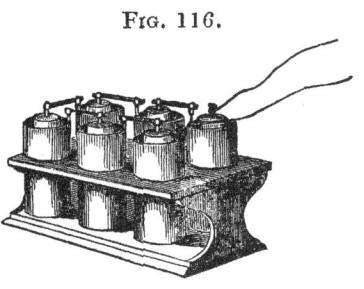
FIG. 116.

minute, and heating to redness twelve or fourteen inches of fine iron wire.

A series of thirty cells of the smaller size, six inches high, and three and a half inches in diameter, forms a very efficient battery for the lecture table; it heats from eighteen inches to two feet of iron-wire, deflagrates mercury most brilliantly, and burns metallic leaves vividly. The cells of

* Golding Bird's Elements of Natural Philosophy.

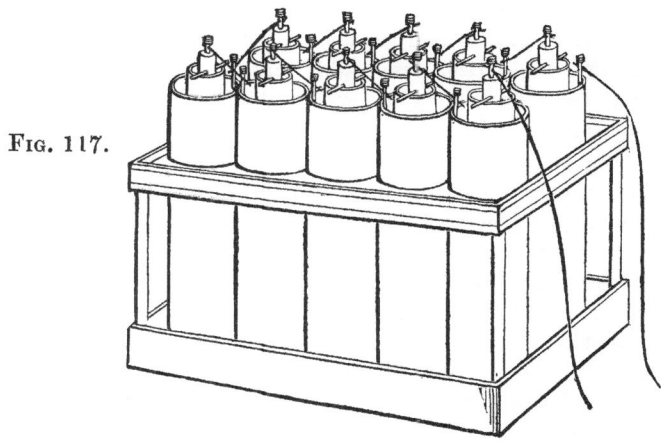

Fig. 117.

the sustaining battery must be plentifully supplied with sulphuric acid, without which the power is but feeble.   Mr. Daniell recommends a mixture of eight parts of water, and one of oil of vitriol, which has been saturated with sulphate of copper, for the copper cell, the internal tube being filled with the same acid mixture without the copper.   The porous cells should be well soaked in dilute sulphuric acid for an hour or two before being used ; and after their removal from the battery they should be repeatedly rinsed, or allowed to soak for some time in warm water, to dissolve out all the metallic salt from their pores.   If this be not attended to they will be soon destroyed.

(268) It was found by Mr. Daniell,* that the action of the constant battery is by no means proportioned to the surfaces of the conducting hemispheres, but approximates to the simple ratio of their diameters ; and hence, he concludes that the circulating force of both simple and compound voltaic circuits increases with the surface of the conducting plates surrounding the active centres.   On these principles he constructed a constant battery, consisting of seventy cells, in a single series, which gave between charcoal points, separated to a distance of three quarters of an inch, a flame of considerable volume, forming a continuous arch, and emitting radiant heat and light of the greatest intensity.   The latter, indeed, proved highly injurious to the eyes of the spectators, in which, although they were protected by grey glasses, of double thickness, a state of very active inflammation was induced.   The whole face of Mr. Daniell became scorched and inflamed, as if it had been exposed for many hours to a bright midsummer's sun.   The rays, when reflected from an imperfect parabolic metallic mirror in a lantern, and collected into a focus by a glass lens, readily burnt a hole in a paper

* Transactions of the Royal Society, May 30th, 1819.

M

at a distance of many feet from their source. The heat was quite intolerable to the hand held near the lantern. Paper steeped in nitrate of silver, and afterwards dried, was speedily turned brown by this light ; and when a piece of fine wire-gauze was held before it, the pattern of the latter appeared in white lines corresponding to the parts which it protected. The phenomenon of the transfer of the charcoal from one electrode to the other, noticed by Dr. Hare, but first observed by Professor Silliman, was abundantly apparent ; taking place from the *zincode* (or positive pole) to the *platinode* (or negative pole). The arch of flame between the electrodes was attracted or repelled by the poles of a magnet, according as the one or other pole was held above or below it : and the repulsion was at times so great as to extinguish the flame. When the flame was drawn from the pole of the magnet itself, including the circuit, it rotated in a beautiful manner.

The heating power of this battery was so great, as to fuse with the utmost readiness, a bar of platinum, one-eighth of an inch square ; and the most infusible metals, such as pure rhodium, iridium, titanium, the native alloy of iridium and osmium, and the native ore of platinum, placed in a cavity, scooped out of a hard carbon, freely melted in considerable quantities.

(269) Mr. Gassiot, afterwards with the view of ascertaining the possibility of obtaining a spark before the circuit of the voltaic battery is completed, prepared first 160, and then 320 series of the constant battery in half-pint porcelain cells, excited with solutions of sulphate of copper and muriate of soda ; but although the effects, after the contact had been completed, were exceedingly brilliant, not the slightest spark could be obtained. He was equally unsuccessful with a water battery, of 150 series, each series being placed in a quart-glass vessel ; and also with a water battery belonging to Mr. Daniell, consisting of 1020 series; but when a Leyden battery of nine jars was introduced into the circuit of the latter, sparks passed to the extent, in one instance, of $\frac{6}{5000}$ of an inch. We have, however, seen (249), that with a water battery of 2400 pairs, Mr. Crosse rarely fails to produce not only a spark, but a stream of light between platinum wires, not having previously been brought into contact.

Mr. Gassiot mentions in his paper,* that having been present at the experiments of Professor Daniell, above alluded to, he was induced to prepare 100 series of the large constant battery ; but although this powerful apparatus was used under every advantage, and the other effects produced were in every respect in accordance with the extent of the elements employed, still no spark could be obtained until the circuit was completed : even a *single fold of a silk handkerchief*, or a piece of

* Read December 19th, 1839.

dry tissue paper was sufficient to insulate the power of the battery, though after the circuit had been once completed, it fused titanium, and heated sixteen feet four inches of No. 20 platinum wire.

Fig. 118.   Fig. 119.

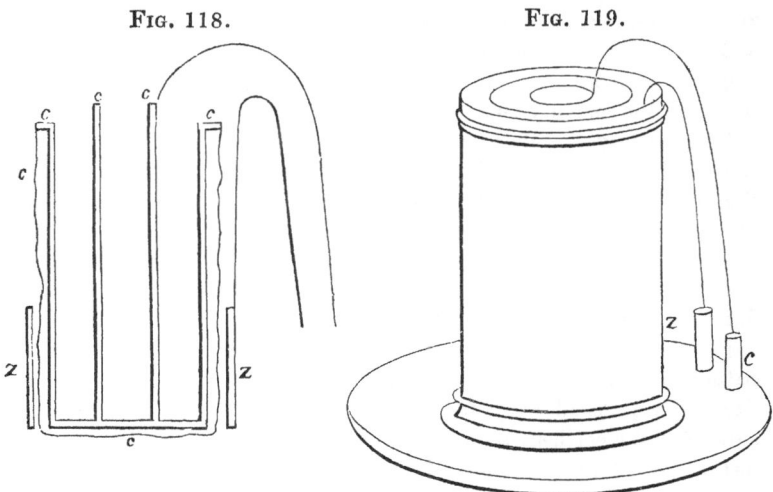

(270) The form of constant battery shown in section in Fig. 118, and in perspective in Fig. 119, was introduced by F. W. Mullins, Esq. In an earthenware pot, six inches deep, and four wide, is placed a cylinder of amalgamated zinc, standing on three legs half an inch long, cut out of the cylinder, the depth of which, including the legs, is only two inches. Within this cylinder, and at three-eighths of an inch distant, stands a copper vessel, having a rim a quarter of an inch wide, surrounding its outer edge, round which a thin clear bladder is tied ; the bottom of the vessel rests on a circular piece of baked box-wood, which projects one-fourth of an inch beyond the cylinder; the bladder, well cleaned and moistened, is drawn all over, and fastened round the upper rim by a string, the wood at the bottom preventing it from contact with the copper, which would otherwise injure the membrane. This cylinder, which is the depth of the pot, is pierced with six holes equidistant from the top and bottom, which communicate with an inner cylinder, separated from the outside one by a space of three-fourths of an inch, the bottom being on a level with the lower edge of the holes, and soldered to the larger cylinder. This chamber is intended to hold crystals of sulphate of copper, when required, and to receive the solution, which should not rise higher than the upper edges of the holes. A small quantity of muriate of ammonia (sal ammoniac), in the proportion of five parts of the saturated solution, to one hundred of water, is to be poured *outside* the membrane, until it rises to the upper edge of

the zinc ; the latter solution does not require renewal, the former will require the addition of a few crystals of the sulphate every four hours. The connexions are made by strips of copper, which are soldered to the zinc and inside copper cylinder, and which bend over the edge of the jar, and enter two cups holding mercury, from which the wires for transmitting the electrical current through any apparatus may proceed, or by fastening the mercury cups *directly* on wires soldered to the two metals.

This battery continues active as long as a particle of metallic salt remains in solution ; for if to the nearly exhausted and colourless solution, you add about six drops of the saturated salt, the original power of the battery is instantly restored, and continues for some time. In order then to keep up a varying current, it is only necessary to keep a few crystals of sulphate of copper on the shelf, which will be gradually dissolved, and carried in solution to the external surface of the copper.

(271) Mr. Mullins has constructed a somewhat different form of battery for intensity effects. " I have first," he says, " as in the quantity battery, a shallow cylinder of zinc within, and close to the internal surface of the earthenware pot next the copper cylinder as before ; but, instead of letting the inside of this cylinder go for nothing, the internal surface of the copper is lined with very thin caoutchouc for insulation : then comes another small cylinder of zinc ; then a copper one, lined as the last ; then a zinc : and lastly, a copper, each copper of course enveloped in membrane. This battery is one of extraordinary power in decomposition and other effects of intensity, which, in my opinion depends upon a new principle, which is developed in this mode of construction and arrangement, that is, the *restricting* the electric current to gradually *diminishing metallic surfaces* as it advances, so that, as the quantity accumulates, the conducting surfaces are *reduced,* and of course a much higher degree of intensity is a necessary consequence."

(272) In Sturgeon's " Annals of Electricity," for April, 1837, page 224, there is the following account of a sustaining battery of great power compared with its size, by the Rev. John Shillibeer. It is represented in Fig. 120.

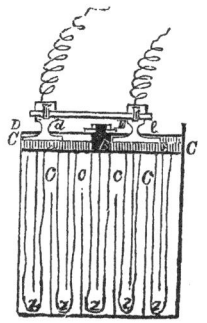

FIG. 120.

" Finding on trial that a single piece of zinc, (surface three inches) surrounded by a membrane, when placed in a copper vessel containing a solution of sulphate of copper, could be soon erected, and that a rotatory motion might be kept up during pleasure, I formed the apparatus about to be described, which consists of a well made copper trough, about three inches deep, and two and a quarter wide, divided into compartments according to the number of zinc plates employed. Fig. 121 represents a section of the copper trough. The plates are soldered firmly to a copper bar, and by the aid of a screw are fastened to a piece of hard wood, answering for a cover to the trough, which with the zinc plates and movement for directing the course of the electric fluid, may be seen by reference to Fig. 120. In a groove cut out of the cover on each side the screw, is fitted a copper slide; and these slides are joined by an elbow, to a piece of ivory forming a handle, by which, passing immediately over the screw, each wing of the slide may be brought readily into contact with the copper or zinc.

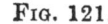
FIG. 121.

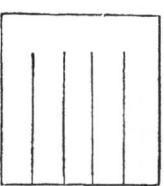

" The use of this movement or pole director, I will endeavour to make apparent.

" In Fig. 120 let *AA* represent a section of the wood cover with a groove for the slide to stand in, *B* the screw in connection with the zinc, *CC*, the opposite sides of the copper trough *Dd Ee*, the two wings of the slide. Now let the wing *Dd* be in contact with *C*, and *Ee* with *B;* it is evident that the stream of electricity is going out from the wire in connection with the wing *Dd*, and is returning by the wire appended to the wing *Ee*, into the zinc plates *via B*. Shift the slide so that *Ee* be in contact with *C*, and *Dd* with *B*, the stream will be reversed, making its exit by *Ee*, and returning by *Dd* to the zinc, via *B*. The zinc plates should be covered tightly with a membrane, so as to prevent any possible precipitation of sulphate of copper upon them. When used for sustaining a weight, the membrane should be slipped off, and instead of sulphate of copper, diluted nitrous acid should be used."

(273) Fig. 122 represents a single cell of Mr. Smee's voltaic arrangement, which, considering its advantages to arise from a mechanical help to the evolution of the hydrogen gas, he calls the *chemico-mechanical* battery. The circumstances which led the author to the construction of this admirable battery, are detailed in a paper inserted in the 16th volume of the L. & E. Phil. Mag. He observes, that " the influence of different conditions of surfaces is a subject which has escaped all experimenters, which is singular, as many must have noticed, that in a circuit the greatest quantity of gas is given off at the corners, edges, and points. Following this hint, a piece of spongy platinum, consisting as it does, of an infinity of points, was placed in contact with amalgamated zinc, when a most violent action ensued, so that but little doubt could be

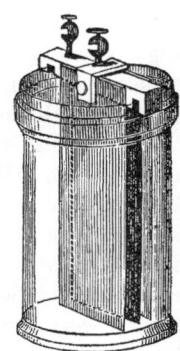
FIG. 122.

entertained of its forming a very powerful battery. The fragile nature of this material precludes it from being thus used, and therefore it was determined that another piece of platinum should be coated with the finely divided metal. This experiment was attended with a similar good result, and the energy of the metal thus coated, was found to be surprising. After a variety of experiments, Mr. Smee found that silver plates were preferable for receiving the precipitated platinum, and he gives the following directions for preparing them :—" Each piece of metal is to be placed in water, to which a little dilute sulphuric acid, and nitro-muriate of platinum is to be added. A simple current is then to be formed by zinc placed in a porous tube with dilute acid ; when, after the lapse of a short time, the metal will be coated with a fine black powder of metallic platinum. The trouble of this operation is most trifling, only requiring a little time after the arrangement of the apparatus, which takes even less than the description." The cost is about 6d. a plate, of 4 inches each way, or 32 inches of surface. It is necessary to make the surface of the silver rough, by brushing it over with a little strong nitric acid, which gives it instantly a frosted appearance, and after being washed it is ready for the platinizing process ; but the finely divided platinum does not adhere firmly to very smooth metals.

(274) The arrangement of the platinized silver battery will be immediately understood from the figure. A piece of the platinized silver has a beam of wood fixed on the top to prevent contact with the zinc, and is furnished with a binding-screw. A strip of stout and well amalgamated zinc, varying from one half to the entire width of the silver, is placed on each side of the wood, and both are held in their place by a binding screw sufficiently wide to embrace the zincs and the wood. This arrangement is immersed in a jar or glass, containing dilute sulphuric acid, (1 oil of vitriol and 7 water), and not the slightest effect is produced till a communication is made between the metals, when it instantly hisses and bubbles, and an active voltaic battery is obtained. For intensity effects it may be arranged as an ordinary Wollaston's battery with advantage, as shewn in Fig. 123 or a series of glass tumblers may be connected together ; 10 or 12 form a very efficient battery, having a very elegant appearance, and well adapted for the lecture table, as the action in

Fig. 123.

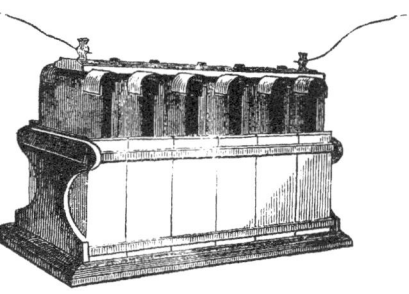

each cell may thus be very clearly seen. On account of the rapid removal of the hydrogen gas, there is, in this form of galvanic battery, but little tendency for the zinc to be deposited in a metallic state upon the negative metal; nevertheless, when it is required in action for a long period, it may be advisable to separate the metals by a porous earthenware vessel; or what answers the purpose equally well, by a thick paper bag, the joinings of which must be effected by shell-lac dissolved in alcohol. By these means the sulphate of zinc is retained on the zinc side of the battery. It may also be arranged as a circular disc battery, or as a Cruickshank, each cell being divided or not by a flat porous diaphragm; but whatever arrangement is adopted, the closer the zinc is brought to the platinized metal the greater will be the power.

In using the chemico-mechanical battery, it is important that no salt of copper, lead, or other base metal be dropped into the exciting liquid, as by that means there is a chance of getting a deposit on the negative metal, copper in particular is apt to get precipitated, in which case the platinized silver should be immersed in dilute sulphuric acid, to which a few drops of nitro-muriate of platinum should be previously added, by this process the baser metals are dissolved, and metallic platinum thrown down.

The platinized silver battery has become a great favourite with the public, and it has probably done already a greater amount of work than all the other batteries put together. It is simple in its construction, remarkably manageable in its applications, and elegant in its appearance. It is soon set in action, and as quickly cleaned and put aside; and although it has not the constancy of the admirable battery of Daniell, or the wonderful energy of the battery of Grove, it may be kept in active operation for six, eight, ten, or more days, when a sufficiency of acid is supplied to it; hence, its extensive application in the new art of electro-metallurgy.

(275) The most powerful voltaic battery, however, that has yet been brought before the public, is that of Professor Grove, which we will now proceed to describe. In a paper read before the Royal Academy of Sciences of Paris, April 15, 1839, Mr. Grove alludes to the powerful development of Electricity which would be occasioned by the combination of four elements instead of three; as, by this means, we should have nearly the sum of chemical affinities instead of their difference. He then describes some experiments which he considers as possessing a high interest, as they prove a well-known chemical phenomenon to depend on Electricity, and thus tighten the link which binds these two sciences; and they led to the discovery of a voltaic combination much more powerful than any previously known. Gold-

leaf is well-known to be unaffected by either nitric or by muriatic acid *alone*, though in a mixture of the two acids the metal dissolves. Mr. Grove cemented the bowl of a tobacco-pipe into the bottom of a wine-glass; into this he poured pure nitric acid, while the wine-glass was filled with muriatic acid to the same level; in this latter acid two strips of gold-leaf were allowed to remain for an hour, at the end of which time they were found as bright as when first immersed. A gold-wire was now made to touch the nitric acid and the extremity of one of the strips of gold-leaf; this was instantly dissolved while the other strip remained unaltered. Two strips of gold-leaf were afterwards made the electrodes of a single pair of voltaic metals in muriatic acid; the acid was decomposed, and the positive electrode was dissolved.

(276) The action is evidently this : as soon as the Electric current is established, both the acids are decomposed, the hydrogen of the muriatic acid unites with the oxygen of the nitric, and the chlorine attacks the gold. By the test of the galvanometer, the gold which was dissolved, was found to represent the zinc of an ordinary voltaic combination ; and reasoning on the phenomena, it occurred to Mr. Grove to substitute zinc for the gold; and on submitting it to the test of experiment, he found that a single pair, composed of a strip of amalgamated zinc, an inch long, and a quarter of an inch wide, a cylinder of platinum, three quarters of an inch high, with a tobacco-pipe bowl, and an egg-cup, readily decomposed acidulated water. He then substituted for the muriatic acid, caustic potash, and found the action equally powerful ; then, sulphuric acid, with four or five times its volume of water ; and, although with this the intensity was a little diminished, yet, from its exercising less local action on the zinc, he was eventually induced to give it the preference.

Mr. Grove then constructed a small battery, of a circular shape, consisting of seven liqueur glasses, and seven pipe bowls : the diameter was four inches; the height one inch and a quarter. This pocket battery gave about a cubic inch of mixed gases in two minutes.

(277) The sectional diagram, Fig. 124, exhibits the mode of fitting up four pairs of zinc and platinum foil

FIG. 124.

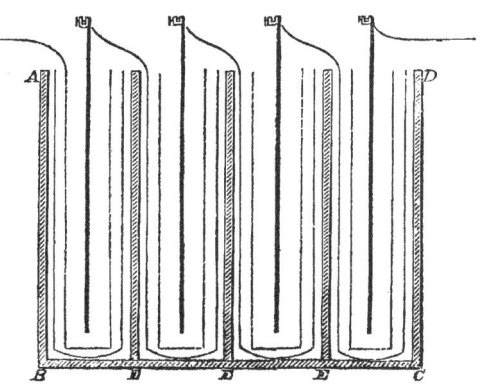

plates, as recommended by the inventor. *A B C D* is a trough of stoneware or glass, with partitions *E, E, E,* dividing it into four acid proof cells. The dotted lines represent four porous vessels, of a parallelopiped shape, so much narrower than the cells as to allow the liquid which they contain, to be double the volume of that which surrounds them ; the four dark central lines represent the zinc plates, and the five lines which curve under the porous vessels, the sheets of platinum foil, which are fixed to the zinc by little clamp screws. Common rolled zinc, about one-thirtieth of an inch thick and well amalgamated, may be employed. On the zinc side, or into the porous vessels is poured a solution of either muriatic acid, diluted with from two to two-and-a-half water, or, if the battery be intended to remain a long time in action, of sulphuric acid, diluted with four to five water ; and on the platinum side, concentrated nitro-sulphuric acid, formed by previous mixture of equal measures of the two acids. The apparatus should be provided with a cover containing lime, to absorb the nitrous vapour. Fig. 125

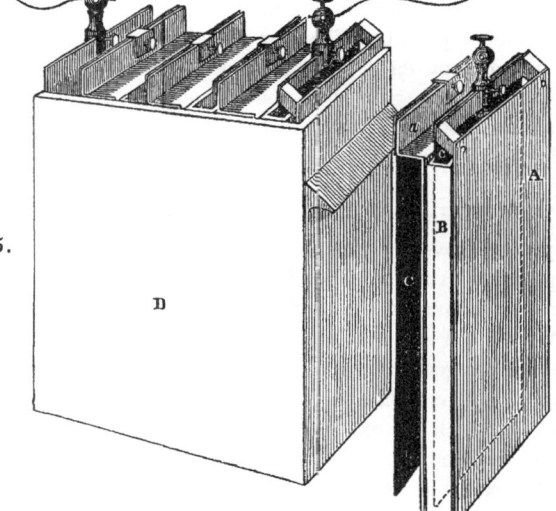

FIG. 125.

represents the complete battery ; and the first set of plates, removed from the porcelain trough D, showing very clearly the arrangement.— A *a* is the bent zinc plate ; B, the insulated platinum plate, in its porous cell ; C, the next platinum plate connected by means of a binding screw with the zinc at *a*.

(278) On the evening of March 13, 1840, Mr. Grove delivered at the Royal Institution, a lecture on voltaic reaction and polarization, and afterwards exhibited two batteries, constructed as above described.— They were charged some time previously to the lecture ; and up to the

period of its conclusion, remained in perfect inactivity, until the circuit was completed. One of these was arranged as a series of five plates, and contained altogether about four square feet of platinum foil : with this the mixed gases were liberated from water, at the surprising rate of 110 cubic inches per minute. A sheet of platinum, one inch wide, and twelve inches long, was heated in the open air through its whole extent, and the usual class of effects were produced in corresponding proportion. With the other arrangement, consisting of fifty plates, of two inches by four, arranged in single series, a voluminous flame of one inch and a quarter long, was exhibited by charcoal points, which showed beautifully the magnetic properties of the voltaic arc, and bars of different metals were instantly run into globules, and dissipated in oxide. These surprising effects were produced, it must be remembered, by a battery which did not cover a space of sixteen inches square, and was only four inches high. In a paper inserted in the 16th vol. of the L. and E. Phil. Mag., Mr. Grove describes a battery of thirty-six elements, each consisting of a square inch of platinum foil and zinc, and charged with concentrated nitric, and diluted sulphuric acid, of each of which it took a pound, so that for the expense of about a shilling, he could experiment for eight or nine hours without fresh charge, with a battery which gave between charcoal points an arc of light 0·4 of an inch long. Professor Jacobi states, that he has readily fused Iridium, with a nitric acid battery, after it has been at work a whole day.*

(279) The following explanation of the superior power of this battery is given by Mr. Grove.† "In the common zinc and copper battery the resulting power is as the affinity of the anion‡ of the generating Electrolyte for zinc, minus its affinity for copper. In the common constant battery, it is as the same affinity plus that of oxygen for hydrogen, minus that of oxygen for copper : in the combination in question, the same order of positive affinities minus that of oxygen for azote. As nitric acid parts with its oxygen more readily than sulphate of copper, resistance is lessened, and the power correlatively increased. With regard to the second material question, that of cross precipitation : in

---

* On Monday evening, May 22d, the Author had the pleasure of attending a scientific soirèe at the house of Mr. Gassiot, on which occasion an arrangement of 100 pairs of the platinum nitric acid battery was set in action. The performances of this battery were brilliant in the extreme : the flame between charcoal terminals was exceedingly voluminous, and so brilliant as to be almost insupportable to the naked eye ; upwards of two feet of stout iron wire were heated to whiteness, and ultimately fused, and sulphuret of antimony was decomposed, and the metal brilliantly deflagrated.

† L. & E. Phil. Mag. Vol. 15, p. 289.

‡ The terms *anion, cation, Electrolyte,* &c., will be explained in the next Lecture.

the common combination, zinc is precipitated on the negative metal, and a powerful opposed force created : in the constant battery, copper is precipitated, and the opposition is lessened : in this there is no precipitation, and consequently no counteraction.

" If the operation of the battery be watched, the nitric acid changes colour, assuming first a yellow ; then a green ; then a blue colour ; and lastly, becomes aqueous ; after some time nitrous gas, and ultimately hydrogen, is evolved from the surface of the platinum."

In the paper from which the above extract is taken, Mr. Grove describes an arrangement of his battery, which, *theoretically*, should evolve 213 cubic inches of mixed gases per minute, or, nearly seven and a half cubic feet per hour; and, should the period arrive when Electricity shall supersede steam, and become a means of loco-motion, the form of battery which he describes would probably be the best that could be devised. An excellent method of economising space, was proposed by Mr. Spencer, of Liverpool, viz., to plait or crimp the negative metal : by this means, in a given space, the surface may be doubled without increasing the mean distance between the metals.

(280) In the L. & E. Phil. Mag. for December, 1842, a most extraordinary and perfectly novel voltaic battery is described by Mr. Grove. It consists of a series of 50 pairs of platinized platinum plates, each about a quarter of an inch wide, enclosed in tubes partially filled alternately with oxygen and hydrogen gases, as shewn in Fig. 126. The tubes were charged with dilute sulphuric acid, sp. gr. 1·2, and the following effects were produced :—

Fig. 126.

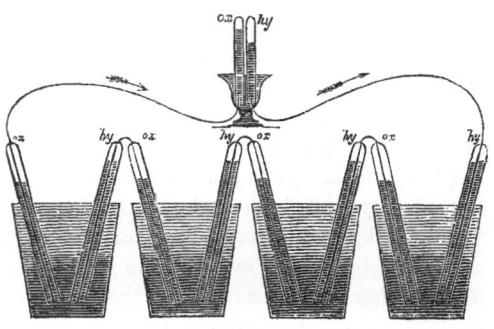

1st, A shock was given which could be felt by five persons joining hands, and which when taken by a single person was painful.

2nd, The needle of a galvanometer was whirled round, and stood at about 60°; with one person interposed in the circuit it stood at 40°, and was slightly deflected when two were interposed.

3rd, A brilliant spark visible in broad daylight was given between charcoal points.

4th, Iodide of potassium, hydrochloric acid, and water acidulated with sulphuric acid, were severally decomposed : the gas from the decomposed water was eliminated in sufficient quantity to be collected and detonated. The gases were evolved in the direction denoted in the figure, *i. e.* as the chemical theory and experience would indicate, the hydrogen, travelling in one direction throughout the circuit, and the oxygen in the reverse. It was found that twenty-six pairs were the smallest number which would decompose water, but that four pairs would decompose iodide of potassium.

5th, A gold leaf electroscope was notably affected.

When the tubes were charged with atmospheric air, no effect was produced, nor was any current determined when the gases employed were carbonic acid and nitrogen, or oxygen and nitrogen : when hydrogen and nitrogen gases were used, a slight effect was observed, which Mr. Grove is inclined to refer to the *oxygen* absorbed by the liquid when exposed to the air, which, with the hydrogen, would give rise to a current.   I shall have occasion to refer to this most remarkable battery in the next Lecture ; both as furnishing a powerful argument for the chemical theory of the voltaic pile, and as exhibiting a beautiful instance of the reciprocal action of natural forces, establishing, as was pointed out by Mr. Grove, the interesting fact that gases, in combining and acquiring a liquid form, evolve sufficient force to decompose a similar liquid, and cause it to acquire a gaseous form.

The original experiment was not made with sufficient exactness to show with accuracy the proportional diminution of gas in each tube, though it was plain that the hydrogen diminished much more rapidly than the oxygen.   A graduated battery is, however, being prepared by that liberal promoter of science, Mr. Gassiot, by which the point will be accurately determined.

(281) In the opinion of Professor Schœnbein, the oxygen does not immediately contribute to the production of the current in the gaseous voltaic battery ; he ascribes the augmentation of the current which is occasioned by having the alternate glass tubes filled with oxygen, to the *depolarizing* action exerted by that gas upon the negative platinum electrodes which are inserted in the tubes containing it, and perhaps also to some other electromotive force called into play.   This electrician found that an aqueous solution of hydrogen combined chemically with pure water, and the circuit thus formed completed by platinum, produced a current which passed from the solution to the water, though no trace of free oxygen was contained in either fluid.   By letting either oxygen or atmospheric air pass into the solution of hydrogen, he did not find the current sensibly increased, from which he drew the conclusion that the current generated cannot be due to the combination of free hydrogen with isolated oxygen ; an inference, he observes, which may

also be drawn from Mr. Grove's arrangement itself; for the oxygen contained in one tube cannot be supposed to combine with the hydrogen enclosed in another tube. M. Schœnbein recommends using chlorine gas instead of oxygen, by which he thinks the effects will be greatly increased.

(282) Fig. 127 represents an useful arrangement of copper and zinc plates for a voltaic battery, the contrivance of J. A. Van Melsen, of Maestricht. The copper soldered to the zinc in each pair envelopes the zinc of the following pair, so as to be exposed to the two surfaces of this plate, but without being in contact with it. It differs from Wollaston's pile, in having the metallic plates much nearer to each other : they are only about $\frac{1}{13}$ inch apart, and are maintained thus by small pieces of cork interposed between the plates of zinc and those of copper, whilst the plates of copper of the consecutive

FIG. 128.

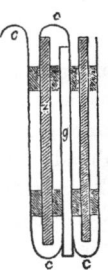

elements are separated by squares of glass of the same size as the plates. Fig. 128 represents two elements of the series. All the pairs are placed in a kind of wooden frame, Fig. 129, carefully varnished, in which they are easily retained without its being necessary to attach them by screws to a bar of wood, as is the case in the Wollaston combination. This arrangement presents the additional advantage of greatly facilitating the taking to pieces of the elements. The pairs united in the frame are at once immersed into the acidulated liquid contained in the trough: the plates of zinc are carefully amalgamated. Van Melsen describes a battery* on this

FIG. 129.

FIG. 127.

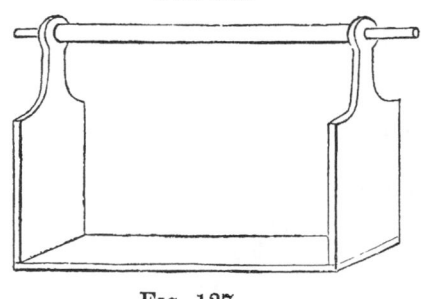

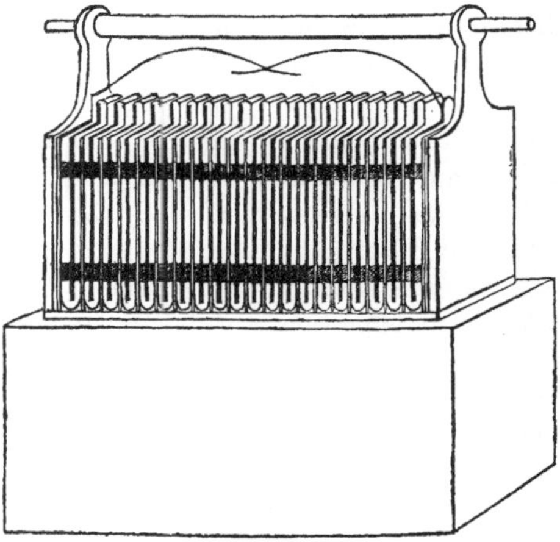

* Proceedings of the Electrical Society, p. 186.

plan, which he constructed for the Maestricht University, consisting of
52 pairs, of which the plates of zinc are $6\frac{1}{2}$ inches wide, and $7\frac{7}{8}$ inches
high. By its means a platinum wire $\frac{1}{60}$ of an inch thick, and $17\frac{3}{4}$
inches long, was reduced to incandescence with an extraordinary brilliancy, and fell into seven pieces, at the extremities of which the melted
metal arranged itself in globules. A silver wire $\frac{1}{30}$ of an inch thick,
and $15\frac{3}{4}$ inches long, became intensely red, and fell into fragments. An
iron wire $\frac{1}{20}$ of an inch thick, and $15\frac{3}{4}$ inches long, was speedily
brought to the most vivid state of ignition, and was reduced into four
pieces, in which, in many places, the melted iron was gathered into
large globules. At the period of this latter experiment the battery had
already been a long time in action, and was much weakened. When
the battery was first excited, in order to produce a spark, the two slips
of copper which serve as conductors were brought into contact. The
parts in contact became immediately soldered together, so that it was
necessary to employ a certain effort to separate them.

(283) The author has much pleasure in adducing his testimony in
favour of this form of voltaic battery,—where intense effects are required for a short time it is a most excellent arrangement. The battery constructed by the writer consists of twenty-five pairs, each zinc
plate being $6\frac{1}{2}$ inches long, by $5\frac{1}{2}$ inches wide, and when charged with
a mixture of 60 parts in volume of water, 1 of nitric and 1 of sul-
phuric acid, as recommended by Van Melsen, it heats to intense whiteness from eight to ten inches of stout platinum wire, evolves in a voltameter, the plates of which are two inches long, and $1\frac{1}{4}$ inch wide,
forty cubic inches of mixed gases per minute, and burns metallic leaves
with great brilliancy. It is far better, however, the zinc plates being
amalgamated, to use pure dilute sulphuric acid (1 acid + 8 water)
alone; by this means we avoid local action, and maintain the action of
the battery for a much longer period, though the first effects are
certainly not nearly so energetic. A great improvement would be
also, to employ as the negative element copper-wire gauze, on which
has been deposited from an acid solution, copper in a state of minute
sub-division, as recommended by Mr. Walker, the secretary of the
Electrical Society.

(284) The occasion of mentioning this gentleman's name shall be
taken as an opportunity of describing his simple and ingenious form of
the constant battery. It consists in coating the interior of an earthen
jar with copper released from a solution, and is effected in the following manner:*—" Take a large jelly pot—one to contain a quart will
do very well—place in it a few pieces of wax. Then stand it by the
fire till the wax is entirely melted, and the vessel has become exceed-

* Proceedings of the Electrical Society, page 29.

ingly hot; after this turn it about so that the wax shall adhere to every part of the interior; pour off the remaining wax and place aside the jar to cool. When it is quite cold, brush the wax coating with black lead until the whole is well covered. It is requisite for this purpose to be acquainted with the nature of the black lead employed; some specimens are far inferior to others; it is not unfrequently found that the common article used for domestic purposes, is better for this than the finer and purer sorts. When the vessel is coated well, pour in it a solution of sulphate of copper, and place within this a porous diaphragm containing acidulated water. In this put a strip of amalgamated zinc, and bend its wire so that it shall press upon the blackened surface of the jar; after a few hours' action the jar will be coated with copper, and will form a very efficient generating cell, to be excited in the usual way." Now, the surface of the copper, as thus precipitated, is somewhat analagous to the platinized surface of the silver in Mr. Smee's battery; it consists of myriads of minute glittering points; or, if the deposition of the metal be allowed to continue until the solution is exhausted, the surface will become black instead of bright, and will consist of much more minutely divided particles. The escape of hydrogen gas is thus facilitated, as in Mr. Smee's arrangement, and the battery may be used either as a constant *acid*, or a constant Daniell's battery. When employed as the latter, Mr. Walker finds it convenient, particularly when the action is required to be kept up for a long time, to use diaphragms formed of good *fresh* common plaster of Paris, cast in a mould of sheet copper.

When the action of a battery is required to be kept up for many months, as in experiments on electro-crystallization, the porous cells should be at least $\frac{1}{4}$ of an inch thick, and the bottoms formed of pipe-clay, well rammed down, while in a plastic state, to about three-fourths of an inch in thickness; this method is found to answer much better than a solid, earthen baked bottom, the latter being sure to crack even with a very moderate degree of use.

Some electricians have proposed to use wooden cylinders as diaphragms for the sulphate of copper battery, but they do not answer well. The author experimented with them very considerably when they were first spoken of, (he believes by Mr. Mullins); large ones he could never get to answer at all; he constructed, however, a battery of 40 pairs, with small wooden cylinders, each 3 inches high, and $2\frac{1}{2}$ inches wide, and with it strong shocks could be given, and a brilliant spark obtained between charcoal points.

(285) In the following series the metals are ranged according to their electrical characters, and in the same relation to each other as zinc has to copper, so that one of them operates as *zinc* to all those above it,

and the more distant from one another any two metals stand in the series, the greater the galvanic action they will develope.

Platinum.	Mercury.	Tin.
Gold.	Copper.	Iron.
Silver.	Lead.	Zinc.

Hence, as we have already seen, a galvanic series of platinum and zinc is more powerful than one of copper and zinc; and the latter again more powerful than one of lead and zinc, &c. It is not, however, to be understood, that the power of any two metals in the table depends upon the *number* of intermediate ones, because a series of platinum and iron is much feebler than a series of copper and zinc; although in the former case there are six intermediate metals, and in the latter, there are only three. Charcoal and plumbago stand higher in the scale of electric bodies than platinum, so that a galvanic series of plumbago and zinc is very powerful. Now, plumbago or graphite is a combination of iron and carbon, and the hint has been thrown out by Jacobi,* that by adding more carbon to that which usually enters into the composition of cast-iron, we should probably arrive at a compound whose galvanic properties would be equal to those of platinum. The object may be obtained by a species of cementation, or by remelting cast-iron with additional carbon in closed vessels.

(286) This high *negative* character of carbon, enables us to understand how it is, that cast-iron and zinc form so effective a voltaic circle, standing as iron and zinc do, immediately next each other in the above series. It was Mr. Sturgeon who first formed a large battery of these metals; it is described in the 5th volume of his Annals of Electricity. It consists of 10 cast-iron cylindrical vessels, and the same number of cylinders of amalgamated, rolled zinc, with dilute sulphuric acid. The cast-iron vessels are 8 inches high, and $3\frac{1}{2}$ inches in diameter. The zinc cylinders are the same height as the iron ones, about 2 inches in diameter and open throughout. The iron and zinc cylinders are attached in pairs to each other by means of a stout copper wire. The zinc of one pair is placed in the iron of the next, and so on throughout the series; contact being prevented by discs of mill-board placed in the bottom parts of the iron vessels.

With ten pairs in series, Mr. Sturgeon states, that he has usually obtained fourteen cubic inches of the mixed gases per minute, and ten and a half cubic inches, when the battery has been in action an hour and a half. On one occasion he states, that he obtained twenty-two cubic inches per minute, fused one inch of copper wire, one-twenty-fifth of an inch in diameter; kept four inches white hot, and eighteen inches red hot, in broad daylight. Eight inches of watch main-springs

* See his " Galvanoplastic Art," translated by Mr. Sturgeon, p. 4.

were kept red hot, and two inches white hot, for several successive minutes.

(287) Immediately after reading the above account of the performances of the cast-iron battery, the writer, struck with its extraordinary power, proceeded to construct a similar arrangement of ten pairs ; and, though he cannot say that he has ever obtained, under any circumstances, anything like the effects above described, he has, nevertheless, found it very convenient as a stationary battery, though, from its necessary weight and bulk, it can never be put in competition with the elegant arrangement of Smee, or the beautiful contrivance of Daniell.

(288) The anomalous electric condition of iron, and its efficiency when employed as the negative element in a voltaic arrangement, was pointed out by Mr. Martyn Roberts in the L. & E. Phil. Mag. :[*] when iron is associated with copper, it is highly positive to that metal ; but when it is associated with zinc, it is more highly *negative* to the zinc than copper would be under similar circumstances. When two similar-sized galvanic combinations—one of iron and zinc, and the other of copper and zinc—were compared, a deviation of the needle of the galvanometer, amounting to twenty-five degrees, was found in favour of the iron and zinc pair : and when two equal-sized batteries were applied under similar circumstances, to the decomposition of water, the iron battery, with a measured quantity of acid, gave four cubic inches of mixed gases in 104 minutes, while the copper battery gave only one and a-half cubic inch in 125 minutes.

(289) This surprising intensity of the zinc-iron circuit is thus explained by Professor Poggendorff :[†]—

" The intensity of the voltaic circuit depends on two things,—the electromotive force, and the resistance. It is the quotient from the division of the former by the latter. Now though the electromotive force between zinc and iron is smaller than between zinc and copper, silver, or platinum : nevertheless, the current of the zinc-iron circuit is stronger, because the iron offers less resistance to the transition of the current than copper does. The current, however, possesses less tension than that of the copper circuit ; or, in other words, it is weakened by the insertion of a foreign resistance in a greater proportion than that of the copper-zinc ; and it was found that the interposition of a wire of German silver, fifty feet long, weakened the current from the iron-zinc more than that of the copper-zinc ; and further it was supposed, that by a continued increase of the inserted resistance, it would be possible to make the current of the iron circuit, not only as weak, but even

* Vol. xvi. p. 142.   † Poggendorff's Annalen, vol. i. p. 255.

weaker than that of the copper circuit.   Professor Poggendorff did not, however, succeed practically in effecting this.

(290) Mr. Roberts has, however, offered an explanation,* which Electricians in this country will, I think, be inclined to adopt in preference to that given by the learned German, who is one of the most powerful and strenuous supporters of the Contact Theory of Galvanism.† It is simply,—that copper, when immersed in an acidulated solution, does not retain so clean a metallic surface as iron does, when exposed to a like action.   When a copper-zinc pair is placed in dilute sulphuric acid, an action takes place upon both the metals, and the balance of their affinities for the acid determines the direction of intensity of the electric current : but an obstacle to its free circulation arises by the resistance offered to its passage from the acid into the copper, because this metal has in a measure been acted upon by the acid, and its surface partially oxidated : but as the affinity of the base for the acid, under these circumstances, is not sufficient to cause the solution of the oxide, it therefore remains upon the surface of the copper-plate ; and as oxides are worse conductors of Electricity than their metallic bases, we have here a resistance presented by the oxidated surface to the entrance of the electric current into the copper plate. On the other hand, when an iron-zinc pair is immersed in dilute acid, we have also an action on both metals ; but the balance of affinities is here not so much in favour of the zinc, as when it is in combination with copper, and therefore the *intensity* or *electro-motive force* generated by the iron-zinc, is not so great as in that of the copper-zinc battery : but the *quantity* circulated by the iron-zinc is greater, because the surface of the iron not only oxidates, as did the copper, but in consequence of its greater affinity for the acid, this oxide becomes dissolved in the liquid, and it is thus removed from the surface of the metal, which remains purely metallic, bright, and far more fitted to conduct Electricity than would be the oxidated surface of a copper plate : it therefore offers less resistance to its entrance, and a larger *quantity* is thus circulated, although (in consequence of the balance of affinities) in less intensity or *electromotive force* by an iron-zinc, than by a copper-zinc galvanic pair.

(291) Mr. Roberts has introduced a form of battery on the above principles, which, as it has come into very general use for blasting purposes, we shall here describe.‡   For general purposes it consists of twenty single negative iron, and twenty single positive zinc plates, of six inches square, arranged alternately in a frame of wood, and connected in the fol-

* L. & E. Phil. Mag. vol. xix., p. 196.
† See next Lecture.   ‡ Proc. Elec. Soc. p. 357.

lowing peculiar manner. Let the numbers and *z*, Fig. 130, represent the zinc,
and the letters and *i*, the iron plates;
let *a* and *b* be joined together, and stand
free as a double terminal plate or pole,
having of course a wire proceeding from
them as a conductor; then join 1 to

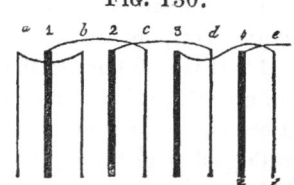

Fig. 130.

c, 2 to *d*, 3 to *e*, and so on, terminating the other end of the battery
by a positive plate, but having both its surfaces opposed to a negative
plate, as is the condition of 4.

In a battery of this construction, there is no cross play of Electricity, because two plates intervene between every positive plate, and
the negative plate in metallic connection with it. Its power is very
great in consequence of the closeness of the plates one to the other. It
is very compact; and the absence of insulating cells, renders it very
convenient, as it can with no trouble be put into, or taken out of, its
box. The plates are put into a frame
made of bars of wood, as in Fig. 131.
The plates are kept from touching
each other by strips or rods of wood
about $\frac{1}{4}$ or $\frac{1}{8}$ of an inch square, and long
enough to extend from the top to the
bottom of each plate, one rod to each
side of a plate; or if the plate be very

Fig. 131.

large, another in the middle. The box containing the exciting liquid
(dilute sulphuric acid, one part acid to thirty of water), is put together
with white-lead joints, as these are perfectly water-tight. A battery
of this construction is found to be far more powerful, and constant in
its action, than an equal sized one of copper and zinc.

(292) While speaking of the electrical properties of iron, we may
take the opportunity to detail some peculiar voltaic conditions of that
metal. In the L. & E. Phil. Mag., vols. 9, 10, and 11, several papers
on this curious subject will be found, by Schoenbein, Faraday, and
others; but we must confine ourselves here to the simple facts, referring
to the original papers, for a theoretical explanation of them.

If one of the ends of an iron wire be made red hot, and, after cooling,
be immersed in nitric acid, sp. gr. 1·35, neither the end that has been
heated, nor any other part of the wire will be affected, whilst acid of
this strength is well known to act rather violently upon common iron.
By immersing an iron wire in nitric acid of sp. gr. 1·5, it becomes likewise indifferent to the same acid of 1·35.

(293) The principal facts that the writer has experimentally verified,

and the observations which he has made, in repeating Schoenbein's experiments, are as follow :—

1°. It is well known, that when iron wire is immersed in nitric acid, sp. gr. 1·35, it is attacked with violence ; but Sir John Herschel was, it seems, the first person who noticed that if the wire was associated with gold or platinum, it was quite inactive in acid of that strength.— When an iron wire, one-sixteenth of an inch in diameter, was touched at a given point with platinum, and dipped into nitric acid sp. gr. 1·37, it was not at all acted upon, but remained, for any length of time, perfectly bright. Once touching it in the acid with the platinum was sufficient to render it inactive when the platinum was removed, as long as it remained in the acid ; but if it were taken out, wiped, and then again immersed, action commenced, but soon again ceased.

2°. If the acid was diluted with an equal bulk of water, platinum did not preserve iron wire from its action, even when coiled thickly round it : it appeared, indeed, rather to quicken the action ; but, although it did not protect the iron under these circumstances, it did under others which will be mentioned presently.

3°. If a wire, having been made inactive, by being touched with a piece of platinum, was touched, while in the acid, with a piece of zinc, or another common iron wire, it was immediately thrown into violent action. Half a wire, four inches long, was heated to dull redness, the blue tinge was visible through three inches : when the wire was cold, these three inches were quite inactive in nitric acid sp. gr. 1·39, the other end was active ; but when the heated end was made bright by filing, it was rendered active likewise.

4°. When an inactive wire and one that was active, were dipped into the same vessel, and made to touch at their parts *above* the fluid, action was excited in the indifferent wire. A common wire was made to touch an indifferent one, and both dipped into nitric acid, the indifferent one going in *first :* by this means the common wire was rendered indifferent, not being in the slightest degree acted on by the acid ; the second wire rendered indifferent a third ; the third, a fourth ; and so on. This experiment was found to succeed best with a wire that had been made indifferent by platinum ; but with care, it will answer equally well with a wire that has been made indifferent in the fire, the conditions appearing to be, perfect contact and gradual immersion. When these wires were taken out of the acid, and wiped, they always returned to the active state, but were again made indifferent by repeating the process.

5°. A wire, polished very bright, and protected by platinum, was immersed in a solution of nitrate of copper and nitric acid, which acted very strongly on common iron, copper being deposited on the metal ; the protected wire remained, however, bright ; after a few seconds, the platinum was removed—the iron became instantly as common iron ;

but when the platinum was allowed to remain in contact an hour or two, and then removed, the wire was left in the peculiar state, exhibiting the curious phenomenon of a piece of polished iron remaining untarnished in a solution of acid nitrate of copper. The wire thus inactive, on being touched with a piece of common iron, was instantaneously rendered active, undergoing rapid solution and becoming covered with a coating of copper.

6°. A piece of common wire was bent into the form of a fork, and slipped down an inactive wire into nitric acid, by which it was itself rendered inactive ; now, if another piece of wire was made to touch the fork, before being introduced into the acid, it was rendered itself inactive ; but if it was first thrown into action, and then made to touch either end of the fork, it threw all the wires into action, unless the first wire was one *rendered inactive by the fire,* in which case it was not thrown into action : the writer could not, in this experiment, succeed in making one end of the fork active and the other passive, as described by Schoenbein ; he tried it many times, and in every case, every wire was thrown into action, when either was touched in the acid with an active wire.

7°. In order to observe the electrical phenomena, a galvanometer was used in the manner described by Faraday : a platinum wire was connected with one of the cups, and the other end dipped into a glass containing nitric acid, of the above strength ; if now, an iron-wire was connected *first* with the other cup of the galvanometer, and then the other end immersed in the acid, it was inactive, and no deflection of the needle took place ; but if it was first put into the acid, and afterwards connected with the galvanometer, it was active, and the needle was deflected in the same manner as if it had been zinc, i. e., whichever pole of the needle the wire of the galvanometer with which it was joined passed *immediately over* moved *west.*

8°. If an inactive wire was in this experiment substituted for the platinum, it acted precisely as platinum, both with regard to its preserving action, and to the direction of the electrical current produced ; and here it may be observed that a striking proof is by this experiment afforded, that voltaic action is due to chemical action, for, when the wires were so arranged that both should be inactive, there was not the slightest electrical current evinced by the galvanometer ; but when either was thrown into action by being touched by a common wire, that wire became instantly as zinc, and the needle was strongly deflected.

9°. If the iron-wire had a piece of platinum foil attached to it, the moment the circuit was closed, bubbles of gas made their appearance on the platinum, but none on the iron ; but when the platinum was removed the gas rose rapidly from the iron, which was not, however, thrown into action.

10°. When two glasses were filled with acid, and connected by a compound platinum and iron-wire, all the phenomena which took place in a *single* glass, were observed, and the platinum or inactive wire in one glass, exerted a protecting influence on the iron on the other, provided the communication was first made through the galvanometer; a touch from a common wire also threw the iron into action, producing a strong electrical current; the same was the case with three or four glasses connected by a compound wire.

11°. When the acid was diluted, so as to have a sp. gr. of 1·2, platinum, as was before observed, could not protect iron from its action, neither when it was connected with the galvanometer did it, if the iron was dipped into the acid first; but if it was first connected with the galvanometer, and then put into the acid, *no action whatever took place in any length of time, even when the platinum was removed;* but it always commenced when the inactive wire was once touched in the acid with a common iron-wire, or with a piece of copper; but the iron thus made inactive did not as in strong acid possess the power of rendering other wire inactive, but was always thrown into action itself when a piece of common wire was substituted for the platinum, whether it was connected with the galvanometer first or not :—the first wire in this case acted as platinum to the second.

12°. When two cups were employed, and connected by a piece of bent wire, and so arranged that the iron-wire should be active, on removing the connecting wire, and taking a fresh piece, if it were dipped first into the cup containing the iron-wire, and then the other end brought into the platinum cup, *that end was inactive, and there was no passage for the electrical current, the needle of the galvanometer being quiescent; but when it was put into an active state the electrical current passed.* Here then we have the iron made inactive without any *metallic* communication with the platinum, and when inactive it is found incapable of conducting, or, at any rate, it obstructs very considerably the passage of an electrical current.

13°. If the iron-wire was inactive it was impossible to make either end of the connecting bent wire so, neither could it be, if it were dipped into the platinum cup *first;* the action of nitric acid of this strength, viz. 1·2, is not an effervescing action, the iron is slowly dissolved ; when a piece of clean metal is dipped in it, it speedily becomes covered with a brown substance, which is gradually deposited, but dissolved by agitation.

14°. When iron-wire is made the positive electrode of a galvanic battery, consisting of fifteen or twenty pairs, and dilute nitric, sulphuric or phosphoric acid the subject of experiment, the negative electrode consisting of a platinum wire, if that pole be first dipped into strong nitric acid, and the circuit closed by a common iron-wire, that wire is immediately inactive, as regards the action of the acid on it, and it behaves precisely as platinum or gold in *giving off oxygen* from the decomposed water, while the platinum wire becomes surrounded with a greenish fluid (nitrous acid) : any other mode of closing the circuit will not answer, and if, while oxygen is given off from the iron-wire, it is once brought into contact with the platinum, it ceases to give off oxygen when separated from it, and will not again do so till exposed to the air.

15°. The same phenomena occur with diluted acid, only hydrogen gas is given off in great abundance from the platinum, and as before, when the wires are made to touch in the liquid the iron ceases to perform the office of platinum, and becomes gradually dissolved ; exposure to the air, however, brings it again to the peculiar state.

16°. Diluted sulphuric and phosphoric acid exhibit similar phenomena, but the iron cannot be made inactive in muriatic acid with that or any other voltaic power ; it is always converted into muriate. When diluted nitric acid is employed, and when two cups are connected by a common iron-wire, the effects are the same, and if the connecting wire be removed, and the cups joined by another, in the manner before described, that end in the cup in which the platinum negative electrode was, gives off oxygen, while the other end undergoes solution, and the iron-wire which acted the part of the positive electrode gives off oxygen also ; if four cups be employed a similar result is obtained ; but the quantity of oxygen liberated diminishes as the number of elements increases : if either of the ends of the wires be now touched with a common iron-wire, its peculiar state is destroyed, and it becomes as the other end, while the oxygen it gave off appeared to be divided between the two inactive wires, and if the iron-wire in immediate connection with the battery be made active, and all the others but the middle one made active also, then the middle wire performed the office of the positive electrode.

Much more might be said on this curious subject ; the above must, however, suffice here, and those who are anxious to see the matter fully discussed may be referred to the 9th, 10th, and 11th volumes of the L. & E. Phil. Mag. A voltaic battery, consisting of zinc and passive iron, or of active and passive iron, in either case excited after the manner of a Grove's battery, was described, Jan. 17, in a communication from Professor Schoenbein to the London Electrical Society. The power of the arrangement is said to be very great. Its economy is also a matter of importance, and the value of the salt produced (sulph. ferri) is not to be overlooked.

(294) The electrical character of an alloy of metals does not, it must be observed, always take a place between those metals of which it consists, but more frequently it stands either much higher or much lower in the series. Such is the case with brass, which mostly acts in galvanic arrangements, either quite as well, or even better than copper, which is one of its constituents. (Jacobi.) On the other hand, either amalgamated zinc, or a compound of zinc and quicksilver, acts even better than zinc alone, although quicksilver itself stands high in the galvanic series. A compound is described by Jacobi, which is still better than quicksilver and zinc; it consists of 38 parts of quicksilver, 22 parts of tin, and 12 parts of zinc. Nevertheless, he observes, in such alloys as these, where too much quicksilver is introduced, the disadvantage is, that they are extremely brittle, and have but little coherence.

(295) The inaction of amalgamated zinc in acidulated water, (236) is considered by Mr. Grove* as being the effect of polarization; but of one which differs from ordinary cases of polarization, in that the *cations†* of the electrolyte, instead of being precipitated on the negative metal, combine with it, and render it so completely positive, that the current is nullified, and not merely reduced in intensity as in other cases.

The experiments made by Mr. Grove, to verify this idea, are curious and striking.

1°. Half the surface of a strip of copper was amalgamated and immersed with a strip of zinc in water, acidulated with $\frac{1}{4}$th of sulphuric or phosphoric acid; on making the plates touch there was a rapid evolution of gas from the *unamalgamated* part of the copper, while only a few detached bubbles appeared on the amalgamated portion.

2°. A large globule of mercury was placed in the bottom of a glass of acidulated water, and by means of a copper wire, the whole surface of which was amalgamated, it was made to communicate with one extremity of a galvanometer, while a strip of amalgamated zinc, immersed in the same liquid, communicated with the other extremity; at the instant of communication an energetic current was indicated, which, however, immediately diminished in intensity, and at the end of a few minutes the needle returned to zero; scarcely any gas was evolved, and of the few bubbles which appeared, as much could be detected on the surface of the zinc as of the mercury.

3°. With the same arrangement a strip of platinum, well amalgamated, was substituted for the mercury. In a few minutes the current became null or very feeble, and if, after the cessation of the current, the zinc was changed for unamalgamated platinum, this latter evolved

* L. & E. Phil. Mag. vol. xv. p. 81.   † See next Lecture.

torrents of hydrogen, and the needle indicated a violent current in a contrary direction.

1 4°. With things arranged as in 2°, sulphate of copper was substituted for acidulated water,—a constant current was produced, and copper was precipitated on the mercury, as long as crystals of the sulphate were added to the solution.

(296) In these experiments it is shown that mercury, which, in its normal state, is well known to be inefficient as the positive metal of a voltaic combination, is in many cases equally inefficient as a negative metal from its faculty of combining with the cations of electrolytes, which rendering it equally positive with the metal with which it is voltaically associated, the opposed forces neutralize each other. But if, as in 4°, the cation of the electrolyte is not of a highly electro-positive character, the zinc (or other associated metals) retains its superior oxidability, and the voltaic current is not arrested.

(297) The application of these experiments to the phenomena presented by amalgamated zinc, Mr. Grove thinks evident ; all the heterogeneous metals with which the zinc may be adulterated, and which form minute negative elements, being amalgamated, become by polarization equally positive with the particles of zinc, and consequently without the presence of another metal to complete the circuit, all action is arrested as in the case of pure zinc. The fact of amalgamated zinc being positive with respect to common zinc, of its precipitating copper from its solutions, and other anomalies, are also explained by these experiments.

(298) The last form of voltaic battery that I shall describe, is the arrangement of Dr. Leeson, in which, instead of sulphate of copper, a solution of bichromate of potash (ten parts water to one of bichromate), is employed as the exciting agent. AA, Fig. 132, shows a vertical,

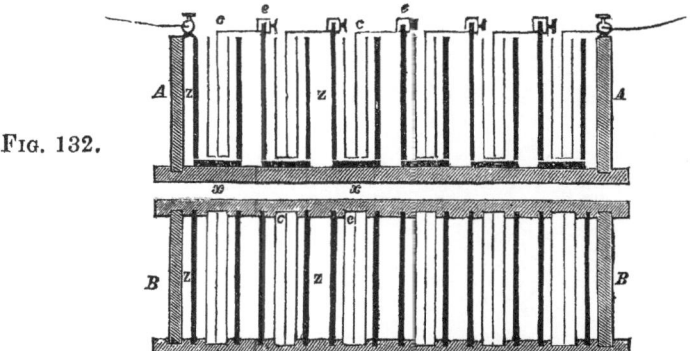

FIG. 132.

and BB, an horizontal section of the wooden trough rendered watertight. It is grooved at the sides, as seen in BB, so as to receive the zinc plates ZZ : between each pair is a groove to receive the flat porous cell, containing the copper plate C. Each zinc plate rests on a piece of zinc, which forms as it were the bottom of a cell: one of each pair of zinc plates, Z, is higher than the other, as seen in the vertical section for the convenience of forming the connection, which is effected by binding over the copper plate, and attaching it to the tall zinc one by a small binding screw, as seen at e. The trough is charged with acid solution, and the porous cell containing the copper with the solution of bichromate. Each trough contains ten or twelve cells. By having the zinc which surrounds the copper, in three pieces, the trouble of binding is avoided, and it is much easier of manipulation. It will be seen by this, that the expensive plan of employing actual partitions between the respective pairs is avoided, each arrangement of zinc forming its own cell. It is scarcely requisite to mention, that the zinc is not of necessity to be accurately fitted in its groove, under the idea of making each cell water-tight, the fallacy of this idea having been long since developed.

(299) Having now described all the forms of the galvanic battery which appear to have any claim to public notice, this lecture shall be here brought to a close : and in the following one the effects of voltaic action, and some of the results which have been furnished by the aid of this powerful instrument of research, shall be described.

LECTURE V.

VOLTAIC ELECTRICITY *(continued)*.

Heat and light—voltaic flame between charcoal points—practical applications of the heating power of the voltaic current—General Pasley's operations on the wreck of the Royal George—the Round Down Cliff explosion—Grove's experiments on the voltaic disruptive discharge — Chemical phenomena — Faraday's new terms—decomposition of water, and saline solutions—experiments with sulphate of magnesia and chloride of silver—Davy's experiments—Definite electro chemical action — Volta-Electrometer — quantity of electric force in matter—secondary results—Crosse's experiments on electro crystallization—the author's experiments—the Electrical "Acarus"—Crosse's "Electrical" springs—Daniell's experiments on the Electrolysis of secondary compounds—Electro metallurgy—Metallo-chromes—the odour accompanying the oxygen from electrolyzed water—Experiments of M. Schoenbein and Mr. Gann.

(300) On comparing the Electricity of the Voltaic battery with that of the electrical battery, we find a difference between the two which may be expressed in the three following particulars :—1. the *intensity* of voltaic Electricity, as compared with statical, is exceedingly *low :* 2. the quantity of Electricity set in motion by the smallest voltaic circle, is almost infinitely greater than that from the electrical machine ; indeed, it has been shown by Faraday,[*] that two wires—one of platinum and one of zinc, each one-eighteenth of an inch in diameter—placed five-sixteenths of an inch apart, and immersed to the depth of five-eighths of an inch in acid, consisting of one drop of oil of vitriol and four ounces of distilled water, at a temperature of about 60°, and connected at the other extremities by a copper wire, eighteen feet long and one-eighteenth of an inch thick, yield as much Electricity in eight beats of a watch, or $\frac{8}{150}$ of a minute, as an electrical battery, consisting of fifteen jars, each containing 184 square inches of glass, coated on both sides, and charged by thirty turns of a fifty-inch plate machine : 3. while the discharge of the electrical battery is instantaneous, in the voltaic battery a current circulates in an uninterrupted and continuous stream, although the wire uniting the opposite ends is constantly tending to restore the electric equilibrium.

(301) In considering the effects of voltaic Electricity, it will be convenient to do so in relation to these three circumstances, as contrasting

[*] Experimental Researches, 371, et seq.

it with ordinary Electricity. In a former Lecture it has been shown,
that a piece of glass or sealing wax rubbed with flannel, and held near
the cap of the gold-leaf electroscope, causes an immediate divergence of
the leaves; but the largest calorimotor that has ever been constructed,
is incapable of producing an equal effect : indeed, it is only by the application of the condenser, that any indications of Electricity can be
obtained from it ; but with a battery of many pairs the effect is very
distinct, though water be the sole exciting agent, as we have already
seen (249). And it matters not what the size of the plates may be ;
pairs of copper and zinc, one quarter of an inch square, being quite as
effectual as plates four inches square, numerous alternations being the
only requisite. Here then we see a remarkable difference between the
simple and the compound voltaic circle, and between quantity and
intensity. From the largest calorimotor that was ever constructed, we
can obtain no direct shock, and only feeble electro-chemical effects,
while thirty or forty pairs of zinc and copper, four inches square, excited
by the same acid, will diverge gold leaves, give shocks, and decompose
acidulated water very rapidly : in general terms it may be stated *cœteris
paribus*, that the *quantity* of the Electric current bears a relation to the
size of the plates, and the *intensity* to the number of the alternations.

(302) *Heat and light.* The wonderful heating powers of an extensive voltaic battery, and the intense light emitted between charcoal
points, were noticed in the last Lecture (278). In the Proceedings of the Electrical Society (4to. volume), a series of experiments
performed with a sulphate of copper battery, consisting of 160 cells, are
detailed. The deflagration of mercury is described as most brilliant ;
and the length of the flame between charcoal points, was three-fourths
of an inch. Zinc turnings were speedily deflagrated, and their oxide
was seen floating about the room. In these experiments, the following
interesting result was *first* obtained :—When the ends of the main wires
were placed across each other (at about one or two inches from their
extremities, not touching, but with an intervening stratum of air, the
striking distance through which the Electricity passed, producing a brilliant light), that wire connected with the *positive* end of the battery,
became red-hot from the point of crossing to its extremity. The corresponding portion of the other wire remained comparatively cold. The
wires were removed from the battery : that which had been made the
positive was made the negative, and that which had been *negative* was
made positive. The results were still the same :—the *positive* wire
becoming in all cases heated from its end to the point of crossing, and
finally bending beneath its own weight. When a piece of sulphuret of
barytes was placed on the table, with one wire resting on it ; upon
bringing the other to within the striking distance, the portion conti-

guous to the wire was fused, but could not be collected. When sulphuret of lead was similarly placed, the metal was released in small quantities; but when sulphuret of antimony was placed in circuit, the most brilliant effects were obtained. The negative wire was firmly held on the sulphuret, and the positive brought to within one-eighth of an inch of it, the heat of the flame immediately disengaged the elements combined with the metal, and they were dissipated in the form of vapour, leaving a small portion of fused metal, in a state of intense heat. When the main wires were crossed, and their ends placed in two separate jars, containing distilled water, in about two minutes the water in the positive cell boiled; that in the other presenting no such appearance. On applying a powerful magnet, the flame from the charcoal points obeyed the known laws of Electro-magnetism (236), being attracted or repelled as the case may be; or following the motion of the magnet, if the latter were revolved. But when a powerful horse-shoe magnet was held horizontally, with its *north* or marked side uppermost, and the wire from the negative side of the battery firmly pressed on the magnet, the positive wire being brought to within the striking distance, a brilliant circular flame of Electrical light *was seen to revolve from left to right, as the hands of a watch.* When the position of the magnet was reversed, the flame revolved from *right to left.* The appearance of the flame was not unlike that of the brush from an Electrical machine, received on a large surface, only much more brilliant. (67)

(303) Fig. 133, represents the appearance presented by the voltaic

FIG. 133.

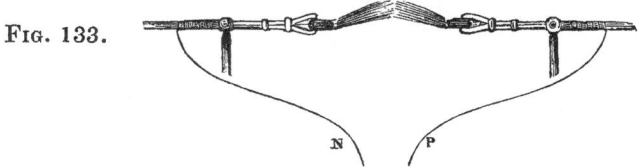

flame between pencils of well-burnt boxwood charcoal, or, which answers better, between pencils formed of that plumbago-like substance found lining the interior of long-used coal-gas retorts. The arched form of the flame is owing to the ascensional force of the heated air. With respect to the charcoal light, the following observations were made by De la Rive, with a Grove's battery of forty pairs.* The luminous arch cannot be obtained between two charcoal points, until after the two points have been in contact, and are heated around this point of contact. We may then, by separating them gradually, succeed in having between them a luminous arch, an inch or more in length. Wood charcoal, which, after having been powerfully ignited, has been quenched by means of water, is that which gives the most beautiful light, on account

* Proceedings of the Electrical Society, page 75.

of its conducting power being increased. Coke, though it succeeds as well as charcoal, does not give so brilliant and white a light : it is always rather blueish, and sometimes red. The transfer of particles of carbon from the positive to the negative pole, whilst the luminous arch is produced, is evident ; but it is especially sensible in vacuo. A cavity is observed to be formed in the point of the positive charcoal, presenting the appearance of a hollow cone, in which the solid cone, formed by the deposition of particles of carbon, might penetrate almost exactly. The phenomenon is almost the same in the air, except, that the accumulation of carbon on the negative point is less, because a portion of the molecules burn in the transfer : and the positive point presents only a flat, instead of a hollow surface. This latter result probably arises from the combustion of the thin exterior of the hollow cone, which must be formed in air as well as in vacuo.

It appears very probable, that the luminous arch itself is only the result of the incandescence of the particles of carbon transferred from one hole to the other,—incandescence, which in the air is accompanied by a partial combustion, and which takes place without combustion in vacuo. When the current is not very powerful, the eye can almost see these particles move in the direction indicated. This transfer can only take place, as we have seen, when the charcoal has been previously strongly heated,—a circumstance which, probably diminishing its cohesion, facilitates the disjunction of the particles. De la Rive also found that spongy platinum and copper, reduced by hydrogen into an impalpable powder, heaped up in glass tubes, produce the same effect as charcoal ; and that the same takes place when a piece of solid metal is placed at the negative end, though, if the solid metal be at the positive, and the spongy metal at the negative, only a very short spark is obtained. De la Rive also repeated the beautiful experiment of the influence of the magnet on the luminous arch, and observed that the light is not polarized : he also remarked, that it can be employed as solar light, to illuminate an object to be daguerréotyped.

(304) The colour of the light which attends the voltaic disruptive discharge, varies with the substances between which the discharge passes. If thin metallic leaves be employed, they are deflagrated with considerable brilliancy. The beautiful effects are not, however, owing to the combustion of the metals, though in some cases increased by this cause, but arise from a dispersion of their particles analogous to that of the more momentary explosion of the Leyden battery. (Daniell) Gold leaf emits a white light, tinged with blue; silver, a beautiful emerald green light ; copper, a bluish white light, with red sparks; lead, a purple; and zinc, a brilliant white light, tinged with red. The experiments are best performed by fixing a plate of polished tinned iron to

one wire of the battery, and taking up a leaf of any metal, on the point of the other wire, bringing it in contact with the tin plate. Even under distilled water, the disruptive discharge of the voltaic battery takes place in a stream of brilliant light.

(305) The temperature of metallic conductors is sometimes actually *reduced* by the passage of Electricity through them. Thus it has been shown by Lenz,* that if two bars of bismuth and antimony be soldered across each other, at right angles, and they be touched with the conducting wires of the battery, so that the positive Electricity will have to pass from the antimony to the bismuth, the temperature of the metals will be elevated; but when the current moves in the opposite direction, viz:—from the bismuth to the antimony, the metals become cooled at their point of contact. If a cavity be excavated at this point, and a drop of water, previously cooled to nearly 32°, be placed therein, on the current passing, *the water will become rapidly frozen.*

(306) The best method of showing the power of the voltaic current to heat metallic wire, is to roll about eighteen inches of wire into a long spiral, and place it in the interior of a glass tube, its ends passing through corks, so as to be readily twisted round the terminal wires of the battery : by this means, a high temperature may be communicated to the glass tube, though the wire may not be ignited ; and by immersing it in a small quantity of water, that fluid may speedily be raised to its boiling point. When a wire in the voltaic circuit is heated, the temperature frequently rises first, or most at one end; but it was shown by Faraday, that this depends on adventitious circumstances, and is not due to any relation of positive or negative, as respects the current. Faraday has also shown,† that the same quantity of Electricity which, passed in a given time, can heat an inch of platinum wire of a certain diameter red hot, can also heat a hundred, a thousand, or any length of the same wire, to the same degree, provided the cooling circumstances are the same for every part in all cases.

(307) It was Lieut.-General Pasley who first applied the heating power of the galvanic battery to a useful practical purpose. While engaged in operations on the river Thames, he was written to by Mr. Palmer,‡ who recommended him to employ the galvanic battery instead of the long fuse then in use, and after being put in possession of the method of operating, he immediately adopted it, and has since turned it to excellent account in the removal of the wreck of the Royal George at Spithead, as is well known.

The destruction by gunpowder of the Round Down Cliff on the line of the South Eastern Railway, on Thursday, 26th January, 1843,

* Pog. Annal. Vol. xliv. p. 342.   † Experimental Researches, 853, note.
‡ Smee's Electro-metallurgy, p. 297.

is a splendid example of the successful application of a scientific principle to a great and important practical purpose. In this grand experiment, by a single blast, through the instrumentality of the galvanic battery, 1,000,000 tons of chalk were in less than five minutes detached and removed, and 10,000*l.* and twelve months' labour saved.

(308) The following account is abridged from the Report in the *Times* newspaper :—" The experiment has succeeded to admiration, and as a specimen of engineering skill, confers the highest credit on Mr. Cubitt who planned, and on his colleagues who assisted in carrying it into execution. Everybody has heard of the Shakspeare Cliff, it would be superfluous therefore to speak of its vast height, were not the next cliff to it on the west somewhat higher : that cliff is Round-Down Cliff, the scene and subject of this day's operations. It rises to the height of 375 feet above high-water mark, and was, till this afternoon, of a singularly bold and picturesque character. As a projection on this cliff prevented a direct line being taken from the eastern mouth of Abbot's Cliff Tunnel to the western mouth of the Shakspeare Tunnel, it was resolved to remove, yesterday, no inconsiderable portion of it from the rugged base on which it has defied the winds and waves of centuries. Three different galleries, and three different shafts connected with them, were constructed in the cliff. The length of the galleries or passages was about 300 feet. At the bottom of each shaft was a chamber 11 feet long, 5 feet high, and 4 feet six inches wide. In each of the eastern and western chambers, 5,500 lbs. of gunpowder were placed, and in the centre chamber 7,500 lbs., making in the whole, 18,000 lbs. The gunpowder was in bags, placed in boxes : loose powder was sprinkled over the bags, of which the mouths were opened, and the bursting charges were in the centre of the main charges. The distance of the charges from the face of the cliff was from 60 to 70 feet. It was calculated that the powder, before it could find a vent, must move 100,000 yards of chalk, or 200,000 tons. It was confidently expected that it would move one million.

" The following preparations were made to ignite this enormous quantity of powder :—At the back of the cliff a wooden shed was constructed, in which three galvanic batteries were erected. Each battery consisted of 18 Daniell's cylinders, and two common batteries of 20 plates each. To these batteries were attached wires which communicated at the end of the charge by means of a very fine wire of platinum, which the electric current as it passed over it made red hot to fire the powder. The wires, covered with ropes, were spread upon the platinum, which became red-hot when the electrical current traversed it. The wires, covered with ropes, were spread upon the grass to the top of the cliff, and then falling over it, were carried to the eastern, the centre, and the western chambers. Lieutenant Hutchinson, of the Royal Engineers, had the command of the three batteries, and it was arranged that when he fired the centre, Mr. Hodges and Mr. Wright should simultaneously fire the eastern and western batteries. The wires were each 1000 feet in length, and it was ascertained by experiment that the current will heat platinum wire sufficiently hot to ignite gunpowder to a distance of 2,300 feet of wire.

" Exactly at twenty-six minutes past two o'clock, a low, faint, indistinct, indescribable, moaning subterranean rumble was heard, and immediately afterwards the bottom of the cliff began to belly out, and then almost simultaneously about 500 feet in breadth of the summit began gradually but rapidly to sink. There was no roaring explosion, no bursting out of fire, no violent and crashing splitting of rocks, and comparatively speaking very little smoke : for a proceeding of mighty and irrepressible force, it had little or nothing of the appearance of force. The rock seemed as if it had exchanged its solid for a fluid nature, for it glided like a stream into the sea, which was at a distance of 100 yards, perhaps more from its base, filling up several large pools of water which had been left by the receding tide. As the chalk, which crumbled into fragments, flowed into the sea without splash or noise, it discoloured the water around with a dark, thick, inky-looking fluid ; and when the sinking mass had finally reached its resting place, a dark brown colour was seen on different parts of it which had not been carried off the land."

(309) The circumstance of so little smoke being seen attendant on the combustion of such a prodigious quantity of gunpowder, occasioned to many a good deal of surprise, and induced a belief that the whole of the gunpowder had not been fired ; but when we consider that the smoke owes its visibility principally to the solid and finely divided charcoal * which is suspended in it, and that in passing through such an immense mass of limestone, it must have been *filtered* as it were from this solid matter, our wonder at the absence of smoke on this occasion will cease.

(310) The following table exhibits the results of the experiments of M. Becquerel on the conducting power of metals for voltaic electricity :—

Copper	. . .	100	Platinum .	. .	16·4
Gold	. . .	93·6	Iron	. .	15·8
Silver	. . .	73·6	Tin .	. . .	15·5
Zinc	. . .	28·5	Lead	. . .	8·3

* The principal gaseous results of the combustion of gunpowder are carbonic oxide, carbonic acid, nitrogen, and sulphurous acid ; the solid residue consists of carbonate and sulphate of potassa, sulphuret of potassium and charcoal.

In consequence of its good conducting power, a voltaic battery will
heat to redness a greater length of silver wire than of platinum ; nevertheless, if a compound wire be formed of several alternate links of platinum and silver, as shewn   FIG. 134.
in Fig. 134, and disposed
between the poles of a powerful battery, the platinum
links will become red hot
during the passage of the

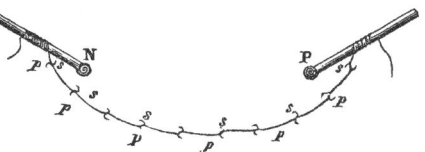

current, while the alternate silver links will remain dark.   The charge,
which passes freely along the silver, meets with resistance enough in
the platinum to produce ignition.

(311) On reflecting on the remarkable difference between the heating
effects of the positive and negative wires of the voltaic battery (302),
it occurred to Mr. Grove * that it might be due to the interposed
medium, and that were there any analogy between the state assumed
by voltaic *Electrodes* (316) in elastic media, and that which they assume
in *Electrolytes* (317), it would follow that the chemical action at the
positive electrode in atmospheric air would be more violent than that
at the negative, and that if the chemical action were more violent, the
heat would necessarily be more intense.

By experiments performed with an arrangement of thirty-six pairs
of his nitric acid battery, Mr. Grove established the following points :—

1°. If zinc, mercury, or any oxidable metal constitute the positive
electrode, and platinum the negative one, in atmospheric air, while the
disruptive discharge is taken between them, a voltameter (335) inclosed
in the circuit yields considerably more gas than with the reverse
arrangement.

2°. In an oxidating medium, the brilliancy and length of the arc are
(with certain conditions) directly as the oxidability of the metals between
which the discharge is taken.

3°. In an oxidating medium, the heat and consumption of metal is
incomparably greater at the *anode* (316) than at the *cathode* (316).

4°. If the disruptive discharge be taken in dry hydrogen, in azote,
or in a vacuum, no difference is observable between the heat and light,
whether the metals be oxidable or inoxidable, or whether the oxidable
metal constitute the positive or negative electrode.

5°. The volume of oxygen absorbed by the disruptive discharge
taken between a positive electrode of zinc and a negative one of platinum
in a vessel of atmospheric air, is equal to that evolved by a voltameter
included in the same circuit.

(312) A remarkable analogy between the electrolytic and disrup-

* L. and E. Phil. Mag. vol. xvi. p. 478.

tive discharges is here presented, but there are two elements which obtain in the latter which have little or no influence on the former, viz. the volatility and state of aggregation of the conducting body. This was shown remarkably in the case of iron, which in air or in oxygen gave a most brilliant voltaic arc, while in hydrogen, or a vacuum, with the same power a feeble spark only was perceptible at the moment of disruption. Mercury, on the other hand, gave a tolerably brilliant spark in hydrogen, azote, or a vacuum, and one more nearly approaching to that which it gives in air.

(313) It has been established by Faraday, that in electrolysis, a voltaic current can only pass by the derangement of the molecules of matter; that the quantity of the current which passes is directly proportional to the atomic disturbance it occasions : he deduces from this, that the quantity of electricity united with the atoms of bodies is as their equivalent numbers, or in other words, that the equivalent numbers of different bodies serve as the *exponents* of the comparative quantities of electricity associated with them.* " Now," observes Mr. Grove, " what takes place in the disruptive discharge? When we see dazzling flame between the terminals of a voltaic battery, do we see electricity, or do we not rather see matter, detached, as Davy supposed, by the mysterious agency of electricity, and thrown into a state of intense chemical or mechanical action ? Matter is undoubtedly detached during the disruptive discharge, and this discharge takes its tone and colour from the matter employed. Now, as this separation is effected by electricity, electricity must convey with it either the identical quantity of matter with which it is associated, or more or less ; more it can hardly convey, and if less, some portion of electricity must pass in an insulated state or unassociated with matter, and some with it." Mr. Grove proceeded to institute some experiments with a view of determining whether the quantity of matter detached by the voltaic disruptive discharge was definite for a definite current, or bore a direct equivalent relation to the quantity electrolyzed in the liquid portions of the same circuit. The great difficulties attending such an enquiry, defied accurate results : but sufficient was gathered to afford strong grounds for presumption, that the separation of matter in the voltaic arc *is* definite for a definite quantity of electricity, and that the all important law of Faraday is capable of much extension ; and uniting this view with the experiments of Faraday on the identity of electricity from different sources, and with those of Fusinieri on the statical electrical discharge, it would follow as a corollary, that every disturbance of electrical equilibrium is inseparably connected with an equivalent disturbance of the molecules of matter.

* Experimental Researches, 518, 524, 732, 783, 836, 839.

(314) *Chemical Phenomena.*—Before entering upon this interesting branch of our subject, it will be necessary that we describe the new terms introduced by Faraday, and state his reasons for adopting them. According to the views of this celebrated philosopher,* electro-chemical decomposition is occasioned by an *internal corpuscular action,* exerted according to the direction of the electric current, and is due to a force either *superadded to, or giving a direction to the ordinary chemical affinity* of the bodies present. He conceives the effects to arise from *forces* which are *internal,* relative to the matter under decomposition, and not *external* as they might be considered if directly dependent upon the poles. He supposes that the effects are due to a modification, by the electric current, of the chemical affinity of the particles through or by which that current is passing, giving them the power of acting more forcibly in one direction than in another, and consequently making them travel by a series of successive decompositions and recompositions in opposite directions, and finally causing their expulsion or exclusion at the boundaries of the body under decomposition, in the direction of the current, *and that* in larger or smaller quantities according as the current is more or less powerful.

Thus, in Fig. 135, the particles *a a,* could not be transferred, or travel from one pole N, towards the other P, unless they found particles of the opposite kind *b b,* ready to pass in the contrary direction ; for it is by virtue of their increased affinity for those particles, combined with their

Fig. 135.

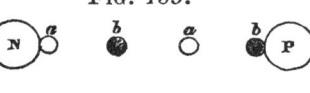

Fig. 136.

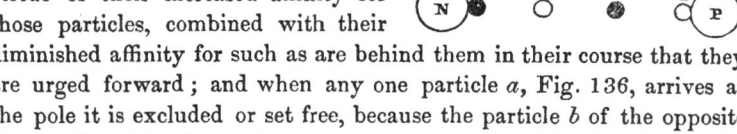

diminished affinity for such as are behind them in their course that they are urged forward ; and when any one particle *a,* Fig. 136, arrives at the pole it is excluded or set free, because the particle *b* of the opposite kind, with which it was the moment before in combination, has, under the super-inducing influence of the current, a greater attraction for the particle *a′,* which is before it in its course, than for the particle *a,* towards which its affinity has been weakened : *a* may be conceived to be expelled from the compound *a b,* by the superior attraction of *a′* for *b,* that superior attraction belonging to it in consequence of the relative position of *a′b* and *a,* to the direction of the axis of electric power super-induced by the current. The electric current is looked upon by Faraday as *an axis of power, having contrary forces, exactly equal in amount, in contrary directions.*

(315) According to Faraday's views then, the determining force is

* Experimental Researches, 518, 524.

not at the so called *poles* of the voltaic battery, but *within* the body under decomposition (132) : to avoid, therefore, confusion and circumlocution, and for the sake of greater precision of expression, he, with the assistance of two friends, framed the following new terms, which have since been almost universally adopted.

What are called the *poles* of the voltaic battery are merely the surfaces or doors by which the Electricity enters into, or passes out of, the substance suffering decomposition ; Faraday hence proposes for them the term *electrodes*, from ἤλεκτρον and ὁδὸs a way, meaning thereby, the substance, or surface, whether of air, water, metal, or any other substance which serves to convey an electric current into, and from the decomposing matter, and which bounds its extent in that direction.

(316) The surfaces at which the electric current enters, and leaves a decomposing body, he calls the *anode*, and the *cathode ;* from ἄνα upwards, and ὁδὸs a way,—*the way which the sun rises ;* and κατὰ downwards, and ὁδὸς a way,—*the way which the sun sets.* The idea being taken from the earth, the magnetism of which is supposed to be due to electric currents, passing round it in a constant direction from *east* to *west*,—if, therefore, the decomposing body be considered as placed, so that the current passing through it shall be in the same direction, and parallel to that supposed to exist in the earth, then the surfaces at which the Electricity is passing into and out of the substance, would have an invariable reference, and exhibit constantly the same relations of powers. The *anode* is therefore, that surface at which the electric current enters : it is the *negative* extremity of the decomposing body ; is where oxygen, chlorine, acids, &c., are evolved ; and is against or opposite the positive electrode. The *cathode* is that surface at which the current leaves the decomposing body, and is its *positive* extremity : the combustible bodies, metals, alkalies, and bases are evolved there, and it is in contact with the negative electrode. Thus,

in Fig. 137, if we suppose a current of Electricity traversing a wire in the direction of the darts, and entering at E, then, on separating the

FIG. 137.

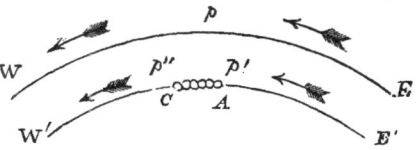

wires at p, p' p'' would become its electrodes : *p'* would be the *anelectrode* or emitting electrode, and *p''* the *cathelectrode*, or receiving electrode; E being the wire connected with the last *active* copper plate, and W the wire connected with the last *active* zinc plate of a battery ; and if we suppose the chain of small circles to represent the fluid under decomposition, A will be its anode, and C its cathode.

(317) Compounds directly decomposable by the electric current are called *electrolytes*, from ἤλεκτρον and λύω to set free,—to *electrolyze* a

body is to decompose it electro-chemically : the elements of an electrolyte are termed *iöns*, from ἰὼν, participle of the verb εἶμι to go ; *anions* are the iöns which make their appearance at the anode, and were formerly termed the electro-negative elements of the compound, and *cations* are the iöns which make their appearance at the cathode, and were termed the electro-positive elements. Thus, chloride of lead is an *electrolyte*, and when *electrolyzed* evolves two *iöns*, chlorine and lead, the former being an *anion*, and the latter a *cation :* water is an electrolyte, evolving likewise two iöns, of which oxygen is the anion, and hydrogen the cation : muriatic acid is likewise electrolytical, boracic acid on the other hand is not.

(318) Mr. Daniell proposes further to distinguish the doors by which the current enters and departs, by the terms *zincode* and *platinode*, the former being the plate which occupies the position of the generating plate in the battery, and the latter of the conducting plate ; when water is decomposed, therefore the last particle of oxygen gives up its charge to the zincode, and the last particle of hydrogen gives up its charge to the platinode, each passing off in its own elastic form.

(319) The chemical power of the voltaic pile was discovered and described by Messrs. Nicholson and Carlisle, in the year 1800. Water was the first substance decomposed. In 1806, Davy communicated to the Royal Society his celebrated lecture " on some chemical agencies of Electricity," and in 1807, he announced the grand discovery of the decomposition of the fixed alkalies. The years from 1831 to 1840, are marked in science by the publication of the masterly Researches of Faraday, in which, much that was before unintelligible has not only been explained and enlightened, but a new character has been stamped on electrical, as connected with chemical science. Of these remarkable essays, it has been said that in point of originality of talent, and perspicuity, they rank among the first efforts of philosophy of the age, if indeed they do not surpass all others.

(320) When water and certain saline solutions are made part of the electric circuit, so that a current of Electricity may pass through them, they are decomposed, that is, they yield up their elements in obedience to certain laws. Water is resolved into oxygen and hydrogen gases, and the acid and alkaline matters of the neutral salts, which it holds in solution, are separated, *not* in an indiscriminate manner, but the oxygen and acids are always developed at the anode, and the hydrogen and alkaline bases are given off at the cathode. If pure water be submitted to the action of the current, in consequence of its bad conducting power it is decomposed with great difficulty ; but a greatly increased conducting power is however given to it by the addition of salts, and particularly by sulphuric acid, though that compound is *not* itself capable of electro-

lysis. One essential condition of electrolysis is liquidity : and the current of a powerful battery will be completely stopped by a film of ice, not more than one-sixteenth of an inch in thickness.* To decompose acidulated water, it may be confined in a glass tube, sealed hermetically at one extremity, and made part of the electrical circuit by means of gold or platinum wires, or arranged as in Fig. 138,

FIG. 138.

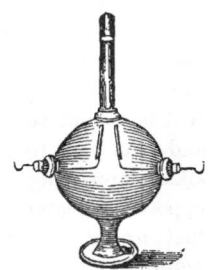

the wires being about a quarter of an inch apart. When the tube is about half full of the mixed gases, if a spark from the electrophorus (Fig. 10,) be passed between the wires, an explosion will take place, and if care be taken to prevent any escape, by the expansion, the tube will be re-filled with water, that fluid having been re-produced by the explosion. If two glass receiving tubes be employed, one over either electrode, gas will be collected in each ; but that in the tube over the cathode will be rather more than double in volume than even the anode, the former being hydrogen, and the latter oxygen. Now the hydrogen and oxygen gases are to each other in water exactly as two to one, by volume : and the reason they do not appear precisely in this proportion in the electrolysis of that fluid is because oxygen is partially soluble. By referring to Fig. 135, it will be immediately seen how it is that there is no visible transfer of the oxygen and hydrogen : if the electrodes were several inches apart there would be no appearance of decomposition between them. The oxygen a', of the atom of water a' b', under the super-inducing influence of the current, is transferred to the hydrogen b, of the second atom of water a b : the oxygen of this second atom is in like manner transferred to the hydrogen of a third, and so on till the electrode P is arrived at, against which the oxygen of the last particles is evolved, having nothing to combine with.

(321) Take a syphon-shaped tube, and, placing its bent part in a wine glass for support, fill it with the blue infusion of red cabbage; then put a few crystals of some known salt, such as sulphate of soda, into the tube, and electrize the solution. In a short time the liquid nearest the cathode of the battery, will become green, indicating the presence of free alkali ; and the liquid nearest the anode will become red, showing that an acid is present : reverse the direction of the current, and the colours will also gradually be reversed. Thus sulphate of soda is an Electrolyte, the anion of which is sulphuric acid and the cation, soda ; and in all salts decomposable by the voltaic current, the acid passes to the anode, and the base to the cathode.

(322) If two glasses be taken, both being filled with the blue infu-

* Faraday's Experimental Researches, 381, et seq.

sion, holding in solution sulphate of soda, and an inverted glass tube, in
which two platinum wires are sealed, be immersed in each, as shown in
Fig. 139, and the two glasses connected to-

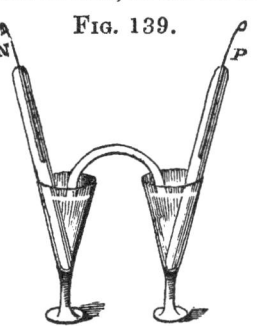

FIG. 139.

gether by a glass syphon filled with the liquid,
on electrizing the solution, it will be found
that notwithstanding they are in separate vessels, the blue liquor will, as before, be turned
red and green; and if the experiment be continued sufficiently long, the alkali of the salt
will be found to have passed from P to N,
and the acid from N to P, the acid and
alkali appearing to traverse the syphon in
opposite directions.   It was hence inferred,
that *under the influence of electrical attraction, the usual chemical
affinities are suspended;* but the same explanation which accounted
for the phenomenon of the decomposition of water, will serve here: a
line of particles of sulphate of soda extend from one electrode to the
other; and it is by a series of decompositions and recompositions that
the effect is produced.

(323) In various experiments of decomposition, the little form of
apparatus of which Fig. 140, is a sketch, will

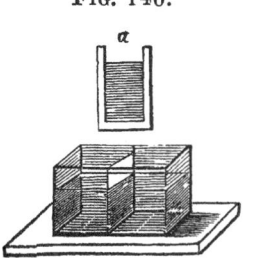

FIG. 140.

be found exceedingly useful.   It is a cell of
plate glass, made by cementing five pieces together with transparent varnish, and supporting them upon a wooden foot, in which they
are fastened.   The cell is about five or six
inches long, and an inch broad, and may be
divided into two parts by the insertion of the
temporary diaphragm *a*, which is a small frame
of cane with muslin stretched over it.   When
this is in its place, a separate electrode may be introduced on each side
of it; they may most conveniently consist of two pieces of thin platinum foil, about four inches long and half an inch broad.

To show the evolution of chlorine at the anode or positive pole, fill
the glass cell with weak salt and water, acidulated with muriatic acid,
and coloured blue by a few drops of the sulphuric solution of indigo;
then introduce the electrodes.   In a few minutes the *anodic* division
will be found to lose colour, and will finally become colourless, owing
to the separation of chlorine, which by its bleaching powers destroys
the blue of the indigo.

The presence of uncombined iodine may be demonstrated, by filling
the cell with a weak solution of *starch*, to which a little common salt
and iodide of potassium have been added; then on passing the electric

current through the liquid, the iodine will show itself at the anode by a beautiful blue colour, it being the property of this singular substance to strike a fine deep blue colour with starch.

(324) A substance cannot be transferred in the electric current beyond the point where it ceases to find particles with which it can combine ; and it cannot be too strongly impressed, that electro-chemical decomposition does not depend upon any direct attraction or repulsion exerted by the metallic terminations either of the voltaic battery, or of the ordinary electrical machine. The beautiful experiments of Faraday, in which *air* was shown to act as a pole, have been quoted (128—132) : in the following equally beautiful experiments,* the decomposition of sulphate of magnesia against a surface of water, is most satisfactorily shown.

(325) A glass basin, four inches in diameter, and four inches deep, had a division of mica, *a*, (Fig. 141), fixed across the upper part, so as to descend one inch and a half below the edge, and to be perfectly water-tight at the sides. A plate of platinum, *b*, three inches wide, was put into the basin on one side of the division *a*, and retained there by a glass block below, so that any gas produced by it in a future stage of the experiment, should not ascend beyond the mica, and cause currents in the liquid on that side. A strong solution of sulphate of magnesia was carefully poured without splashing into the basin until it rose a little above the lower edge of the mica division *a*, great care being taken

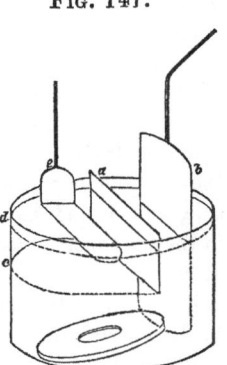

FIG. 141.

that the glass or mica, on the unoccupied or *c* side of the division in the figure, should not be moistened by agitation of the solution above the level to which it rose. A thin piece of clean cork, well wetted in distilled water, was then carefully and lightly placed on the solution at the *c* side, and distilled water poured gently on to it, until a stratum, the eighth of an inch in thickness, appeared over the sulphate of magnesia. All was then left for a few minutes, that any solution adhering to the cork might sink away from it, or be removed from the water on which it now floated ; and then more distilled water was added in a similar manner, until it reached nearly to the top of the glass. In this way, solution of the sulphate occupied the lower part of the glass, and also the upper on the right-hand side of the mica ; but on the left-hand side of the division, a stratum of water from *c* to *d*, one inch and a-half in depth, reposed upon it, the two presenting, when looked through

* Faraday's Exp. Researches, 494.

horizontally, a comparatively definite plane of contact. A second platinum pole e, was arranged so as to be just under the surface of the water, in a position nearly horizontal, a little inclination being given to it, that gas evolved during decomposition, might escape. The part immersed was three inches and a-half long by one inch wide ; and about seven-eighths of an inch of water intervened between it and the solution of sulphate of magnesia.

(326) The latter pole e, was now connected with the negative end of a voltaic battery, of forty pairs of plates, four inches square; whilst the former pole b, was connected with the positive end. There was action and gas evolved at both poles ; but, from the intervention of the pure water, the decomposition was very feeble, compared to what the battery would have effected in an uniform solution. After a while (less than a minute) magnesia also appeared at the negative side. *It did not make its appearance at the negative metallic pole, but in the water*, at the place where the solution and the water met; and on looking at it horizontally, it could be there perceived lying in the water upon the solution, not rising more than a fourth of an inch above the latter ; whilst the water between it and the negative pole was perfectly clear. On continuing the action, the bubbles of hydrogen, rising upwards from the negative pole, impressed a circulatory movement on the stratum of water, upwards in the middle, and downwards at the side, which gradually gave an ascending form to the cloud of magnesia in the part just under the pole, having an appearance as if it were there attracted to it ; but this was altogether an effect of the currents, and did not occur till long after the phenomena looked for, were satisfactorily ascertained.

(327) After a little while the voltaic communication was broken, and the platinum poles removed with as little agitation as possible from the water and solution, for the purpose of examining the liquid adhering to them. The pole e, when touched by turmeric paper, gave no traces of alkali ; nor could anything but pure water be found upon it. The pole b, though drawn from a much greater depth and quantity of fluid, was found so acid as to give abundant evidence to litmus paper, the tongue, and other tests. Hence, there had been no interference of alkaline salts in any way, undergoing first decomposition, and then causing the separation of the magnesia at a distance from the pole by mere chemical agencies. This experiment was repeated again and again, and always satisfactorily.

(328) Thus it is clearly shown, that both water and air may officiate as a *pole*, and that one *element or principle* only has no power of transference, or of passing towards either pole ; and hence there appears but little reason to consider the phenomena of electro-chemical decomposi-

tion, as due to the *attraction* or attractive powers of the metallic terminations of the battery.   Indeed, if, in accordance with the usual theory,
a piece of platinum be supposed to have sufficient power to attract a
particle of hydrogen from the particle of oxygen with which it was the
instant before combined, there seems, as Faraday has observed, no sufficient reason, nor any fact, except those to be explained, which show
why it should not, according to analogy with all ordinary attractive
forces, as those of gravitation, magnetism, cohesion, chemical affinity, &c. *retain* that particle which it had just before taken from a distance, and from previous combination.   Yet it does not do so, but
allows it to escape freely.

(329) It would not be possible, perhaps, to bring forward a more
instructive, or a more beautiful instance of the *transfer of elements*, and
their progress in opposite directions, parallel to the electric current,
than is furnished by *chloride of silver* when decomposed by silver-wire
poles.   Upon fusing a portion of this compound on a piece of glass, and
bringing the poles into contact with it, there is abundance of silver
evolved at the negative pole, and an equal abundance absorbed at the
positive pole, for no chlorine is set free; and by careful management
the negative wire may be withdrawn from the fused globules as the
silver is reduced there, the latter serving as the continuation of the pole,
until a wire or thread of revived silver, five or six inches in length, is
produced.   At the same time, the silver at the positive pole is as rapidly dissolved by the chlorine, which seizes upon it, so that the wire
has to be continually advanced as it is melted away.   The whole experiment includes the action of only two elements—silver and chlorine.

(330) According to the theory of Faraday, no element or substance
can be transferred, or pass from pole to pole, unless it be in chemical
relation to some other element or substance tending to pass in the opposite direction, the effect being essentially due to the mutual relation of
such particles.   Thus, pulverized charcoal, or sublimed sulphur, diffused
through dilute sulphuric acid, exhibits no tendency to pass to the negative pole, neither do spongy platinum, or gold precipitated by sulphate
of iron, yet in these cases the attraction of cohesion is almost perfectly
overcome ; the particles are so small as to remain for hours in suspension,
and are perfectly free to move by the slightest impulse towards either
pole ; and *if in relation* by chemical affinity to any substance present,
are powerfully determined to the negative pole.

(331) In Davy's celebrated paper on "Some Chemical Agencies of Electricity," read before the Royal Society, November 20th,
1806, the following experiments on the passage of acids, alkalies,
and other substances through various attracting chemical menstrua, are described : " An arrangement was made, consisting of

three vessels, as shown in Fig. 142:
solution of sulphate of potash was
placed in contact with the negatively
electrified point, pure water was
placed in contact with the positively

Fig. 142.

electrified point, and a weak solution of ammonia was made the middle
link of the conducting chain: so that no sulphuric acid could pass to
the positive point in the distilled water without passing through the
solution of ammonia: the three glasses were connected together by
pieces of amianthus. A power of 150 pairs was used: in less than five
minutes it was found, by means of litmus paper, that acid was collecting
round the positive point: in half an hour the result was sufficiently
distinct for accurate examination.

" The water was sour to the taste, and precipitated solution of nitrate
of barytes.

" Similar experiments were made with solution of lime, and weak
solutions of potash and soda, and the results were analogous. With
strong solutions of potash and soda a much longer time was required for
the exhibition of the acid; but even with the most saturated alkaline
lixivium, it always appeared in a certain period. Muriatic acid, from
muriate of soda, and nitric acid, from nitrate of potash, were transmitted
through concentrated alkaline menstrua under similar circumstances.
When distilled water was placed in the negative part of the circuit,
and a solution of sulphuric, muriatic, or nitric acid, in the middle, and
any neutral salt with a base of lime, soda, potash, ammonia, or magnesia, in the positive part, the alkaline matter was transmitted through
the acid matter to the negative surface, with similar circumstances to
those occurring during the passage of the acid through alkaline menstrua; and the less concentrated the solution the greater seemed to be
the facility of transmission.

" I tried in this way muriate of lime with sulphuric acid, nitrate of
potash with muriatic acid, sulphate of soda with muriatic acid, and
muriate of magnesia with sulphuric acid; I employed the power of
150; and in less than forty-eight hours, I gained in all these cases
decided results; and the magnesia came over like the rest.

" Strontites and barytes passed like the other alkaline substances,
readily through muriatic and nitric acids; and vice versâ, these acids
passed with facility through aqueous solutions of barytes and strontites;
but in experiments in which it was attempted to pass sulphuric acid
through the *same menstrua,* or to pass barytes or strontites through
this acid, the results were very different.

" When solution of sulphate of potash was in the negative part of the
circuit, distilled water in the positive part, and a saturated solution of
barytes in the middle, no sensible quantity of sulphuric acid existed in

the distilled water after thirty hours, the power of 150 being used : after four days sulphuric acid appeared, but the quantity was extremely minute : much sulphate of barytes had formed in the intermediate vessel : the solution of barytes was so weak as barely to tinge litmus, and a thick film of carbonate of barytes had formed on the surface of the fluid. With solution of strontites the result was very analogous, but the sulphuric acid was sensible in three days.

" When solution of muriate of barytes was made positive by the power of 150, concentrated sulphuric acid intermediate, and distilled water negative, no barytes appeared in the distilled water, when the experiment had been carried on for four days ; but much oxymuriatic acid had formed in the positive vessel, and much sulphate of barytes had been deposited in the sulphuric acid."

Fig. 143.

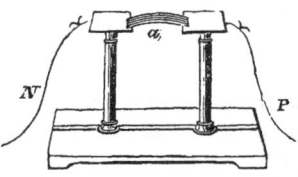

(332) Sulphate of barytes may be decomposed by employing two insulated discs of platinum, as in Fig. 143, one of which is to be put in communication with the negative, and the other with the positive end of the pile : on each of these a few grains of finely powdered sulphate of barytes, moistened by a drop or two of water, is placed, and the discs connected by a few filaments of wet cotton, they should be within half an inch of each other : in a few minutes barytes will be evinced by test papers at the negative disc, and sulphuric acid at the positive.

(333) These experiments of Davy's excited at the time they were announced the utmost astonishment, and the only way by which they could be at all explained, was by supposing that throughout the whole circuit the natural affinities of substances were suspended and destroyed ; but that which made the *wonder*, is, in fact, the *essential condition* of transfer and decomposition, and the more alkali there is in the course of an acid, the more will the transfer of that acid be facilitated from pole to pole. The instances in which sulphuric acid could not be passed through barytes, or barytes through sulphuric acid, enter within the pale of the law established by Faraday, by which liquidity is so generally required for conduction and decomposition. In assuming the solid state of *sulphate of barytes* these bodies became virtually non-conductors to Electricity of so low a tension as that of the voltaic battery, and the power of the latter over them was almost infinitely diminished.

(334) When the material out of which the poles are formed is liable to the chemical action of the substances evolved, either simply in consequence of their natural relation to them, or of that relation aided by the influence of the current, then they suffer corrosion, and the parts dissolved are subject to transference in the same manner as the particles of the body originally under decomposition. Thus platinum being made

the positive and negative poles, in a solution of sulphate of soda, has no affinity for the oxygen, hydrogen, acid, or alkali evolved, and refuses to combine with, or retain them. Zinc can combine with the oxygen and acid : at the positive pole it does combine, and immediately begins to travel as oxide towards the negative pole. Charcoal, which cannot combine with the metals, if made the negative pole in a metallic solution, refuses to unite to the bodies which are ejected from the solution upon its surface; but if made the positive pole in a dilute solution of sulphuric acid, it is capable of combining with the oxygen evolved there, and consequently unites with it, producing both carbonic acid and carbonic oxide in abundance.

(335) Among the many grand discoveries with which Faraday has enriched electrical science, that of *definite electro-chemical action*, is one of the most important. In the investigation of this question, it was necessary to construct an instrument which should measure out the Electricity passing through it, and which, being interposed in the course of the current used in any particular experiment, should serve at pleasure, either as a *comparative standard* of effect, or as a positive measurer of this subtile agent. Water, acidulated with sulphuric acid, was the electrolyte chosen ; and Fig. 144, exhibits one of the forms of apparatus employed : *d* is a straight tube, closed at the other extremity and graduated : through the sides pass the platinum wires *bb'*, being fused into the glass and connected with two plates within : the tube is fitted by grinding into one mouth of a double-necked bottle, one-half or two-thirds full of water, acidulated with sulphuric acid. The tube is filled by inclining the bottle ; and when an electric current is passed through it, the gases evolved, collect in the upper part of the tube, and displace the diluted acid, the stopper *c* being left open.

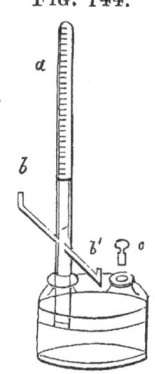

FIG. 144.

When the graduated part of the tube *a*, is filled with the mixed gases, the electric circuit may be broken, by removing the wires connected with *bb'*, the stopper *c* replaced, and the metre tube refilled, by properly inclining the instrument : a second measure of gas is then collected, on re-establishing the circuit, and so on. Fig. 145, is another very useful form of this instrument, to which its inventor has given the name of the Volta-Electrometer.

FIG. 145.

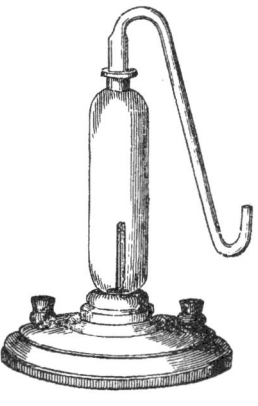

(336) By a series of experiments made with this apparatus, under a variety of forms, with different-sized platinum electrodes, and with

solutions of various degrees of strength, it was proved that *water, when subjected to the influence of the electric current is decomposed in a quantity exactly proportionate to the quantity of Electricity which passes through it,* whatever may be the variations in the conditions and circumstances under which it may be placed; and hence, that the instrument may be employed with confidence, as an exact measurer of Voltaic Electricity.

(337) A detail of one experiment with protochloride of tin,* will be sufficient as an example, both of definite electro-chemical decomposition, and of the masterly method of examining the question which was adopted by this celebrated electrician. A piece of platinum wire had one extremity coiled into a small knob; and having been carefully weighed, was sealed hermetically into a piece of bottle-glass tube, so that the knob should be at the bottom of the tube within. The tube was suspended by a piece of platinum wire, so that the heat of a spirit-lamp could be applied to it. Recently-fused protochloride of tin was introduced in sufficient quantity to occupy, when melted, about one-half of the tube. The wire of the tube was connected with a volta-electrometer, which was itself connected with the negative end of a voltaic battery; and a platinum wire connected with the positive end of the same battery, was dipped into the fused chloride in the tube, being, however, so bent, that it could not, by any shake of the hand or apparatus touch the negative electrode at the bottom of the vessel. The whole arrangement is delineated in Fig. 146.

FIG 146.

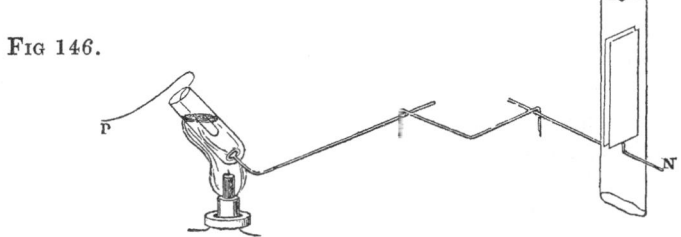

(338) Under these circumstances, the chloride of tin was decomposed; the chlorine evolved at the positive electrode formed *bichloride* of tin, which passed away in fumes; and the tin evolved at the negative electrode combined with the platinum, forming an alloy fusible at the temperature to which the tube was subjected, and therefore, never occasioning metallic communication through the decomposing chloride.— When the experiment had been continued so long as to yield a reasonable quantity of gas in the volta-electrometer, the battery connexion was broken, and the positive electrode removed, and the tube and remaining

* Faraday's Experimental Researches, 789.

chloride allowed to cool. When cold, the tube was broken open, the rest of the chloride and the glass being easily separable from the platinum wire and its button of alloy. The latter, when washed, was then re-weighed, and the increase gave the weight of the tin reduced.

(339) The following are the particular results of one experiment:— The negative electrode weighed at first 20 grains : after the experiment, it, with the button of alloy, weighed 23·2 grs. The tin evolved by the electric current, at the cathode, weighed therefore 3·2 grains. The quantity of oxygen and hydrogen collected in the volta-electrometer = 3·85 cubic inches. As 100 cubic inches of oxygen and hydrogen, in the proportions to form water, may be considered as weighing 12·92 grains, the 3·85 cubic inches would weigh 0·49742 of a grain : that being therefore the weight of water decomposed, by the same electric current, as was able to decompose such weight of protochloride of tin as could yield 3·2 grains of metal. Now 0·49742 : 3·2 :: 9 (the equivalent of water) : 57·9, which should therefore be the equivalent of tin, if the experiment had been made without error, and if the electro-chemical decomposition *is in this case also definite*. In some chemical works, 58 is given as the chemical equivalent of tin ; in others, 57·9. Both are so near to the result of the experiment, and the experiment itself is so subject to slight causes of variation, that the numbers leave little doubt of the applicability of the *law of definite action* in this and all similar cases of decomposition. Chloride of lead was experimented upon in a manner exactly similar, except that *plumbago* was substituted for platinum, as the positive electrode. The mean of three experiments gave 100·85, as the equivalent for lead : the chemical equivalent is 103·5, the deficiency being probably attributable to the solution of part of the gas in the volta-electrometer.

(340) In some experiments several substances were placed in succession, and decomposed simultaneously by the same electric current : thus protochloride of tin, chloride of lead, and water, were acted on at once, the results were in harmony with each other : the tin, lead, chlorine, oxygen and hydrogen evolved, being *definite in quantity*, and electro-chemical equivalents to each other.

(341) By these and numerous other experiments, an irresistible mass of evidence was produced to prove the truth of the important proposition, that the chemical power of a current of Electricity is in direct proportion to the absolute quantity of Electricity which passes (336), which also is not merely true with one substance, as water, but generally with all electrolytic bodies ; and farther, that the results obtained with any *one substance*, do not merely agree amongst themselves, but also with those obtained from *other substances*, the whole combining together into *one series of definite electro-chemical actions*.

(342) The following is a summary of certain points respecting electrolytes, iöns, and electro-chemical equivalents, developed by Dr. Faraday, and given in the seventh series of his Experimental Researches :*

i. A single iön, that is, one not in combination with another, will have no tendency to pass to either of the electrodes, and will be perfectly indifferent to the passing current, unless it be itself a compound of more elementary iöns, and itself subject to decomposition. Upon this fact is founded much of the proof adduced in favour of the new theory of electro-chemical decomposition put forward in a former series of these Researches.

ii. If one iön be combined in right proportions with another strongly opposed to it in its ordinary chemical relations, that is, if an aniön be combined with a cathiön, then both will travel, the one to the anode, and the other to the cathode of the decomposing body.

iii. If therefore an iön pass towards one of the electrodes, another iön must be also passing simultaneously to the other electrode, though from secondary action it may not make its appearance.

iv. A body decomposable directly by the electric current, that is, an electrolyte must consist of two iöns, and must give them up during the process of decomposition.

v. There is but one electrolyte composed of the same two elementary iöns, at least such appears to be the fact, dependent upon a law, *that only single electro-chemical equivalents of elementary iöns can go to the electrodes, and not multiples.*

vi. A body not decomposable when alone, as boracic acid, is not directly decomposable by the electric current when in combination ; it may act as an iön, going wholly to the anode or cathode ; but it does not give up its elements, except occasionally by chemical action.

vii. The nature of the substance of which the electrode is formed, provided it be a conductor, causes no difference in the electro-decomposition, either in kind or degree; but it seriously influences by secondary action, the state in which the iöns finally appear. Advantage may be taken of this principle, in combining and collecting such iöns, as, if evolved in their free state, would be unmanageable.

viii. A substance which, being used as the electrode, can combine altogether with the iön evolved against it, is also an iön, and combines in such cases in the quantity represented by its electro-chemical equivalent. All the experiments agree with this view, and it seems, at present, to result as a necessary consequence. Whether in the secondary action that takes place where the iön acts, not upon the matter of the electrode, but upon that which is round it in the liquid, the same consequence follows, will require more extended investigation to determine.

* 826, et seq.

P

ix. Compound iöns are not necessarily composed of electro-chemical equivalents of simple iöns. For instance—sulphuric, phosphoric, and boracic acids, are iöns, but not electrolytes, that is, not composed of electro-chemical equivalents of simple iöns.

x. Electro-chemical equivalents are always consistent, that is, the same number which represents the equivalent of a substance A, when separating from a substance B, will also represent A when separating from a third substance C. Thus 8 is the electro-chemical equivalent of oxygen, whether separating from hydrogen, or tin, or lead ; and 104 is the electro-chemical equivalent of lead, whether separating from oxygen, chlorine, or iodine.

xi. Electro-chemical equivalents coincide, or are the same with ordinary chemical equivalents.

(343) The theory of definite electro-chemical action led Faraday to the consideration of the absolute quantity of electric force in matter : for although, as he observes, we are utterly ignorant of what an atom really is, we cannot resist forming some idea of a small particle, which represents it to the mind, and there is an immensity of facts which justify us in believing that the atoms of matter are in some way endowed or associated with electrical powers, to which they owe their most striking qualities, and amongst them their mutual chemical affinity. Now, to decompose a single grain of acidulated water, an electric current, powerful enough to retain a platinum wire $\frac{1}{104}$ of an inch in thickness, redhot, must be sent through it for three minutes and three quarters, and this quantity of Electricity is *equal to a very powerful flash of lightning:* yet the electrical power which holds the elements of a grain of water in combination, or which makes a grain of oxygen and hydrogen in the right proportions, unite into water when they are made to combine, equals in all probability the current required for the separation of that grain of water into its elements again ; and this Faraday has shown to be equal to 800,000 charges of a Leyden battery, of fifteen jars, each containing one hundred and eighty-four square inches of glass, coated on both sides : indeed, a beautiful experiment is described by Faraday, in which the chemical action of dilute sulphuric acid on 32·31 parts, or one equivalent of amalgamated zinc, in a simple voltaic circle, was shown to be able to evolve such quantity of Electricity in the form of a current, as passing through water could decompose nine parts, or one equivalent of that substance ; thus rendering the proof complete, (bearing in mind the definite relations of Electricity,) *that the Electricity which decomposes, and that which is evolved by the decomposition of a certain quantity of matter, are alike.*

(344) *Secondary Results :*—In investigating the action of the voltaic current on chemical compounds, it is important to distinguish care-

fully between *primary* and *secondary* results. When a substance yields uncombined and unaltered at the electrodes, those bodies which have been separated by the electric current, then the results may be considered as primary; but when any second reaction takes place, by which the substances, which appear at the electrodes, are *not* those which the *immediate* decomposition of the compounds would produce, then the results are secondary, although the bodies evolved may be elementary.

These secondary results occur in two ways, being sometimes due to the mutual action of the evolved substance on the matter of the electrode, and sometimes to its action on the substances contained in the body itself, under decomposition. Thus, when carbon is made the positive electrode in dilute sulphuric acid, carbonic oxide, and carbonic acid occasionally appear there instead of oxygen : for the latter acting on the matter of the electrode, produces these secondary results. Or if the positive electrode, in a solution of nitrate, or acetate of lead, be platinum, then peroxide of lead appears there equally a secondary result with the former; but now depending upon an action of the oxygen on a substance in the solution. Again, when ammonia is decomposed by platinum electrodes, nitrogen appears at the anode ; but though an *elementary* body, it is a *secondary* result in this case, being derived from the chemical action of the oxygen, electrically evolved there upon the ammonia in the surrounding solution. In the same manner, when aqueous solutions of metallic salts are electrolyzed, the metals evolved at the cathode, though elements, are always secondary results, and not immediate consequences of the decomposing power of the electric current.

(345) In like manner, those interesting compounds which M. Becquerel has obtained by feeble electric currents, are of a secondary character, and in the theory of electrolytic action must be regarded as essentially chemical, and carefully distinguished from those which are directly due to the action of the electric current.

(346) While on the subject of secondary results, I may take occasion to mention some of the beautiful results which Mr. Crosse has obtained, by long continued voltaic actions. This gentleman has for many years been engaged with experiments, on the conversion and production of mineral substances, and has succeeded in forming quartz, arragonite, malachite, and a great variety of most interesting minerals, by the slow and long continued action of the voltaic battery, charged with pure water only. It was not, however, till the meeting of the British Association at Bristol, in 1836, that the results of his experiments were first announced to the public by Mr. Crosse ; and it is probable that his modesty would have prevented him from coming for-

ward then, had he not been urged by the scientific enthusiasm of Dr.
Buckland, who introduced him to the Geological section as "a philosopher who had made great discoveries by the use of a brick with a
hole in it, immersed in a pail of water."

(347) When the author had the pleasure of visiting Broomfield in
the summer of the year 1842, the following experiments were in progress :—

1°. In an oval glass dish, of the capacity of about two quarts, was
placed on the bottom horizontally, a flat piece of clay-slate, a few inches
square, with a platinum wire round its middle, and connected with the
negative pole of a *sulphate of copper battery*, of eight pairs of plates.
Upon this was placed a piece of mountain limestone, of a few ounces
weight, round the middle of which passed a platinum wire, connecting
it with the positive pole.   This stone was prevented from touching the
slate below by three small wedges of deal, placed as supports.   The
glass dish was filled with spring water, and a stream of gas was rising
from each wire.   After two months' action the negative platinum wire
was entirely covered throughout its whole length, under water, with
crystalline carbonate of lime, and the positive wire had produced a
great effect upon that part of the limestone with which it was in contact, having *eaten into* it so as to form a neck round it.   In another
month the effects greatly increased, and carbonate of lime began to form
rapidly over the whole of the slate, as well as over the greater part of
the inner surface of the glass basin.   It so happened that the most
elevated part of the limestone stood perpendicularly above a part of the
negative wire, from which a constant stream of hydrogen gas, in minute
bubbles, was playing against the little wall of limestone above it.
Exactly where this line of bubbles exists, about half an inch in width,
is a line of most beautiful translucent crystals of carbonate of lime upon
the limestone, and occupying the whole surface of that part of it which
is exposed to the current of hydrogen gas.   The actual size of these
crystals when the writer saw them, was about one-eighth of an inch
in length ; but as it was Mr. Crosse's intention not to interfere with
the progress of the experiment for a considerable time, they have doubtless increased much in size and beauty.

2°. In a glass jar of spring water were placed two pieces of clay
slate, and between them a piece of crystallized carbonate of strontia,
connected with the positive pole of a sulphate of copper battery, of six
pairs of plates, the lower slate being in connexion with the negative
pole : both slates were thickly covered with pearly-white carbonate of
strontia in a botryoidal formation : the glass was also partially covered
with stalactitic carbonate of strontia.   This experiment had been going
on eight months.

3°. In a similar jar, carbonate of barytes was being positively electrified : the negative wire and a portion of the slate was covered with a beautiful mamillated formation of carbonate of barytes.

4°. In a similar jar, sulphate of barytes was being positively electrified : the slate was studded with brilliantly transparent crystals of sulphate of barytes.

5°. A piece of solid opaque white quartz, suspended in a glass basin, filled with solution of pure carbonate of potash, was kept positively electrified by a similar battery, a similar piece of quartz being in the same manner kept negative. Some small pieces of quartz were placed between the two : there was a considerable formation of minute crystals.

(348) Among other experiments suggested to the writer, by his visit to Broomfield, the following are worthy of being recorded :—

1°. Two pieces of white marble placed horizontally in a glass basin, are connected by platinum wires with the positive and negative terminations of a battery of twenty pairs, in glass jars, charged with salt and water. The basin is filled with spring water; the experiment has been going on six months; the positive marble is cut nearly a third through its thickness; and the edges of the negative marble, and the negative side of the basin are covered with crystals. There is a strong smell of chlorine proceeding from the water, evidently occasioned by the decomposition of the chlorides contained in it; and there is no doubt, as Mr. Crosse remarks, that this small quantity of chlorine which is evolved at the positive pole, lends very material assistance to the transference.

2°. A similar battery is acting on a piece of mountain limestone attached to the positive pole, a strip of slate being connected with the negative; after the lapse of six months, the action on the limestone has been very great, the platinum wire being completely buried, while the negative slate is beginning to be studded with crystals.

3°. To the positive pole of a battery of twenty pairs, charged with salt and water, is attached a crystal of sulphate of barytes. This rests on the bottom of an inverted gallipot, which is placed in a large glass jar filled with spring water : a piece of common slate, attached to the negative pole, rests also on the pot, at about half an inch distant from the sulphate of barytes. After six months' action, the negative side of the gallipot has become studded with *beautiful transparent crystals*, many of which can be distinctly pronounced to be four-sided, and tabular. These crystals are rapidly increasing both in size and number : the glass jar itself is also beginning to be crystallized upon : there are no crystals yet formed on the slate; but there is an amorphous deposit on the negative wire.

4°. Under similar circumstances (except that instead of a gallipot, a small inverted tumbler is employed), a crystal of sulphate of strontia is kept positively electrified : there is a similar formation of transparent crystals over the negative side of the inverted jar, and also over the side of the large jar in which the whole is contained. The odour of chlorine is very distinct in both these experiments.

5°. The carbonates of barytes and strontia kept positively electrified in vessels of spring water : after four months' action, have transferred beautiful crystals to the negative sides of the basin ; and in the case of carbonate of strontia, the negative slate is very thickly studded : the evolution of chlorine is in both cases very powerful.

6°. A common large garden flower-pot *without* a hole in the bottom, is filled with fragments of common red-brick, and placed on two pieces of brick standing in a common salting pan: the pot is kept filled with spring-water, the droppings being poured back every morning. Three platinum wires from the positive extremity of a salt-and-water battery, of sixty pairs of cylinders, in three series of twenty pairs each, envelop two of the pieces of brick, about three inches beneath the surface ; and three silver wires from the negative extremity, are twisted round two other pieces at the opposite side of the pot. A few days after the commencement of the experiment, a strong odour of chlorine rose from the positive side ; and now, after the lapse of four months, there is a large accumulation of carbonate of lime on the negative side of the pot, not only over the fragments of brick, but all over the *outside* of the pot, and between the bottom of the pot and the crucible under the negative side. With the aid of a lens, a large accumulation of small crystals of carbonate of lime, can be seen between the interstices of the bricks.

This experiment is a modified repetition of one I saw at Broomfield, which was as follows :—

In a large, common, glazed salting-pan, filled with the spring-water of the country, a common red-brick was laid horizontally, each end resting on a half-brick of the same sort. The two ends of the brick were connected respectively with the positive and negative terminations of a sulphate of copper battery, of nine pairs of nine-inch plates : the upper surface of the brick was covered with clear river-sand. At the termination of a quarter of a year, the apparatus was taken apart, and the following observations were made :—

On attempting to lift the whole brick from the two half-bricks that supported it, it was found, that while the positive end was easily removed from the brick below it, the negative end required some little force to separate it from its support ; and when the two were wrenched asunder, it was observed that they had been partially cemented together by a tolerably large surface of beautiful snow-white crystals of *arragonite*, thickly studding that part of the brick in groups, the crystals of each radiating from their respective centres. Here and there were formed in some of the little recesses in the brick, elevated groups of needle arragonite, meeting together in a pyramidal form in the centre ; while in the open spaces between, were some exquisitely-formed crystals of carbonate of lime, in cubes, rhomboids, and more particularly, in short six-sided prisms, with flat terminations, translucent and opaque, sufficiently large to determine their form, without the use of a lens. The positive end of the brick, and that which supported it, were also covered with crystals, much smaller, and apparently of a different nature. On emptying the water from the pan, there was found at its negative end, at the bottom, a very large quantity of snow-white carbonate of lime to the extent of some ounces in weight, in the form of a gritty powder in minute crystals. Three-fourths of the whole interior of the pan were covered with myriads of crystals of carbonate of lime, so firmly adhering to the pan, as not to be separated without the aid of an acid.

(349) Of the action of a weak acid on limestone, when concentrated at the positive pole, Mr. Crosse showed me the following pretty application :—

In a saucer, filled with a concentrated solution of nitrate of potash, a flat, polished piece of white marble was placed ; and upon the middle of the marble, a common sovereign, with its reverse in contact with the marble, and having a stout glass rod supported perpendicularly on the coin, to keep it in its place. Between the rod and the coin was affixed a platinum wire, which was connected with the positive pole of a sulphate of copper battery, of eight pairs of plates ; while round the marble, but not touching it, was a coil of similar wire connected with the negative pole. The nitric acid was soon separated from the potassa, and attacked the marble in contact with the sovereign ; and at the expiration of three days, *the coin was perfectly embedded in the marble.* The experiment was then put an end to ; and the marble being taken out and inverted, the sovereign fell out of its stony receptacle, leaving a tolerably perfect impression on the marble. A very singular result took place, when the water in which the nitrate of potash was dissolved had a very minute quantity of hydrochloric acid added to it, the end of the glass rod, for about an inch in length, being ground. In the course of three days the gold was slightly acted on, and the whole ground surface of the glass rod was evenly and tolerably permanently gilded, which gilding eventually extended above the level of the fluid in the basin.

(350) It was in the course of his experiments on electro-crystallization, that that extraordinary insect about which so much public curiosity has been expended, was first noticed by Mr. Crosse. In

justice to this talented individual, who was most shamefully and absurdly assailed by some ignorant individuals, on account of this insect, and who underwent much calumny and misrepresentation in consequence of experiments, which, " in this nineteenth century, it seems a crime to have made," I shall give a detailed account of that experiment in which the " *acarus* " first made its appearance. A wooden frame was constructed, of about two feet in height, consisting of four legs proceeding from a shelf at the bottom, supporting

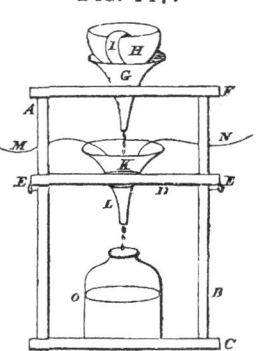

FIG. 147.

another at the top, and containing a third in the middle. *A B*, Fig. 147, represents two of the four uprights, or legs, issuing from the base *C*, supporting the moveable shelf *D*, which shelf is kept in its place by four pins *E*, passing through the four uprights, and may be raised or lowered at pleasure. Each of these shelves was about seven inches square. The upper shelf was pierced with an aperture in which was fixed a funnel of wedgewood ware, *G*, within which rested a quart basin, on a circular piece of mahogany placed within the funnel. When this basin was filled with fluid, a strip of flannel, wetted with the same, was suspended over the edge of the basin, and inside the funnel, which, acting as a syphon, conveyed the fluid out of the basin through the funnel in successive drops. The middle shelf of the frame was likewise pierced with an aperture in which was fixed a smaller funnel of glass *L*, containing a piece of somewhat porous red oxide of iron from Vesuvius, *K*, immediately under the dropping of the upper funnel. This stone was kept constantly electrified by means of two platinum wires *M N*, on either side of it, connected with the poles of a voltaic battery, of nineteen pairs of five-inch zinc and copper plates, excited by water only. The lower shelf supported a wide-mouthed bottle *o*, to receive the drops as they fell from the second funnel. When the basin was nearly emptied, the fluid was poured back again from the bottle below into the basin above, without disturbing the position of the stone. The fluid with which the basin was filled, was made as follows :—A piece of black flint was reduced to powder, having been first exposed to a red heat, and quenched in water. Of this powder, two ounces were taken and fused with six ounces of carbonate of potash : the soluble glass formed, was dissolved in boiling water, diluted, and hydrochloric acid added to supersaturation, the object being to form, if possible, crystals of silica at one of the poles of the battery. On the fourteenth day from the commencement of the experiment, Mr. Crosse observed, through a lens, a few small whitish excrescences or nipples, projecting from about the middle of the electrified stone, and nearly under the dropping of the fluid above. On the eighteenth day these projections enlarged, and seven or eight filaments, each of them longer than the excrescence from which it grew, made their appearance on each of the nipples. On the twenty-second day, these appearances were more elevated and distinct ; and on the twenty-sixth day, each figure assumed the form of a *perfect insect*, standing erect on a few bristles which formed its tail. Till this period, Mr. Crosse had no notion that these appearances were any other than an incipient mineral formation ; but it was not until the twenty-eighth day when he plainly perceived these little creatures move their legs, that he felt any surprise. When an attempt was made to detach them from the stone, they immediately died ; but in a few days they separated themselves, and moved about at pleasure. In the course of a few weeks, about a hundred of them made their appearance on the stone : at first, each of them fixed itself for a considerable time in one spot, appearing to feed by suction ; but when a ray of light from the sun was directed upon it, it seemed disturbed, and removed itself to the shaded part of the stone. . . . . . Mr. Crosse adds, " *I have never ventured an opinion as to the cause of their birth ;* and for a very good reason—I was unable to form one. The most simple solution of the problem, which occurred to me, was, that they arose from *ova* deposited by insects floating in the atmosphere, and that they might possibly be *hatched* by the electric action. I next imagined, as others have done, that they might have originated from the water, and consequently made a close examination of several hundred vessels filled with the same water as that which held in solution the silicate of potassa. In none of these vessels could I perceive the trace of an insect of that description. I likewise closely examined the cre-

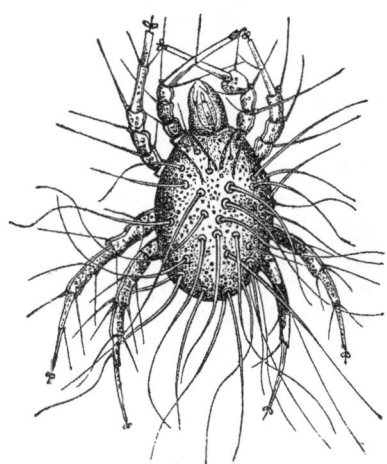

FIG. 148.

vices and most dusty parts of the room with no better success."

(351) In subsequent experiments, this same insect (which it appears is of the genus acarus, but of a species not hitherto observed, and of which a magnified representation is given in Fig. 148), made its appearance in electrified solutions of *nitrate and sulphate of copper, of sulphates of iron, and sulphate of zinc ;* also on the wires attached to the poles of a battery working in a concentrated solution of silicate of

potassa, as shewn in Fig. 149; also in *fluo-silicic*
acid, in the arrangement shown in Fig. 150,
in which a glass bason is shown partly filled with
fluo-silicic acid, to the level 1 : 2, a small porous
pan made of the same materials as a garden-pot
partly filled with the same acid to the level 2,
with an earthen cover 3, placed upon it to keep
out the light, dust, &c. : 4, a platinum wire connected with the positive pole of the battery, with
the other end plunged into the acid in the pan, and
twisted round a piece of common quartz.—
The platinum wire passes under the cover of
the pan : 5, a platinum wire connected with
the negative pole of the same battery, with the
other end dipping into the bason an inch or
two below the fluid, and, as well as the other,
twisted round a piece of quartz.   After eight
months' action, Mr. Crosse perceived two or three insects in their incipient state, appearing on the naked platinum wire at the bottom of
the quartz *in the glass basin at the negative*
pole.   In Fig. 151, a magnified view is given
of the wire, &c., 1 being the platinum wire ;
2, the quartz ; 3, the incipient insects.   At
the suggestion of Mr. Crosse, that indefatigable electrician Mr. Weekes, of Sandwich, in
Kent, repeated some of these experiments, by
passing currents of Electricity through vessels filled with solutions of
silicate of potash under glass receivers inverted over mercury, the
greatest possible care being taken to shut out extraneous matter ; and
in some cases previously filling the receivers with oxygen gas. The result
has been, that after an uninterrupted action of upwards of a year *insects*
have made their appearance, in every respect perfectly resembling those
which occurred in the Broomfield experiments, as the writer can testify,
having had many opportunities of examining each through a powerful
microscope.   In some recent experiments of Mr. Weekes', the acarus
has made its appearance in solution of ferrocyanuret of potassium.

(352) In the Transactions of the Electrical Society, the reader will
find a full and most explicit account of the arrangements of Mr. Weekes,
together with a statement of all the precautions adopted to avoid ambiguity ; and it may here be added, that the writer has experiments in
progress, the silica employed in which has been derived from three
sources : 1, from pure transparent rock crystal : 2, from the decomposition of silicated hydro-fluoric acid : and 3, from window-glass.   It

FIG. 149.

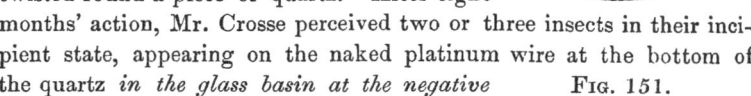

FIG. 150,

FIG. 151.

would be premature to enter into a detail of the arrangements here; but should the results be of a *positive* character, they will not be withheld from the public.

(353) Not among the least interesting of Mr. Crosse's experiments, are those in which he has imitated in a most extraordinary manner, "constant" and "intermittent" springs with the acid of the voltaic battery. The experiments were made in the following manner:—

1°. A common garden-pot full of moistened pipe-clay was placed in a basin full of water : a platinum wire connected with the negative extremity of a sulphate of copper battery, of twelve pairs of plates, each two inches long, by one inch wide, was placed three inches deep into the middle of the clay ; and a second platinum wire connected with the positive pole, was plunged into the water in the basin, to the same depth. Within a fortnight *fissures* took place in the clay in contact with the negative wire ; and in six or eight weeks, these fissures filled with water, which was drawn up two inches above the level of the water in the basin. A small pool of water was formed round the negative wire, which at last overflowed and trickled constantly into the basin below. Here, then, was a *constant electrical spring*.

2°. Here the experiment was varied ; but the apparatus was precisely similar. In this, both wires were plunged three inches deep into the same pot of moist pipe-clay, at the opposite sides, but about three-quarters of an inch from each side. Within a fortnight, fissures took place at the negative, but none at the positive wire. In a month or six weeks more, these fissures filled with water which overflowed, and after a day or two *ebbed*, and then again overflowed, and so on, being apparently acted on by change of weather. Mr. Crosse generally found the spring overflowing when the barometer was *very low ;* and the reverse, when it was *high*. Here then was an electrical *intermittent spring*.

(354) In subsequent experiments, Mr. Crosse found it better to employ porous earthen-pots, open at the top and bottom, filled within an inch of the top with pipe clay kneaded with water to the consistence of putty, and plunged into a basin ; three platinum wires issuing from one stout wire connected with the negative extremity of the battery, being plunged three inches deep into the clay ; and a group of six platinum wires issuing from one connected with the positive pole, being immersed to the same depth in the water. With this arrangement, if the battery is active, the water will rise in one night half an inch above the surface of the clay in the pot, the lip of which, together with the whole rim, to the depth of an inch, is glazed. Under the lip is placed a small shoot of sheet copper, to convey the water, as it falls drop by drop from the lip, to a graduated glass vessel. In one experiment, Mr. Crosse mixed dilute sulphuric acid with the pipe-clay, instead of distilled water. *Not*

one drop of water was raised upwards to the negative wire; but the
water in the basin, which was also acidulated with the same acid, was
changed to a most beautiful *rose-red.* In a letter received from Mr.
Crosse, addressed to the author, in the beginning of the year 1840, he
says, " My two springs—the one *constant,* the other *intermittent*—are
in as good action as ever. The intermittent one overflows generally
when the barometer is somewhat below 29° ; and is generally dry when
the barometer is above that point. A row of open porous pitchers
being filled with pipe-clay, all their lips turned the same way and all
negatively electrified, may yield a succession of drops, which being collected in a shoot, may be used *to turn a small water-wheel,* thus producing perpetual motion ; and *provided the power be found equivalent to
produce such increased effect,* it may be applied in the most important
ways. Also, the fissures formed in the clay at the negative pole, may
be converted into *metallic lodes,* by mineralizing the water in the basin
with metallic and other solutions :—*this I have already done.*"

(355) The author's first repetition of these extraordinary experiments
were not attended with successful results. By employing, however, a
salt-and-water battery, of forty pairs, the observations of Mr. Crosse
have been verified in a most satisfactory manner. After a continued
action of about eight weeks, several ounces of water were drawn to the
negative wire upwards of three inches above the level of the water in
the exterior basin ; and now, after the lapse of thirteen weeks, there is
a continual flow of water over the top and sides of the porous jar,
amounting to upwards of an ounce daily. Common river-water was
employed to fill the basin, and to knead the pipe-clay. The odour of
chlorine from the positive wire is very powerful.

(356) To return to the consideration of the secondary results of
decomposition : it appears that there are two modes by which substances may be decomposed by the voltaic battery : 1st, by the direct
force of the current ; and 2nd, by the action of bodies which that current may evolve. There are also two modes by which new compounds
may be formed, i. e. by combination of the evolving substances whilst
in their *nascent* state directly with the matter of the electrodes ; or else
their combination with those bodies, which being contained in, or associated with, the body suffering decomposition, are necessarily present at
the anode and cathode. When *aqueous* solutions of bodies are used,
secondary results are exceedingly frequent. They are not, however,
confined to aqueous solutions, or cases where water is present. Whenever hydrogen does *not* appear at the *cathode,* in an aqueous solution, it
always indicates that a secondary action has taken place there.

(357) Professor Daniell, of King's College, who has greatly distinguished himself by his researches in electro-chemistry, has published

in the Philosophical Transactions, a series of admirable papers on the electrolysis of secondary compounds, of which, however, I regret that the popular character of this work will deprive me of the pleasure of giving anything more than a very general account. The primary object of these researches was, the determination of the relative proportions of the decompositions both of water and salt, when various saline solutions were subjected to the action of the voltaic current, and their relation to the amount of electrolytic force in action, with a view to increase our knowledge of the constitution of saline bodies in general.

(358) From an elaborate series of experiments on the sulphates of soda, potash, and ammonia, phosphate of soda, nitrate of potash, &c., it appeared " that in the electrolysis of a solution of a neutral salt in water, a current which is just sufficient to separate single equivalents of oxygen and hydrogen from a mixture of sulphuric acid and water, will separate single equivalents of oxygen and hydrogen from the saline solution, while single equivalents of acid and alkali will make their appearance at the same time at the respective electrodes ;" and further experiments showed, that whenever dilute sulphuric acid is used, there is a transfer of acid towards the *zincode* or anode, the quantity scarcely ever exceeding the proportion of one-fourth of an equivalent, as compared with the hydrogen evolved. Mr. Daniell thought possibly this might be owing to the acid being mechanically carried back to the platinode (cathode), as in all cases there is a mechanical convection of the liquid from the anode to the cathode ; and this is greater in proportion to the inferiority of its conducting power. If, however, this deficiency of acid were owing to the mechanical *re-transfer*, mechanical means, such as increasing the number of diaphragms, would stop it : the proportion, however, was, even under these circumstances, still maintained. No difference was observed, whether the oxygen was allowed to escape from a platinum anode, or whether it was absorbed by copper or zinc : the metals, of course, being dissolved in proportions equivalent to the hydrogen developed at the cathode. Solution of potash, baryta, or strontia, similarly treated, exhibited a transfer of about one-fourth of an equivalent towards the cathode.

(359) In order to remove the ambiguity which might thus possibly be conceived to arise from the employment of dilute sulphuric acid, as the measurer of the electrolytic force, the following arrangement was substituted for the Voltameter :—a green glass tube (into the bottom of which, as a cathode, was welded a weighed platinum wire) was filled with chloride of lead, maintained in a state of fusion by a spirit-lamp : the corresponding anode was made of plumbago. At the termination of the experiment, the tube was broken, the wire and adhering button of lead weighed ; and the result showed that " the same current which

is just sufficient to resolve an equivalent of chloride of lead, which is a
simple electrolyte unaffected by any associated composition, into its
equivalent *iöns*, produces the apparent phenomena of the re-solution of
water into its elements, and at the same time of an equivalent of sulphate of soda, into its proximate principles."

(360) Aqueous solutions of the chlorides were next tried, as the simple
constitution of this class of salts promised to throw light upon the nature
of the electrolysis of secondary compounds. A weighed plate of pure tin
was made the anode of a double cell of peculiar construction, which was
charged with a strong solution of chloride of sodium, and a tube of fused
chloride of lead, as before, included in the circuit. Not a bubble of gas
appeared on the tin electrode, and no smell of chlorine was perceptible;
but hydrogen in equivalent proportion to the quantity of tin dissolved,
was given off at the cathode ; and the cell contained an equivalent proportion of free soda. One equivalent of lead was reduced in the voltameter tube. Muriate of ammonia treated in the same way, gave precisely similar results, proving it to be " an electrolyte" whose simple
anion was chlorine and compound cathion nitrogen, with four equivalents of hydrogen. Its electrolytic symbol, therefore, instead of being
$(Cl + H) + (N + 3 H)$ is $Cl + (N + 4 H)$, confirming in a striking
manner the hypothesis of Berzelius, of the base $(N + 4 H)$ called ammonium.

(361) In discussing the results of all these experiments, we must
bear in mind the fundamental principle, " that the force which we have
measured by its definite action at any one point of a circuit, cannot
perform more than an equivalent proportion at any other point of the
same circuit." " The sum of the forces which held together any number of *iöns* in a compound electrolyte, could, moreover, only have been
equal to the force which held together the elements of a single electrolyte, electrolyzed at the same moment in one circuit.'

(362) In the electrolysis of the solution of sulphate of soda, and many
of the other salts, water seemed to be electrolysed ; at the same time acid
and alkali appeared in equivalent proportion with the oxygen and hydrogen at the respective electrodes. " We must conclude," says Mr. Daniell,
from the above-mentioned principle, " that the only electrolyte which
yielded was the sulphate of soda, the iöns of which were not, however,
the acid and alkali of the salt, but an anion, composed of an equivalent
of sulphur and *four* equivalents of oxygen and the metallic cathion
sodium. From the former, sulphuric acid was formed at the anode, by
the secondary action and evolution of one equivalent of oxygen ; and
from the latter, soda at the cathode, by the secondary action of the
metal and the evolution of an equivalent of hydrogen."

(363) To avoid circumlocution (but only when speaking of electro-

lytic decomposition), Mr. Daniell proposes to adopt the word iön, introduced by Dr. Faraday, as a general termination, to denote the compounds which in the electrolysis of a salt, pass to the anode; and that
they should be specifically distinguished by prefixing the name of the
acid slightly modified, as is shown in the following table:—

Ordinary Chemical Formula.   Electrolytic Formula.

Sulphate of copper $(S + 3 O) + (Cu + O) = (S + 4 O) +$ Cu oxysulphion of copper.

Sulphate of soda $(S + 3 O) + (Na + O) = (S + 4 O) +$ Na oxysulphion of sodium.

Nitrate of potassa $(N + 5 O) + (Ka + O) = (N + 6 O) +$ Ka oxynitrion of potassium.

Phosphate of soda $(P + 3\frac{1}{2}O) + (Na + O) = (P + 4\frac{1}{2}O) +$ Na oxyphosph. of sodium.

(364) The following experiments, strongly favouring the above view,
were made by Professor Daniell:—

A small glass bell with an aperture at top, had its mouth closed, by
tying a piece of membrane over it. It was half filled with a dilute
solution of caustic potassa, and suspended in a glass vessel containing
a strong neutral solution of sulphate of copper, below the surface of
which it just dipped. A platinum electrode connected with the last
zinc rod of a large constant battery of twenty cells, was placed in the
solution of potassa; and another connected with the copper of the first
cell was placed in the sulphate of copper immediately under the diaphragm which separated the two solutions. The circuit conducted
very readily, and the action was very energetic. Hydrogen was given
off at the cathode in the solution of potassa, and oxygen at the anode in
the sulphate of copper. A small quantity of gas was also seen to rise
from the surface of the diaphragm. In about ten minutes, the lower
surface of the membrane was found beautifully coated with metallic
copper, interspersed with oxide of copper, of a black colour, and hydrated oxide of copper, of a light blue. The explanation of these phenomena is this:—In the experimental cell we have two electrolytes separated by a membrane through both of which the current must pass to
complete the circuit. The sulphate of copper is resolved into its compound anion, sulphuric acid + oxygen (oxysulphion), and its simple
cathion, copper. The oxygen of the former escapes at the zincode,
but the copper in its passage to the platinode, is stopped at the surface
of the second electrolyte, which for the present we may regard as water,
improved in its conducting power by potassa. The metal here finds
nothing by combining with which it can complete its course; but,
being forced to stop, yields up its charge to the hydrogen of the second
electrolyte, which passes on to the cathode, and is evolved. The corresponding oxygen stops also at the diaphragm, giving up its charge
to the anion of the sulphate of copper. The copper and oxygen thus
meeting at the intermediate point, partly enter into combination, and
form the black oxide; but from the rapidity of the action, there is not

time for the whole to combine, and a portion of the copper remains in a metallic state, and a portion of the gaseous oxygen escapes. The precipitation of blue hydrated oxide doubtless arose from the mixing of a small portion of the two solutions. Nitrate of silver, nitrate of lead, proto-sulphate of iron, sulphate of palladium, and proto-nitrate of mercury, were similarly treated, and afforded analagous results somewhat modified by the nature of the metallic base. Sulphate of magnesia was subjected to the same process in hopes of finding magnesium, but magnesia alone was deposited.

(365) The theory of ammonium, as proposed by Berzelius, and the hypothesis of Davy developing the general analogy of all salts, whether derived from oxyacids or hydracids, may by this evidence, especially when taken in conjunction with the recent researches on the constitution of organic bodies, be considered as almost experimentally demonstrated.

(366) The bisalts yield results which at first sight do not accord with the preceding deductions : a strong solution, for example, of pure crystallized bisulphate of potassa was made, and its neutralizing power carefully ascertained by the alkalimeter. Evaporation and ignition with carbonate of ammonia, gave the quantity of neutral sulphate yielded by a certain measure of the solution. An equal measure was then placed in each arm of the double diaphragm cell, and the current passed through till 70·8 cubic inches of mixed gases were collected : half the solutions from the anode and cathode were then separately neutralized, and half evaporated and ignited in the vapour of carbonate of ammonia. It was then found that the anode had gained eighteen grains, and the cathode lost nineteen of free acid : of potassa, the anode had lost 9·9 grains, and the cathode gained an equal quantity. Thus, though the solution conducted very well, not more than one-fifth of an equivalent of the potassa was transferred to the cathode, as compared with the hydrogen evolved, while half an equivalent of acid was transferred to the anode, while a whole equivalent of oxygen was evolved. On this experiment Mr. Daniell remarks :—

" I think we cannot hesitate to admit that in this case the current divided itself between two electrolytes, that a part was conducted by the neutral sulphate of potassa, and a larger part by the sulphuric acid and water. It is a well known fact, that the voltaic current will divide itself between two or more metallic conductors in inverse proportion to the resistance which each may offer to its course : and that it does not in such cases choose *alone* the path of *least* resistance. Analogy would lead me to expect a similar division of a current between two electrolytes ; but I am not aware whether such a division has ever before been pointed out."

(367) These considerations enable us to explain the apparent anomalies in the electrolysis of diluted sulphuric acid and alkaline solutions alluded to (358). The results are explained by supposing that the solution is a mixture of two electrolytes : with sulphuric acid, they are H + (S + 4 O) oxysulphion of hydrogen ; (H + O) water. The current so divides itself, that three equivalents of water are decomposed, and one equivalent of oxysulphion of hydrogen. Analagous changes occur with the alkaline solutions, the alkaline metal passing as usual to the cathode.*

(368) According to Professor Daniell's view of Faraday's beautiful experiments with sulphate of magnesia (325), the first electrolyte was resolved into a compound anion, sulphuric acid + oxygen, which passed to the anode, and the simple cathion, magnesium, which on its passage to the cathode was stopped at the surface of the water from not finding any iön, by temporarily combining with which it could be further transferred according to the laws of electrolysis. At this point, therefore, it gave up its charge to the hydrogen of the water, which passed in the usual manner to the cathode, and the circuit was completed by the decomposition of this second electrolyte. The corresponding oxygen, of course, met the magnesium at the point where it was arrested in its progress, and, combining with it, magnesia was precipitated. This combination of the oxygen and metal is looked upon by Professor Daniell, as a *secondary* result due to the local affinity of the elements thus brought into juxta-position, and in no way connected with the primary phenomena of the current, which would have completed its course, whether this combination had taken place or not; i. e. whether magnesium and oxygen had been separately evolved, or whether magnesia had been formed by the combination of the two. It also seemed probable that, although in the very slow action of this experiment this combination invariably took place, by varying the experiment so as to evolve metals possessing different degrees of affinity for oxygen, and particularly by shortening the time in which the evolution might take place, instances might be found of some portion of the metal escaping this combination, which would thus afford the most incontrovertible proof of the point to be established. Professor Daniell was thus led to the experiment above detailed.

(369) *Electro-metallurgy.*—On this subject it will be necessary to say but little here, the works of Messrs. Smee and Walker having supplied the experimentalist with the fullest details; and every individual who is desirous of becoming practically acquainted with this new and

* The scientific reader will do well to study carefully the original papers of Professor Daniell, in the Philosophical Transactions : for the above abstract, the writer is principally indebted to the pages of the Philosophical Magazine.

captivating branch of electrical science, will not fail to provide himself
with one or both of these publications. Nevertheless, as the whole art
of Electro-metallurgy is based on the chemical powers of the voltaic
current, it would be impossible to omit a short outline in this division
of our subject.

(370) In our historical account of the sulphate of copper battery of
Daniell (262, et seq.,) it was stated, that on completing the circuit the
electrical current passes freely through the metallic solution ; that no hydrogen makes its appearance on the conducting plate, but that a beautiful pink coating of pure copper is deposited on it, and thus perpetually
renews its surface.   In the discovery of this battery, then, we find the
origin of electro-metallurgy ; for it appears that in his earlier experiments it was noticed by Mr. Daniell, that on removing a piece of the
reduced copper from a platinum electrode, scratches on the latter were
copied with accuracy on the copper, and M. De la Rue later, in a
paper in the Phil. Mag.,* detailing some experiments with a voltaic
battery of ordinary construction, charged with sulphate of copper, made
the observation that " the copper-plate is covered with a coating of
metallic copper, which is continually being deposited ;" and he proceeds
to remark, " so perfect is the sheet of copper thus formed, that on being
stripped off it has the polish and even a counterpart of every scratch of
the plate on which it is deposited."   On reading this passage at the
present time, when the art of Electro-metallurgy is so extensively practised, we can hardly resist a feeling of surprise that the application of
the *facts* discovered by Daniell and De la Rue did not occur to either
of these gentlemen.   They were, however, too intent on the battery
itself, to attend to any collateral circumstances ; and it was left for
Jacobi, in Russia, and Spencer, in this country, to do so.   The process
of the former distinguished philosopher was called " Galvano-plastic :"
that of Mr. Spencer, " Electrography."   And though it is quite certain
that the discovery was made by each, independent of the other, the
priority must be given to Jacobi, who states in the preface of his
" Galvanoplastik,"† that it was in the month of February, 1837,
while prosecuting his galvanic investigations, that he discovered a
striking phenomenon which presented itself in his experiments, and furnished him with perfectly novel views ; and Mr. Spencer, in his pamphlet,‡ informs us, that his first results were obtained in 1838.

(371) The description of an original experiment is generally interesting : it is always so when connected with a subject of much practical
importance.   I shall therefore insert Mr. Spencer's account of his first
successful experiment in electrography :||—" I selected a very prominent

* Vol. ix. p. 484.   † Translated from the German edition, by Wm. Sturgeon.
‡ Griffin's Scientific Miscellany, No. iv. p. 33. || See his Pamph. p. 33.

copper medal. It was placed in a voltaic circuit, and a surface of copper deposited on one of its sides to about the thickness of a shilling. I then proceeded to get the deposition off. In this I experienced some difficulty, but ultimately succeeded. On examination with a lens, every line was as perfect as the coin was from which it was taken. I was then induced to use the same piece again, and let it remain a much longer time in action, that I might have a thicker and more substantial mould, in order to test fairly the strength of the metal. It was accordingly put again in action, and let remain until it had acquired a much thicker coat of the metallic deposition; but on attempting to remove it from the medal, I found I was unable. It had apparently completely adhered to it. I had often practised, with some degree of success, a method of preventing the oxidation of polished steel, by slightly heating it until it would melt fine bees' wax : it was then wiped apparently completely off; but the pores or surfaces of the metal became impregnated with the wax. I thought of this method, and applied it to a copper coin. I first heated it, applied wax, and then wiped it off so completely that the sharpness of the coin was not at all interfered with. I proceeded as before, and deposited a thick coating of copper on its surface. Being desirous to take it off, I applied the heat of a spirit-lamp to the back, when a sharp crackling noise took place, and I had the satisfaction of perceiving that the coin was completely loosened. In short, I had a most complete and perfect copper mould of one side of a halfpenny."

(372) The first kind of apparatus employed by Mr. Spencer was simply a common tumbler to hold the copper solution, and a gas-glass, having one end closed with brown paper, or plaster of Paris, to contain the saline solution; the coin to be copied, and a piece of zinc of equal size, were attached to the extremities of a piece of copper wire. The gas-glass being fixed in the axis of the tumbler, the zinc was placed in it, and the copper wire bent in such a manner as to bring the coin immediately under it in the copper solution. The battery process is subsequently described by Mr. Spencer; but he gives no method of depositing copper on any surface but a metallic one. In Jacobi's pamphlet, however, which was published at St. Petersburg, in March, 1840, the use of plumbago, for giving a conducting surface to non-metallic substances, and so enabling them to answer all the purposes of metallic originals, is distinctly alluded to. It appears, however, that Mr. Murray has the merit of having introduced this discovery into this country; and the Society of Arts have recorded their sense of its value, by presenting this gentleman with a silver medal. The employment of the battery was first suggested by Mr. Mason, who, by connecting a piece of copper with the anode in a second cell, the object to be copied

being connected with the cathode, showed that the quality of the copper was much better than when reduced in the single cell apparatus,
besides the great advantage that was gained by the unlimited number
of operations that may be going on at the same time.

(373) Fig. 152 represents the single cell apparatus.   FIG. 152.
Z is a rod of amalgamated zinc ; m, the mould ; w,

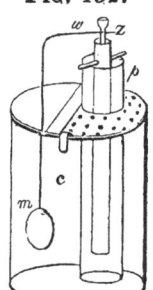

the wire joining them ; c, the copper solution ; p, a
tube of porous earthenware, containing a solution of
acid and water.   To put this in action, pour in the
copper solution, fill the tube with the acid water, and
place it as shown in the figure.   *Last* of all put in the
bent wire, having the zinc at one end, and the mould
at the other.   It is essential that the copper solution be
kept saturated, or nearly so ; with which view the perforated shelf must be kept well furnished with crystals
of sulphate of copper.   The mould must not be too small in proportion
to the size of the zinc, and the concentrated part of the solution must
not be allowed to remain at the bottom, or the copy will be irregular
in thickness.

Fig. 153 represents the battery apparatus.   A is a cell of Daniell's
battery (or Smee's may be used) ; B is the decomposition cell, filled

FIG. 153.

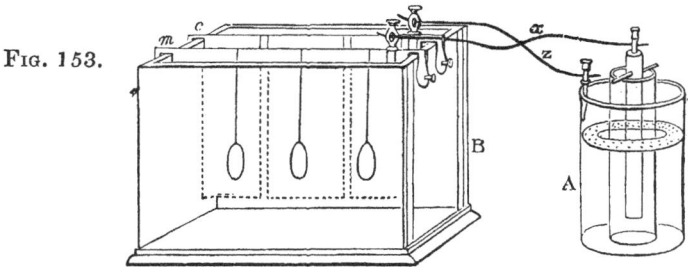

with the dilute acid solution of sulphate of copper ; c, the sheet of
copper, to furnish a supply ; m, the mould to receive the deposit.   To
charge this, pour in the several solutions, and connect the wire z with
the copper sheet and the copper of the battery.   *Last* of all, attach the
wire x to the zinc and the moulds.   The charging liquid is a mixture
of one part *sulphuric acid*, two parts *saturated* solution of sulphate of
copper, and eight parts water.   When the circuit is complete, the copper from the solution is transferred to the mould, and the copper sheet
is dissolved, being converted, with the sulphuric acid, into sulphate of
copper ; thus keeping up the strength of the solution.   Rather a longer
time is required by this method than with the single cell ; but two
days will produce a medal of very good substance, firm and *pliable* :

the time required, however, for these experiments, depends much on the *temperature*. If the solutions are kept boiling, a medal may be made in a few hours : in severe weather, the action of the battery almost ceases, and it is necessary to carry on the operations before a good fire.

Fig. 154 represents a form of electrotype trough, to be used without acid or mercury. It consists of a wooden box, well varnished in the interior, and divided into two unequal cells by a partition of porous wood. The larger cell is filled with a saturated solution of sulphate of copper; the smaller, with a half saturated solution of muriate of ammonia. In the former is a shelf, containing a supply of crystals. The zinc plates employed for this, are *pure*. This is a matter of some importance, because there then is no need of amalgamation to destroy local action ; and the instant the circuit is interrupted, all action ceases. Ordinary zinc

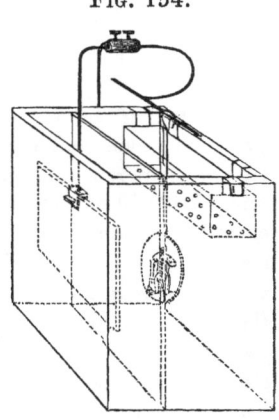

FIG. 154.

may be used ; but less power is obtained. This form of apparatus may be used as a single cell, or with a decomposition trough. It must not be expected that the action will be equally quick with that resulting from the addition of acid ; but it will be sure—perhaps more so than in the other instance; for it must be borne in mind, that in " electrotype manipulation the failure in nine cases out of ten results from the power of the battery being too strong, and not from its being too weak."— *Walker.*

(374) For taking off copper plates, the apparatus shown in Fig. 155, may be employed ; and by adopting the arrangement shown in Fig. 156, six electrotypes may be taken off at the same time. A is the battery; B, the trough ; *z*, the

Fig. 155.

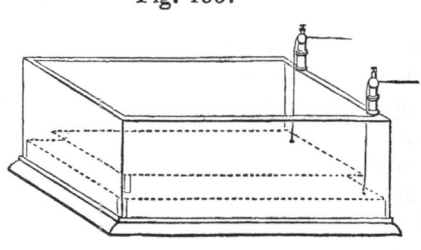

wire connecting the copper plate C with the copper cell of the battery ; *x*, the wire connecting the mould with the zinc of the battery ; *a, a, a, a, a,* five bent wires, each having a mould at one end, and a piece of copper at the other. The following directions are given by Mr. Walker, for charging this trough. " Connect the copper plate C, with the battery : place a *wire* with its extreme

FIG. 156.

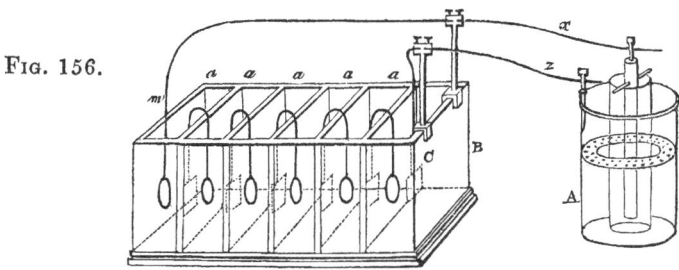

ends dipping in the extreme cells of the trough; then, having previously connected the zinc and mould with the wire x, place the zinc in the porous cell, and the mould in its place at m; in about two minutes it will be covered with copper. After this there is no fear of chemical action; then remove the end of the copper wire from the cell containing m, and place it in the next cell; complete the circuit with the bent wire a, having a mould at one end, and a sheet of copper at the other. After waiting two minutes for a deposit of copper, remove the end of the wire one cell further forward; and so continue, till the six moulds are placed in." The advantage, in point of economy, from using this form of decomposition trough, is at once apparent, when it is remembered that for every *ounce* of copper released from the solution in the generating cell, an ounce will be deposited on each mould, and about *an ounce* of zinc will be consumed in effecting this. Whether, there-fore, one, or six, or even twenty moulds be placed in series, the *same quantity* of zinc will be required; and hence, an ounce of zinc may be made to furnish Electricity enough to produce, according to the will of the experimenter, one, or six, or more medals, each weighing an ounce. This follows immediately from the laws of definite electro-chemical action, developed by Faraday (335 et seq).

(375) To enter into a detail of the numerous precautions which re-quire to be attended to, in order to ensure certain and perfect results— to describe the various methods of taking good impressions from coins, medallions, and plaster-casts, the preparation of moulds, &c.;—and to describe the methods of electro-plating, electro-gilding, and electro-etching, would lead me far beyond the limits of my present design. I, therefore, refer my readers with confidence to the very pleasingly written little work on "Electrotype Manipulation," by Mr. C. V. Walker; a work the sterling value of which may, I presume, be inferred, from its having already reached a twelfth edition. Here will be found every particular connected with *mould making*, soldering, and gilding, mi-nutely detailed; and instructions in the art of working in metals, through the agency of the voltaic battery, plainly given. Mr. Smee's "Elements of Electro-metallurgy," in which the subject is pursued

more at length, may also be studied with much advantage ; and the student will derive much information from the pamphlets of Jacobi and Spencer.

(376) It may perhaps be almost unnecessary to remark that these metallic deposits are all *secondary* results of the voltaic battery. Water is the compound electrolyzed, and hydrogen the element disengaged at the cathode by the *direct* action of the current ; but this element re-acts on the metallic solution, combining with its oxygen, and setting free the metal. Oxygen is also disengaged at the anode ; but it is not set free, because it immediately meets with an element with which it can combine. So also, in the processes for gilding and silvering, where *cyanides* of gold and silver are employed, the anodes being pure gold and pure silver. Cyanagon is the substance disengaged at the anode ; and though itself a compound body, it is capable of combining with these noble metals, when presented to them in its nascent state ; and thus a quantity of metallic salt is retained in the solution, equivalent to that which is decomposed.

(377) When acetate of lead is electrolyzed, under peculiar circumstances, it gives rise to secondary results of a particularly beautiful character : peroxide of lead is deposited at the anode ; and by carefully regulating the thickness of this compound, a series of most magnificent colours may be produced on a plate of highly polished steel. The process recommended by Mr. Gassiot, to form these *metallo-chromes*, is this :—Place the polished steel plate in a glass basin, containing a clear solution of acetate of lead, and over it a piece of card with some regular device cut out, as shewn in Fig. 157. A small rim of wood should be placed over

FIG. 157.

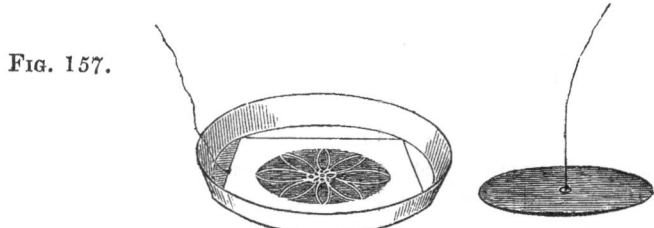

the card, and on that a circular copper disc. On contact being made from 5° to 20°, with two or three cells of a small constant battery, the steel plate being connected with the copper or silver, and the copper disc with the zinc, the deposit will be effected, and a series of exquisite colours will appear on the steel plate. These colours are not films of oxygen and acid, as some have imagined, but lead, in a high state of oxidation, thrown down on the surface of the steel, and the varied tints are occasioned by the varying thickness of the precipitated film, the light being reflected through them from the polished metallic surface below.* By reflected light, every prismatic colour is seen; and by transmitting light, a series of prismatic colours complimentary to the first series will appear, occupying the place of the former series. The colours are seen in the greatest perfection, by placing the plate before a window, and inclining a sheet of white paper at an angle of 45° over it.

(378) Allusion was made in a former Lecture (161) to a peculiar odorous principle developed during the passage of a torrent of dense sparks of Electricity between two metallic plates enclosed in a cylinder of glass.   Every person who has been in the habit of experimenting with a large electrical machine, must have remarked this peculiar odour, particularly during the escape of positive Electricity from a point : the same is perceptible when lightning has struck an object, and when a discharge from a powerful battery is sent through several sheets of paper it is exceedingly strong.   It was, however, M. Schoenbein who first noticed that it accompanies the electrolyzation of water : he has investigated the circumstances attending the phenomena, and has obtained results which will, he thinks, afford a clue to the discovery of their cause.   The odour which accompanies the electrolyzation of water, he observes, is only disengaged at the positive electrode.   He also finds that the odoriferous principle can be preserved in well closed glass vessels for any length of time.   The only metals which he found to yield this odour, were gold and platinum ; but dilute sulphuric, phosphoric, and nitric acids, and aqueous solutions of several of the salts, also disengage it.   Raising the temperature of the fluid to the boiling point prevents the odour from arising ; and the addition of comparatively small quantities of powdered charcoal, iron, zinc, tin, lead, antimony, bismuth, or arsenic, or a few drops of mercury, to the odorous principle contained in a bottle, immediately destroys the smell ; and the same happens when platinum or gold, heated red hot, is introduced into the vessel containing that volatile substance.

(379) A series of experiments relating to this subject were made by Mr. Gann, and communicated to the Electrical Society.†   The electrodes employed were copper, iron, silver, platinum, zinc, charcoal, and plumbago ; the first four afforded the odour abundantly, either when the electrodes were both of the same metal or of dissimilar metals.   If, however, both the electrodes were of zinc, it was not perceptible ; but when copper, iron, silver, or platinum were *positive* in a jar of atmospheric air, and zinc *negative*, it was developed ; not so, however, when such arrangement was reversed ; neither was it obtained from charcoal or plumbago ; but in lieu of it, a very disagreeable odour was emitted, similar to that given off by charcoal points under water.   When dilute

* See a paper by Mr. Warrington, on this subject : Phil. Mag. vol. xvi. p. 52.
† Proceedings of the Electrical Society, p. 160.

sulphuric acid was electrolyzed with platinum electrodes, in an open voltameter, the odour was very powerful; but it was not produced when copper, zinc, or iron electrodes were employed. With muriate of soda the odour was not perceptible, until the gas obtained from platinum electrodes at the positive pole was placed over ammonia and water, to absorb the chlorine; then the tube became hot, and the residual gas emitted the peculiar odour.

Mr. Gann next procured gas from voltaic action with platinum electrodes placed in a solution of sulphuric acid water; and having filled several jars from the positive and negative electrodes, they were preserved over water. Two measures of the hydrogen, and one of the oxygen, were detonated in a eudiometer over mercury. There was no residual gas, nor any odour remaining; and when a series of explosions from a Leyden phial were passed through a portion of the oxygen over mercury, the peculiar odour was entirely destroyed.

(380) From his experiments, Mr. Gann is induced to think that this peculiar odour may be emitted from all metallic bodies, if they are treated in such a manner as to prevent oxidation or combination with other bodies; and that it is not confined to the *noble* metals, as supposed by M. Schoenbein.

LECTURE VI.

VOLTAIC ELECTRICITY *(continued)*.

Physiological effects of Galvanism—Dr. Ure's experiments on the body of a recently executed criminal—Dr. Wilson Philip's experiments—Animal Electricity—The Raia Torpedo—The Gymnotus—Faraday's experiments with—Schoenbein's ex·periments—Electricity of effluent steam—Mr. Armstrong's and Mr. Pattinson's experiments—Theory of the Voltaic pile.

(381) *Physiological Effects of Galvanism.*—The action of Galvanic Electricity on the living animal is the same as that of the common electric current, account being taken of the *intensity* of the one, and the *duration* of the other. When any part of the body is caused to form part of the circuit of the voltaic pile, a distinct shock resembling that of a large electrical battery weakly charged is experienced every time the connection with the extremities is made ; and besides this, if the pile be a large one, a continued aching pain is frequently felt as long as the current is passing through the body, and if the slightest excoriation or cut happen to be in the path of the current, the pain is very severe. The intensity of galvanic electricity is so low that it requires good conductors for its transmission ; unless therefore the skin be previously moistened, it will not force its way through the badly conducting cuticle, or outer skin. The most effectual mode of receiving the whole force of the battery, is to wet both hands with water, or with a solution of common salt, and to grasp a silver spoon in each, and then to make the connection between the poles of the battery. Another method is to plunge both hands into two separate vessels of water into which the extremities of the wires from the battery have been immersed. Volta has remarked, that the pain is of a sharper kind on those sensible parts of the body included in the circuit, which are on the negative side of the pile ; this is particularly remarkable with the water-battery, and the same has been noticed with regard to the pungency of the common electrical spark.

(382) It does not require a voltaic pile to exhibit the effects of Galvanic Electricity on the animal, whether living or dead. The simple

application of a piece of zinc and one cf silver to the tongue and lips, frequently gives rise, at the moment of the contact of the metals, to the perception of a luminous flash ; but the most certain way of obtaining this result, is to press a piece of silver as high as possible between the upper lip and the gums, and to insert a silver probe into the nostrils, while at the same time a piece of zinc is laid upon the tongue, and then to bring the two metals into contact. Another mode is to introduce some tin foil within the eyelid, so as to cover part of the globe of the eye, and place a silver spoon in the mouth, whic'ı must then be made to communicate with the tin foil, by a wire of sufficient length ; or, conversely, the foil may be placed on the tongue, and the rounded end of a silver probe applied to the inner corner of the eye, and the contact established as before. This phenomenon is evidently produced by an impression communicated to the retina or optic nerve, and is analogous to the effect of a blow on the eye, which is well known to occasion the sensation of a bright luminous coruscation, totally independent of the actual presence of light. In the like manner the flash from galvanism is felt, whether the eyes are open or closed, or whether the experiment is made in day light or in the dark. If the pupil of the eye is watched by another person when this effect is produced, it will be seen to contract at the moment the metals are brought into contact ; a flash is also perceived the moment the metals are separated from each other. When different metals are applied to different parts of the tongue, and made to touch each other, a peculiar taste is perceived : in order that this experiment should succeed, the torgue must be moist, the effect is materially diminished if it be previously wiped, and cannot be produced at all if the surface be quite dry. The quality of the metal laid upon the tongue influences the kind of taste which is communicated ; the more oxidable metal giving rise to an acid, and the less oxidable metal to a distinct alkaline taste. Similar differences have been observed by Berzelius, with regard to the sensations excited in the tongue, by common Electricity, directed in a stream upon that organ from a pointed conductor ; the taste of positive Electricity being acid, and that of negative Electricity caustic and alkaline.

(383) If the hind legs of a frog be placed upon a glass plate, and the crural nerve dissected out of one, made to communicate with the other, it will be found on making occasional contacts with the remaining crural nerve, that the limbs of the animal will be agitated at each contact. Aldini, the nephew of the original discoverer of galvanism, produced very powerful muscular contraction, by bringing a part of a warm-blooded and of a cold-blooded animal into contact with each other, as the nerve and muscle of a frog, with the bloody flesh of

the neck of a newly decapitated ox, and also by bringing the nerve of one animal into contact with the muscle of another.

(384) If a crown piece be laid upon a piece of zinc of larger size, and a living leech placed upon the silver coin, it suffers no inconvenience as long as it remains in contact with the silver only, but the moment it has stretched itself out and touched the zinc, it suddenly recoils, as if from a violent shock. An earth worm exhibits the same kind of sensitiveness. The convulsive movements excited in the muscles of animals after death, by a powerful galvanic battery, are extremely striking, if the power is applied before the muscles have lost their contractility. Thus if two wires connected with the poles of a battery of a hundred pair of plates, are inserted into the ears of an ox or sheep, when the head is removed from the body of the animal recently killed, very strong actions will be excited in the muscles of the face every time the circuit is completed. The convulsions are so general, as often to induce a belief that the animal has been restored to life, and that he is enduring the most cruel sufferings. The eyes are seen to open and shut spontaneously ; they roll in the sockets as if again endued with vision ; the pupils are at the same time widely dilated ; the nostrils vibrate as in the act of smelling ; and the movements of mastication are imitated by the jaws. The struggles of the limbs of a horse galvanised, soon after it has been killed, are so powerful, as to require the strength of several persons to restrain them.

(385) The following account of some experiments made by Dr. Ure on the body of a recently executed criminal, will serve to convey a tolerably accurate idea of the wonderful physiological effects of this agent, and will be impressive from their conveying the most terrific expressions of human passion and human agony.

" The subject of these experiments was a middle-sized, athletic, and extremely muscular man, about thirty years of age. He was suspended from the gallows nearly an hour, and made no convulsive struggle after he dropped. While a thief who was executed along with him was violently agitated for a long time. He was brought into the anatomical theatre of our university, about ten minutes after he was cut down. His face had a perfectly natural aspect, being neither livid, nor tumefied, and there was no dislocation of the neck.

" Dr. Jeffray, the distinguished professor of anatomy, having on the preceding day requested me to perform the galvanic experiments, I sent to his theatre the next morning with this view my minor voltaic battery, consisting of two hundred and seventy pairs of four-inch plates, with wires of communication, and pointed metallic rods with insulating handles, for the more commodious application of the electric power.

About five minute before the police-officers arrived with the body, the battery was charged with dilute nitro-sulphuric acid, which speedily brought it into a state of intense action. The dissections were skilfully executed by Mr. Marshall, under the superintendence of the professor.

"*Experiment* 1.—A large incision was made in the nape of the neck just below the occiput; the posterior half of the atlas vertebra was then removed by bone forceps; when the spinal marrow was brought into view, a profuse flow of fluid blood gushed from the wound, inundating the floor. A considerable incision was at the same time made in the left hip, through the great gluteal muscle, so as to bring the sciatic nerve into sight, and a small cut was made in the heel, from neither of these did any blood flow. The pointed rod connected with one end of the battery, was now placed in contact with the spinal marrow, while the other rod was applied to the sciatic nerve; every muscle of the body was immediately agitated with convulsive movements, resembling a violent shuddering from cold. The left side was most powerfully convulsed. On moving the second rod from the hip to the heel, the knee being previously bent, the leg was thrown out with such violence as nearly to overturn one of the assistants, who in vain attempted to prevent its extension.

"*Experiment* 2.—The left phrenic nerve was now laid bare at the outer edge of the *sternothyroideus* muscle, from three to four inches above the clavicle; the cutaneous incision having been made by the side of the sterno-cleido-mastoideus. Since this nerve is distributed to the diaphragm, and since it communicates with the heart through the eighth pair, it was expected by transmitting the galvanic current along it, that the respiratory process would be renewed. Accordingly a small incision having been made under the cartilage of the seventh rib, the point of the one insulating rod was brought into contact with the great head of the diaphragm, while the other point was applied to the phrenic nerve in the neck. This muscle, the main agent of respiration, was immediately contracted, but with less force than was expected. Satisfied from ample experience on the living body, that more powerful effects can be produced by galvanic excitation, by leaving the extreme communicating rod in close contact with the parts to be operated on, while the electric chain or circuit is completed by running the end of the wires along the top of the plates in the last trough of either pole, the other wire being steadily immersed in the last cell of the opposite pole, I had immediate recourse to this method. The success of it was truly wonderful; full, nay, laborious breathing, instantly commenced, the chest heaved and fell, the belly protruded, and again collapsed with the relaxing and retiring diaphragm. This process was continued without interruption, as long as I continued the electric discharges. In the judgment of many scientific friends who witnessed the scene, this respiratory experiment was perhaps the most striking ever made with philosophical apparatus.

" Let it also be remembered, that for full half an hour before this period, the body had been well nigh drained of its blood, and the spinal marrow severely lacerated. No pulsation could be perceived meanwhile at the heart or wrist ; but it may be supposed that but for the evacuation of blood, the essential stimulus of that organ, this phenomenon might also have occurred.

" *Experiment* 3.—The super-orbital nerve was laid bare in the forehead, as it issues through the supra-ciliary foramen in the eyebrow ; the one conducting rod being applied to it, and the other to the heels, most extraordinary grimaces were exhibited every time the electric discharges were made, by running the wire in my hand over the edges of the plates in the last trough, from the two hundred and twentieth to the two hundred and seventieth pair, thus fifty shocks, each greater than the preceding ones, were given in two seconds. Every muscle of his countenance was simultaneously thrown into fearful action, rage, horror, despair, and anguish, and ghastly smiles united their hideous expression in the murderer's face, surpassing far the wildest representations of a Fuseli or a Kean. At this period several of the spectators were obliged to leave the room from terror or sickness, and one gentleman fainted.

" *Experiment* 4.—The last galvanic experiment consisted in transmitting the electric power from the spinal marrow to the ulnar nerve, as it passes by the internal condyle at the elbow ; the fingers now moved nimbly, like those of a violin performer ; an assistant who tried to close the fist, found the hand to open forcibly in spite of his efforts. When one rod was applied to a slight incision on the top of the forefinger, the fist being previously clenched, the fingers extended instantly, and from the convulsive agitation of the arm, he seemed to point to the different spectators, some of whom thought he had come to life. About an hour was spent in these operations."

(386) In these experiments the positive wire was always applied to the *nerve*, and the negative to the *muscles*, that this is important, will appear from the following facts :

Let the posterior nerve of a frog be prepared for electrization, and let it remain till its voltaic susceptibility is considerably blunted, the crural nerves being connected with a detached portion of the spine; plunge the limbs into one wine glass full of water, and the crural nerves, &c. into another glass ; then take a rod of zinc in one hand, and a silver tea-spoon in the other, plunge the former into the water of the limbs glass, and the latter into the water of the nerves glass, without touching the frog itself, and gently strike the dry parts of the metals together; feeble convulsive movements or mere twitching of the fibres will be perceived at each contact; reverse now the position of the metal rods, and on renewing the contact between them, very lively convulsions will take place, and if the limbs are skilfully disposed in a narrow conical glass, they will probably spring out to some distance. Or, let an assistant seize in his moistened left hand the spine and nervous cords of the prepared frog, and in his right a silver rod, and let another person lay hold of one of the limbs with his right hand, and a zinc rod in the moist fingers of the left; on making the contact, feeble convulsive twitching will be perceived as before; now let the metals be reversed; on renewing the contact, lively movements will take place, which become very conspicuous; if one limb be held nearly horizontal, while the other hangs freely down, at each touch of the voltaic pair, the drooping limb will start up and strike the hand of the experimenter.

Hence for the purposes of resuscitating the dormant irritability of the nerves, as Dr. Ure remarks, or the contractility of their subordinate muscles, the positive pole must be applied to the former, and the negative to the latter.

(387) Some interesting researches, on the relation between Voltaic Electricity, and the Phenomena of Life, were published in the Philosophical Transactions by Dr. Wilson Philip.

In his earlier researches, he endeavoured to prove that the circulation of the blood, and the action of the involuntary muscles, are independent of the nervous influence. In a paper, read in January, 1816, he showed the immediate dependence of the secretory function on the nervous influence. The eighth pair of nerves distributed to the stomach, and subservient to digestion, were divided by incisions in the necks of several rabbits; after the operation, the parsley which they ate remained without alteration in their stomachs, and the animals, after evincing much difficulty in breathing, appeared to die of suffocation. But when in other rabbits similarly treated, the galvanic power was distributed along the nerve below its section, to a disc of silver placed closely in contact with the skin of the animal opposite to its stomach, no difficulty of breathing occurred. The voltaic action being kept up for twenty-six hours, the rabbits were then killed, and the parsley was found in as perfectly digested a state as that in healthy rabbits fed at the same time; and their stomachs evolved the smell peculiar to that of rabbits during digestion. These experiments were several times repeated with similar results.

Thus a remarkable analogy is shown to exist between the galvanic energy and the nervous influence, the former of which may be made to supply the place of the latter, so that while under it, the stomach, otherwise inactive, digests food as usual.

(388) Dr. Philip was next led to try galvanism as a remedy in asthma. By transmitting its influence from the nape of the neck to the pit of the stomach, he gave decided relief in every one of twenty-two cases, of which four were in private practice, and eighteen in the Worcester infirmary. The power employed varied from ten to twenty-five pairs.

(389) These results of Dr. Philip have since been confirmed by Dr. Clarke Abel, of Brighton (Journ. Sc. ix.): this gentleman employed in one of the repetitions of the experiments, a comparatively small and in the other a considerable power. In the former, although the galvanism was not of sufficient power to occasion evident digestion of the food, yet the efforts to vomit and the difficulty of breathing, constant effects of dividing the eighth pair of nerves, were prevented by it. The symptoms recurred when it was discontinued, but vanished on its re-application. "The respiration of the animal," he observes, "continued quite free during the experiment, except when the disengagement of the nerves from the tin foil, rendered a short suspension of the galvanism necessary during their re-adjustment. The non-galvanized rabbit wheezed audibly, and made frequent attempts to vomit. In the latter experiment, in which greater power of galvanism was employed, digestion went on as in Dr. Philip's experiments.

(390) It had been suggested by an eminent French physiologist, M. Gallois, that the motion of the heart depends entirely upon the spinal marrow, and immediately ceases when the spinal marrow is removed or destroyed. But Dr. Philip rendered rabbits insensible by a blow on the occiput, the spinal marrow and brain were then removed, and the respiration kept up by artificial means; the motion of the heart and circulation were carried on as usual. When spirit of wine or opium was applied to the spinal marrow or brain, the rate of circulation was accelerated. These experiments appear to confute the notion of M. Gallois.

(391) The general inferences deduced by Dr. Philip from his numerous experiments are, that Voltaic Electricity is capable of effecting the formation of the secreted fluids when applied to the blood, in the same way in which the nervous influence is applied to it; and that it is capable of occasioning an evolution of caloric from arterial blood, when the lungs are deprived of the nervous influence, by which their function is impeded, and even destroyed; when digestion is interrupted by withdrawing this influence from the stomach, these two vital functions are renewed by exposing them to the influence of a galvanic trough.

" Hence," says he, " galvanism seems capable of performing all the functions of the nervous influence in the animal economy; but obviously, *it cannot excite the functions of animal life, unless when acting on parts endowed with the living principle.*"

(392) *Animal Electricity.*—There are some remarkable instances of the generation of Electricity in living animals, to whom the power seems principally to be given as a means of defence. Of these animals the Raia Torpedo appears to have been noticed at a very early period, since we find a description of its properties in the writings of Pliny, Appian, and others. It inhabits the Mediterranean and North Seas; its weight when full grown is about eighteen or twenty pounds.

It is frequently met with also upon the Atlantic coast of France, as well as along our southern shores, especially in Torbay, where it attains a great size. It is taken by the *trawl* in company with its congenitors, the rays. At Malta it is known by the name of *Haddayla*, a term which has reference to its benumbing power : in France it is called *La Tremble*, and with us it has various designations, as the cramp or numb fish, and the Electric Ray. Dr. Davy has recently reduced the four previous recognised species of torpedo into two species, viz. : the torpedo *diversicolor*, and the torpedo *oculata*, a term having reference to certain markings on the back which have been likened to eyes.

The generic characters of the torpedo (Henry Letheby) are :—The disc of the body is nearly circular; the pectoral fins large; the two dorsal fins are placed so far back as to be on the tail ; the surface of the body is smooth ; the tail short and rather thick, and the mouth armed with small sharp teeth.

It is an ovo-viviparous animal, the young being matured in their descent along the oviduct, where they are retained for several months ; and during this detention, the yolk bag disappears and the fish are perfected; the electrical organs also become gradually developed. Dr. Davy thinks these are formed from matters which are absorbed by the branchial filaments; an opinion which he deduces from the fact of these not attaining such great size and length in other rays, while they also drop off when the organs are complete.

When touched it communicates a benumbing sensation, and by repeated contacts gives a series of electric shocks. In the Philosophical Transactions, 1773 and 1775, there are accounts of some experiments of Mr. Walsh on this animal. He placed a living torpedo on a wet napkin, and formed a communication through five persons, all of whom were insulated. The person at one extremity touched some water, in which a wire, proceeding from the wet napkin, terminated ; the last person in the series having a similar mode of communication with a

wire, which at intervals could be brought into contact with the back of
the animal.   In this manner shocks were communicated to the five,
and afterwards to eight persons.   Mr. Walsh could not succeed in
affecting the electroscope, or in obtaining a spark by this electricity.
But he observed that every time the animal gave a shock, which was
not generally perceptible beyond the finger with which it was touched,
a contortion of the body followed, as if the animal were anxious to
make its escape ; its eyes were also depressed, so that he could tell by
observing the eyes, when the animal attempted to make a discharge,
even upon non-conducting bodies.   Mr. Cavendish constructed an
artificial torpedo of wood, connected with glass tubes and wires, and
covered with a piece of sheep-skin leather.   To render the effect of this
instrument more like that of the animal, with regard to the difference of
the shock in and out of water, it was necessary to substitute thick
leather in the place of the wood ; and with this improvement the
apparatus succeeded admirably.   In air the sensation of the shock was
experienced at the elbows ; but under water it was confined chiefly to
the hands.   On touching this artificial torpedo under water, a shock
was obtained as powerful as if it had been touched by both hands.
Being touched under water with two metallic spoons, it gave no shock ;
but in air, the shock was very strong.   Cavendish also made an esti-
mate between the strength of his artificial torpedo and that dissected
by Hunter, with reference to surface.   His own battery consisted of
seventy-six feet of coated surface, and he calculated that the animal
retained a charge fourteen times as great as that of the battery, or was
equivalent to one thousand and sixty-four feet of coated glass.

(393) In the Philosophical Transactions for 1773, there is a detail
of the anatomical structure of this curious fish, from the pen of the
celebrated Hunter.[*]

" The nerves," he says, " inserted into each electric organ, arise by
three large trunks from the latter and posterior parts of the brain.
The first of these, in its passage outwards, turns round a cartilage of
the cranium, and sends a few branches to the first gill, and to the anterior part of the head, and then passes into the organ at its anterior
extremity.   The second trunk enters the gills between the first and
second openings, and furnishes it with small branches, passing into the
organ near the middle.   The third trunk, after leaving the skull,
divides into two branches, which pass to the electric organ through the
gills, one between the second and third openings, the other between the
third and fourth, giving small branches to the gill itself.   These nerves

* For a detailed account of the anatomy of this fish, see also Proceedings of the
Electrical Society, p. 512.

having entered the organs, ramify in every direction between the columns, and send in small branches on each partition where they are lost.

"The magnitude and number of the nerves bestowed on these organs, in proportion to their size, must, on reflection, appear as extraordinary as the phenomena they afford. Nerves are given to parts either for sensation or for action. Now if we except the more important senses of seeing, hearing, smelling, and tasting, which do not belong to the electric organs, there is no part, even of the most perfect animal, which, in proportion to its size, is so liberally supplied with nerves, nor do the nerves seem necessary for any sensation which can be supposed to belong to the electric organs. And with respect to action, there is no part of any animal with which I am acquainted, however strong and constant its natural actions may be, which has so great a proportion of nerves. If, then, it be probable that these nerves are not necessary for the purposes of sensation or action, may we not conclude that they are subservient to the formation, collection, or management of the electric fluid, especially as it appears evident from Mr. Walsh's experiments, that the will of the animal does absolutely control the electric powers of the body which must depend upon the energy of the nerves."

(394) At the request of his illustrious brother, Dr. John Davy undertook a series of experiments at Malta, on this form of electricity. He succeeded in communicating distinct magnetism to a needle by the electricity of a small torpedo, and in throwing into violent motion the needle of a magnetic multiplier; but he failed in obtaining a spark, or any igniting power, though by using Harris's air-electrometer he obtained evidence of *heating* powers; nor could he affect the electroscope, or obtain any indications of attraction and repulsion in air.*

His electro-chemical results were highly satisfactory; he decomposed strong solutions of common salt, nitrate of silver, and superacetate of lead; and he inferred that the *under* surface of the organ corresponds to the zinc, and the upper surface to the copper extremity of the voltaic battery.

(395) The Gymnotus is another fish possessed of electrical properties; it is a native of the warmer regions of America, and Africa, inhabiting large rivers, especially those of Surinam. In Africa it chiefly occurs in the branches of the Senegal. It is so named from the absence of the dorsal fin. There are several species, all inhabiting fresh water lakes and rivers; but one species alone is electrical. In general aspect it very much resembles an eel,—the body is smooth, without scales; a

---

* Prior to Dr. Davy's experiments, MM. de Blainville and Flourens had obtained magnetic deflection; and recently, Linari and Matteucci have obtained the spark from the torpedo.

long ventral fin extends from just behind the head, to the very extremity of the tail; around the mouth are many papillæ lodged in crypts, which are merely mucous glands.   The mouth is armed with sharp teeth, and projecting into it are numerous fringes, which, from their vascularity, doubtless serve a purpose in respiration.   The œsophagus is short, terminating in a capacious stomach with thick rugose parietes. The rest of the alimentary canal is short, doubled on itself, and terminates in the *cloaca*, which is situated in the mesial line, a few inches from the under jaw.   The whole cavity of the abdomen is not more than seven inches long, and contained, in the female specimen examined by Mr. Letheby, besides the alimentary canal, ovaries filled with ova of a bright orange colour, the heart, the liver, and upper part of the air bladder.   The rest of the animal is made up of the electrical organs and muscles of progression, together with an air sac, which runs beneath the spine the whole length of its body.*   The Gymnotus was first described by Richer, in 1677, and its anatomical structure by Mr. Hunter, in the 63rd and 65th volumes of the Philosophical Transactions.   The electric organs consist of alternations of different substances, and are most abundantly supplied by nerves; their too frequent use is succeeded by debility and death.   That these organs are not essential to the animals, is proved by their thriving after they have been removed.

By touching this fish with one hand a shock is felt in the fingers, wrist, and elbow.   The hand of one person being held in water at the distance of three feet from the animal, upon a second person's irritating the eel, the former will receive a shock.   Dr. Williamson placed a catfish in the same vessel of water with the eel, and then dipped his own hand in the water.   The Gymnotus swam up to the fish, but turned away without offering any violence to it ; it soon returned, and after regarding attentively the cat-fish for some seconds, gave it a shock, which made it turn up its belly, and continue motionless.   The shock was perceptible to Dr. Williamson at the same time.   Whenever fish that had been thus rendered motionless were removed to another vessel, they recovered.

(396) In his Tableau Physique des Régiones Equitoriales, &c., Humboldt has given some curious details respecting the electrical eel, which inhabits the rivers and lakes of the low provinces of Venezuela and the Caraccas.   It is met with most frequently in the stagnant pools, dispersed at intervals over the plains which extend from Oronoco to the Apuré.   The old road near Urutica has even been abandoned on account of the danger experienced in crossing a ford, where the mules were, from the effects of the shocks, often paralyzed and drowned.

* See Mr. Letheby's elaborate paper on the dissection of a Gymnotus Electricus.— Proceedings of the London Electrical Society, p. 367.

Even the angler sometimes receives a stroke conveyed along his rod and line. These eels are about six feet in length, and occasion a highly painful sensation, more resembling the effect of a blow on the head, than the shock of a common electric discharge ; a peculiarity of effect referable perhaps to a great quantity of electricity of small intensity. The following particulars given by Humboldt, and taken from the Edinburgh Review, vol. xvi. p. 250, are worth inserting :—

" The Indians entertain such a dread of the Gymnotus, and show so much reluctance to approach it when alive and active, that Humboldt found extreme difficulty in procuring a few to serve as the subjects of experiment. For this express purpose he stopped some days on his journey across the Llanos to the river Apuré, at the small town of Calaboze, in the neighbourhood of which he was informed they were very numerous. But though his landlord took the utmost pains to gratify his wish, he was constantly unsuccessful. At last he determined to proceed himself to the spot, and was conducted to a piece of shallow water, stagnant and muddy, but of the temperature of 79°, surrounded by a rich vegetation of the great Indian fig trees and the odoriferous sensitive plants. Here he soon witnessed a spectacle of the most novel and extraordinary kind. About thirty horses and mules were quickly collected from the adjacent savannahs, where they run half wild ; these the Indians drove into the marsh. The gymnoti, roused from their slumbers by the noise and tumult, mount near the surface, and swimming like so many livid water-serpents, briskly pursue the intruders, and gliding under their bellies, discharge through them the most violent and repeated shocks. The horses, convulsed and terrified, their mane erect, and their eyes staring with pain and anguish,. make unavailing struggles to escape. In less than five minutes two of them sunk under the water and were drowned. Victory seemed to declare for the eels ; but their activity now began to relax ; fatigued by such expense of nervous energy, they shot their electric discharges with less frequency and effect. The surviving horses gradually recovered from the shock, and became more composed and vigorous. In a quarter of an hour the gymnoti finally retired from the contest, and in such a state of languor and complete exhaustion, that they were easily drawn on shore by help of small harpoons fixed on cords."

(397) A fine specimen of the Gymnotus (Fig. 158 being a correct

FIG. 158.

representation) was for some time in the possession of the proprietors of the Gallery of Practical Science in Adelaide-street. It was brought to this country by Mr. Porter, and deposited in the Gallery in August 1838, where it remained in a healthy and vigorous condition till March 14th, 1842, when it died from the effects of a rupture of a blood-vessel consequent upon expansion of the ovarium.

The length of this fish was forty inches. At first it was fed with blood, which was nightly put into the water, which was changed for fresh water in the morning; subsequently it was supplied with small fish, such as gudgeons, carp, and perch, one of which, on an average, it consumed daily.

(398) It will not be uninteresting to give a brief account of some of the experiments made by Dr. Faraday with this fish, the results having afforded every proof of the identity of its power with common Electricity.*

1st, *Shock.*—This was very powerful when one hand was placed on the body near the head, and the other near the tail. When the *dry* hands grasped metallic conductors in contact with the fish, scarcely any shock was felt; but when the hands were wetted, smart shocks were experienced.

2nd, *Galvanometer.*—A pair of collectors were thus constructed: a plate of copper eight inches long, by two inches and a half wide, was bent into a saddle shape, that it might pass over the fish, and inclose a certain extent of the back and sides, and a thick copper wire was brazed to it, to convey the electric force to the experimental apparatus; a jacket of sheet caoutchouc was put over the saddle, the edges projecting at the bottom and the ends; the ends were made to converge so as to fit in some degree the body of the fish, and the bottom edges were made to spring against any horizontal surface on which the saddles were placed. The part of the wire liable to be in the water was covered with caoutchouc.

(399) By causing the fish to send a powerful discharge through these collectors, a galvanometer of no great delicacy being included in the circuit, a deflection of the needle amounting to 30° was produced; the deflection was constantly in a given direction, the electric current being always from the anterior parts of the animal through the galvanometer wire to the posterior parts. The former were, therefore, for the time, externally positive, and the latter negative.

(400) *Making a Magnet.*—When a little helix, containing twenty-two feet of silked wire wound on a quill, was put into the circuit, and an annealed steel needle placed in the helix; the needle became a magnet, and the direction of its polarity in every case indicated a cur-

* Experimental Researches, 15th Series.

rent from the anterior to the posterior parts of the Gymnotus through the conductors used.

(401) *Chemical Decomposition.*—Polar decomposition of iodide of potassium was obtained by moistening three or four folds of paper in the solution, placing them between a platinum plate and the end of a platinum wire connected respectively with the two saddle conductors. Whenever the wire was in conjunction with the conductor at the fore part of the Gymnotus, iodine appeared at its extremity; but when connected with the other conductor, none was evolved at the place on the paper where it before appeared. By this test Dr. Faraday compared the middle part of the fish with other portions before and behind it, and found that the conductor A, which being applied to the middle, was negative to the conductor B applied to the anterior parts, was, on the contrary, positive to it when B was applied to places near the tail. So that within certain limits the condition of the fish externally at the time of the shock appears to be such that any given part is negative to other parts anterior to it, and positive to such as are behind it.

(402) *Evolution of Heat.*—The experiments were not decisive on this point, as might be expected; the instrument employed was a Harris's thermo-electrometer.*

(403) *Spark.*—The electric spark was first obtained in the following manner. A good magneto-electric coil, with a core of soft iron wire, had one extremity made fast to the end of one of the saddle collectors, and the other fixed to a new steel file; another file was made fast to the end of the other collector. One person then rubbed the point of one of these files over the face of the other, whilst another person put the collectors over the fish, and endeavoured to excite it to action. By the friction of the files contact was made, and broken very frequently; and the object was to catch the moment of the current through the wire and helix, and by breaking contact *during the current* to make the Electricity sensible as a spark. The spark was obtained four times: a revolving steel plate, cut *file-fashion* on its surface, was afterwards substituted for the lower file; and for the upper file wires of iron, copper, and silver, with all of which the spark was obtained.

---

* Mr. Gassiot, however, by employing an electrometer of peculiar construction, having, instead of one straight wire, two separate wires, one of fine silver and the other of fine platinum, made under the personal inspection of Mr. Harris, has succeeded in developing the heating power of the Gymnotus in the first experiment, (made on the 21st of May, 1839,) the circuit was completed through the platinum wire, when the liquid in the electrometer rose *one* degree.

In the second experiment the circuit was completed through the silver wire, when the liquid rose *two* degrees.

In subsequent experiments the spark was obtained *directly* between fixed surfaces, the inductive coil being removed, and only short wires used. The apparatus employed was a glass globe, through the upper cap of which a copper wire, slightly bent at its lower extremity, and carrying a slip of gold leaf, was passed; a similar wire terminating with a brass ball within the globe was passed through the lower cap. The gold leaf and brass ball were brought into all but actual contact, and when the wires were connected with the saddle collectors, and the fish provoked to discharge a current of Electricity, the gold leaf was attracted to the ball, and a spark passed.*

(404) When the shock is strong, it is like that of a large Leyden battery charged to a low degree, or that of a good voltaic battery of perhaps one hundred or more pairs of plates, of which the circuit is completed for a moment only; and great as is the force of a single discharge, the fish is able to give a double, and even a triple shock, with scarcely a sensible interval of time. Dr. Faraday endeavoured to form some idea of the *quantity* of Electricity, by connecting a large Leyden battery with two brass balls above three inches in diameter, placed seven inches apart in a tube of water, so that they might represent the parts of the Gymnotus to which the collectors had been applied; but to lower the intensity of the discharge, eight inches in length of six-fold wetted string were interposed elsewhere in the circuit, this being found necessary to prevent the easy occurrence of the spark at the ends of the collectors when they were applied to the water near to the balls, as they had been before to the fish. Being thus arranged, when the battery was strongly charged and discharged, and the hands put into the water near the balls, a shock was felt much resembling that from the fish; and though the experiments have no pretension to accuracy, yet as the tension could be in some degree imitated by reference to the more or less ready production of a spark, and after that, the shock be used to indicate whether the quantity was about the same, Dr. Faraday thought that it may be concluded that a single medium discharge of the fish was at least equal to the Electricity of a Leyden battery of fifteen jars containing 3500 square inches of glass coated on both sides, and charged to its highest degree.

* It was Mr. Gassiot, I believe, who first obtained attractions of gold leaves in the following manner:—

A common glass tumbler, having two small holes drilled on each side, was inverted on a wooden stand: two copper wires, with small brass balls attached, were passed through the holes; to each ball a strip of gold leaf was fixed about 1 inch long and $\frac{1}{8}$ of an inch wide:—the leaves being placed parallel to each other, were then approximated to within about $\frac{1}{30}$ or $\frac{1}{40}$ of an inch. On making contact with the eel, the leaves were not only attracted, but were actually fused, scintillating in the most beautiful manner.

(405) Numerous other interesting experiments were made by Dr. Faraday with this fine specimen of the Gymnotus, from all of which it was rendered evident that all the water, and all the conducting matter around the fish, through which a discharge circuit can in any way be completed, is filled at the moment with circulating electric power, and this state might be easily represented by drawing the lines of inductive action upon it. In the case of a Gymnotus surrounded equally in all directions by water, these would resemble generally in disposition the magnetic curves of a magnet having the same straight or curved shape as the animal, i. e., provided he in such cases employed, as may be expected, his four electric organs at once. That all the conducting matter around the fish is filled at the moment with circulating electric power, was proved by the fact, that a number of persons all dipping their hands at the same time into the tub, the diameter of which was forty-six inches, received a shock of greater or less intensity according as they were more or less favourably situated with regard to the direction of the current.

(406) The Gymnotus can stun and kill fish which are in very various positions to its own body. Dr. Faraday describes the behaviour of the eel on one occasion when he saw it eat, as follows :—
A live fish, about five inches in length, caught not half a minute before, was dropped into the tub. The Gymnotus instantly turned round in such a manner as to form a coil inclosing the fish, the latter representing a diameter across it : a shock passed, and there in an instant was the fish struck motionless, as if by lightning, in the midst of the water, its side floating to the light. The Gymnotus made a turn or two to look for its prey, which having found, he bolted, and then went searching about for more. Living as this animal does in the midst of such a good conductor as water, it seems at first surprising that it can sensibly electrify anything ; but in fact it is the very conducting power of the water which favours and increases the shock by moistening the skin of the animal through which the Gymnotus discharges its *battery*. This is illustrated by the fate of a Gymnotus which had been caught and confined for the purpose of transmission to this country. Notwithstanding its wonderful powers, it was destroyed by a *water rat*, and when we consider the perfect manner in which the body of the rat is insulated, and that even when he dives beneath the water not a particle of the liquid adheres to him, we shall not feel surprised at the catastrophe.

(407) The Gymnotus appears to be sensible when he has shocked an animal, being made conscious of it, probably, by the *mechanical impulse* he receives, caused by the spasms into which he is thrown. When Dr. Faraday touched him with his hands, he gave him shock

after shock; but when he touched him with glass rods, or insulated conductors, he gave one or two shocks felt by others having their hands in at a distance, but then ceased to exert the influence, as if made aware it had not the desired effect. Again, when he was touched with the conductor several times for experiment on the galvanometer, &c., and appeared to be languid or indifferent, and not willing to give shocks, yet, being touched by the hands, they, by convulsive motion, informed him that a sensitive thing was present, and he as quickly showed his power and willingness to astonish the experimenter.

(408) In these most wonderful animals then we behold the power of converting the *nervous* into the *electric* force. Is the converse of this possible? Possessing, as we do, an electric power far beyond that of the fish itself, is it irrational, or unphilosophical, to anticipate the time when we shall be able to reconvert the electric into the nervous force? Seebeck taught us how to commute heat into Electricity, and Peltier, more recently, has shown us how to convert the Electricity into heat. By Œrsted we were shown how to convert the electric into the magnetic force, and Faraday has the honour of having added the other member of the full relation, by re-acting back again and converting magnetic into electric forces.

(409) The following are the experiments suggested by Faraday, as being rational in their performance and promising in anticipation.

1°. If a gymnotus or torpedo has been fatigued by frequent exertion of the electric organs, would the sending of currents of similar force to those he emits, or of other degrees of force, either continuously or intermittingly, in the same direction as those he sends forth, restore him his powers and strength more rapidly than if he were left to his natural repose?

2°. Would sending currents through, in the contrary direction, exhaust the animal rapidly?

3°. When, in the torpedo, a current is sent in the natural direction, *i. e.* from below upwards, through the organ on one side of the fish, will it excite the organ on the other side into action? or if sent through in the contrary direction, will it produce the same, or any, effect on that organ?

4°. Will it do so if the nerves proceeding to the organ or organs be tied? and will it do so after the animal has been so far exhausted by previous shocks, as to be unable to throw the organ into action in any, or in a similar, degree of his own will?

(410) It is for the physiologist to pursue this inquiry: to him it belongs to connect these two branches of physical philosophy, a minute acquaintance with practical anatomy being quite as indispensable as a thorough knowledge of the laws of Electricity. " Never, however," as Daniell observes, " was there a more tempting field of research, or a

higher reward offered for its successful cultivation, than that which is presented by *animal Electricity*."

(411) In the autumn of 1839, Professor Schoenbein, of Bale, went through a series of experiments with the London Gymnotus, and obtained results entirely in accordance with those just described. One fact, however, was observed during the decomposition of iodide of potassium, which greatly surprised those who witnessed it. At the instant when the paper, impregnated with the iodide, was put in communication with the fish, a visible spark was observed : this spark did not occur every time, but in an exceptional manner, although the experiments were repeated in circumstances as similar as possible. " So far as I myself have been able to observe," says Schoenbein ; " we never obtained a spark, either at the moment when we complete the circuit of a galvanic pile, by means of an electrolytic body, or at the moment when this latter is put out of the action of the current. I dare not then express an opinion upon the nature and cause of the phenomenon just mentioned, especially as I fear to decide whether the spark really occurred at the opening of the circuit, or at the instant of its being closed."

(412) In summing up some exceedingly interesting remarks on the electrical powers of the Gymnotus,* Schoenbein declares it to be his opinion, that the true cause of the phenomena is still completely obscure, and must neither be sought for in the physical or chemical constitution, nor in a fixed organization of certain parts of the animal : but that there exists, without our being able at present to determine how, an intimate connection between the vital actions dependent on the will of the fish, and the physical phenomena which these vital actions produce. Until we know more exactly the nature of Electricity, we shall be unable to detect this intimate relation which exists between electric and vital action, until we know whether Electricity is only a particular condition of what we call matter, or whether it arises from particular vibrations of the ether, or, in fine, whether like gravity it must be regarded as a primitive and specific force of nature. So long as we are without an exact idea of what Electricity is, the different modes of its development will, of course, be incomprehensible to us, and we shall scarcely be able to say anything upon the cause of animal Electricity, even though anatomists and physiologists should have very carefully studied the structure of the fish, and should have made us most intimately acquainted with all its fibres and its most minute nerves."

(413) *Electricity of Effluent Steam.*—The first account we have of an observation on the Electricity of a jet of steam, while issuing from a boiler, is contained in a letter addressed by H. G. Armstrong, Esq., to Professor Faraday, and published in 17th vol. of the L. and E. Phil.

* See a Translation of his paper in Proceedings of London Elec. Soc., p. 133.

Mag., with the date of Oct. 14th, 1840. It appears that the phenomenon was first noticed by the engine-man entrusted with the care of
a steam engine at Seghill, about six miles from Newcastle : it happened
that the cement, by which the safety valve was secured to the boiler,
had a crack in it, and through this fissure a copious horizontal jet of
steam continually issued. Soon after this took place, the engine-man
having one of his hands accidentally immersed in the issuing steam,
presented the other to the lever of the valve, with the view of adjusting
the weight, when he was greatly surprised by the appearance of a brilliant spark, which passed between the lever and his hand, and was
accompanied by a violent wrench in his arms, wholly unlike what he
had ever experienced before. The same effect was repeated when he
attempted to touch any part of the boiler, or any iron work connected
with it, provided his other hand was exposed to the steam. He next
found that while he held one hand in the jet of steam, he communicated
a shock to every person whom he touched with the other, whether such
person was in contact with the boiler, or merely standing on the brick-
work which supported it ; but that a person touching the boiler, received
a much stronger shock than one who merely stood on the bricks.

(414) In following up these experiments, Mr. Armstrong provided
himself with a brass plate having a copper wire attached to it, which
terminated in a round brass knob. When this plate was held in the
steam, by means of an insulated handle, and the brass knob brought
within about a quarter of an inch from the boiler, the number of sparks
which passed in a minute was from sixty to seventy, and when the
knob was advanced about one-sixteenth of an inch nearer to the boiler,
the stream of Electricity became quite continuous. The greatest distance between the knob and the boiler, at which a spark would pass
from one to the other, was fully an inch. A Florence flask, coated with
brass filings on both surfaces, was charged to such a degree with the
sparks from the knob, as to cause a spontaneous discharge through the
glass,—and several robust men received a severe shock from a small
Leyden jar charged by the same process. The strength of the sparks
was quite as great when the knob was presented to any conductor communicating with the ground, as when it was held to the boiler.

(415) A long and well conducted series of experiments have been
made by Mr. Armstrong, on the Electricity evolved under these peculiar circumstances.* By standing on an insulated stool, and holding
with one hand a light iron rod immediately above the safety valve of a

* See L. and E. Phil. Mag. vol. xvii. pp. 370, 452 ; vol. xviii. pp. 50, 133, 328 ;
vol. xix. p. 25 ; vol. xx. p. 5 ; vol. xxii. p. 1. See also papers on the same subject
by Mr. Pattinson, vol. xvii. pp. 375, 457 ; and by Dr. Schafhaeutl, vol. xvii. p.
449 ; vol. xviii. pp. 14, 95, 265.

locomotive engine, while the steam was freely escaping, and then advancing the other hand towards any conducting body, he obtained sparks of an inch in length : when the rod was held five or six feet above the valve, the length of the sparks was two inches : and when a bunch of pointed wires, attached to the rod, were held, points downwards in the issuing steam, sparks *four inches long* were drawn from a round knob, on the opposite extremity of the iron rod. On insulating the boiler, large and brilliant *negative* sparks an inch long were drawn from it—the Electricity of the steam being positive.

(416) A small boiler was constructed by Mr. Armstrong—it was arranged on a stove which was insulated ; when the rate of evaporation was about a gallon in an hour, and the pressure in the boiler 100 lbs. on the square inch, by connecting the knob of a Leyden phial with the boiler or stove, he was able to give it a charge, and he found that Electricity could be collected in much greater abundance from the evaporating vessel, than from the issuing steam. The Electricity of the steam was generally positive, that of the insulated boiler being negative; occasionally, however, these conditions were reversed, and after the boiler had been in use for some time, positive Electricity rarely appeared in the jet, even when circumstances were most favourable to its development. No alteration was effected by washing out the boiler with water, but when it was washed with solution of *potash* or *soda*, the *positive* condition of the steam jet was restored, and, by dissolving a little potash in the water from which the steam was generated, the quantity of Electricity was amazingly increased ; on the other hand, when a small quantity of *nitric acid*, or *nitrate of copper*, was added to the water, the Electricity of the steam became negative.

(417) Subsequent experiments led Mr. Armstrong to the conclusion, that the excitation of Electricity takes place at the point where the steam is subjected to friction; and, in a paper recently read before the Royal Society by Professor Faraday, it is shown that the steam itself has nothing to do with the phenomenon. By means of a suitable apparatus, Faraday found that Electricity is never excited by the passage of pure steam, and is manifested only when water is at the same time present; and hence, he concludes, that it is altogether the effect of the friction of globules of water against the sides of the opening, or against the substances opposed to its passage, as the water is rapidly moved onwards by the current of steam. Accordingly, it was found to be increased in quantity by increasing the pressure and impelling force of the steam. The immediate effect of this friction was, in all cases, to render the steam or water positive, and the solids, of whatever nature they might be, negative. In certain circumstances, however, as when a wire is placed in the current of steam, at some distance from the

orifice whence it has issued, the solid exhibits the positive Electricity already acquired by the steam, and of which it is then merely the recipient and the conductor. In like manner the results may be greatly modified by the shape, the nature and the temperature of the passage through which the steam is forced. Heat, by preventing the condensation of the steam into water, likewise prevents the evolution of Electricity, which again speedily appears by cooling the passages, so as to restore the water which is necessary for the production of that effect. The phenomena of the evolution of Electricity, in these circumstances, is dependent also on the quality of the fluid in motion, more especially in relation to its conducting power. Water will not excite Electricity unless it be pure: the addition of any soluble salt or acid, even in minute quantity, is sufficient to destroy this property. The addition of *oil of turpentine*, on the other hand, occasions the development of Electricity of an opposite kind to that which is excited by water; and this Faraday explains by the particles and minute globules of the water having each received a coating of oil, in the form of a thin film, so that the friction takes place only between that external film and the solids, along the surface of which the globules are carried. A similar but more permanent effect is produced by the presence of olive oil, which is not, like oil of turpentine, subject to rapid dissipation. Similar results were obtained when a stream of compressed air was substituted for steam in these experiments. When moisture was present, the solid exhibited negative, and the stream of air positive Electricity; but when the air was perfectly dry, no Electricity of any kind was apparent.*

(418) In Mr. Armstrong's last paper† on this subject, he states that his recent experimennts have confirmed the conclusion, that the excitation of Electricity takes place at the point where the steam is subjected to friction, and he describes several improvements in his apparatus by which the energy of the effects is amazingly increased. By means of a boiler furnished with a stop cock and discharging jet of peculiar construction, he has produced effects upwards of *seven* times greater than those from a plate electrical machine of three feet in diameter, worked at the rate of seventy revolutions per minute. This boiler is a wrought iron cylinder, with rounded ends, and measures three feet six inches in

---

* See an excellent abstract of a Lecture on this interesting subject delivered by Dr. Faraday at the Royal Institution, in the *Electrical Magazine*, (a new scientific periodical, the first number of which made its appearance on the 1st of July, under the able superintendence of Mr. C. V. Walker, Hon. Secretary to the London Electrical Society.) In this Lecture it was demonstrated that the effects are not due to evaporation, or to the mere transition of water into vapour; they do not therefore afford any clue to elucidate the mysteries of atmospheric electricity and lightning.

† L. and E. Phil. Mag. vol. xxii. p. 1.

length, and one foot six inches in diameter. It rests on an iron frame, containing the fire, and the whole apparatus is supported on glass legs to insulate it. It is found much more convenient and effectual to collect Electricity from the boiler than from the steam cloud, but, in order to obtain the highest effect from the boiler, the Electricity of the steam must be carried to the earth by means of proper conductors.

(419) The following report of some experiments with this apparatus was furnished by L. Boscawen Ibbetson, Esq.* To the boiler (Fig. 159)

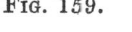

Fig. 159.

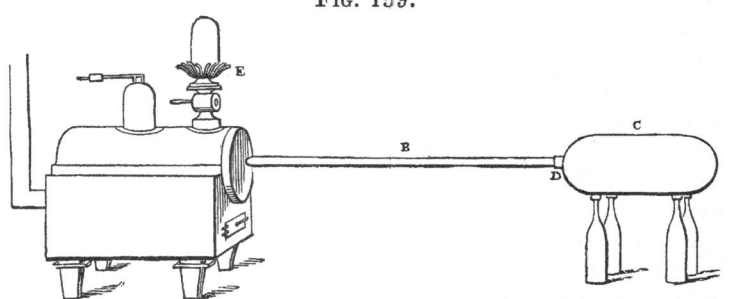

was attached a conductor B, fourteen feet long, and a prime conductor C, which latter was supported by four long glass bottles, filled with warm water and corked. When the steam was let off through a series of jets, as at E, all the bottles were covered with beautiful streams of Electricity ; and one of the bottles was so charged, that it was fractured and burst. A spark was procured from the prime conductor 15 *inches in length*. Ether and spirits of turpentine were ignited through the knuckles, at a distance of six inches from the point D, the conductor being removed ; and *gunpowder* was exploded by the intervention of wetted string, *without the aid of a jar*. A Leyden jar, 5 inches in diameter, with 6¾ inches in height, coated, placed in communication with the point D, gave 120 spontaneous discharges in one minute ; and a jar 9 inches in diameter, with 13½ inches in height, coated, gave 28 spontaneous discharges in one minute. When the small boiler (Fig. 160) was used, the Electricity from A was *positive*, that from the steam being negative ; but when the water in the boiler was reduced in quantity, the Electricity from the boiler became negative, and that of the steam positive. The change could be traced taking place. When the boiler was washed with a weak alkaline solution, the boiler always remained

Fig. 160.

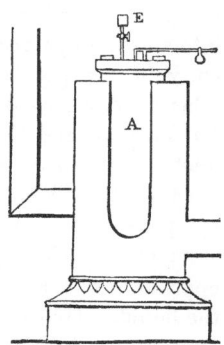

* Proceedings of the London Electrical Society, p. 527.

negative and the steam positive ; afterwards, when it was washed with very dilute nitric acid, the Electricity from the boiler became positive, and that from the steam negative. But when at the orifice E, a wooden jet was placed inside the copper one, the character was again changed, the boiler becoming negative and the steam positive. The colour of the spark from the steam Electricity seemed to Mr. Ibbetson to differ from that obtained by means of ordinary Electricity, and to be much more intense.

(420) In Mr. Pattinson's experiments, on one of the locomotive engines belonging to the Newcastle and Carlisle Railway, sparks *four inches* long were given off from the person of an individual standing on an insulating stool, and holding a copper rod, terminated by sharp-pointed wires, in the current of steam, blowing forcibly out of the safety-valve at a pressure of 52 lbs. per inch. The Electricity was ascertained to be positive. It is certainly, as Mr. Pattinson observes, a novel and curious light in which to view the splendid locomotive engine in its rapid passage along the railway line, viz., that of an enormous electrical machine,—the steam analogous to the glass plate of an ordinary machine, and the boiler to the rubber ; while torrents of electricity might continually be collected, by properly disposing conductors in the escaping steam.

(421) *Theory of the Voltaic Pile.*—Is the proximate cause of the voltaic current the contact of the two dissimilar metals, or is it the action of the oxidizable metal on the water of the acid solution ? This question has been the subject of much profound discussion. It has already been stated that the first view of the subject was adopted by Volta, who, attributing the Electricity of the pile to the contact of dis-similar metals, regarded the interposed solutions merely as imperfect conductors, admitting the transfer of Electricity when the circuit was completed ; and when incomplete, throwing the whole by induction into an *electro-polar* state. This view has been adopted and reasoned on, with their peculiar ingenuity, by the German philosophers ; on the other hand, a powerful mass of evidence has been brought against it by Dr. Faraday, and the chemical theory has obtained, in this country at least, almost universal assent.

(422) It will be proper, however, to attempt a popular account of the present state of this interesting question. By Davy the electric state of the pile was considered as due partly to the contact of the opposed metals, and partly to the chemical action exerted on them by the liquid. He concluded, to use his own words,* that " chemical and electrical attractions are produced by the same cause ; acting, in one case, on *particles*, in the other on masses of matter : and that the same

* Philosophical Transactions, 1826, p. 389.

property, under different modifications, is the cause of all the phenomena exhibited by different voltaic combinations." By Dr. Wollaston the phenomena were referred solely to chemical action ; and he even attributed the Electricity of the common machine to the oxidizement of the amalgam, and found, contrary to the experiments of his great contemporary, that the electrical machine was not active in atmospheres of hydrogen, nitrogen, or carbonic acid. The first suggestion, however, of the chemical origin of voltaic Electricity, is to be found in a paper communicated by Fabroni, in 1792, to the Florentine Academy. This philosopher ascribed the convulsions in the limbs of the frog, in the experiments of Galvani and Volta, to a chemical change made by the contact of one of the metals with the liquid matter on the parts of the animal body ; to a decomposition of this liquid ; and to the transition of oxygen from a state of combination with it, to combination with the metal. He maintained that the convulsions were chiefly due to the chemical changes, and not to the Electricity incidental to them, which he considered, if operating at all, to do so in a secondary way. Pepys placed a pile in an atmosphere of oxygen, and found that in the course of a night, 200 cubic inches of the gas had been absorbed : while in an atmosphere of azote, it had no action. MM. Biot and Cuvier also observed the quantity of oxygen absorbed, and inferred from their experiments that, "although strictly speaking the evolution of Electricity in the pile was produced by oxidation, the share which this had in producing the effects of the instrument, bore no comparison with that which was due to the contact of the metals, the extremity of the series being in communication with the ground.

(423) The source of the Electricity of the voltaic pile was made by Dr. Faraday the subject of the 8th, 16th, and 17th series of his Experimental Researches. By the arrangement shown in Fig. 160 he succeeded in producing Electricity *quite independent of* FIG. 161.
contact. A plate of zinc (Fig. 161) was cleaned and bent in the middle to a right angle ; a piece of platinum, about three inches long and half an inch wide, *b*, was fastened to a platinum wire, and the latter bent as in the figure. These two pieces of metal were arranged as shown in the sketch ; at *x* a piece of folded bibulous paper, moistened in a solution of iodide of potassium, was placed on the zinc, and was pressed upon by the end of the platinum wire ; when, under these circumstances, the plates were dipped in the diluted nitric and sulphuric acids, or even in solution of caustic potash, contained in the vessel *c*, there was an immediate effect at *x*, the iodide being decomposed, and iodine appearing at the anode, that is, against the end of the platinum wire. As long as the lower

ends of the plates remained in the acid, the electric current proceeded, and the decomposition proceeded at x. On removing the end of the wire from place to place on the paper, the effect was evidently very powerful ; and on placing a piece of turmeric paper between the white paper and the zinc, both papers being moistened with a solution of iodide of potassium, alkali was evolved at the *cathode* against the zinc, in proportion to the evolution of iodine at the *anode ;* the galvanometer also showed the passage of an electrical current ; and we have thus a simple circle of the same construction and action as those described in the last Lecture, except in the absence of metallic contact.

(424) It is shown by Faraday that *metallic contact* favours the passage of the electrical current, by diminishing the opposing affinities. When an amalgamated zinc plate is dipped into dilute sulphuric acid, the force of chemical affinity exerted between the metal and the fluid is not sufficiently powerful to cause sensible action at the surfaces of contact, and occasion the decomposition of the water by the oxidation of the metal, though it *is* sufficient to produce such a condition of Electricity as would produce a current, if there were a path open for it ; and that current would complete the conditions necessary, under the circumstances, for the decomposition of water. Now when the zinc is *touched* by a piece of platinum, the path required for the Electricity is opened, and it is evident that this must be far more effectual than when the two metals are connected through the medium of an *electrolyte ;* because a *contrary and opposing action* to that which is influential in the dilute sulphuric acid is then introduced, or at any rate the affinity of the component parts of the electrolyte has to be overcome, since it cannot conduct *without decomposition,* and this decomposition reacts upon, and sometimes neutralizes, the forces which tend to produce the current.

(425) The mutual dependence and state of the chemical affinities of two distant portions of acting fluids, is well shown in the following experiments. Let P (Fig. 162) be a plate of platinum, Z a plate of amalgamated zinc, and y a drop of dilute sulphuric acid ; no sensible chemical action takes place till the points $P\ Z$ are con-

FIG. 162.

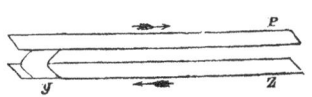

nected by some body capable of conducting Electricity : *then* a current passes ; and as it circulates through the fluid at y, decomposition ensues.

In Fig. 163, a drop of solution of iodide of potassium is substituted, at x, for the acid : the same set of effects occur ; but the electric current is in the opposite direction, as shown by the arrows.

FIG. 163.

In Fig. 164, the dilute sulphuric acid, and the iodide of potassium, are opposed to each other at y and x: there is no metallic contact between the zinc and platinum; but there is an opposition of forces; the stronger (that brought into play by the acid) overcomes the weaker, and determines the formation and direction of the current; not merely making that current pass through the weaker liquid, but actually *reversing* the tendency which the elements of the latter have in relation to the zinc and platinum if not thus counteracted, and forcing them in a contrary direction to that they are induced to follow, that its own current may have free course.

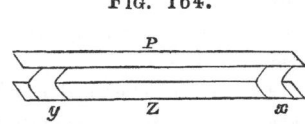

FIG. 164.

(426) To decompose a compound by the current from a single pair of plates was considered impossible : by some beautiful experiments, however, Faraday proved that iodide of potassium, protochloride of tin, and chloride of silver, may be decomposed by a single pair, excited with dilute sulphuric acid; and thereby showed the direct opposition and relation of the chemical affinities concerned at the two points of action. Where the sum of the *opposing* affinities was sufficiently beneath the sum of the *acting* affinities, decomposition took place; but in those cases where the opposing affinities preponderated, decomposition was effectually resisted, and the current ceased to pass.

(427) By increasing the *intensity* of the current, without, however, causing *more* Electricity to be evolved, solution of sulphate of soda, muriatic acid, nitrate of silver, fused nitre, and fused iodide and chloride of lead, were decomposed by a single pair. This increase in *intensity* was effected by adding a *little nitric acid* to the dilute sulphuric acid, with which the battery was charged; and that this addition caused no increase in the *quantity* of Electricity, was rendered evident from the fact, that mere wires of platinum and zinc evolved sufficient Electricity to decompose muriatic acid, which compound would not, however, yield to a large pair of plates, excited by dilute sulphuric acid alone.

(428) The source of the Electricity in the voltaic circuit, is the chemical action which takes place between the metal and the body with which it combines. As *volta-electro-generation* is a case of mere chemical action, so *volta-electro-decomposition* is simply a case of the preponderance of one set of chemical affinities, more powerful in their nature, over another set which are less powerful; and the forces termed chemical affinity and Electricity are one and the same. It is the union of the zinc with the oxygen of the water, that determines the current in the common voltaic battery; and the quantity of Electricity is dependent on the quantity of zinc oxidized and in a definite proportion to it. The intensity of the current is in proportion to the intensity of

the chemical affinity of the zinc for the oxygen, under the circumstances, and is scarcely (if ever) affected by the use of either strong or weak acid.

(429) Not chemical *combination* alone, but chemical *decomposition* also, is requisite to generate a current of Electricity : the simple union of oxygen with zinc will not produce Electricity. The oxygen must be in combination, and the compound of which it is an element must be an *electrolyte*. A pair of plates of zinc and platinum may be heated in an atmosphere of oxygen gas, sufficiently high to cause rapid oxidation of the zinc, and yet a voltaic circle will not be formed, neither will any current be excited by immersing the plates in liquid chlorine. Strong chemical action and high ignition are known to attend the combination of platinum and tin ; nevertheless, no development of Electricity was found by Faraday to attend the union of these metals ; for, though a good conductor, and capable of exerting a chemical action on tin, platinum is *not* an electrolyte, was *not* decomposed, and therefore there was no Electricity.

(430) When a fluid amalgam of potassium, containing not more than a hundredth of that metal, was put into pure water, and connected through the galvanometer with a plate of platinum in the same water, an electric current was determined from the amalgam, through the water, to the platinum ; so also, when a plate of clean lead and a plate of platinum were put into pure water, there was a current sufficiently intense to decompose iodide of potassium, produced from the lead, through the fluid, to the platinum. The Electricity in both these cases must be referred solely to the oxidation of the metals ; as in neither was there either acid or alkali present to combine with, or in any way act on, the body produced.

(431) Although a piece of amalgamated zinc has not, when alone, power enough to take the oxygen and expel the hydrogen from water, it would appear that it has the power so far to act by its attraction for the oxygen of the particles in contact with it, as to place the similar forces already active between these and the other particles of oxygen, and the particles of hydrogen in the water, in a peculiar state of tension or polarity ; and probably also, at the same time, to throw those of its own particles which are in contact with the water, into a similar, but *opposed* state. Practically, this state of tension is best relieved by touching the zinc in the dilute acid with a metal having a less attraction for oxygen than the zinc ; the force of chemical affinity is then transferred in a most extraordinary manner through the two metals, and it appears impossible to resist the idea, that the voltaic current must be preceded by a *state of tension* in the fluid, and between the fluid and the zinc. Faraday endeavoured to make this state of tension in the

electrolytic conductor evident, by transmitting a ray of polarised light through it; but he did not succeed, either with solution of sulphate of soda or nitrate of lead.

(432) By a series of beautiful experiments, Faraday has shown that electrolytes can conduct a current of Electricity of an intensity too low to decompose them; and in the case of water, when the current is reduced in intensity below the point required for decomposition, then the degree of conduction is the same, whether sulphuric acid or any other of the many bodies which can effect its transferring power as an electrolyte, are present or not; or, in other words, that the *necessary* electrolytic intensity for water is the same, whether it be pure, or rendered a better conductor by the addition of these substances; and that for currents of less intensity than this, the water, whether pure or acidulated, has equal conducting power. The following remarkable conclusion is also pointed out, viz., that when a voltaic current is produced having a certain intensity, dependent upon the strength of the chemical affinities by which that current is excited, it can decompose a particular electrolyte, without relation to the quantity of Electricity passed; the *intensity* deciding whether the electrolyte shall give way or not; and if this be confirmed, circumstances may be so arranged that the *same* quantity of Electricity may pass in the *same time, in at* the *same surface*, into the *same decomposing body in the same state*, and yet differing in intensity, will *decompose in one case and not in the other :* for, taking a source of too low an intensity to decompose, and ascertaining the quantity passed in a given time, it is easy to take another source, having a sufficient intensity, and reducing the quantity of Electricity from it, by the intervention of bad conductors, to the same proportion as the former current, and then all the conditions will be fulfilled which are required to produce the result desired.

(433) What follows is exceedingly important. From the principles of electrolytic action, it is evident that the *quantity* of Electricity in the current cannot be increased with the *quantity* of metal oxidized and dissolved at *each new place of chemical action :* hence in the compound voltaic battery the action of the number of pairs of plates is only to urge forward that quantity of Electricity which is generated by the first pair in the series, and this is effected by the amount of decomposition of water and oxidation of zinc being *equal* in each cell. A little consideration will render this evident; for if we consider that by the decomposition of a certain quantity of water in the first cell a certain quantity of Electricity (equivalent to that associated with the water decomposed) is evolved, it is clear that before this Electricity can pass through the second cell an equal quantity of water must be decomposed, and this can only be effected by the oxidation of an equal weight

of zinc : and so for each succeeding cell the electrochemical equivalent of water must be decomposed in each before the current can pass through it, and this theoretical deduction Faraday has proved by direct experiment. Each cell, then, gives a fresh impulse to the Electricity generated in the first cell ; or, in other words, increases the *intensity* of the current ; and though we may not know what *intensity* really is, being ignorant of the real nature of Electricity itself, it is not difficult to imagine that the *degree* of intensity at which a current of Electricity is evolved by a first voltaic element, shall be increased when that current is subjected to the action of a second voltaic element, acting in conformity and possessing equal power with the first.

(434) It is argued by Poggendorf, who has published in his "Annalen," Jan. 1840, a most profound paper on the theory of the voltaic pile, that the electrolytic law is no proof of the chemical origin of the Electricity of the voltaic apparatus, inasmuch as it is the property of *all* currents, voltaic, frictional, magnetic, thermal, and animal, to decompose on their passage through a series of different fluids equivalent quantities of each. But the greater part of this elaborate memoir is directed against that experiment (424) on which so much stress is laid by the supporters of the chemical theory, in which, as we have seen, two strips, one of zinc and the other of platinum, are separated at their extremities, on the one side by sulphuric acid, and on the other side by a solution of iodide of potassium. An electric current then occurs in a direction which indicates the preponderance of the sulphuric acid circuit over that of the iodide of potassium. Now, in this experiment there is, *first*, the affinity of the oxygen for the zinc ; and *second*, the affinity of iodine for the same metal ; both endeavour to excite a current, but that of the oxygen being the strongest, sets more Electricity in movement than that of the iodine ; the latter is therefore overpowered, and a current thus originates in the direction of the affinity of the oxygen, which, at the same time, since the two metals do not touch, is considered as affording a proof of the non-necessity of metallic contact to excite voltaic Electricity.

(435) On this experiment the following remark is made by Poggendorf : — " The experiment is so remarkable, and the explanation given has in appearance so much plausibility, that it is not to be wondered at if the supporters of the chemical theory have regarded it as the main prop of their opinion. Upon the defenders of the contact theory, however, it made but little impression, probably from their believing that no regard need be had to an isolated fact, speaking apparently in favour of the chemical theory, considering the numerous objections which may be urged against it. In general, they may have contented themselves with this otherwise perfectly correct position, that one metal as soon as

it is in contact with two fluids, can no longer be regarded as a single metal; so that in Faraday's experiment that end of the zinc bar which touched the sulphuric acid would be positive towards that which was moistened by the solution of iodide of potassium." The German professor then details a vast number of experiments made with two metals and two fluids not in contact, and states the following as the main results :—*That the magnitude of the electromotive force in general is altered, sometimes increased, sometimes diminished, by any substances added to water, be it an electrolyte or not, and indeed, (which should be well observed,) increased for one metal combination, and diminished for another, by the same substance, added to the water in the same proportion. Nor has he been able to find that this force stands in direct ratio to the energy of the affinity between the positive metal and the negative constituent of the fluid. It is weak in cases where this energy must be considered as strong, and, on the contrary, strong where but a weak affinity can be admitted. Frequently indeed a current originates, and at times a powerful one, where, to judge from the affinity, not the slightest action should be expected.*

(436) As another result of his experiments, Poggendorf submits that the position that those bodies which, brought between the metallic elements of the voltaic pile, render it active, are all electrolytes,[*] must be thus altered, "that the fluids between the metallic plates must, it is true, be electrolytes, *i. e.,* decomposable bodies, since at least with aqueous fluids, and with a certain intensity of current, no conduction can take place without decomposition ; but that the electromotive force which is developed on the contact of these fluids with the metals, is not in any necessary connection with the conductivity or decomposability, and can be increased or diminished by bodies which are *not* electrolytes, *i. e.* not directly decomposable."

(437) Professor Poggendorf is not satisfied with the *passive* part which the chemical theory assigns to the negative metal in a voltaic combination. He thinks his experiments warrant the conclusion, that it is essential to the *generation* of the current. If the negative metal in a circuit has merely to act a passive part, to perform merely the function of *conducting,* then the best conductor should produce the strongest current, or rather the greatest electromotive force, and as copper is a better conductor of Electricity than platinum, a copper-zinc circuit ought to be more efficacious than a platinum-zinc circuit, which is contrary to fact.

(438) With respect to the experiment with sulphuric acid and iodide of potassium, the German electrician states that it is only at

* Faraday's Exp. Researches, 858, 921.

first and *transitorily* that the sulphuric acid has the ascendancy, and that subsequently, although unquestionably it attacks the zinc more violently than the iodide of potassium, it gives way to the latter salt : a fact he thinks conclusive against the chemical theory. The reason that Faraday obtained different results is to be accounted for from the sulphuric acid not being pure, but mixed with nitric acid, in which case it always maintains a high degree of superiority over iodide of potassium. The addition of the nitric acid, according to the Faradayan theory, increases the *intensity* without interfering with the *quantity* of the Electricity produced ; but if the intensity of a chemical action is to be measured by the quantity of metal dissolved from a *unity* of surface in the *unity* of time, then if sulphuric and nitric acids be taken of such a degree of concentration that they both dissolve just the same quantity of a like zinc surface in the same time, there is, says the German professor, no reason why the nitric acid should enjoy any single advantage over the sulphuric acid, more particularly as both are non-electrolytes. Nevertheless, as nitric acid does develope a greater degree of electromotive force than sulphuric acid, the chemical theorists must suppose that the *quality* of the chemical action produces a specific difference in the excited Electricity ; but Poggendorf declares that he has convinced himself in the most positive manner *that the result of the addition of the nitric acid does decidedly not arise from the chemical attack of this acid on the zinc, but solely from an action of it on the platinum.* The acids were separated by animal membrane, a zinc (amalgamated) plate being immersed in the sulphuric, and a platinum in the nitric acid; the two other plates, zinc and platinum, standing in solution of iodide of potassium. The result was, *that the separated acids not only excite an electromotive force quite as great as the mixed, but have a slight superiority over these :* a fact, in the professor's opinion, perfectly destroying the chemical theory of Galvanism. Finally, not only the cases examined in the memoir, but others in previous ones by Fechner and others, are considered as proving in the most evident manner that the energy of the direct chemical attack of the fluid on the positive metal does in no way stand in any connection with the intensity of the excited electromotive force; and, on the other hand, it is *not proved* that the local action is ever converted into circulating, or weakened by it.* What has been advanced as such is founded or error. And lastly, it is urged that the decrease of the hydrogen at the zinc, which results on the closing of the circuit, does not happen from a *transfer* of *this* hydrogen to the negative metal, but simply from the oxygen being carried by the current to the zinc, and there combining with the hydrogen.

* Faraday's Exp. Researches, 996.

(439) The theories of voltaic Electricity have been examined with much attention by Professor De la Rive, of Geneva. The following is a brief abstract of an admirable memoir which he published in 1828, entitled, "*Analysis of the circumstances which determine the direction and intensity of the electric current in a voltaic pair.*"

(440) 1st.—He strongly contends that, insuring the absence of calorific and mechanical action, no Electricity can be developed in bodies, when they do not undergo chemical action. All the experiments that have hitherto been brought forward in opposition to this, are unsatisfactory, owing to the very great difficulty of securing the absence of chemical action in their prosecution. Messrs. Pfaff and Becquerel employed a condenser, of which one of the plates was copper, and the other zinc, between which a communication was established by means of an insulated arc of copper : the plates were put in vacuo, in hydrogen, or in azote carefully dried; or the copper plate was gilt, and the zinc plate covered with a thin coating of lac varnish. But De la Rive says, it can easily be shown, that in one case sufficient air (atmospheric) is always present to produce slight oxidation of the zinc ; and in the other the coating of varnish is too thin to prevent oxidation, which took place through the pores the alcohol produced by evaporation.

(441) But chemical action may be entirely excluded. Pairs of platinum and rhodium, and pairs of platinum and gold, give no current in very pure nitric acid ; nor do pairs of platinum and palladium in dilute sulphuric acid ; but a drop of hydrochloric acid in the one, or nitric acid in the other, immediately determines one. The following experiment was made :—

Two plates of perfectly polished steel were immersed in a flask containing solution of potash : one was insulated, and the other metallically fixed by its extremity to a plate of platinum immersed in the same liquid : the two steel plates were fixed in a cork, the upper end of each passing into the air. In three years the immersed surfaces had not lost any degree of polish ; yet, according to Volta, the plate connected with the platinum ought to have been oxidated, particularly, since potash is a good conductor of Electricity. The ends outside the cork were both oxidated—the associated plate by far the most so : hence it follows that oxidation must have commenced in order to the existence of an electric current. The current produced by this oxidation, decomposes water, and in consequence determines a stronger oxidation on the steel plate connected with the platinum ; and this oxidation, in its turn, increases the energy of the current, and is thus both cause and effect.

(442) He says sufficient attention has not been paid to the formation of coatings of *suboxide* on the surfaces of metals, which sometimes take place with great rapidity, and may be seen, by comparing the re-

cently brightened surface of a metal with one that has for some time
been exposed to the air. If a bright surface be rubbed with a cork, the
metal is always negative ; but if the rubbing be deferred for a time, the
metal is always positive, even in dry air. This is evidently occasioned
by the formation of a coating of suboxide, which is removed by the
rubbing substances, the friction afterwards taking place between the
metal and its oxide, which causes the former to be positive. Again, if
a metal which has been brightened and allowed to remain for some time
in dry air, be fixed to the negative pole of a battery, and a plate of platinum to the positive, and both immersed in dilute sulphuric acid,
oxygen appears at the positive pole some seconds before the hydrogen
shows itself at the negative, which shows that the latter must have been
employed in deoxidating the negative metal.

(443) The experiment of Becquerel, of immersing pure oxide of
manganese and platinum in pure water, is also unsatisfactory ; for the
current is not perceptible for more than half an hour, during which
time the Electricity due to chemical action (either from the slight
deoxidation of the peroxide, or the formation of a hydrate), is accumulating—here the platinum is positive.

" All chemical action disengages Electricity ; but the Electricity disengaged, is not, in every case, nor under every form, proportional to the
vivacity of the chemical action. Two principal circumstances may
explain this anomaly :—viz., the immediate recomposition in a larger
or smaller proportion of the two Electricities, at the points at which
they are separated by chemical action ; and the particular nature of this
action, which according to the bodies between which it is exerted, gives
rise to electric effects more or less intense."

(444) It is necessary here carefully to distinguish the Electricity
perceived from the Electricity *produced :* the latter must evidently be
proportional to the extent of the chemical action ; that is, that in a
given time it depends upon the number of *chemical atoms* which are
combined, and consequently upon all the other circumstances which
may have exerted an influence upon the number of these combinations
(the extent of the surface exposed to chemical action, the vivacity of
that action, &c.) The Electricity perceived, is a portion of the Electricity produced, a portion which depends on the relative conductibility
of the bodies entering into the system in which the Electricity is propagated, upon the disposition of the different parts of the system, and
upon the nature of the apparatus to be employed in showing the
presence of the Electricity, &c., circumstances which all have an influence on the degree of facility with which the two electric principles
follow some certain course, or become again immediately united to the
same surface from which by chemical action they are separated.

(445) When a capsule of platinum, filled with sulphuric or diluted nitric acid, is placed upon the plate of a condenser, and a plate of zinc held in the fingers is immersed in it, a very feeble charge is given to the plate of the condenser, although the chemical action may have been very lively : the reason is, not that there has not been an enormous disengagement of Electricity—a fact which may be proved by employing this Electricity in producing a current,—but, in this experiment, the negative Electricity developed in the zinc, unites with the positive with much greater facility than it can pass through the fingers and the body of the experimenter, in order to lose itself in the earth. There will, therefore, be only a very feeble positive tension, often scarcely any ; but if the diluted acid is replaced by concentrated sulphuric acid, though the chemical action will be less lively, the electric tension will be much stronger, this acid being a very bad conductor, and the passage of the Electricity from the liquid to the metal immersed in it, being extremely difficult, the two Electricities uniting, on the surface attacked, in much smaller proportions : if, instead of a piece of metal, a piece of wood rather moist is immersed in the concentrated sulphuric acid, the *positive tension* acquired by the acid is still stronger. If a capsule, made of an oxidable metal, be employed, and after heating it, a few drops of a liquid capable of attacking it at that high temperature, in ever so small a degree (pure water is sufficient), be poured into it, a quantity of negative Electricity is developed, which is sufficiently strong to be sensible without the assistance of the condenser, and even to give sparks. In this case the drop of liquid injected into the heated capsule is converted into vapour while it is attacking the metal, and carries off with it the positive Electricity which cannot then combine immediately with the negative Electricity left in the metal ; but if even the smallest quantity of liquid remains in the capsule, unvaporized, the immediate recomposition takes place, and only very feeble traces of negative Electricity can be obtained. If the Electricity developed by the action of a gas, or by that exerted by a humid body, such as the hand, or a piece of wood, upon the metal with which it is in contact, be often much stronger than the Electricity resulting from the much livelier action of a liquid, the reason is, that in the former case, the *immediate recomposition* of the two electric principles, is almost *null*, in consequence of the imperfect conductibility of the exciting bodies, and that the Electricity produced is almost entirely perceived. There *is*, however, a slight recomposition ; for the negative tension of an insulated metal is sensibly augmented by giving a *translatory* motion to the gas which attacks its surface ; the consequence of which is, that the positive Electricity accumulated in the gas, being removed with it, cannot unite with the negative left in the metal. The principle of the immediate

recomposition of the two Electricities applies also to the production of electric currents in a pair. In very lively chemical actions, the larger proportion of the Electricities developed often undergoes this recomposition; a small part only runs through the whole circuit, especially if it be not a very good conductor, which is the reason that the strongest currents are not always those produced by the most lively chemical actions, and that in a pair the metal most attacked, *is not always the positive one ;* that is, the one whence the current commences. However, the latter case occurs only when each of the two metals of the pair are immersed in different liquids. A single example may be adduced : a plate of zinc is immersed in concentrated sulphuric acid, and a plate of copper in nitric acid : the two acids are immediately in contact, and the two metallic plates communicate by means of the wire of a galvanometer. In this pair the zinc is positive, though it is much less attacked than the copper, because the two Electricities developed by the action of the sulphuric acid on the zinc, can be more easily reunited by making the tour of the circuit, than by passing from the sulphuric acid to the zinc, and reciprocally ; while on the contrary the two Electricities developed by the action of the nitric acid on the copper, reunite immediately with the greatest facility, in consequence of the conductibility of the nitric acid, and the ready passage of the Electricity from that acid to the copper ; while to make the circuit, they would be obliged to traverse the concentrated sulphuric acid, which is a very imperfect conductor, and pass from the zinc to the acid—a very difficult passage. Two circumstances prove the exactitude of this explanation :—1. The same result is obtained in the preceding experiment, by substituting a plate of zinc similar to that which is immersed in the sulphuric acid, for the plate of copper immersed in the nitric acid. 2. If a capsule of platinum be put upon the plate of a condenser, and filled in succession with nitric acid, and concentrated sulphuric acid, and a plate of copper or zinc held between the fingers, be immersed in the former liquid, and a plate of zinc in the latter, a much stronger positive Electricity is obtained in the second case than in the first.

(446) In applying these principles to the explanation of the theory of the voltaic pile, De la Rive remarks, that the use of the pile is to facilitate the passage of the current through imperfect conductors, and *not to increase the quantity of Electricity ;* for the utmost that can be effected by a pile composed of a certain number of similar pairs, is to compel all the Electricity produced by only *one* of its pairs, to pass through the conducting body, which connects its poles. The only means of attaining this object, is to separate the two metals of a pair, by other pairs, as similar to the first as possible. These intermediate

pairs, the number of which should correspond to the more or less imperfect conductibility of the bodies interposed, will each produce as much Electricity, as the extreme pairs. But these Electricities do *not pass through the conductor*, they only compel the Electricities of the extreme pairs to pass through it almost in totality.

(447) Let us see how this effect is produced. " We shall take a pile in activity, and suppose that all the pairs of which it is composed are so exactly similar in every respect, that the free Electricity on each of them has the same intensity. Let *b*, be a pair in the pile taken at hazard, and disposed in such a manner, that its *zinc* is immersed in the same liquid as the copper of the pair *a*, which precedes it; and its copper in the same liquid as the zinc of the pair *c*, which follows it. The chemical action of the liquid upon the zinc of the pair *b*, develops in it a certain quantity of Electricity; the portion of this Electricity, which does not undergo immediate recomposition, remains free, and the same for all the pairs, they being similar and symmetrically disposed, with relation to each other. According to this, the positive Electricity of *b*, developed by chemical action, in the liquid in which the copper of *a* is immersed, neutralizes the negative Electricity of this latter pair, which is equal to it. In the same manner, the negative Electricity of *b*, which by chemical action is carried to the zinc, and thence to the copper, in contact with the zinc, neutralizes the positive Electricity of *c*, which also is perfectly equal to it. There remains then an excess of free positive Electricity, in the liquid in which the zinc of *a* is immersed; and an excess of free negative Electricity, perfectly equal upon the copper of *c*. But these free Electricities are neutralized by the equal and opposite Electricities of the following pairs, with regard to which, we may reason in the same manner as for the pairs *a, b, c*. Thence there results an excess of free positive Electricity at the extremity of the pile, at the side of *a;* and an exactly equal excess of negative Electricity, at the extremity, situated at the side of *b*. Such is found to be the fact, if a communication be established between each of the extremities and an electroscope; and if they be united by a conductor, the two excesses of free Electricity are collected together and form the current. The intensity of this current as experiment has proved, ought to be perfectly equal to that of the current which is established in the pile itself between all the pairs."

(448) M. De la Rive next proceeds to show how it happens, that though the quantity of free Electricity developed upon each pair of the pile be frequently, not *mathematically* the same, yet the current which traverses a conductor, uniting the two extremities, is still *mathematically* equal to that which traverses each of the pairs.

To establish this important result, instead of soldering the zinc and copper of the same pair to each other, an independent conductor must be fixed to each. By means of these two conductors, a metallic communication is established between the two metals of the pair, by the intervention of one of the wires of a double galvanometer, the second wire of which serves as conductor to the current of a second pair of the same pile, or to effect a communication between the two poles.

(449) If these two currents are carefully made to pass in contrary directions, in each of the wires of the galvanometer, their action on the needle will be always found absolutely null, provided they are mathematically equal. This equality is easily explained. Take the most feeble pair in the pile; let b be the pair; the positive Electricity disengaged by b, cannot neutralize all the negative of a; there will remain then, in the copper of a, an excess of negative Electricity, which will retain, by neutralizing it, an equal quantity of positive ; the result will be, that a, though much stronger than b, can only set at liberty a quantity of positive Electricity, equal to that of b. It appears from this analysis, that the current of each pair, and consequently the current of the whole pile should be equal to the current produced by the *weakest pair*. Now experiment fully proves, that if a feeble pair is introduced into a pile composed of energetic pairs, the immediate result is a considerable diminution in the force of the current of the pile, and consequently of the current of each of the other pairs. But this reduction is never sufficient to render this current equal to that, which would be developed by the pair, introduced in an isolated state. Indeed, any pair whatever necessarily produces a greater quantity of Electricity, when it is in the circuit, than when it is isolated. From these valuable remarks, we see how necessary it is, in the construction of compound voltaic batteries, to prepare plates as similar as possible, both in size and quality of metal; for of how many pairs soever the arrangement may consist, and how perfect and alike soever all the other pairs may be, the introduction of one smaller, or faulty pair, will inevitably reduce the power of the battery to that, which would result from an equal number of pairs of plates, of the size and condition of the feeble pair.

(450) The same indefatigable electrician published also in 1836 another essay, embodying a series of experimental arguments against the contact theory. This memoir was afterwards replied to by Fechner, in a paper published in Poggendorf's Annalen,* entitled " Justification of the Contact Theory." I shall give one or two extracts from each of these memoirs, more, however, with a view of exhibiting specimens of the profundity of thought and skill thrown by both parties into the argument, than with an expectation of enabling any of my readers to form a conclusion respecting these hardly contested theories.

* Vol. xiii. p. 481.

(451) Amongst other important experiments, quoted by De la Rive, is the following. A piece of potassium or sodium was fixed, in a solid manner, by one of its ends to a platinum forceps, while the other extremity was held by means of a wooden or ivory one. If, after having well brightened it, it is surrounded by very pure oil of naphtha, and the condenser be touched with the end of the platinum forceps, no electrical sign is observable; while, if the naphtha oil is taken off, and none remain adhering to the metal, this is observed to oxidate rapidly by the contact of the air, and the Electricity indicated by the electroscope is of the most lively kind. The condenser is scarcely necessary to render it perceptible. If, sometimes, some indications of Electricity are obtained when the potassium or sodium is on the oil of naphtha, then a small quantity of humidity has been introduced into the liquid, which had remained adhering to the surfaces of the two metals, and which exercises on them a chemical action, which it is easy to recognise. Immersed in azote and in hydrogen, the two metals still give rise to a development of Electricity, proceeding from the action exerted upon them, either by the gas or by the aqueous vapour, from which it is impossible entirely to free them; and in proof of this chemical action, we see their surfaces lose their metallic brightness and become tarnished, very much as would have taken place in the air.

(452) By a variation in the method of performing this experiment, Fechner brings it forward as furnishing an argument *against* the chemical theory. If the potassium be brought into connexion with the earth, by means of *moist* wood, then powerful action is produced *in* the petroleum, arising, according to the chemical theory, from the chemical action produced through the moisture, and according to the contact theory from the increased conducting power of the wood. If the one-half of the bar of wood, which stood in connexion with the potassium, was moistened, and the other half air-dried, then no effect was produced on the condenser, provided the dry half of the wood was held in the hand; and this was even the case if the potassium was moistened with acidulated water during the contact, so that a violent chemical action took place, a proof that the non-conducting power of the dried wood is sufficient to explain the negative result.

A delicate electrometer was constructed to present the smallest possible surface : it consisted solely of a very thin and short brass wire, which, as the axis of a surrounding gum lac cylinder, traversed the perforated bottom of an inverted drinking-glass, and from which, within the glass, was suspended, between the pole plates of a dry pile, a very small gold leaf, $2\frac{1}{2}$ inches long, while the Electricity could be transferred to the prominent end of the brass, without the glass. Into the potassium ball was inserted a thin platinum wire, as short as the convenience of transfer of the Electricity allowed, and the ball itself, for the purpose of increasing its surface, was pressed between two copper plates, which had been soaked in petroleum, as smooth as was possible, without cutting the potassium ball with the platinum wire. Thus, the entire electrometer might have been somewhat about double the size of the surfaces of the potassium. The potassium disc, with the upwards bent platinum wire proceeding from it, was placed in a small glass, and covered with petroleum to about half an inch high, the platinum wire which projected from the petroleum, and which nowhere touched the glass, was discharged on to the electrometer, the glass being held in the hand. *The divergence to the side which indicates the negative Electricity followed in this case, quite as constantly, evidently, and certainly, as if the potassium had been insulated in the air.* It is true, observes Fechner, that when the potassium is brought from the air into the petroleum, the chemical action of the adhering moisture is shown by the gas bubbles which rise from the liquid ; but this development of gas soon ceases, and, twenty-four hours after, it had *entirely* disappeared : the electrical signs in the petroleum were *of quite the same force* as during the development of gas and even in the air, so that any objection raised on the grounds of chemical action is valueless, and the experiment is entirely in favour of the *contact* theory.

(453) From experiments described by De la Rive, in which two similar plates of zinc are furnished with a brass knob soldered to each, the inner surface of one plate and both exterior and interior surfaces of the other being covered with lac varnish. When these plates are made sometimes to stand in the place of the plates of a condenser, and sometimes using one of them and another brass plate, it was shown that when entirely protected from the action of the air by means of a layer of varnish, a plate of zinc does not become electric in its contact with a brass knob, and indeed that it conducts itself as a homogeneous plate of brass ; for when the brass knob was touched with the copper element of a heterogeneous plate, the zinc of which was held in the hand, it became charged with negative Electricity, though according to the contact theory all kind of action should have been neutralized, from the opposition of two pairs of plates perfectly similar.

(454) These experiments were repeated by Fechner with contrary results. He states, that in order to lay aside the objection which perhaps might be raised respecting the chemical action of the air upon the copper knob, he fixed a platinum wire to it, and then varnished the whole over so as to have the platinum alone exposed. Nevertheless, when the platinum was touched with the finger or with a slip of paper moistened in distilled water, the zinc condensers became quite as well charged with positive Electricity as if it had not been var-

nished. Becquerel and Peltier arrived at similar results;* and Pfaff, who states that he repeated De la Rive's experiments quite in accordance with his own statement, always observed the same action of the zinc condensers *with* as without varnish.

(455) The following experiment is produced by Fechner as an *experimentum crucis* against the chemical theory. Ten pairs of zinc and copper, in every respect as equal to one another as possible, were arranged into a "couronne des tasses," so that half of the said pairs produced a current opposite in its direction to that which was originated by the other half. The exciting fluid used was common water. Such an arrangement being connected with the galvanometer, can, according to either of the two theories, have no effect upon the needle, provided everything in the two systems of cells be equal: muriatic acid was then put into one of the systems, and it was found that in these circumstances the previous equilibrium was in the first instance maintained, but that by degrees the current of the water cells got the ascendancy over the acid system. "According to the contact theory," says Fechner, "the explanation of this experiment is easy." The addition of muriatic acid increases the action only by diminishing the opposition to the conduction present in the circuit, and this diminution is of as great advantage to the Electricity (which is developed by contact in the cells without acid) in its entire circulation throughout the circuit, as the Electricity of the pairs of plates which are in the very acid fluid. " How the result is to be explained according to the chemical theory, I cannot conceive."

(456) I think, however, that Schœnbein has given a very satisfactory explanation of this experiment on the *chemical* principle. From the results of various experiments, it appeared that only in a few instances the chemical difference of the exciting fluids contained in the two systems of cells determines a difference of currents produced by the two sets of pairs, and that the general rule is the production of *current equilibrium*. Now it had been established by De la Rive (445) that the electricities which are set free by chemical action at the two ends of a closed compound circle, unite themselves in two ways : one of which is the pile itself, the other the conductor placed between the poles, the quantities of Electricity recombining within each of the two conducting mediums depending upon the peculiar degree of conducting power in each. If in the case in question we consider the acid cells as originating the current and the water cells merely as the medium placed between the poles, it is evident that by far the larger portion of the Electricity developed must be re-united within the pile, and only a

* Traité d'Electricité, ii. p. 139.

small quantity pass through the water cells and through the galvanometer; but as we know that the water cells also give rise to a
current, which, on account of the peculiarity of the arrangement, would be
in direction opposite to that excited by the acid cells, it appears from the
fact of equilibrium usually taking place, that both currents are generally
equal to one another.   If ten voltaic pairs be taken, half of them
being put into water cells, and the other half into acid ones, and
arranged in the usual way, a current is produced much weaker than
that which is obtained from five pairs alone placed within the acid
fluid.   Why (the contact theorist may ask) should this be?—the extent
of chemical action in the whole arrangement must be greater than that
of only a part!—how then does it happen that the voltaic effect of ten
pairs is smaller than that produced by five?*   The answer to this
question is too obvious to require farther consideration.

The celebrated papers of Faraday on the theory of the voltaic pile
were read before the Royal Society in February and March 1840.   I
shall attempt a brief analysis of these memoirs, as they form the most
powerful series of experimental arguments that have hitherto been
brought together against the contact theory, and are considered by the
great majority of electricians, in this country at least, as quite unan-
swerable.

(457) In the arrangement shown in Fig. 165, the glasses D E are

Fig. 165.

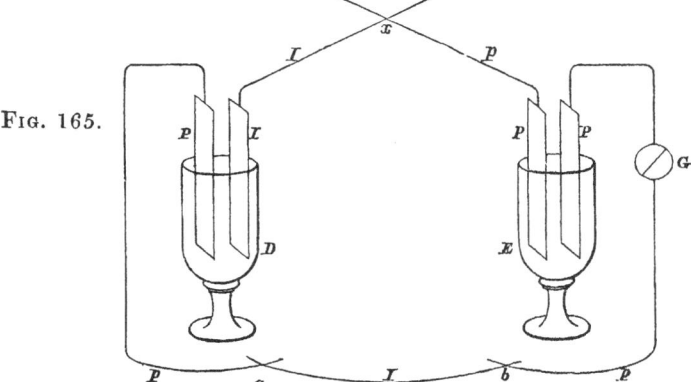

filled with solution of sulphuret of potassium; P, I, in D, are plates of
platinum and iron; and P, P, in E, plates of platinum; G is a galvanometer.   Here it will be observed that there are three metallic
contacts of platinum and iron, viz., at x, a, and b; with certain pre-

* See also, in relation to this subject, the experiments with the water
battery—(253.)

cautions *no* current passed, though heating either of the junctions at *a*, *b*, or *x*, caused a *thermo-current* deflecting the galvanometer from 30° to 50°; and when the tongue or a wet finger was applied at either of the junctions, a strong current passed : contact of platinum and iron therefore in this case produced nothing. Zinc, gold, silver, potassium, and copper, introduced at *x*, produced no current, so no electromotive force exists between these metals and platinum and iron. Various other combinations of metals were tried with similar negative results. In *green nitrous acid*, iron and platinum produced no current ; neither did it in solution of potassium. Now, according to the contact theory, the contact effects between metals and liquids, so far from being balanced, give rise to the phenomena of the pile : it cannot therefore be supposed that in the above cases the effects are balanced without straining the point in a most unphilosophical manner. According to the chemical theory, however, the facts admit of very simple explanation : where there is no chemical action there is no current, and a single experiment shows the operator what he is to expect. The contact theory cannot explain why, substituting *zinc* for *iron*, a powerful current should be produced in sulphuret of potassium with platinum ; but the chemical theory at once recognises a chemical action on the zinc, and the same is the case with copper, silver, tin, &c. Many circuits of three substances, all being conductors, were next tried, but without establishing anything like electromotive force.

(458) To account for the current of the voltaic pile, distinct and important cases ought to be brought forward, and not a case where the current is *infinitesimally* small. To account for the phenomena obtained with sulphuret of potassium, the contact force must be supposed to be balanced in some cases (iron and platinum), and not in others (lead and platinum) : in the latter case, the current ceases when a film of sulphuret has been formed by the chemical action, though the circuit be a good conductor. The case therefore will stand thus :—

Iron . . . . . . platinum	.	sulphuret of potassium	{ Electromotive forces balanced.
Lead . . . . . platinum	.	sulphuret of potassium	{ Electromotive forces *not* balanced.
Lead { with a film of sulphuret a good conductor } platinum	.	sulph. potas. . . .	{ Electromotive forces balanced.

Nothing therefore can be predicted by the contact theory regarding results.

(459) Some active circles excited by the sulphuret of potassium are next examined. *Tin* and platinum produced a strong current, tin being *plus ;* after a time the needle returned to 0, the tin becoming invested with a non-conducting sulphuret. The current here could not

have been produced by the contact force of the sulphuret, because it happens to be a non-conductor.

Lead and platinum produced a strong current, which ceased when the lead became invested with sulphuret ; nevertheless, though chemical action ceased, and therefore no current was called forth, the arrangement conducted a feeble *thermo* current exceedingly well, the *sulphuret of lead being a conductor :* this was an excellent case in point. Lead and gold, lead and palladium, lead and iron, gave similar results.

Bismuth with platinum, gold or palladium, gave active circles, the bismuth being *plus ;* in less than half an hour the current ceased, though the circuit was still an excellent conductor of thermo-currents. Bismuth with iron, nickel, or lead, produced similar results. Copper associated with any metal chemically inactive in the solution of sulphuret gave a current, which did not come to a close as in the former cases, and for this reason, the sulphuret of copper does not adhere to the metal, but falls from it in scales, exposing a fresh surface to the action of the sulphuret of potassium. Antimony, platinum, and sulphuret of potassium produced a powerful and permanent current, but the sulphuret of antimony does not adhere to this metal, which sufficiently explains the phenomenon, showing it to be dependent on chemical action. Sulphuret of antimony is not a conductor. Silver acts like copper ; the current is continuous, and the sulphuret of silver separates from the metal. Sulphuret of silver is a non-conductor. Zinc also gives a permanent current, but sulphuret of zinc is soluble in sulphuret of potassium. Now, sulphuret of zinc is a non-conductor ; how then in this case can the current be produced by contact ? All the phenomena with sulphuret of potassium are decidedly unfavourable to the contact theory : with tin and cadmium it gives an impermeable non-conducting body : with lead and bismuth an impermeable conducting body : with antimony and silver it produces a permeable non-conducting body : with copper a permeable conducting body : and with zinc a soluble non-conducting body. The chemical action and its resulting current are perfectly consistent with all these variations ; but the phenomena can only be explained, on the contact theory, by making special assumptions to suit each particular case.

(460) A series of experiments were then made with different metals in solutions unequally heated, and the results were considered as affording striking proofs of the dependence of the current on chemical action, according perfectly with the known influence of heat, and not cognizable by the theory of contact without fresh assumptions being added to those already composing it. The electric current appeared to be determined not by the amount of chemical action which takes place, but by the intensities of the affinities concerned ; and the intensity of currents is exactly proportional to the degree of affinity which reigns

between the particles, the combination or separation of which produces the currents.

(461) The effect of dilution is next examined. In Fig. 166, the part below *m* is strong acid, and that above diluted; the wires being platinum, and the fluid nitric acid, drawing the end of the wire B upwards above *m*, or depressing it from above *m* downwards, caused great changes in the galvanometer. The wires, silver, iron, lead, tin, cadmium, and zinc, being compared, it was found that the metal in the weaker acid was *plus* to that in the stronger. The fluids being strong and dilute muriatic acid, and the metals silver, copper, lead, tin, cadmium, and zinc, being compared, the metal in the strongest acid was *plus*, and the current in most cases powerful. The fluids being strong and dilute solution of caustic potash, with iron, copper, lead, tin, cadmium, and zinc, the metal in the strong solution was positive. Cases occurred also in which metals in acids of a certain strength were negative to the same metals in the same acid, either stronger or weaker.

FIG. 166.

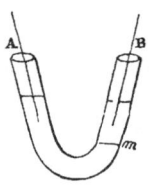

Iron and silver being in the tube C D, Fig. 167, whichever metal was in weak acid was positive to the other in the strong acid; it was merely requisite to raise the one and lower the other metal to make either positive at pleasure. Of the metals, silver, copper, iron, lead, and tin, any one can be made positive or negative to any other, with the exception of silver positive to copper: and such are the wonderful changes that may be brought about by the mere effect of dilution, that the order of these metals may be varied in a hundred different ways by the mere effect of dilution.

FIG. 167.

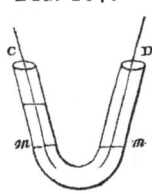

(462) The same metals in the same acid of the same strength at the two sides, may be made to change their order thus:—Copper and nickel being put into strong nitric acid, the copper will be positive: in dilute acid the nickel will be positive. Zinc and cadmium, in strong acid; the cadmium will be positive; in dilute acid, the *zinc* strongly positive. An effective battery may be constructed by employing only *one* metal and *one* fluid; thus,—if the parts of the tubes at *a*, Fig. 168, contain

FIG. 168.

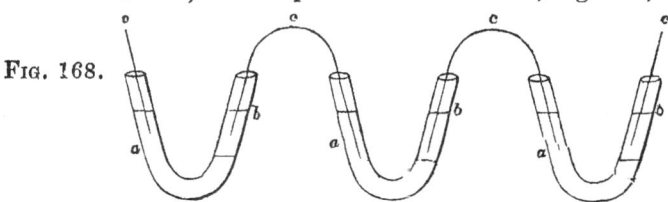

strong nitric or sulphuric acid, and the parts at b, diluted acid of the same kind, then, by connecting these tubes by wires, rods, or plates, of one metal only, such as copper, iron, silver, tin, lead, or any of those metals which become positive and negative by difference of dilution in the acid, we have a voltaic arrangement.

(463) *Where chemical action has been, but diminishes or ceases, the electric current diminishes or ceases also.*   If a piece of tin be put into strong nitric acid, it will generally exert no action in consequence of the film of oxide which is on its surface ; and if two platinum wires connected with a galvanometer be put into the acid, and one of them pressed against the tin, no current will be produced.   If now the metal be scratched under the acid, so as to expose a clean surface of metal, *chemical action takes place, and a current is produced ;* but this is only for a moment ; for oxide of tin is soon formed, chemical action ceases, and the current with it.

(464) *When chemical action changes, the current changes also.*   If copper and silver be associated in dilute solution of sulphuret of potassium, the copper will be chemically active and positive, and the silver will remain clean until of a sudden the copper will cease to act, and the silver will become instantly covered with sulphuret, showing by that, the commencement of chemical action there ; and the needle of the galvanometer will jump through 180°.

(465) *Where no chemical action occurs, no current is produced ; but a current will occur the moment chemical action commences.*   This is well illustrated by the following experiment :—In Fig. 169, let both tubes be filled with the same pure, pale, strong nitric acid, and the two platinum wires pp, being connected by a galvanometer, and the wire i, of iron, no current is produced : now, let a drop of water be put in at b, and stir the water and acid together by means

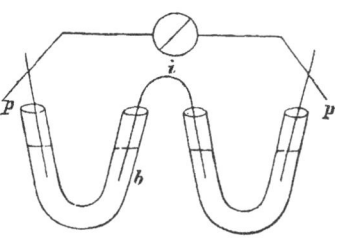

Fig. 169.

of the end of the wire i, chemical action commences, nitrous gas is evolved, and the iron wire acquires a positive condition at b, producing a powerful current.

(466) *When the chemical action which either has, or could have produced a current in one direction, is reversed or undone, the current is reversed or undone also.*   It was shown by Volta, in 1802, that crystallized oxide of manganese was highly *negative* to zinc, and similar metals, giving, according to his theory, Electricity to the zinc at the point of contact.   In 1833, Becquerel examined this subject, and thought the facts favourable to the theory of contact.   According,

however, to De la Rive, *the peroxide is at the time undergoing chemical change and losing oxygen*, a change perfectly in accordance with the direction of the current it produces. Peroxide of manganese associated with platinum in green nitrous acid, originates a current, and is *minus* to the platinum ; but a chemical action is going on, the peroxide giving up oxygen, and converting the nitrous into nitric acid. Peroxide of lead produces similar phenomena in solution of common salt, and in potash it is minus to platinum ; but direct experiments show that there is sufficient chemical action to account for the effects.

(467) Faraday concludes his elaborate defence of the chemical theory of galvanism, with the following remarks on the improbable nature of the assumed contact force :—" It is assumed that where two dissimilar metals touch, the dissimilar particles act on each other, and induce opposite states ; that the particles can discharge these states one to the other, and yet remain unchanged ; and that while thus *plus* and *minus*, they can discharge to particles of like matter with themselves, and so produce a current. But if the acting particles are not changed, it should follow, that the force which causes them to assume a certain state in respect to each other, is unable to make them retain that state, thus denying the equality between cause and effect. If a particle of platinum by contact with a particle of zinc, willingly gives off its own Electricity to the zinc, because this, by its presence, tends to make the platinum assume a negative state, why should the particle of platinum take Electricity from any other particle of platinum behind it, since that would only tend to destroy the very state which the zinc had just forced it into ? This is quite contrary to common induction ; for there a ball rendered negative, not only will not take Electricity from surrounding bodies, but if we force Electricity into it, it will, as it were, be *spurred back again* with a power equal to that of the inducing body. Or, if it be supposed that the zinc particle by its inductive action tends to make the platinum particle positive, and the latter, being in connection with the earth by other platinum particles, calls upon them for Electricity, and so acquires a positive state, why should it discharge that state to the zinc—the very substance which making the platinum assume that condition, ought of course to be able to sustain it ? Or why should not Electricity go from the platinum *to the zinc*, which is as much in contact with it as its neighbouring platinum particles are ? Or if the zinc particle in contact with the platinum particle, tends to become positive, why does not Electricity flow to it from the zinc particles behind, as well as from the platinum ? There is no sufficient, probable, or philosophic cause assigned for the assumed action, or reason given why one or other of the consequent effects above mentioned should not take place. The contact theory assumes that a force which

is able to overcome powerful resistance, *can arise out of nothing* : that without any change in the acting matter, or the consumption of any generating force, a current can be produced, which shall go on for ever against a constant resistance, or only be stopped as in the voltaic trough by the ruins which its exertions have heaped upon its own course. The chemical theory, on the other hand, sets out with a power, the existence of which is *pre-proved*, and then follows its variations, rarely assuming anything which is not supported by some corresponding simple chemical fact. The contact theory sets out with an assumption to which it adds others, as the cases require, until at last the contact force, instead of being the firm unchangeable thing at first supposed by Volta, is as variable as chemical force itself. Were it otherwise than it is, and were the contact theory true, then the quality of cause and effect must be denied. Then would perpetual motion also be true ; and it would not be difficult, upon the first given case of an electric current by contact alone, to produce an electromagnetic arrangement, which, as to its principle, would go on producing mechanical effects for ever."*

(468) It would be difficult to give a satisfactory explanation of the theory of Mr. Grove's gaseous voltaic battery (280), on the contact hypothesis. " Where," says its ingenious author, " is the contact, if not everywhere ? Is it at the points of junction of the liquid, gas, and platinum ? If so, it is there that the chemical action takes place ; and as contact is always necessary for chemical action, all chemistry may be referred to contact ; or, upon the theory of a universal plenum, all natural phenomena may be referred to it. Contact may be necessary ; but how can it stand in the relation of a cause, or of a force ?" In the opinion of Mr. Grove, the most interesting effect of this extraordinary battery is, the fact which it establishes, that gases in combining and acquiring a liquid form, evolve sufficient force to decompose a similar liquid, and cause it to acquire a gaseous form ; for it has been proved, that the gases *evolved* at the electrodes are exactly equal to the quantity absorbed in each pair of tubes.

---

* Some of the experiments brought forward by Faraday, have been commented upon by M. Martin, a translation of whose paper appears in the Trans. Elec. Soc., p. 269 ; but as the main arguments of the English philosopher have not been touched upon in the Essay, I have not thought it worth while to allude to it *particularly*.

LECTURE VII.

MAGNETISM.

Historical Notice—General facts and principles—Induction—Attraction—Repulsion—Fracture, &c.—Haldat's magnetic figures—Universality of magnetism—Magnetism developed by rotation—Methods of making artificial magnets—Terrestrial magnetism—Mariner's compass—Variation and dip—Intensity of the Earth's magnetism—Humboldt's researches—Magnetic stations—Magnetic Observatory at Dublin—Theory of terrestrial magnetism—Dr. Dalton's remarks on the Aurora Borealis in connection with magnetism.

(469) ALTHOUGH the attractive power of the loadstone appears to have been known in times of very remote antiquity, and its properties studied even during the dark ages, yet its *directive* power, and that of a needle touched or rubbed with it, seems to be the discovery of modern times. In an old French poem, called La Bible Guyot, which is contained in a curious quarto manuscript of the thirteenth century, still existing in the Royal Library at Paris, and in several other authors, there are notices which prove that the mariner's compass was known in the twelfth century; and Cardinal James de Vitri, who flourished about the year 1200, mentions the magnetic needle in his History of Jerusalem; and he adds, that it was of indispensable utility to those who travelled by sea.

(470) A Neapolitan named Flavio Gioia, who lived in the thirteenth century, has been regarded by many as the inventor of the compass. Dr. Gilbert affirms that Paulus Venetus brought the compass from China to Italy in 1260; and Ludi Vestomannus asserts, that about 1500, he saw a pilot in the East Indies direct his course by a magnetic needle like those now in use. The variation of the needle was discovered two hundred years before the time of Columbus; but the *variation of the variation*, that is, the fact that the variation was not a constant quantity, but varied in different latitudes, was first noticed by the discoverer of America, as appears from the following extract from Irving's Life and Voyages of Columbus, vol. i. p. 201. "On the 13th of September, 1492, he perceived about night-fall that the needle, instead of pointing to the north-star, varied but half a point, or between five and six

degrees, to the north-west, and still more on the following morning. Struck with this circumstance, he observed it attentively for three days, and found that the variation increased as he advanced. He at first made no mention of this phenomenon, knowing how ready his people were to take alarm ; but it soon attracted the attention of the pilots, and filled them with consternation. It seemed as if the laws of nature were changing as they advanced, and that they were entering into another world, subject to unknown influences. They apprehended that the compass was about to lose its mysterious virtues ; and without this guide, what was to become of them in a vast and trackless ocean? Columbus tasked his science and ingenuity for reasons with which to allay their terrors. He told them that the direction of the needle was not to the polar star, but to some fixed and invisible point. The variation was not caused by any failing in the compass, which, like the other heavenly bodies, had its changes and revolutions, and every day described a circle round the pole. The high opinion that the pilots entertained of Columbus, as a profound astronomer, gave weight to his theory, and their alarm subsided."

(471) That ferruginous substances always possess a greater or less degree of magnetism, has long been known. One Julius Cæsar, a surgeon of Rimini, first observed the conversion of iron into a magnet. In 1590 he noticed this effect on a bar of iron, which had supported a piece of brick work, on the top of a tower of the church of St. Augustin. The very same fact was observed about 1630, by Gassendi, on the cross of the church of St. John, at Aix, which had fallen down in consequence of having been struck with lightning. He found the foot of it wasted with rust, and possessing all the properties of a loadstone.

(472) In the year 1600 Dr. Gilbert, of Colchester, published a work, entitled " Physiologia Nova, seu Tractatus de Magneti et Corporibus Magneticis," which contains almost all the information concerning magnetism which was known during the two following centuries. He regarded the earth as acting on a magnetised bar, and upon iron, like a magnet, the directive power of the needle being produced by the action of magnetism of a contrary kind to that which exists at the extremity of the needle directed towards the pole of the globe. He gave the name of *pole* to the extremities of the needle, which pointed towards the poles of the earth, conformably to his views of terrestrial magnetism ; calling the extremity that pointed towards the north, the south pole of the needle, and that which pointed to the south, the north pole.

(473) Newton, Huygens, and Hooke, with some of the other philosophers who flourished about the end of the seventeenth century, were occupied to a certain extent with the subject of magnetism. Some of

their observations and discoveries are referred to in a manuscript volume of notes and commentaries, written by David Gregory, in 1693, in a copy of Newton's Principia, and used by Newton in improving his second edition. Newton had supposed that the law of magnetic action approaches to the inverse triplicate ratio of the distance; but Gregory did not adopt this opinion, and invalidates the arguments that were used in its support. Newton committed another mistake in asserting that red-hot iron has no magnetic property.

(474) In 1683, Dr. Edmund Halley published his theory of magnetism. He regarded the earth's magnetism, as caused by four poles of attraction, two of them near each pole of the earth ; and he supposes, "that in those parts of the world that lie nearly adjacent to any one of these magnetic poles, the needle is governed thereby, the nearest pole being always predominant over that more remote." He supposes that the magnetic pole, which was, in his time, nearest Britain, was situated near the meridian of the Land's End, and not above 7° from the north pole ; the other north magnetic pole being in the meridian of California, and about 15° from the north pole of the earth. He placed one of the two south poles about 16° from the south pole of the globe, and 95° west from London ; and the other, or the most powerful of the four, about 20° from the south pole, and 120° east of London.

(475) In order to account for the change in the variation, Dr. Halley, some years afterwards, added to these reasonable suppositions the very extraordinary one, that our globe was a hollow shell, and that within it a solid globe revolved, in nearly the same time as the outer one, and about the same centre of gravity, and with a fluid medium between them. To this inner globe he assigned two magnetic poles, and to the outer one, other two : and he conceived the change in the variation of the needle to be caused by a want of coincidence in the times of rotation of the inner globe and the external shell. " Now supposing," says he, such an external sphere, having such a motion, we may solve the two great difficulties in every former hypothesis : for, if this exterior shell of the earth be a magnet, having its poles at a distance from the poles of diurnal rotation, and if the internal nucleus be likewise a magnet, having its poles in two other places, distant also from its axis, and these latter, by a gradual and slow motion, change their places in respect of the external, we may then give a reasonable account of the four magnetic poles, as also of the changes of the needle's variation." From some reasons, which Dr. Halley then states, he concludes " that the two poles of the external globe are fixed in the earth ; and that if the needle were wholly governed by them, the variation would be always the same, with some little irregularities ; but the internal sphere, having such a gradual translation of its poles, influences the needle, and directs

it variously, according to the result of the attractive and directive power of each pole, and consequently there must be a period of revolution of this internal ball, after which the variation will return as before."

(476) Mr. Graham, a celebrated mathematical instrument-maker, in London, discovered in 1722, the daily variation of the needle. While the needle was advancing by a gradual motion to the westward, Mr. Graham found that its north extremity moved westward during the early part of the day, and returned again in the evening to the eastward, to the same position which it occupied in the morning, remaining nearly stationary during the night. Mr. Graham at first, ascribed these changes to defects in the form of his needles; but, by numerous and careful observations, repeated under every variation of the weather, and of the heat and pressure of the atmosphere, he concluded that the daily variation was a regular phenomenon of which he could not find the cause. It was generally a maximum, between 10 o'clock, A.M., and 4 o'clock, P.M. ; and a minimum, between 6 and 7 o'clock, P.M. Between the 6th of February, and the 12th of May, 1722, he made a thousand observations in the same place, from which he found that the greatest westerly variation was $14°\ 45''$, and the least, $13°\ 50''$; but in general, it varied between $14°\ 35''$ and $14°$, giving $35''$ for the amount of the daily variation.

(477) Various speculations respecting the cause of the phenomena of magnetism, had been hazarded by different authors : but it was reserved for M. Epinus to devise a rational hypothesis, which embraced and explained almost all the phenomena which had been observed by previous authors. This hypothesis, which he has explained at great length in his "Tentamen Theoriæ Electricitatis et Magnetismi," published in 1759, may be stated in the following manner :—

i.—In all magnetic bodies there exists a substance which may be called the magnetic fluid, whose particles repel each other with a force inversely as the distance.

ii.—The particles of this fluid attract the principles of iron, and are attracted by them in return, with a similar force.

iii.—The particles of iron repel each other, according to the same law.

iv.—The magnetic fluid moves through the pores of iron and soft steel with very little obstruction ; but its motion is more and more obstructed as the steel increases in hardness or temper, and it moves with the greatest difficulty in hard tempered steel and the ores of iron.

(478) The method of making artificial magnets, which was practised by the philosophers of the seventeenth century, was a very simple, but a very inefficacious one. It consisted in merely rubbing

the steel bar to be magnetised upon one of the poles of a natural or artificial magnet, in a direction at right angles to the line joining the poles of the magnet. Towards the middle of the eighteenth century, however, the art of making artificial magnets had excited general attention ; and it is to Dr. Gowin Knight, an English physician, that we are indebted for the discovery of a method of making powerful magnets. This method he kept secret from the public, but it was afterwards published by Dr. Wilson. Duhamel, Canton, Michell, Antheaume, Savery, Epinus, Robison, Coulomb, Biot, Scoresby, and others, made various improvements on this art ; but for a detailed account of their numerous experiments, I must refer to the article on magnetism, drawn up Sir David Brewster, for the seventh edition of the Encyclopædia Britannica, from which this short sketch of the history of magnetism is, in a great measure, compiled.

(479) One of the ablest cultivators of the science of magnetism was the celebrated Coulomb, who, by the application of the principle of torsion, first used by Michell, determined the correct law of magnetic attraction and repulsion. Provided with such a delicate instrument as the torsion balance, this philosopher was enabled to apply it with singular advantage to almost every branch of science. His first object was to determine the law according to which magnetism is distributed to a magnetic bar. It was, of course, well known, that the magnetism in the middle of a bar was imperceptible, and that it increased according to a regular law, and with great rapidity, towards each of the poles. By suspending a small proof needle, with a silk fibre, and causing it to oscillate horizontally opposite different points of a magnetic bar placed vertically, Coulomb computed the part of the effect that was due to terrestrial magnetism, and the part that was due to the action of the bar ; and in this way he showed the extreme rapidity with which magnetism is increased towards the poles.

(480) In examining the distribution of Electricity, in a circular plane, Coulomb found that the thickness of the electric stratum was almost constant from the centre, to within a very small distance of the circumference, when it increased all on a sudden with great rapidity, (as has been shown in a previous Lecture.) He conceived that a similar distribution of magnetism took place in the transverse section of a magnetic bar ; and by a series of nice experiments with the torsion balance, he found this to be the case, and established the important fact, that the magnetic power resides on the *surface* of iron bodies, and is entirely independent of their mass.

(481) The effect of temperature on magnets was another subject to which Coulomb directed his powerful mind ; but he did not live to give an account of his experiments, which were published after his death by

his friend M. Biot. Coulomb found that the magnetism of a bar, magnetized to saturation, diminished greatly by raising its temperature from 12° of Reaumur to 680°, and that when a magnetic bar was tempered at 780°, 860°, and 950° of Reaumur, the development of its magnetism was gradually increased, being more than double at 900° of what it was at 780°. He found also that the directive force of the bar reached its maximum when it was tempered at a bright cherry red heat at 900°, and that at a higher temperature the force diminished.

(482) Coulomb made many valuable experiments on the best methods of making artificial magnets, and he subjected all the various processes that had previously been employed to the test of accurate measurement. His experiments on the best forms of magnetic needles are equally valuable ; but the most interesting of his researches, and the last to which he devoted his great talents, were those which relate to the action of magnets upon all natural bodies. Hitherto iron, steel, nickel, and cobalt, had been regarded as the only magnetic bodies ; but in the year 1802, Coulomb announced to the institute of France, that all bodies whatever are subject to the magnetic influence, even to such a degree as to be capable of accurate measurement. In order to determine if the phenomena were owing to particles of iron disseminated through the bodies, which he subjected to experiment, he tried a needle of silver, purified by cupellation, and another needle of silver alloyed with $\frac{1}{320}$th part of iron, and he found that the action of a magnet on the former was four hundred and fifteen times less than upon the latter. Hence, as he had previously shown, that the forces exerted by magnets on needles are proportional to the absolute quantities of iron which they contain, there will be four hundred and fifteen times less iron in the pure than in the impure silver ; and since the latter contained $\frac{1}{320}$th part of its weight of iron, the first will contain $\frac{1}{415}$th part of $\frac{1}{320}$th, or $\frac{1}{132800}$th, or it will contain one hundred and thirty-two thousand, seven hundred and ninety-nine parts of pure silver, and one of iron, a quantity of alloy beyond the reach of chemical detection.

(483) Amongst the scientific travellers who have contributed to our knowledge of terrestrial magnetism, Baron Alexander Humboldt was one of the most distinguished. Himself an accurate and scientific observer, and possessed of instruments and methods of research, he made numerous accurate observations on the dip and variation of the needle in various parts of the earth, and particularly near the magnetic equator ; and by means of these valuable data, M. Biot was enabled to throw much light on the subject of terrestrial magnetism.

(484) In the aërostatic ascent of MM. Gay, Lussac and Biot, they

were unable to detect any change in the intensity of terrestrial magnetism at the height of four thousand metres. Saussure, however, had found that the intensity was considerably less on the Col du Géant than at Chamouni and Geneva; the difference in the levels of these places being, in the one case, ten thousand, and in the other, seven thousand eight hundred feet; but his observations contradict his conclusions. M. Kupffer has more recently obtained a similar result, by observations on Mount Elbrouz; having found a decrease in intensity in rising four thousand five hundred feet above his first station; and he explains the result obtained by MM. Gay, Lussac and Biot, by supposing that an increase of intensity was produced by a diminution of temperature Mr. Henwood, on the other hand, has made observations at the surface of Dolcoath mine, at one thousand three hundred and twenty feet beneath its surface, and on a hill, at seven hundred and ten feet above the level of the sea, without being able to detect any difference in the intensity. To the late Captain Foster we are indebted for many valuable observations on the magnetic intensity, made at Spitzbergen and elsewhere. From these he concluded that the diurnal change in the horizontal intensity is principally, if not wholly, owing to a small change in the amount of the *dip*. The maximum took place at about 3h. 30' A. M., and the minimum at 2h. 47' P. M.; its greatest change amounting to $\frac{1}{83}$rd of its mean value. Captain Foster is of opinion, that these changes have the sun for their primary agent, and that his action is such as to produce a constant inflection of the pole towards the sun during the twenty-four hours; an idea which had been previously stated by Mr. Christie.

(485) About the year 1818, Professor Barlow, of Woolwich, turned his attention to the subject of magnetism, with a view principally of calculating the effect of ship's guns on the compass. In trying the effect of different iron balls, he was led to the curious facts—that there exists round every mass of iron, a great circle inclined to the horizon, at an angle equal to the complement of the dip of the needle;—that the plane of this circle is a plane of no attraction upon a needle whose centre is in that plane;—that if we regard this circle as the magnetic equator, the tangent of the deviation of the needle from its north or south pole will be proportional to the rectangle of the sign of the double latitude, and cosine of the longitude;—that when the distance of the needle is variable, the tangent of deviation will be reciprocally proportional to the cube of the distance, and that, all things else being the same, the tangent of deviation will be proportional to the cubes of the diameters of the balls, or shells, whatever be their masses, provided their thickness exceeds a certain quantity. Mr. Barlow was, from these discoveries, enabled to invent a most ingenious method of correct-

ing the error of the compass, arising from the attraction of all the iron on board ships. This source of error had been noticed by Mr. Wales, Mr. Downie in 1794, and by Captain Flinders; but it is to Mr. Bain that we owe the distinct establishment and explanation of this source of error. As a hollow shell of iron, about four pounds in weight, acts as powerfully at the same distance as a solid iron ball of two hundred pounds weight, Mr. Barlow happily conceived that a plate of five or six pounds weight might be made to represent and counteract the amount of the attraction of all the iron on board a vessel, and therefore leave the needle as free to obey the action of terrestrial magnetism as if there were no iron in the ship at all. After this ingenious contrivance had been submitted to the Admiralty, it was tried in every part of the world; and even in the regions which surround the magnetic pole, where the compass becomes useless, it never failed to indicate the true magnetic direction, when the connecting plate was properly applied. "Such an invention as this," says Captain Parry, "so sound in principle, so easy in application, and so universally beneficial in practice, needs no testimony of mine to establish its merits; but when I consider the many anxious days and sleepless nights which the uselessness of the compass in these seas had formerly occasioned me, I really should have esteemed it a kind of ingratitude to Mr. Barlow, as well as great injustice to so memorable a discovery, not to have stated my opinion of its merits, under circumstances so well calculated to put them to a satisfactory trial." For this beautiful invention, the board of longitude conferred upon Mr. Barlow the highest reward of five hundred pounds; and the emperor of Russia, who was never inattentive to the interests of science, sent him a fine gold watch, and a rich dress chain, for the same contrivance.

(486) The late Dr. Morichini, an eminent physician at Rome, first announced it as an experimental fact, that an unmagnetised needle could be rendered magnetic, by the action of the violet rays of the sun. His experiments were successfully repeated by Dr. Carpi, at Rome, and the Marquess Ridolfi, at Florence : but M. d'Hombre Firmas, at Alais, in France, Professor Configliachi, of Pavia, M. Bérard, of Montpellier, failed in obtaining decided magnetic effects from the violet rays. In 1814, Dr. Morichini exhibited the actual experiment to Sir Humphry Davy, and 1817, Dr. Carpi showed it to Professor Playfair. A few months after Sir Humphry witnessed the experiment, Sir David Brewster met him at Geneva, and learned from him the fact, that he had paid the most diligent attention to one of Morichini's experiments, and that he saw an unmagnetised needle rendered magnetic by violet light. The following account of the experiment, made by Dr. Carpi, was given to Sir David, by Professor Playfair :—" The violet light was

obtained in the usual manner, by means of a common prism, and was collected into a focus by a lens of sufficient size. The needle was made of soft wire, and was found upon trial, to possess neither polarity, nor any power of attracting iron filings. It was fixed horizontally upon a support, by means of wax, and in such a direction, as to cut the magnetic meridian at right angles. The focus of violet rays was carried slowly along the needle, proceeding from the centre, towards one of the extremities, care being taken never to go back in the same direction, and never to touch the other half of the needle. At the end of half an hour after the needle had been exposed to the action of the violet rays, it was carefully examined, and it had acquired neither polarity nor any force of attraction : but after continuing the operation twenty-five minutes longer, when it was taken off and placed on its pivot, it traversed with great alacrity, and settled in the direction of the magnetical meridian, with the end over which the rays had passed turned to the north. It also attracted and suspended a fringe of iron filings. The extremity of the needle that was exposed to the action of the violet rays, repelled the north pole of a compass needle. This effect was so distinctly marked, as to leave no doubt in the minds of any who were present, that the needle had received its magnetism from the action of the violet rays." In this state of the subject, Mrs. Somerville made some simple and well conducted experiments, which seemed to set the question at rest, from the distinct and decided character of the results. A sewing needle, an inch long, and devoid of magnetism, had one half of it covered with paper, and the other exposed to the violet rays of the spectrum, five feet distant from the prism. In two hours it acquired magnetism, the exposed end exhibiting north polarity. The *indigo* rays produced an equal effect, and the *blue* and the *green* the same in a less degree. The yellow, orange, and red rays, had no effect even after three days' exposure to their action. Pieces of blue watch spring, received a higher magnetism. When the sun's light fell upon the exposed end through blue-coloured glasses, or through *blue* or *green* riband, the same magnetic effects were produced. The experiments of Mr. Christie, an account of which was read to the Royal Society, a short time before Mrs. Somerville's, confirmed her results to a certain degree, by a different mode of observation. The general opinion seems, however, now to be, that light does not exercise any decided effect in producing magnetism. The experiments of MM. P. Ries and Moser, were made with needles, both polished and oxidated, and also with wires half polished; and polarized as well as common light was made to fall on them in a concentrated state, but no decided effect upon their number of oscillations could be observed; and they state that they think themselves justly entitled *to reject totally a discovery which for seventeen years has at different times disturbed science.**

(487) A valuable series of observations on the influence of the aurora borealis on the magnetic needle, was made by Dr. Dalton, at Kendall and Keswick, during seven years, from May, 1786, to May, 1793, and has been published in his meteorological observations and essays, which appeared in 1793. During these observations, he noticed the effect produced on the magnetic needle, and he was thus led to study the phenomena of the aurora, and to establish, beyond a doubt, the relation of all its phenomena to the magnetic poles and equator. In some cases, however, Dr. Dalton did not observe any perceptible disturbance of the needle.

(488) Professor Hanstein, who has been extensively occupied with the subject of magnetism, observes, that large extraordinary movements of the needle, in which it traverses frequently, with a shivering motion, an arc of several degrees on both sides of its usual position, are seldom, perhaps never exhibited, unless when the aurora borealis is visible, and that this disturbance of the needle seems to operate at the same time in places the most widely separate.

(489) From the extensive series of accurate observations, made by M. Arago, at Paris, since 1818, the needle was almost invariably found to be affected by auroræ that were seen in Scotland; and so striking was the connexion between the two classes of facts, that the existence of the aurora could be inferred from the derangements of the

---

* On 24th February, 1840, the following account of some experiments on this subject was laid before the Royal Irish Academy by Mr. G. J. Knox and the Rev. T. Knox:—" Having procured several hundred needles, of different lengths and thicknesses, and having ascertained that they were perfectly free from magnetism, we enveloped them in white paper, leaving one of their extreme ends uncovered. Taking advantage of a favourable day for trying experiments upon the chemical ray, (known by the few seconds required to blacken chloride of silver,) we placed the needles at right angles to the magnetic meridian, and exposed them for two hours, from eleven till one, to the differently refrangible rays of the sun, under coloured glasses. Those beneath the red, orange, and yellow, showed no trace of magnetism, while those beneath the blue, green, and violet, exhibited, the two first feeble, but the last strong traces of magnetism.

" To determine how far the oxidating power of the violet ray is concerned in the phenomenon, we exposed to the different coloured light needles whose extremities had been previously dipped in nitric acid, and found that they became magnetic (the exposed end having been made a north pole) in much shorter time than the others, and that this effect was produced in a slight degree, under the red, (when exposed a sufficient length of time,) strongly under white glass, and so strong under violet glass, that the effect took place even when the needles were placed in such a position along the magnetic meridian, as would tend to produce, by the earth's influence, a south pole in the exposed extremity.

" Conceiving that the inactive state produced in iron (as observed by Schoenbein, 292 et seq.) when plunged into nitric acid, sp. gr. 9·36, or being made the positive pole of a battery, or by any other means, might throw some light upon the nature of the change produced, experiments were instituted to this effect, which showed that no trace of magnetism could be thereby produced."—See some curious experiments on the magnetic influence of the *lunar spectrum* in 20th vol. of Phil. Mag.

needle. M. Arago has likewise discovered, that early in the morning, often ten or twelve hours before the aurora is developed in a very distant place, its appearance is announced by a particular form of the curve, which exhibits the diurnal variation of the needle, that is, by the value of the morning and evening maxima of elongation.

(490) During the late journey of Captain Back to the polar regions, in 1833, 1834, and 1835, he found that the needle was generally affected by the aurora; and on one occasion the deviation which it produced was 8°; he repeatedly observed that when the aurora was concentrated in individual beams, the needle was powerfully affected; but that it generally returned to its mean position, when the aurora became generally diffused. On several occasions, the needle was restless, and exhibited the vibrating action, produced by the aurora, when this motion was not visible; and Captain Back states that he could not account for this, except by supposing the invisible presence of the aurora in full day.

(491) The only metals which were supposed to have a distinct and decided power, and were therefore called magnetic metals, are iron, cobalt, and nickel. Mr. David Lyon has lately endeavoured to show that these metals resemble one another, not only in their principal qualities, but in the numerical values of their qualities: and, he adds, that whilst these three magnetic substances have the values above referred to, near each other, there are no other substances in which the same values come very near, or fall within those of the three magnetic substances. The values to which Mr. Lyon alludes are the following :—

	Specific gravity.	Atomic weight.	Atoms contained in a given space.
Nickel . .	8·27 . .	739·51 . .	1118
Iron . . .	7·21 . .	678·43 . .	1062
Cobalt . .	7·8 . .	738 . .	1057

These speculations, though ingenious, and deserving of attention, have been recently overturned by some observations of Professor Faraday. " Cobalt and chromium," says he, " are said to be both magnetic metals. I cannot find that either of them is so, in its pure state, at any temperatures. When the property was present in specimens, supposed to be pure, I have traced it to iron or nickel."

(492) Mr. Faraday thinks that all metals are magnetic, in the same manner as iron, though not at common temperatures, or under ordinary circumstances. He does not allude to a feeble magnetism, uncertain in its existence and source, but to a distinct and decided power, such as that possessed by iron and nickel; and his impression is, that there is a certain temperature for each metal (well known in the case of iron), beneath which it is magnetic, and above which it loses all power, and

that there is some relation between this *point* of temperature, and the intensity of the magnetic force, which the body, when reduced beneath it can acquire; iron and nickel would then be no more exceptions from the metals in regard to magnetism, than mercury is in regard to liquefaction.

(493) M. Pouillet, on the other hand, thinks that there are five simple magnetic bodies, viz., iron, manganese, nickel, cobalt, and chrome ; and in consequence of having observed some remarkable analogies, between the distance of the atoms of bodies, and their magnetic properties, he was led to suppose that the magnetic limit of different bodies ought to be found at very different temperatures. " I have, indeed," says he, " demonstrated by experiment, first, that cobalt never ceases to be magnetic, or rather that its magnetic limit is at a temperature higher than the brightest white heat ; second, that chrome has its magnetic limit a little below the temperature of dark blood-red heat ; third, that nickel has its magnetic limit about 350° centigrade, nearly at the melting point of zinc ; and fourth, that manganese has its magnetic limit at the temperature of from 20° to 25° below zero. Experiments on these five magnetic bodies seem to prove, first, that heat acts upon magnetism only, in consequence of the greater or less distance which it occasions between the atoms of bodies ; and second, that all bodies would become magnetic if we could by any action whatever, make their atoms approach within a suitable distance."

(494) Having thus given a short sketch of the history and progress of magnetism, I shall proceed to give an account of its general facts and principles.

The native magnet, or natural loadstone, is an ore of iron, consisting chiefly of the two oxides of that metal, together with a small proportion of quartz and alumina. It is usually of a dark grey hue, and has a dull metallic lustre. It is found in considerable masses in the iron mines of Sweden and Norway, and also in different parts of Arabia, China, Siam, and the Philippine Islands. Small loadstones have occasionally been met with among the iron ores of England. The smallest loadstones have generally a greater attractive power, in proportion to their size, than larger ones. They have been found of such a strength, that, though weighing only about twenty-five grains, they could lift a piece of iron *forty-five* times heavier than themselves. Sir Isaac Newton had a small specimen, set in a ring, which was capable of lifting seven hundred and forty-six grains of iron, or two hundred and fifty times its own weight ; and it is stated by Cavallo, that he has seen a loadstone which weighed only six grains and a half, which lifted a weight of three hundred grains.

(495) If we immerse a natural loadstone—no matter of what shape —in a quantity of clean iron filings, we shall find that there are two points exactly opposite each other, on which the filings are accumulated

more abundantly than on any other place, assuming the form shown in Fig. 170, the lines diverging from the ends of the magnet in *curves*, the centre *a*, being nearly free from them.   These are called its poles; and if we balance a small needle of iron on a pivot, and bring it near either of these poles, we shall find that it will be attracted towards it; or, conversely, if we suspend the loadstone by a fine fibre, and bring into the vicinity of its poles a piece of soft iron, it will be drawn towards the iron ; a reciprocal attraction is exerted between them, action and re-action being equal and opposite.

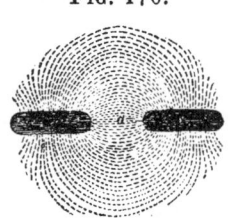

FIG. 170.

(496) The power of the natural magnet is greatly increased, by adapting two pieces of flat iron to its poles, and enclosing it in a silver or brass case.   In Fig. 171, a magnet thus *armed*, is shown : A is the loadstone ; BB, two pieces of soft iron placed against its opposite poles, the lower ends turning inwards, and fastened by transverse bars of copper, *c c*, to the brass case which surrounds the sides and upper part of the stone.   In the top of the box is inserted a ring R, for the purpose of suspending the whole ; and to the lower part of the armature is adapted a piece of soft iron, with a hook, on which is hung as much weight as the strength of the magnet will bear.

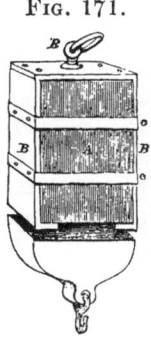

FIG. 171.

(497) When a piece of steel has been rendered magnetic, it exhibits the same properties as the natural loadstone; and since we are in possession of a variety of methods of communicating to it this state, the artificial magnet is always employed in experimental investigations.   I shall describe some of the most approved methods of magnetising iron presently : in the mean time I shall only observe that, for the exhibition of the experiments I shall first have to allude to, the following simple and ready method will be found amply sufficient for communicating to small bars of steel the requisite degree of magnetism.

Let a straight bar of hard tempered steel be held in a position slightly inclined to the perpendicular, the lower end deviating to the north, and struck several hard blows with a hammer, it will be found to have acquired by this process, all the properties of a magnet.*

(498) If a bar which has been thus treated, be supported on a pivot,

---

* A bar of iron heated red-hot, and allowed to cool in the direction of the magnetic dip, (544,) will generally be found magnetic ; a thin piece of wire may be rendered magnetic by forcibly twisting it till it breaks.

in such a manner as to have entire freedom of motion in a horizontal plane, and if all bodies of a ferruginous nature are removed from its vicinity, it will, after a few oscillations, take up a position nearly north and south ; and if it be disturbed from this position, and placed in any other, it will not remain there ; but as soon as it is at liberty to move, it will resume its former position. It will also possess the power of communicating magnetism to hard steel *permanently*, and to soft iron *temporarily*, the degree of strength, of course, depending on its own power ; and with respect to the steel, on the time which it is suspended. If two magnetised bars be poised and placed in different positions respecting each other, it will be found that in some cases they appear to be attracted towards each other ; while in others, they manifest a mutual repulsion. This, however, does not happen capriciously ; the two north poles and the two south poles invariably repel each other ; but the north pole of one magnet always attracts, and is of course attracted by the south pole of the other.[*]

(499) In order to communicate magnetism from a natural or artificial magnet, to unmagnetized iron or steel, it is not necessary that the two bodies should be in contact. The communication is effected as perfectly, though more feebly, when the bodies are separated by space. Thus in Fig. 172, if the *north pole* of an artificial steel magnet A, is placed near the extremity S, of a piece of soft iron B, the

FIG. 172.

end s, will instantly acquire the properties of a *south* pole, and the opposite end n, those of a north pole. The opposite poles would have been produced at n and s, if the south pole s, of the magnet A, had been placed near the iron B.

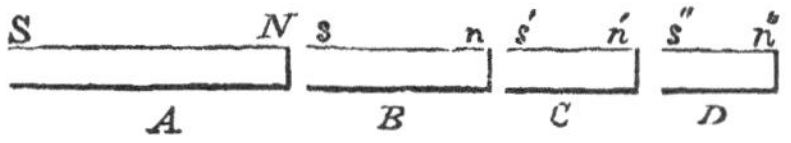

In like manner, the iron B, though only temporarily magnetic, will render another piece of iron C, and this again, another piece D, temporarily magnetic, north and south poles being produced at n', s', and n'', s''.

(500) Here we cannot fail to observe a pointed analogy between the phenomena of magnetic attraction and repulsion, and those of Electricity. In both there exists the same character of double agencies of opposite kind, capable, when separate, of acting with great energy, and being, when combined together, perfectly neutralized, and exhibiting no signs of activity. As there are two electrical, so there are also two magnetic powers ; and both sets of phenomena are governed by the same characteristic laws. So also in the last experiment, the magnetism inherent in B, C, D, is said to be *induced* by the presence of the real magnet A ; and the phenomena are exactly analogous to the com-

[*] Artificial magnets have been constructed by reducing to powder the native magnetic oxide, and forming it into bars with wax and oil.

munication of Electricity to unelectrified bodies by induction, the positive state inducing the negative, and the negative the positive, in the
parts of a conductor placed in a state of insulation, near an electrified body.

(501) A simple experiment will satisfactorily show that soft iron
possesses magnetic properties, while it remains in the vicinity of a magnet. Let A, Fig. 173, be a magnet, and K   FIG. 173.
a key, held either horizontally near one of
its poles, or near its lower edge. Then if
another light piece of iron, such as a small
nail, be applied to the other end of the key,
the nail will hang from the key, and will
continue to do so while the magnet is slowly
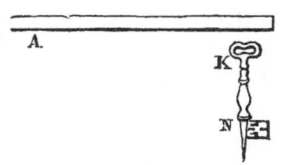
withdrawn; but when it has been removed beyond a certain distance,
the nail will drop from the key, because the magnetism induced on the
key becomes at that distance too weak to support the weight of the
nail. That this is the real cause of its falling off, may be proved, by
taking a still lighter fragment of iron, such as a piece of very slender
wire, and applying it to the key. The magnetism of the key will still
be sufficiently strong to support the wire, though it cannot the nail;
and it will continue to support it, even when the magnet is yet further
removed. It at length, however, drops off.

If the key be held above a portion of iron filings, they will not be
attracted by it; but if the magnet be then brought near the ring of the
key, as in the figure, the iron filings will instantly start up, and be
attracted by the key.

(502) It has been observed, that in all cases where a magnet attracts
iron, a re-action takes place, the iron attracting the magnet; it is the
same with a bar of iron on which magnetism has been induced. It re-
acts upon the magnet, which induces its magnetism, and increases its
magnetic intensity. Hence, we derive a distinct explanation of the
remarkable facts, that a magnet has its power increased by having
a bar of iron placed in contact with one of its poles; and that we can
gradually add more weight to that which is carried by the magnet,
provided we make the addition slowly, and in small quantities, the
power of the magnet being increased by the re-action of each separate
piece of iron that it is made to carry.

These facts enable us to explain the phenomena of magnetic attraction and repulsion. The magnet attracts a piece of iron, by inducing
an opposite polarity at the end in contact with it; and the two opposite principles attract each other. In like manner, the north pole
of one magnet attracts the south pole of the other; and the north and
south poles repel each other, in consequence of the attraction and repulsion of the opposite and similar principles. The attraction of iron
filings is explained in the same way. The particle of iron next the

magnet, has magnetism induced on it, and it becomes a minute magnet, like *B*, Fig. 172. This particle again makes the next particle a magnet, like *C*, and so on, the opposite polarity in each particle of the filings attracting one another, as if they were real magnets.

(503) In comparing the amount of the *attractive* force of two *dissimilar* poles of two magnets, with the amount of the *repulsive* force of the two similar poles, it has been found that the former force is considerably greater than the latter. This result is a necessary consequence of the inductive process above described. When the two attracting poles are in contact, each magnet tends to increase the power of the other, by developing the opposite magnetic states in the adjacent halves, and thus increasing their mutual attraction. But when the two repelling poles are brought into contact, the action of each half brought into contact, has a tendency to develope in that half a magnetism opposite to that which it really possesses, and thus to diminish the two similar principles, and weaken their repulsive power. This injurious influence of opposite poles upon the repulsive power of the magnets in action, is well exhibited when one of the magnets is very powerful, and the other very weak. When the two similar poles are held at a moderate distance, a repulsion is distinctly exhibited; but when they are brought into contact, the stronger *attracts* the weaker magnet, an effect which is produced by its actually destroying the similar weak magnetism in the half next to it, and inducing in that half the opposite magnetism, which of course occasions attraction.

(504) The most favourable position for the bars receiving and sustaining magnetism, is that of parallelism; for in this position, the induced magnetism of magnet *A*, Fig. 174, is the most powerful. Let us first suppose the bars arranged as in Fig. 174, the pole *N*, of the magnet *A*, acts favourably in inducing *south* polar magnetism in *n*, and north polar at *S*; but it is evident that the

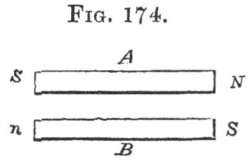

Fig. 174.

remote pole *S*, must tend to weaken the inductive force of *N*, by inducing, though in a feeble degree, north polar magnetism at *n*, and south polar at *S*.

If the soft iron *B*, be placed, as in Fig. 175, the induced magnetism will be nearly as strong as before, the greater proximity of *N*, tending to produce south polar magnetism in *n*, being compensated for by the proximity of *s*, tending to produce north polar magnetism in *n*. In the inclined position *C*, the induced magnetism is still stronger, as *s* acts more

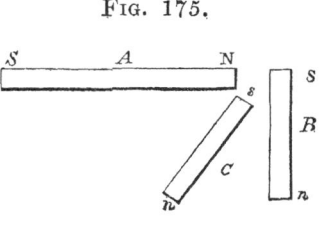

Fig. 175.

powerfully upon n ; and when the two bars are parallel, the inductive power is of course at its maximum.

If a magnetised steel bar be placed with its north pole, opposite the middle of a soft iron bar, the effect will be to induce upon it two north poles at the extremities, and a south pole in the centre ; and so also if the soft iron bar be placed between two magnets, whose north poles are nearest the bar, these north poles will tend to produce two south poles at the extremities, and consequently a northern polarity in the middle.

In like manner a piece of soft iron, ss, ss, Fig. 176, of the form of a cross, will have south poles at s, s, s, s, if the south pole of the magnet A be placed on or near its centre, as it may be conceived to consist of two bars, ss, ss. For the same reason, if a circular piece of soft iron be substituted in place of the cross, and the south pole of the magnet placed on or near its centre, that centre will be a north pole, and every part of the circumference will have a southern polarity.

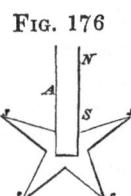

FIG. 176

FIG. 177.

(505) A very instructive experiment, founded on magnetic induction, is exhibited in Fig. 177, where several soft iron wires are suspended from the north pole of a magnet. Each of the ends becomes a south pole by induction, from the action of the north, and consequently the lower ends north poles. The south poles have a natural tendency to repel each other, but are prevented from yielding to their repulsive forces, in consequence of their strong adhesion to the north pole, N. The north poles, n, n, n, are, however, free from this restraint, and exhibit their mutual repulsion, by diverging, as shown in the figure. Hence we see the reason why rows of iron filings, adhering to each other when attracted by a magnet, keep separate from each other by the repulsive forces of the similar poles.

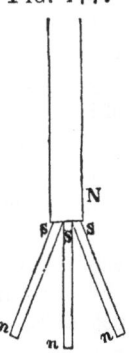

FIG. 178.

In the following experiment of Cavallo, the repulsion of the poles is well illustrated. Let two short pieces of iron wire (Fig. 178) be each fastened to a thread, the threads being joined at their other ends and formed into a loop, by which they are to be suspended from a hook, so as to have full liberty to move. On bringing the pole of a magnet, the north for instance, at a certain distance below the wires, it will

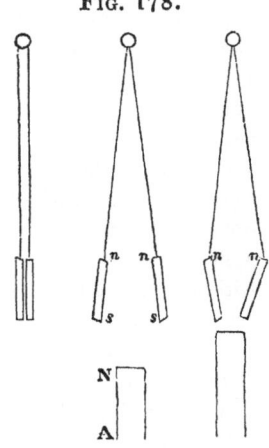

occasion them to recede from each other, as shown in Fig. 178, indicating the repulsion which takes place between the adjacent ends of the wires, in consequence of their being similarly affected by the inductive power of the magnet, the lower ends of both being rendered south poles, and the upper north poles. This divergency of the wires will continue to increase until the magnet has approached to a certain limit; but if the magnet be brought nearer than this limit, its own attractive force becomes so strong as to overpower the repulsion that exists between the lower ends of the wires, and therefore brings them nearer to each other, as shown in Fig. 178, while the repulsion of the upper ends, nn, still continues to manifest itself, by keeping them remote from one another. On removing the magnet entirely, the wires immediately collapse, their magnetism being only of a transitory nature. But if the same experiment be made with sewing needles, instead of soft iron wires, the needles will often continue to repel each other, after the removal of the magnet, having acquired some degree of permanent magnetism, by the circumstances in which they have been placed.

(506) The neutralization or destruction of induced magnetism by two equal and opposite magnetic actions, is shown by the following experiment of Dr. Robinson. If we take a forked piece   FIG. 179. of iron, C, D, E, Fig. 179, and suspend it by the branch D, from the north pole of a magnet, B, it will be magnetised by induction, and will carry a key at its lower end, E, which will be a north pole. If we now apply to the other branch, C, the south pole, S, of another and equal magnet, A, the key will instantly drop off. This obviously arises from the south pole, S, inducing a south pole at E, which either destroys or neutralizes the north polar magnetism previously induced by N.

(507) We have seen the close analogy which exists between the phenomena of Electricity and Magnetism, as far as relates to the law of action and the influence of induction (32 et seq.); but beyond this point it fails us entirely. No natural or artificial magnet has ever been seen with *only one pole*, or one kind of magnetism; electricity, on the other hand, whether positive or negative, is not only capable of being excited by induction, but may be actually transferred from one body to another. A body may without difficulty be electrified, positively or negatively, as has been shown in a former Lecture; but with magnetism there is never any transfer of properties, but only the excitation of those which were already inherent on the body operated on. It became, therefore, interesting to examine experimentally the distribution of magnetism in a part of a magnet, cut from its north or south extremity. The experiment has been often made, both by cutting it through at the middle or neutral point, or by cutting or breaking off

a portion from the end of it. If
N, S, Fig. 180, be a magnet, N,
its north, and S, its south pole, and
A, C, B, the curve representing the
intensity of its magnetism ; then, if
we cut it through the middle, C,
each half, n s, n' s', will be a complete magnet, with a north pole at
n, and a south pole at s, and their

FIG. 180.

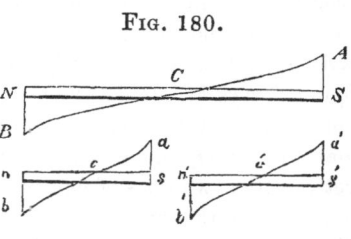

neutral points at c, c'; the curves, a c b, a' c' b', representing the
distribution of their north and south polar magnetism, being similar
to the curve A C B of the large magnet, of which they are halves.

(508) This curious experiment was devised by Epinus ; and when
he first made it, he did not divide the magnet in two, but he set two
steel bars end to end, and magnetised them as one magnet ; so that this
compound magnet had its magnetism distributed as in a single bar,
N S, Fig. 180.   He then separated them and found that each bar was
a perfect magnet, with two poles. To insure success in this experiment,
the united ends of the bars should be ground together, so as to be kept
in perfect contact, and preserved in this state by a powerful pressure,
during the time that they are magnetised.

Upon the separation of the magnets thus united, Epinus found that
two poles were instantly developed in each half, but that the neutral
points, c c', Fig. 180, were nearer the interior poles, s n', or, what is
the same thing, nearer the original neutral point, C, than to n' and s.
In the space of about a quarter of an hour it had, however, advanced
nearer to the middle points, c, c', and continued for some hours, sometimes for days, to advance to these points, which it finally reached, thus
completing the regular distribution of the two opposite magnetisms.

(509) We are indebted to M. Haldat, of Nancy, for the discovery of
magnetic figures, analogous to those first produced with Electricity, by
M. Lichtenberg, and which may easily be exhibited. For this purpose
he employs plates of steel, from eight to twelve inches square, and from
one-twentieth to one-eighth of an inch thick. The plates which he
used were of that kind of steel which is used for the manufacture of
cuirasses ; so that it did not require to be tempered, being sufficiently
hard to preserve the magnetism communicated to it.   Figures of any
kind may be traced on the surface of the steel plate, either by one magnet or by several combined, and the best form for this purpose is that
in which the poles are rounded.   In this way we may write on a steel
plate the name of a friend, or sketch a flower, or a figure, with the
extremity of a magnet.   If it is the south pole that we use, all the
traces that we make will have north polar magnetism ; and if we shake
steel filings on the plate out of a gauze bag, the filings will arrange

themselves in the empty spaces between the lines traced by the pole of the magnet, and thus represent in vacant steel the name which has been written, or the flower or figure which has been sketched. " These figures," says M. Haldat, " have a perfect resemblance to those which are formed on the surface of non-magnetic plates, viz., wood, card, glass, or paper, under which a magnet is placed. The resemblance between the two sorts of figures, when the magnets and the parts magnetised have the same form, is not only exact in the whole figure, but likewise in the smallest details. The filings collect at the parts where the magnetism is most intense, and they arrange themselves in pencils and radii. These curves, and pencils, and rays, so similar at the two poles of the same magnet, have such a resemblance that they do not allow us to distinguish the two parts from one another."

(510) In sifting the iron filings upon the steel plate, a general vibration of the plate, by tapping its edge with the ring of a small key, will assist the filings in taking their proper places ; but we must avoid such vibrations as will produce regular acoustic figures, unless we wish, as M. Haldat has found to be practicable, to unite the magnetic with the acoustic figures, which produces very interesting and varied forms.

In order to remove the magnetism from the steel plates, they may be heated over charcoal, till they become of the straw-coloured temperature ; and to render the re-polishing of them unnecessary, M. Haldat tins them, and the temperature at which the tin melts, when it is required to efface the magnetism, indicates the necessary heat.

(511) As the figures traced on the steel are nothing more than magnets of different forms, and are surrounded on all sides with a substance capable of acquiring the magnetism which may be developed by communication, we might expect, as M. Haldat remarks, that this means of communication between the opposite poles of the magnets would bring them into a neutral state. This, however, is not the case ; and the portion of the metal which surrounds the magnetic figure, performs the part of the *armature* of a loadstone, and the magnetism is thus kept up.

The figures might be rendered permanent, by covering the steel plate either with a gummy or balsamic solution, which will become hard by exposure to the air ; or with a coating of some easily melted substance which becomes fixed at ordinary temperatures. If we sift the iron filings on the steel plate when covered with such a fluid, the filings will take their magnetic position round the traced lines, and will become fixed by the induration or solidification of the fluid coating.

(512) For a long time *iron* was regarded as the only magnetic metal ; for though other substances were found to possess the property of acting and of being acted upon as a magnet, it was imagined that such effects were rather the result of the small quantity of iron entering

into their composition, than to the magnetism residing in the proper substance of the body.

(513) A series of careful experiments were made by Cavallo, on the magnetism of *brass* when hammered. He found that this compound, whether old or new, was made magnetic, when placed between two pieces of card, and hammered, either on an anvil by a hammer, or between two flints. He observes, " It appears that the property of becoming magnetic in brass, by hammering, is rather owing to some peculiar configuration of its parts, than to any admixture of iron ; which is confirmed still further, by observing, that Dutch plate brass, (which is made, not by melting the copper, but by keeping it at a strong degree of heat, whilst surrounded by *lapis calaminaris*,) also possesses that property ; at least such was the case with all the pieces I tried. From this it follows, that when brass is to be used for the construction of instruments wherein a magnetic needle is concerned, as dipping needles, variation compasses, &c., the brass should either be left quite soft, or it should be chosen of such a sort as will not be made magnetic by hammering, which sort, however, does not occur very frequently."

(514) These suggestions of M. Cavallo were not attended to as their importance deserved, and there is no doubt that considerable errors have arisen from their neglect. Many examples have recently occurred in which the errors were detected ; and it is now the invariable practice of well-informed instrument makers, to reject hammered brass bowls for compasses, and to use those which are cast and turned for the purpose.

(515) The existence of magnetism in brass, while there was not the least trace of it either in the copper or zinc of which it is composed, led philosophers to investigate the effects produced by the union of different metals, or by their combination with other substances. Iron itself is a simple chemical body. Steel is a combination of iron and carbon. The loadstone is a combination of iron and oxygen ; and as no magnetism is found either in carbon or in oxygen, we are naturally led to believe, as M. Pouillett has remarked, that the magnetic fluid resides in the substance of the iron, and that it is carried with the atoms of that metal into all the chemical combinations which they form ; we may, therefore, expect to find magnetic properties in all ferruginous bodies, whether the iron be an accidental or a necessary ingredient; and indeed cast-iron, plumbago, and the oxides and sulphurets of iron, exert a sensible action on the needle.

(516) On the other hand, Dr. Matthew Young found that the smallest admixture of antimony was capable of destroying the polarity of iron ; and M. Seebeck states, that an alloy of one part of iron and four parts of antimony were so completely destitute of magnetic action, that even

when put into rotation it exerted no action on the magnetic needle.
The magnetic qualities of nickel, (which stands next to iron in magnetic
susceptibility, and which is usually found to possess considerable
power,) are also destroyed by a mixture with it of other metals.
Chevenix found that a very small portion of arsenic deprived a mass of
nickel, that had previously exhibited a strong magnetic power, of the
whole of its magnetism ; and Dr. Seebeck found that an alloy of two
parts of copper with one of nickel was entirely devoid of magnetism,
and on this account he recommends it as well suited for the manufacture of compass boxes.   On the other hand, Mr. Hatchett ascertained,
that when a large proportion of carbon, or sulphur, or phosphorus, was
combined with iron, the iron was enabled fully to receive and retain its
magnetic properties ; but he at the same time found that there was a
limit, beyond which an excess of any of these three substances rendered
the compound wholly incapable of receiving magnetism.

(517) Animal and vegetable substances, after combination, are said
to be attracted by the magnet.   The flesh, and particularly the blood,
are acted on more powerfully than other parts, and bone less powerfully.   Burned vegetables have the same property, and also soot and
atmospheric dust ; and M. Cavallo has maintained that brisk chemical
effervescences acted upon the magnetic needle.

(518) In 1802, the supposition of universal magnetism was put to
the test of rigorous experiment by Coulomb.   He employed a glass
receiver, from the top of which was suspended, by a silk fibre, the
needle of the substance to be examined about an inch long, and one-
thirtieth thick.   The receiver was then placed so as to enclose the
opposite poles of two powerful bar magnets, each formed of four bars of
steel tempered to a white heat, and the number of oscillations of the
needle between these poles were noted.   It was found that *all substances whatever, when formed into small needles, turned themselves
in the direction of the poles of the magnets, and after a few oscillations,
finally settled in that position.*   When these bodies were moved a very
little way out of their position of equilibrium, they immediately began
to oscillate round it, the oscillations being always performed more
rapidly in the presence of the magnets than when they were removed
out of their influence.   Gold, silver, brass, wood, and all other substances,
whether organic or inorganic, thus obey the power of magnets.   Hence
we cannot avoid the conclusion, either that all bodies are susceptible of
magnetism, or that they contain minute quantities of iron, or other
magnetic metals, which give them their susceptibility.   Various other
methods have been employed in developing magnetism in all bodies
whatever, since the time of Coulomb ; but I must refer for an account of them to the excellent treatise on magnetism, drawn up for

the Encyclopædia Britannica, by Sir D. Brewster. Recently, the universal prevalence of magnetism has been fully established by a beautiful discovery of M. Arago. This distinguished philosopher conceived the idea of studying the oscillations of a magnetic needle, when placed above or near any body whatever. Having suspended a magnetic needle above metal, or even water, and caused it to deviate a certain number of degrees from its position, it began, when left to itself, to oscillate in arcs of less and less amplitude, as if it had been placed in a resisting medium ; and what was peculiarly curious in these experiments, this diminution in the amplitude of the oscillations did not alter the number of oscillations which were performed in a given time. Dr. Seebeck found, that in alloying magnetic with non-magnetic substances, *he formed compounds which exercised no action on the needle.* The alloys which had particularly this singular property, were those consisting of four parts of antimony, and one of iron, or two parts of copper, and one of nickel. In these cases the magnetism of the two ingredients must have been neutralized by their opposite actions.

(519) In consequence of these experiments of M. Arago, which were announced at the sitting of the French Institute, on the 7th of March, 1825, philosophers in every part of Europe turned their attention to the development of magnetism by rotation. The most important results were obtained by Messrs. Babbage and Herschel. A horse-shoe magnet, which lifted twenty pounds, was made to revolve rapidly round its axis of symmetry, placed vertically with its poles uppermost. A circular disc of copper, six inches in diameter, and one-twenty-fourth of an inch thick, was suspended above the revolving magnet. As soon as the rotation of the magnet commenced, the copper began to turn in the same direction, at first slowly, but afterwards with an increased velocity. When the magnet was made to turn in an opposite direction, the disc of copper changed the direction of its motion also, and exhibited the same phenomena. Metallic plates ten inches in diameter, and half an inch thick, when interposed between the magnet and the copper disc, did not sensibly modify the results, as M. Arago had observed. Glass produced no effect ; but a sheet of tinned plate iron diminished greatly the influence of the magnet, while two such plates almost destroyed it. They found also that a disc of copper, ten inches in diameter, and half an inch thick, and revolving with a velocity of seven revolutions in a second, did not communicate any motion to a similar disc freely suspended above it.

(520) Messrs. Babbage and Herschel next sought to determine the effect produced by a solution of continuity in the metallic disc, upon which the revolving magnet acted. For this purpose a disc of lead

twelve inches in diameter, and one-tenth of an inch thick, was suspended at a given distance from a horse-shoe magnet, revolving with the ordinary rapidity, first in its entire state, and afterwards in the state shown in the annexed figures, the black lines in the direction of the radii being the planes where the lead was cut through, Fig. 181, 182, 183, 184, 185. The accelerating forces, represented by $\frac{s}{t^2}$, where

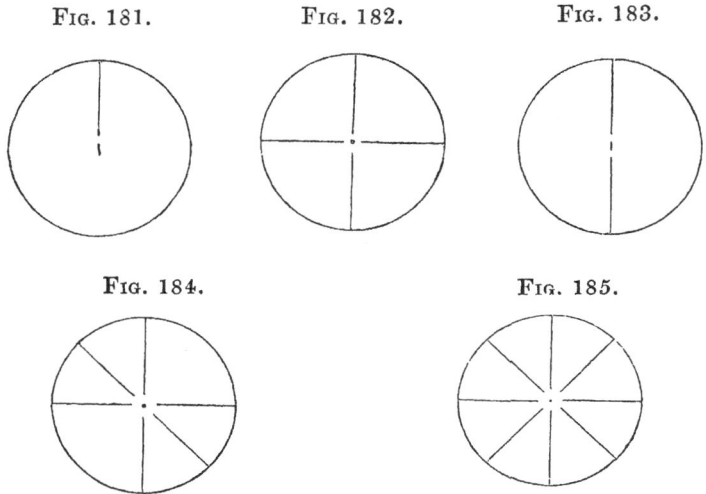

FIG. 181.   FIG. 182.   FIG. 183.

FIG. 184.   FIG. 185.

s is the number of the revolutions, and t, the time employed, are as follow :—

Uncut disc.	Disc as in Fig. 181.	Disc as in Fig. 182.	Disc as in Fig. 183.	Disc as in Fig. 184.	Disc as in Fig. 185.
1258	1047	913	564	432	324

Effects similar, but differing in degree, were obtained with other metals : with soft tinned iron, the cutting produced a very slight diminution of effect, while in copper the same operation reduced the accelerating force in the ratio of five to one.

They next tried the effect of filling up the cuts with different metals. A light upper disc suspended at a given distance above a revolving magnet, performed six revolutions in $54''$ 8. When cut as in Fig. 185, its magnetic action was so weakened, that it took $121''$ 3, to perform six revolutions ; when the right open radial spaces were filled up with tin, its magnetic action was restored to such a degree, that it made six revolutions in $57''$ 3. This fact is very interesting, as tin has less than half the energy of copper.

(521) M. Haldat made some very interesting experiments on this subject. He found that every needle, however weak was its mag-

netism, obeyed the action of the revolving disc; but that this action disappeared entirely when its polarity disappeared. He found it impossible to magnetise needles by the action of the revolving disc, however rapid; and in consequence of ascribing this effect to the want of coercive power, he employed discs of iron and steel, both soft and hardened.

A disc of soft iron acted with more energy than one of copper, and with the same velocity, it dragged the needle twice the distance that a disc of brass did. Iron strongly hammered acted like soft iron, and was unable to give polarity to a steel needle ; but a disc of untempered steel, one twenty-fifth of an inch thick, did not produce any appreciable effect on the magnetic needle, which, after a few irregular oscillations, maintained its ordinary position of equilibrium. Hence he concluded, that the force which acted upon it was in the inverse ratio of the coercive force. M. Haldat also found, that discs in a state of incandescence exercised the same action as those at the ordinary temperature.

(522) Mr. Snow Harris has since shown that several substances not supposed to contain iron, have the power of intercepting the influence of a revolving magnet, contrary to the observations of Messrs. Babbage and Herschel. A circular magnetic disc being delicately balanced on a fine central point by means of a rim of lead, was put into a state of rotation on a small agate cup, at the rate of six hundred revolutions in a minute ; and a light ring of tinned iron also finely balanced on a central pivot was placed immediately over it at about four inches distant, by means of a thin plate of glass, on which its pivot rested. When the ring of tinned iron began to move slowly on its pivot by the influence of the magnet revolving below, a large mass of *copper* about *three* inches thick, and consisting of plates a foot square, was carefully interposed between the magnet and the iron ring. The interposition of the copper soon sensibly diminished the motion of the iron disc, and at length *arrested it altogether*. On again withdrawing the copper, the motion of the disc was restored ; and the same effects were repeatedly obtained. In this experiment both the magnet and the disc were enclosed by glass shades, and supported on a firm base.

(523) The same effects were produced by a mass of silver and zinc : but when their thickness was considerably diminished by removing the central plates, the motion of the disc was not impeded. A very great thickness of *lead* was necessary to stop the disc, in consequence, as Mr. Harris supposes, of its magnetic energy being so much less than that of copper.

(524) With regard to the influence of *heat* on magnetism, Mr. Christie, from a number of experiments made with a torsion balance,

x

the needle being suspended by a brass wire one-four-hundred and fiftieth of an inch in diameter, ascertained the following facts :—

i. Beginning with—3° Fahr. up to 127°, the intensity of magnets decreased as their temperature increased.

ii. With a certain increment of temperature the decrement of intensity is not constant at all temperatures, but increases as the temperature increases.

iii. From a temperature of about 80°, the intensity decreases very rapidly, as the temperature increases; so that, if up to this temperature, the differences of the decrements are nearly constant, the differences of the decrements also increase.

iv. Beyond the temperature of 100°, a portion of the power of the magnet is permanently destroyed.

v. On a change of temperature, the most considerable portion of the effect on the intensity of the magnet is produced instantaneously, showing that the magnetic power resides on, or very near the surface.

vi. The effects produced on soft iron by changes of temperature, are directly the reverse of those produced on a magnet ; an increase of the temperature causing an increase in the magnetic power of the iron. This was observed between the temperatures of 50° and 100° Fahr. Mr. Christie regards this as a strong argument against the hypothesis, that the action of iron upon the needle arises from the polarity which it receives from the earth.

(525) *Methods of making Artificial Magnets.*—Simple induction, by juxta-position with one or more powerful magnets, may suffice for the impregnation of very small bars, and in the infancy of the science, this was the only method employed. For this purpose the bar to be magnetised was simply placed between the opposite poles of two strong magnetic bars of nearly equal power ; but as the principles of magnetic induction came to be better understood, these simple methods were discontinued, and a great variety of ingenious and highly effective processes invented. The first of these was that of Mr. Knight. This method, which was kept a secret during his life-time, but which was made public after his death by Mr. Wilson, consisted in placing the bar to be magnetised, after having tempered it at a cherry-red heat, under the opposite poles $N S$, Fig. 186, of two equal magnets. These magnets were then separated in opposite directions, SA, NA, so that the south pole of the one passes over the

FIG. 186.

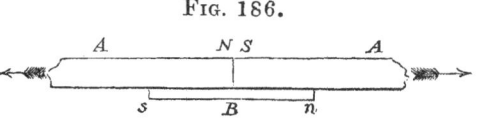

north *polar* half Bn, of the bar B, and the north pole N of the other half, over the south polar half *Bs*, of *B* : this operation is repeated several times till the magnetism of the bar *B* is fully developed.

In this process, the north pole *N*, while it attracts to the half *Bn*, all the south polar magnetism in *Bs*, repels at the same time into *Bs* all the north polar magnetism of *Bn*. The same is true *mutatis mutandis* with the south pole *S*. When the bars *A A* are large and powerful, it has been found that this process is capable of communicating to small bars all the magnetism of which they are susceptible.

(526) Soon after the publication of Dr. Knight's method, small bars thus magnetised were distributed all over Europe, and were eagerly sought after by the cultivators of natural philosophy. When the process, however, was applied to bars of large size, it was found to be defective; philosophers therefore renewed their efforts to devise methods of greater and more universal efficacy. The next improvement was made by M. Duhamel, of the Academy of Sciences, in conjunction with M. Antheaume. It is represented in Fig. 187. The bars *B B* to be magnetised, are placed parallel to each other, and have their extremities united by two pieces *M m* of soft iron, at right angles to the bars; two strong magnets, or two bundles of small bar magnets *A A'*, having their similar poles together,

FIG. 187.

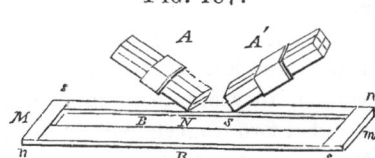

are placed as in the figure, at an angle of about 90°, or inclined 45° to the bar *B*, and then separated from each other, as in Dr. Knight's method; the same operation is repeated on the other bar *B*, and continued alternately on both, till the magnetism is supposed to be completely developed in both bars. When *A A'* are placed on the second bar *B*, the disposition of the poles must be reversed; the pole that was formerly to the right hand being now turned to the left. The two bars *B B* are then turned, so that the undermost faces are uppermost, and the same process carried on as before.

The peculiarity of Duhamel's process consists in the employment of the pieces of iron *M m*, and in the use of bundles of small bars, which are more efficacious than two single ones of the same size. In proportion also as the steel bars acquire magnetism, the connecting pieces participate in the acquisition of a similar power, and serve to retain it in the bars themselves; just as the Electricity which is imparted to the inner coating of a Leyden jar, is retained by the reciprocal influence of the induced, and contrary Electricity of the outer coating. The magnetism of the bars is retained by a similar influence, and greater facility is thus afforded to increase its amount, by the subsequent additions it is receiving from the action of the magnets, as they pass along the surface.

(527) About the same time that Duhamel was occupied with this subject, Mr. Michell, of Cambridge, and Mr. Canton, were separately engaged in the same inquiry. Mr. Michell published his method in 1750, to which he gave the name of *method by double touch*. Having joined together, at the distance of a quarter of an inch, two bundles of strongly magnetised bars, $A A'$, Fig. 188, their opposite poles, $N S$, being together, he placed five or more equal steel bars, $B B' B' B'' B''$, in the same straight line, and resting the extremity of the bundle of magnets, $A A'$, upon the middle of the central bar, B, he moved them *backwards and forwards* throughout the whole length of the line of bars, repeating the operation on each side of the bars, till the greatest possible effect was produced. By this method Mr. Michell found that the middle steel bars, $B B' B'$, acquired a very high degree of magnetic virtue, and greater than the outer bars, $B'' B''$; but by placing these last bars in the middle of the series, and repeating the operation, they acquired the same power as the rest. Mr. Michell states that two magnets will, by his process of double touch, communicate as strong a magnetic virtue to a steel bar, as a single magnet of five times the strength, when used in the process of single touch. The bars $A A'$ act with the sum of their powers in developing magnetism in all parts of the line of the bars between them, and with the difference of their powers in all parts of the line beyond them. The external bars act the same part in this process as the two pieces of soft iron in the method of Duhamel.

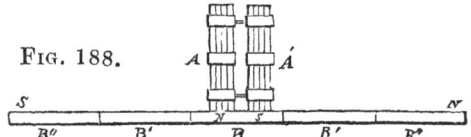

FIG. 188.

(528) In the year 1831 Mr. Canton published his process, which he regarded as superior to the preceding ones. He placed the bars as in Duhamel's method, joined by pieces of soft iron. He then applied Michell's method of double touch, and afterwards he separated the two bundles of magnets, $A A'$; and having inclined them to each other, as in Duhamel's method, he made them rub upon the bar from the middle to its extremities. The peculiarity of Canton's method is the union of these two processes; but Coulomb and others are of opinion that the latter part of the process is the only effectual one.

(529) In order to make artificial magnets, without the aid either of natural loadstones or artificial magnets, Mr. Canton gives the following detailed process :—

He takes six bars of soft, and six of hard steel; the former being smaller than the latter. The bars of *soft* steel should be three inches

long, one-fourth of an inch broad, and one-twentieth thick; and two
pieces of iron must be provided, each having half the length of one of
the bars, and the same breadth and thickness. The bars of *hard* steel
should be each five and a half inches long, half an inch broad, and
three-twentieths of an inch thick, with two pieces of iron of half the
length, and of the same breadth and thickness.

All the bars being marked with a
line quite round them at one end, take
an iron poker and tongs, or two bars
of iron (the larger and the older the
better), and fixing the poker upright,
as in Fig. 189, hold to it with the
left hand near the top, P, by a silk
thread, one of the soft bars, B, having
its marked end downwards; then
grasping the tongs, T, with the right
hand a little below their middle, and
keeping them nearly in a vertical line,
let the bar, B, be rubbed with the
lower end, L, of the tongs, from the
marked end of the bar to its upper
end, about ten times on each side of
it. By this means the bar, B, will
receive as much magnetism as will
enable it to lift a small key at the
marked end ; and when suspended

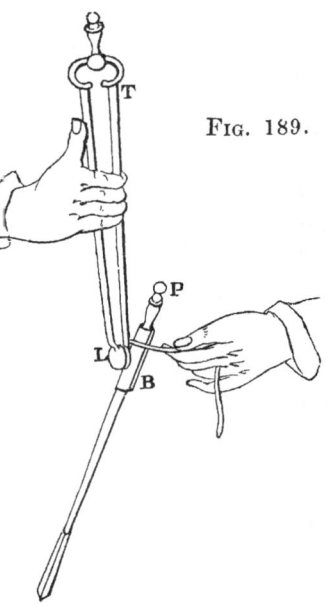

FIG. 189.

by its middle, or made to rest on a point, this end will turn to the
north, and is called its *north* pole, the unmarked end being the *south*
pole.

When four of the soft steel bars are thus rendered magnetic, the other
two, *A C*, *B D*, Fig. 190, must
be laid parallel to each other, at
the distance of about one-fourth
of an inch, having their dissimilar
poles united by the smallest pieces
of iron, *A B*, *C D*. Two of the
magnetised bars are then to be
placed together, as at *G*, with
their similar poles united, and
when separated by a piece of

FIG. 190.

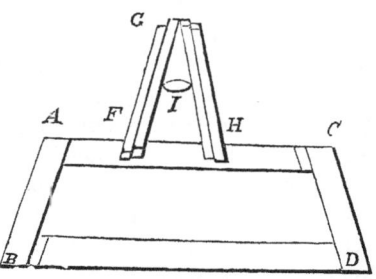

wood, at *I*, they are slid four or five times backwards and forwards
along the whole length of the bar, *A C*, so that the marked end, *F*, of
G is nearest the unmarked end of *A C*, and *vice versa*.

This operation is carefully repeated on *B D*, and on the other sides of both *A C* and *B D*. When this is done, the bars *A C* and *D* are to be taken up and substituted for the two outer bars of the bundles *G K ;* these last being laid down in the place of the former, and magnetised in a similar manner. This operation must be repeated, till each pair of the soft bars has been magnetised three or four times.

When the six soft bars are thus magnetised, they must be formed into two bundles of three each, with their similar poles together, and must be used to magnetise two of the *hard* bars in the manner already described ; and when they are magnetised, other two of the hard bars must be touched in a similar manner. The soft bars are now to be laid aside, and the remaining two *hard* bars magnetised by the *four* hard bars already rendered magnetic ; and when this is done, the operation should be repeated by interchanging the hard bars, till they are impregnated with the greatest degree of permanent magnetism which this method is capable of communicating to them.

(530) In performing the above operations, which may be completed in about half an hour, the bars, *A C*, *B D*, and the pieces, *A B*, *C D*, should be placed in grooves or fixed between pins of wood or brass, to keep them steady during the successive frictions which are applied to them. According to Canton, each of the six artificial magnets thus made will lift about twenty-eight ounces troy. They should be kept in a wooden box, and placed so that no two poles of the same name may be together,—the pieces of iron being placed beside them.

(531) The following is an account of Coulomb's method of making artificial magnets, which consists of the most efficacious parts of the preceding processes, improved and extended by long experience. The apparatus which he uses consists of *fixed* and *moving* bundles of mag - nets. Each of the fixed bundles consists of *ten* bars of steel, tempered at a cherry-red heat, their length being about twenty-one inches, their breadth six-tenths of an inch, and their thickness one-fifth of an inch. Having rendered them as strongly magnetic as possible, with a natural or artificial magnet, he joined them with their similar poles together, and formed them into two beds of four bars each, these beds being separated by small rectangular parallelopipeds, *m n*, of soft iron, pro- jecting beyond their extremities, as shown in Fig. 191. The *moving* bundles consist of four bars tempered at a cherry-red heat, each being about sixteen inches long, six-tenths of an inch wide, and two-tenths of an inch thick. When these bars were magnetised in the same way as the other bars, he united two of them by their width, and two of them by their thickness, so that each bundle was one inch and two-tenths

FIG. 191.

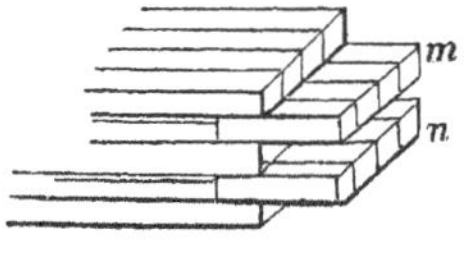

wide, and four-tenths thick.   The bars being separated as before, by
pieces of soft iron, Coulomb found that all kinds of steel, provided
the quality was good, were capable of receiving the same degree of
magnetism.   In order to magnetise a bar, he placed the large fixed
bundles, M N, Fig. 192, in the same straight line, and at a distance a

Fig. 192.

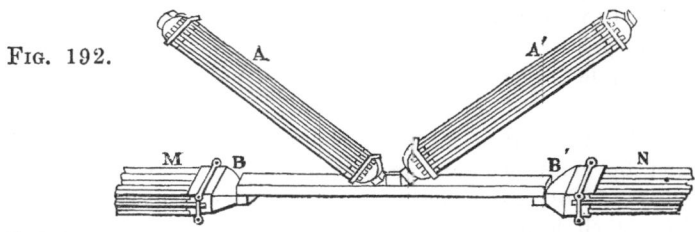

little less than the length of the bar to be magnetised ; and this bar,
B B, was placed as in the figure, so as to rest on the projecting pieces
of iron, so that the contact took place only over a length of one-fifth of
an inch : the two moving bundles, A A′, having their dissimilar poles
separated by a small piece of wood or copper, about one-fifth of an inch
wide, between them ; and each being inclined at an angle of 20° or 30°
to the bar, B B′.   The united poles of the moving bundles are then
moved successively from the centre to each extremity of the bar, B B′,
so that the number of frictions upon each half of the bar may be equal.
When the last friction has been given, the united poles are brought to
the middle point of the bar, B B′, and then withdrawn perpendicularly.
The same operation is then repeated on the other side of the bar, B B′.
If we wish to employ the method of Duhamel, we do not require the
piece of wood or copper, but have only to separate the bars when the
united poles are in the middle of the bar, B B′, making each pole pass
to the extremity of it.

    (532) If the pieces composing the moveable **magnets** have not
received their full power, they will, notwithstanding, communicate to
the bars subjected to their action, a *greater* degree of magnetism than
they possess themselves.   We may therefore now increase *their* power,
by repeating the process on them with the bars which they have themselves impregnated : by so doing three or four times, we shall succeed
in effecting their complete saturation.   If the bars to be magnetised be
very large, Coulomb recommends an increase of the number of the
moveable magnets, each of the bars projecting beyond the last as shown
in Fig. 193.   Thus the pole of each, which   Fig. 193.
Coulomb supposes generally to reside at the
very extremity of the bar, will come immediately in contact with the bar to be magnetised, when the compound magnet is ap-

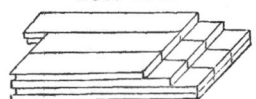

plied to it with the proper inclination, and the whole will powerfully conspire to produce the same effect.

(533) *Horse-shoe magnets.*—The form of a horse-shoe is generally given to magnetic bars when both poles are wanted to act together, which frequently happens in various experiments; such as for lifting weights by the force of magnetic attraction, and for magnetising steel bars by the process of double touch, for which they are exceedingly convenient; fulfilling in this operation all the purposes of compound magnets.

(534) The following is the method of making a powerful magnetic battery of the horse-shoe form, recommended by Professor Barlow :— " Take bars of steel, twelve inches long, and bend them into a horse-shoe shape, their length being six inches, their breadth one inch at the curved part, and three-fourths of an inch at their extremities, and their thickness one-fourth of an inch. Let them be filed nicely, so as to correspond, and lie flatly upon each other. Then drill two holes in each as shown in Fig. 194, and by means of screws V V, passing through these holes, let *nine* horse-shoe bars be bound together. When the heads and ends of the screws are constructed, so as to leave the outer surfaces smooth, the mass of bars must be filed as if they were one piece, and the surface made flat and smooth. When the bars are separated, let them be carefully hardened so as not to warp ; and when they are cleaned and rendered bright, but not polished, magnetise them separately in the following manner :—When the two extremities of the bar are connected by a piece of soft iron, M, the magnetism may be developed in the two halves by Duhamel's method, as in Fig. 195; or, following Epinus, a strong magnet

FIG. 194.

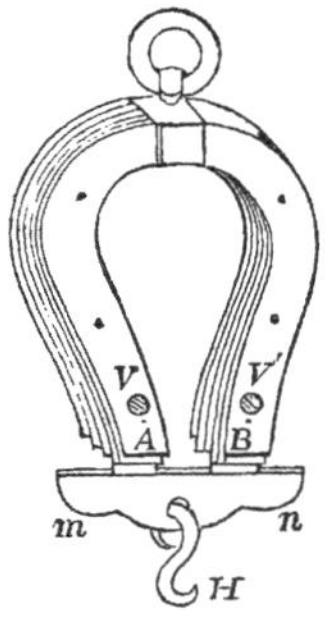

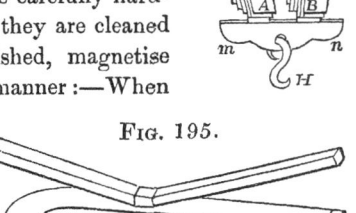

FIG. 195.

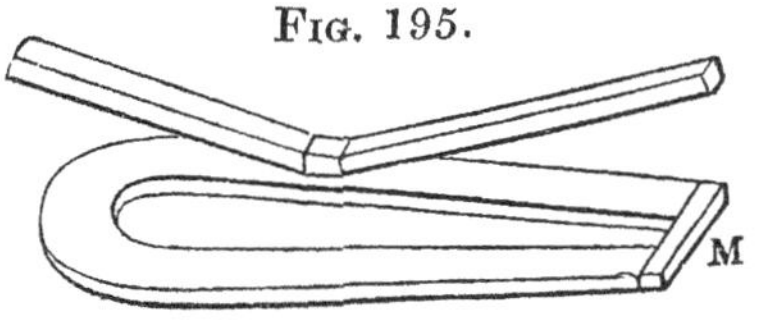

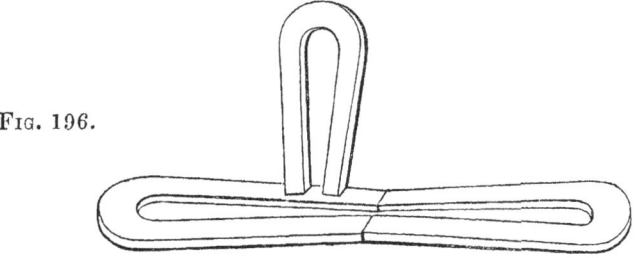

FIG. 196.

may be applied to each pole, and their extremities connected either with a piece of soft iron or another magnet, or two horse-shoe magnets may be applied to each other, as in Fig. 196, uniting the poles which are to be of contrary names. When the magnets are prepared in any of these ways, they are then to be magnetised with another horse-shoe magnet, by placing its north pole next to what is to be the south pole of one of the horse-shoe bars, and then carrying the moveable magnet round and round, but always in the same direction. In this way, a very high degree of magnetic virtue may be communicated to each of the *nine* bars. When this is done, they are to be re-united by the three screws; and their poles or extremities connected by a piece of soft iron, or lifter, as in Fig. 194, having in its middle a hook H, for suspending any weight. As the lifting power depends on the accurate contact of the poles of the magnet with the lifter, the extremities should, after hardening, be properly rubbed down with putty on a flat surface.

A magnet of this size and form was found by Professor Barlow to suspend *forty pounds ;* but he afterwards found, that a greater proportional power could be obtained by using bars that were long in comparison with their breadth.

(535) The following is another simple and efficacious method of making artificial magnets, which has been successfully practised by Mr. Barlow. Having occasion for thirty-six magnets, twelve inches long, one and a fourth broad, and seven-sixteenths of an inch thick, he placed thirty-six bars of steel of these dimensions on a table, so as to form a square, having *nine* bars on each side, the marked or north pole of each bar being in contact with the unmarked or south pole. At the angular points of the square, the under edges of the bars were brought into contact, and the external opening thus left was filled up by a piece of iron one inch and a quarter square, and seven-sixteenths of an inch thick. The horse-shoe magnet described in the preceding section, was set upon one of the bars, so that its *north* pole was towards the unmarked end of the bar, and was then carried or rubbed along the four sides of the bars, and the operation was continued till the compound magnet had gone twelve times round the square. Without removing the magnet, each bar was turned one by one, so as to bring their lower sides uppermost, and the horse-shoe magnet was made to rub along the four sides of the square other twelve times. The bars were then highly magnetised, and the whole process did not occupy more than half an hour.

(536) This last process is the simplest, quickest, and most efficacious of all the methods that have been described ; so that when a person is in possession of one good horse-shoe magnet, consisting of three or four

bars joined together, he may afterwards make any number at the same time : indeed, the more the better. In removing the bars from each other after they have undergone the operation, it is advisable to place small pieces of soft iron on the poles of each before they are separated ; for it is a fact well known to experimentalists, that a very considerable portion of the power of a bar is lost at the moment of its separation from others that have been impregnated with it ; nor is it possible by any means to secure the whole of the magnetism that has been given to it. By following the plan recommended above, however, it will be found that a much larger portion is retained. The whole of the bars become in fact, as one single magnet ; and the act of separation is, of course, analogous to that of fracture.

(537) Magnets should, when laid aside, be placed as nearly as possible in the position which they would assume, in consequence of the action of terrestrial magnetism. If this be neglected, in process of time they will become gradually weaker ; and this deterioration is most accelerated when its poles have a position the *reverse* of the natural one. Under these circumstances indeed, unless the magnet be made of the hardest steel, it will eventually lose the whole of its magnetic power. Two magnets may also very much weaken each other, if they be kept, even for a short time, with their *similar* poles fronting each other. This will readily be understood from what has been said with regard to magnetic induction. The polarity of the weaker magnet is rapidly impaired, and sometimes actually reversed. All rough and violent treatment of a magnet should also be carefully avoided : every concussion or vibration amongst its particles tends to weaken its power.

(538) Horse-shoe magnets should have a short bar of soft iron, adapted to connect the two poles; and should never be laid by, without such a piece of iron adhering to them. Bar magnets should be kept in pairs, with their poles turned in contrary directions, and the dissimilar poles on each side connected by a bar of soft iron, so that the whole may form a parallelogram. They should fit into a box when thus arranged, so as to guard against accidental concussion, and to preserve them from the dampness of the atmosphere. They should be polished, not with a view of increasing their magnetism, but because they are then less liable to contract rust. Both single magnets and needles have their powers not only preserved but increased, by keeping them surrounded with a mass of dry filings of soft iron, each particle of which will re-act, by its induced magnetism, upon the point of the magnet to which it adheres, and maintain in that point its primitive magnetic state.

(539) In the *Comptes Rendus*, for January 2d, 1838, there is an important notice of a communication received from M. De la Rive, relative

to the magnetising of needles by the *nerves*.   The following is an extract :—" Dr. Prevost, of Geneva, has succeeded in magnetising very delicate soft iron needles, by placing them near the nerves, and perpendicular to the direction which he supposed the electric current took. The magnetising took place at the moment when, on irritating the spinal marrow, a muscular contraction was effected in the animal."

(540) There is another method of forming very powerful magnets, which I shall only briefly notice here, as it will form a part of a future Lecture ; viz., by induced electrical currents : soft iron is employed, and the magnetism is only temporary. The process consists in winding spirally round a horse-shoe bar of iron, a copper wire covered with silk thread. A galvanic current is then made to pass through the wire ; and at the moment the contact between the wires and the single voltaic circle is completed, the soft iron becomes powerfully magnetic. Professor Henry mentions a horse-shoe bar of soft iron, weighing twenty-one pounds, which, when magnetised by this process, lifted more than thirty-five times its own weight; whereas, the largest natural magnet known, and in the possession of Mr. Peale, of Philadelphia, lifts three hundred and ten pounds, or about six times its own weight.

(541) *Theory of Magnetism.*—A magnet is considered as composed of minute invisible particles or filaments of iron, each of which has individually the properties of a separate magnet.   It is assumed that there are two distinct fluids—the *austral* and *boreal ;* and under the influence of either in a *free* state, the bar of iron or other metal will point to the north or south poles of the earth, according to circumstances. It is within these small particles or metallic elements that the displacement or separation of the two attractive powers take place ; and the particles may be the ultimate atoms of the iron.

A magnetic bar may therefore be represented (as in Fig. 197), as composed of minute portions, the right-hand extremities of each of which possess one species of magnetism, and the left-hand extremities the other.

FIG. 197.

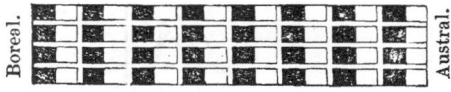

The shaded ends being supposed to possess *boreal,* and the light ends *austral* magnetism.   Then the ends of the bar itself, of which these sides of the elementary magnets form the faces, possess respectively boreal and austral magnetism, and are the boreal and austral poles of the magnet.

In ordinary iron, these fluids exist in a *combined* state, and are therefore perfectly latent, the metal appearing to be destitute of magnetism.   They exist in certain proportions united to each molecule or atom of the metal, and from which they *can never be disunited,* the only change which they are capable of undergoing being their decomposition into the separate fluids, one of which in a permanent magnet is always collected on one, and the other on the opposite side of each molecule.

This theory, if not absolutely correct, is very convenient, and offers a satisfactory explanation of all the ordinary magnetic phenomena.

(542) *Terrestrial Magnetism.* — The tendency of the magnetised needle or bar to turn *nearly* to the north and south when left at liberty to move freely on a pivot, or otherwise suspended, so as to allow of freedom of motion, in a horizontal plane, is derived from a force naturally supposed to reside in the earth. It is found that in this country, as well as throughout Europe, the north pole of the compass deviates a certain number of degrees (about 25°) to the westward of the true north : this deviation is called the magnetic *declination* or *variation.* The vertical plane which passes through the direction of the horizontal needle at any particular place, is called the *magnetic meridian* or the place in contra-distinction to the *geographical* or *true meridian,* which is a vertical plane passing through the poles of the earth. On this property of the needle, is founded that instrument which is of such indispensable utility to the sailor, the mariner's compass. It consists of a needle fixed to a circular card, containing upon its surface the thirty-two points of the compass. This card is balanced upon a pivot fixed in the bottom of a circular box, and the top of the box is a plate of glass for protecting the needle from the motion of the air. In Fig. 198, AB is the compass-box suspended within a larger

FIG. 198.

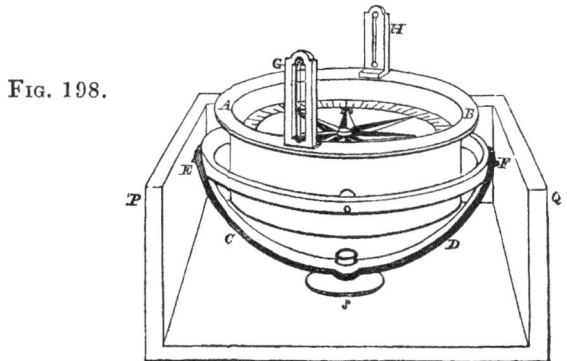

box PQ, upon two concentric brass circles, or gimbals, the outer circles being fixed by horizontal pivots both to the inner circle, which carries the compass-box, and likewise to the outer box, the two axes upon which the gimbals move being at right angles to each other. The effect of this construction is, that the compass-box AB, will retain a horizontal position during the motions of the vessel. The instrument shown in the figure, is the Azimuth Compass : it is furnished with sights GH, through which any object may be seen, and its angle with the magnetic meridian increased. For this purpose, the whole box is hung in detached gimbals CD, EF, which turn upon a stout vertical pin seen above S. In this compass the card is divided on its rim into 360° ; but the divisions are more frequently placed on a light metallic rim which it carries. The eye is applied to the sight H, which is a slip of brass, containing a narrow slit. The other sight G, which is turned towards the object, contains an oblong aperture, along the axis of which is stretched a fine wire, which is made to pass over the object whose angular distance or *azimuth* from the magnetic meridian is to be determined.

(543) The force of terrestrial magnetism is not only exerted on bodies possessed of magnetic power, but also on unmagnetised bodies : a piece of iron may be rendered permanently magnetic by it. It is found that when a bar of soft iron is held in the direction of the magnetic meridian, and inclined to the horizon, it is converted into a temporary magnet possessed of a north and south pole; and if it be kept for a very long time in that position, it acquires permanent magnetism. Old kitchen pokers that have been in use for a number of years, and which when laid aside have frequently the exact position required, are often found to be powerfully magnetic, the lower end exhibiting all the properties of a north pole, repelling the north, and attracting the south pole of the needle; while the upper end is a true, though much weaker south pole, repelling the south and attracting the north pole of the needle.

(544) But not only does the balanced magnetised needle point to the north and south, in consequence of terrestrial magnetism; it is found to be endued with another property also. If we take a needle and balance it on a pivot with the greatest accuracy, it will, of course, lie horizontally on its axis; but if we magnetise it, it will no longer remain in a horizontal position, but will be found to take an inclined position, the north end appearing to preponderate : this arises from the effects of the *dip*, which in this country is about 70°.

(545) Thus the magnetic force may, by the ordinary dynamic mode of resolution of forces, be resolved into two forces, the one acting vertically, and the other horizontally. The latter of these forces is the only one with the action of which gravitation does not interfere ; and accordingly, the mariner's compass indicates by its motions the effects of this part of the terrestrial magnetic force, and this only. In order to ascertain the vertical force, we must proceed in a different manner.

The needle must be furnished with an axis at right angles to its length, and adjusted very carefully, so that it may pass as exactly as possible through its centre of gravity. This of course can only be done when the needle is wholly free from magnetism, and secured from all magnetic influence which the earth might exert upon it. The axis should be supported horizontally, in such a manner as to allow the needle complete freedom of motion in a vertical plane. The needle being balanced, will have no tendency to incline to one side rather than to another, and will remain at rest in any position in which it may happen to be left, as long as no extraneous force is applied to it, and when it is magnetised it will be in a condition to illustrate the property of the dip.

The dipping needle is shown in Fig. 199. It consists of a brass plate supported on a flat stand by three screws. On this plate is fixed a spirit level for adjusting the plate horizontally. A stout hollow brass pillar rising from the centre of the plate carries a circular box, forming the case of the dipping needle, which turns freely on two finely polished planes of agate. In order to obtain an accurate measure of the dip, several measures of it should be taken ; *first*, with the face of the instrument to the east; *secondly*, with the face to the west ; and the same observations repeated after the polarity of the needle has been invested, or the north pole converted into a south pole, and a south pole into a north one. The mean of these four sets of observations will be the true dip required.

FIG. 199.

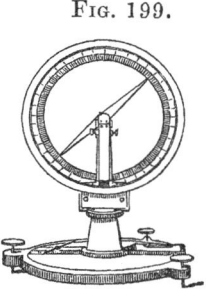

I shall consider these two important classes of phenomena in reference to terrestrial magnetism in their order. 1st, The variation ; and 2ndly, The dip of the needle. Measures of the variation of the needle have been taken by navigators and travellers in all parts of the globe. In some places the magnetic and terrestrial meridian exactly correspond ; in these situations the needle points to the true north and south ; but in most parts of the earth's surface its direction deviates, sometimes to the east, and sometimes to the west. We are indebted to Professor Hansteen for the most satisfactory collection of observations on the variation of the needle, and for the most philosophical generalization of them. He has published a variation chart for 1787, in which it is shown how irregular are the causes on which terrestrial magnetism depends, by the total want of symmetry, and the irregularities and inflexions of the magnetic curves. He makes the *western line of no variation*, or that which passes through all the places on the globe where the needle points to the true north, to begin in latitude 60° to the west of Hudson's Bay, proceeding in a south-east direction through the North American Lake, passing the Antilles and Cape St. Roque, till it reaches the South Atlantic Ocean, where it cuts the meridian of Greenwich at about 65° of south latitude. This line of no variation is extremely regular, being almost straight till it bends round the eastern part of South America, a little south of the equator. On the other hand, his chart shews that the *eastern line of no variation* is extremely irregular, being full of loops and inflexions of the most extraordinary kind, indicating the action of local magnetic forces. It begins in latitude 60° south, below New Holland, crosses that immense island through its centre, extends through the Indian Archipelago with a double sinuosity, so as to cross the equator three times ; first passing north of it to the east of Borneo, then returning to it and passing south through Sumatra and Borneo, and then crossing it again beneath Ceylon, from which it passes to the east through the Yellow Sea. It then stretches along the coast of China, making a semicircular sweep to the west till it reaches the latitude of 70°, where it descends again to the south, and returns northward with a great semicircular bend, which terminates in the White Sea.

These lines of no variation are accompanied through all their windings by other lines, where the variation is 5°, 10°, 15°, &c. ; these lines becoming more irregular as they recede from those of no variation.

(546) In 1833, Professor Barlow constructed a new variation chart, in which he has inserted the magnetic observations of Commander Ross ; and he remarks, that the very spot where this officer found the needle perpendicular, " that is, the pole itself, is precisely that point in my globe and chart in which, by supposing all the lines to meet, the several curves would best preserve their unity of character, both separately and conjointly, as a system."

(547) Hansteen is of opinion, that there are *four points of convergency* in each hemisphere ; and that these points are to be considered as the *four magnetic poles of the globe.* He concludes that they have a regular motion round the globe ; the two northern ones from west to east, in an oblique direction ; and the two southern ones from east to west, also obliquely. The following are the calculated periods of their revolution :—

> The strongest *north* pole in 1740 years.
> The strongest *south* pole in 4609 years.
> The weakest *north* pole in 860 years.
> The weakest *south* pole in 1304 years.

Hansteen considers the four poles as originating in *two* magnetic axes, the two strongest being the termination of one axis, and the two weakest of the other ; and he conceives that they may have been produced, either along with the earth itself, or at a later epoch. According to the first supposition, it is not easy to account for their change of position; but according to the last, they must have originated either from the earth alone, or from some external cause. If they originated in the earth, their change of position is still unsusceptible of explanation; and hence Hansteen conceives that they have their origin from the action of the sun, heating and illuminating the earth, and producing a magnetic tension as it produces electrical phenomena.

(548) The following table exhibits the progressive changes in the variation of the needle, which have taken place in London.

Years.	Observers.	Variations.			
1576	.... Norman	11°	15'		Easterly.
1580	.... Burroughs	11°	17'		Maximum.
1622	.... Gunter	6°	12'		
1634	.... Gellibrand	4°	5'		
1657) 1662)	....	0	0		No variation.
1666	....	0	34'		Westerly.
1670	....	2°	6'		
1672	....	2°	30'		
1700	....	9°	40'		Westerly.
1720	....	13°	0'		
1740	....	16°	10'		
1760	....	19°	30'		
1774	....	22°	20'		
1778	.... Phil. Trans.	22°	11'		
1790	....	23°	39'		
1800	....	24°	36'		
1806	.... Phil. Trans.	24°	8'		
1813	.... Col. Beaufoy	24°	20'	17"	
1815	.... Ditto	24°	27'	18"	Maximum.
1816	....	24°	17'	9"	
1820	....	24°	11'	7"	
1823	....	24°	9'	40"	
1831	....	24°	0'	0"	

These progressive changes are referred by Hansteen to the motion of four magnetic poles.*

(549) An *annual* change in the variation of the needle, depending on the position of the sun, in reference to the equinoctial and solstitial points, has been observed by M. Cassini.

* The greatest variations ever observed were by the Chev. de Langle, between Greenland and Labrador, amounting to 45° west; and by Capt. Cook in 60° south lat., and 92° 35' long., when the variation amounted to 43° 6' east of the geographic meridian.—(Golding Bird.)

Between the months of January and April, the magnetic needle recedes from the north pole of the globe, so that its western declination increases.

From April to the beginning of July, that is, from the vernal equinox to the summer solstice, the declination diminishes, or the needle approaches the north pole of the globe.

From the summer solstice to the vernal equinox, the needle receding from the north pole, returns to the west, so that in October it has nearly the same position as in May ; and between October and March, the western motion is smaller than in the three preceding months. Hence it follows that during the three months between the vernal equinox and the summer solstice, the needle retrogrades towards the east ; and during the following *nine* months its general motion is towards the west.

(550) There is also a daily change in the variation of the needle ; this was first observed in 1724, by Mr. Graham, and has been confirmed with the most accurate instruments in almost every part of the world. When it was first discovered, the needle was supposed to have only two changes in its movements during the day. About seven A.M. its north end began to deviate to the west ; and about two P. M. it reached its maximum westerly deviation ; it then returned eastward to its first position, and remained stationary till it again resumed its westerly course in the following morning. It was afterwards found that the diurnal motion commences much earlier than seven A. M., but its motion is to the east. At half-past seven A.M. it reaches its greatest easterly deviation, and then begins its movement to the west till two P. M. It then returns to the eastward till evening, when it has again a slight westerly motion ; and in the course of the night, or early in the morning, it reaches the point from which it set out twenty-four hours before.

(551) At Paris, the needle is nearly stationary during the night. At sun-rise its north extremity moves westward, as if it were avoiding the solar influence. Towards noon, or more generally from noon to three o'clock, it attains its maximum westerly deviation, and then it returns eastward till nine, ten, or eleven o'clock in the evening ; and then, having reached its original position, it remains stationary during the night. The amount of this daily variation is, for April, May, June, July, August, and September, from 13′ to 15′ ; and for the other six months of the year, from 8′ to 10′ ; on some days it rises to 25′, and on others it does not exceed 5′ or 6′.

(552) The dipping needle also undergoes daily variations, but their amplitude is of less amount ; and there is no doubt that a needle capable of moving in any given azimuth, will experience daily changes ;

and that a needle moveable in every direction round its centre of gravity, would describe every day a cone, whose base would be an ellipse, or some other curve, more or less elongated, in different parts of the earth.

(553) The sun is now universally allowed to be the cause of the diurnal variations of the needle. Canton ascribed them to the action of solar heat, having ascertained that heat tends to diminish the attractive powers of a magnet, and assuming that the direction of the needle was due to the resultant of all the magnetic forces of the terrestrial sphere. When the sun was to the eastward of the needle, the forces lying to the eastward suffered a diminution of power; in consequence of which, the westerly force prevailed, and the north end deviated to the west. When the sun, on the other hand, was to the westward of the needle, the power on that side diminished, and the needle returned again to the eastward.   Canton did not, however, give any explanation of the morning easterly variation of the needle.

(554) MM. Morlet and Hansteen have determined the true form of the magnetic equator (an irregular line crossing the equator at *four* points) with great care ; they have traced it over the whole globe, and have found that its motion is from east to west, in so far as can be determined by direct observations on the position of its nodes.   They both place the magnetic equator wholly to the south of the terrestrial equator, between Africa and America ; its greatest southern latitude being at 25°.   One node is in Africa, in about 22° of east longitude, or in 18° according to Morlet.   In setting out towards the east from this node, which is nearly in the centre of that part of the African continent, the magnetic equator advances rapidly to the north of the terrestrial equator, quits Africa a little to the south of Cape Guardafui, and in the Arabian Sea it attains its most northerly latitude of about 12°, in 62° of east longitude ; between this meridian, and 174° east, the magnetic equator is constantly to the north of the equinoctial line.   It cuts the Indian peninsula a little to the north of Cape Cormorim, traverses the gulf of Bengal, making a slight advance to the equinoctial, from which it is only 8° distant at the entry of the Gulf of Siam.   It then re-ascends a little to the north, almost touches the north point of Borneo, traverses the Isle of Paragua, the strait which separates the most southern of the Philippines, from the Isle of Mindanao, and under the meridian of Naigiou, it again reaches the north latitude of 9°.   From this point it traverses the Archipelago of the Caroline Islands, and descends rapidly to the equinoctial line, which it cuts, according to Morlet in 174°, and according to Hansteen in 187° east longitude.   There is much less uncertainty respecting the position of a second node, also situated in the Pacific Ocean.   Its west longitude ought to be about 120° ; but while

M. Morlet's inquiries lead him to conclude that the magnetic equator merely touches the equinoctial at that point, and then bends again to the south—M. Hansteen makes it cross the line into the northern hemisphere, and continue there through an extent of 15° of longitude, and then return southward, and cross the equinoctial again in about 108° of west longitude, or 23° from the west coast of America. This discrepancy between the results of Morlet and Hansteen is, after all, very trivial; for, in the case just mentioned, the magnetic equator does not go more than $1\frac{1}{2}$° to the equinoctial; and in general the magnetic equator of Morlet differs in no part so much as 2° in latitude from that of Hansteen.

(555) The observations of Captain Duperrey, made on board the Coquille, in the years 1822-25, have contributed considerably to our stock of knowledge on the subject of terrestrial magnetism, and particularly on the form and motion of the magnetic equator. This vessel crossed the magnetic meridian six times, and M. Duperrey was enabled to determine directly two of its points, situated in the Atlantic Ocean. On the chart of Morlet, and in that of Hansteen, the latitudes of those parts, which correspond to the same longitudes, are greater by 1°43′ and 1°50′; and hence M. Arago has concluded that the magnetic equator has approached the terrestrial equator by the same quantities. In the South Sea, near the coast of America, Duperrey has determined two points of the magnetic equator. On the charts of Morlet and Hansteen, the latitudes of these points are about a degree smaller, but the difference is in a direction contrary to that which was found in the Atlantic Ocean; from which it follows, that near the coast of Peru, the magnetic equator has removed from the equinoctial line.

(556) In a chart of the equatorial regions, which M. Duperrey has drawn up and published in the Annales de Chimie, for 1830, these results are laid down :—"That the magnetic equator will meet the equinoctial line only in two points, which are diametrically opposite, the one situated in the Atlantic Ocean, and the other in the great ocean, nearly in the plane of the meridian of Paris. When this equator meets only some scattered islands, it recedes only a little from the equinoctial line ; when the islands are more numerous, it recedes farther; and it reaches its maximum deviation in both hemispheres, only in the two great continents which it traverses. He found also, that between the northern and southern halves of the magnetic equator, there is a symmetry very remarkable, and much more perfect than had previously been believed.

" The dip of the needle increases on each side of the magnetic equator; and Hansteen has projected lines of equal dip in his chart. These lines are nearly parallel to the magnetic equator, till we reach 60° of north latitude, and they then begin to bend round the American magnetic pole, which Commander Ross found to be situated in north latitude, 70° 5' 17", and west longitude, 96° 45' 48", the needle having at this point, in Boothia Felix, lost wholly its directive power, and the dip being 89° 59', within *one minute* of 90°, or vertical.   Had we inferred the position of the needle from the form of the magnetic equator, we should have placed it in 25° of west longitude, viz., the meridian in which the magnetic equator advances farthest to the south, or about $13\frac{1}{2}°$ and $76\frac{1}{2}°$ of north latitude, or $90°—13\frac{1}{2}°$.   This, however, as all arctic observations prove, is not the case; and we are led by the phenomena of the dip, as well as by those of the variation in different parts of the globe, to conclude that every place has its own magnetic axis, with its own pole, and its own equator, as stated by Mr. Barlow."

(557) The dip of the needle, like the variation, undergoes a continual change, increasing in some parts of the world, and diminishing in others. The following tables show the changes of the dip at London, since 1720 ; and at Paris, from 1671 to 1829 :—

AT LONDON.

Year.	Observed.	Observer.	Computed.
1720	74°42'	Graham	76°27'
1773	72 19	Heberden	73 40
1780	72 8	Gilpin	73 18
1790	71 53	ditto	72 39
1800	70 35	ditto	71 58
1810			71 15
1818	70 34	Kater	70 34
1821	70 3	Sabine	....
1828	69 47	ditto	69 43
1830	69 38	ditto	....
1833			69 21

AT PARIS.

Year.	Dip.	Year.	Dip.
1671	75° 0'	1818	68°35'
1754	72 15	1819	68 25
1776	72 25	1820	68 20
1780	71 48	1821	68 14
1791	70 52	1822	68 11
1798	69 51	1823	68 8
1806	69 12	1824	68 7
1810	68 50	1825	68 0
1814	68 36	1826	68 0
1816	68 40	1829	67 41
1817	68 38		ARAGO.

M. Humboldt has shown that the progressive variation in the dip of
the needle is the necessary consequence of a change in the magnetic
latitude, arising from the motion of the nodes of the magnetic equator,
modified by the form of this curve.   And Morlet has applied the same
principle to account for the variations of the dip in different parts of the
globe.   From a series of observations made with a dipping needle by
Dollond, Hansteen has found that the dip during the summer was
about fifteen minutes greater than during the winter, and about four or
five minutes greater in the forenoon than in the afternoon.

(558) It has become a most important practical problem, connected
with the physical condition of the globe, to determine the intensity of
its magnetism at different points on its surface, and the changes which
it undergoes at different seasons of the year, and at different times of
the day.

Mr. Graham first suggested a method of determining this, by the
number of oscillations of the magnetic needle.   And this plan has been
since much improved by Coulomb, Humboldt, and others.

If a needle, whose axis of suspension passes through its centre of
gravity, and which has its north and south polar magnetism equal, and
similarly distributed, be made to vibrate, by turning it from its position,
and allowing it to recover that position by a series of oscillations, it is
evident that the magnetism of the earth will act with equal force on
each half, and that the needle will be drawn into the magnetic meridian
by the combined action of both forces.   The greater the magnetic force,
the more quickly will the needle oscillate, and recover its primitive
position.   The needle is, in short, in the same circumstances as a pendulum, oscillating by the action of gravity; and as in this case, the
forces are as the squares of the number of oscillations made in the same
time.

Suppose the dipping needle be made to oscillate in the plane of the
magnetic meridian, round the line of the dip, and that when an experiment is made at the equator, the number of oscillations in a second is
24, while in another place it is 25 ; then the intensity of the magnetic
force at these places, is as 25^2 to 24^2, or as 625 to 576, or as $1 \cdot 085$ to
$1 \cdot 000$.   By carrying the same needle to different parts of the earth, the
magnetic intensity at these places will be found from the number of its
oscillations.

(559) In the application of this method there are various practical
difficulties, particularly the necessity of the needle resting on knobs,
edges of steel, or agate, during its oscillations ; these difficulties are
avoided by suspending it by a fibre of silk, and allowing it to oscillate
horizontally.   This method is therefore the one adopted, though a little

calculation is necessary to obtain the intensity of terrestrial magnetism, from the number of oscillations that are performed.

Hansteen has drawn up a table, which is too long for insertion here, exhibiting the magnetic intensity in almost every part of the world, from observations made principally by himself and his friends. He has projected, on a map of the globe, the lines passing through all the places in which the intensity has the same value. These lines he calls *isodynamic lines*, or those of equal force, and they are, generally speaking, nearly parallel to each other, and to the lines of equal dip.

(560) The same indefatigable philosopher, not satisfied with the many valuable observations which were made during the various arctic expeditions which were sent out by the British government, and being exceedingly anxious to establish, by direct observations of his own, the existence of the secondary magnetic pole, which he believed existed in Siberia, undertook a journey at the expense of the Norwegian Storthing, and with every encouragement and assistance from the Russian government. The results of his expedition were highly satisfactory ; and in consequence, the Russian Academy of Sciences have been induced to take a new interest in the subject of terrestrial magnetism, which exhibits such important features throughout the Russian empire ; and the Russian government has established regular observatories in various parts of its vast dominions for making magnetic experiments. The Russian empire is actually traversed by *two* lines of no variation, and it is proposed to determine with great precision, every ten years, the exact position of these two lines. Near the first of them, which traverses European Russia, Petersburgh, Moscow, and Kasan are situated ; and near the second, which passes through Siberia, are situated Kiachta and Nizni Oudinsk. Observations are yet wanting to determine in what manner the intensity varies with the height. Humboldt is of opinion that it decreases, confirming the deductions of Kupffer.

By combining all the observations of intensity from 179° to 183°, M. Hansteen has drawn the conclusion, *that the total magnetic intensity is smaller in the southern than in the northern hemisphere.* M. Duperrey has confirmed this result.

(561) The magnetic intensity, like the variation and dip of the needle, undergoes monthly and diurnal changes. Hansteen found by means of the vibrations of a needle delicately suspended, that the *minimum* of daily change of intensity is between ten and eleven in the forenoon, and the *maximum* between four and seven in the afternoon in May, and about seven in June. The intensity is a maximum in December, and a minimum in June. The greatest monthly change in the intensity is a maximum in the months of December and June,

about the time when the earth is in its perihelion and aphelion. It is a minimum near the equinoxes, or when the earth is at its mean distance from the sun. The greatest daily change is least in the winter, and greatest in the summer. The greatest difference of the annual intensity is 0·0359. M. Hansteen has likewise found that the magnetic intensity is diminishing in Europe, and that the decrease is greater in the northern and eastern, than in the southern and western part; an effect which he conceives to be produced by the motion of the Siberian pole towards the east. At Port Bowen, Captain Parry observed an augmentation of the magnetic intensity to take place from the morning till the afternoon, and a diminution of it from the afternoon till the morning. These results of M. Hansteen have been confirmed by Mr. Christie,* who has shown that the terrestrial magnetic intensity is a minimum between ten and eleven o'clock in the morning; the time nearly when the sun is in the magnetic meridian; that it increases from this time until between nine and ten o'clock in the evening, after which it decreases, and continues decreasing during the morning, till it reaches its minimum between ten and eleven. These results were deduced from observations made in May, *within doors*, to determine the positions of the points of equilibrium at which a magnetic needle was retained, at different hours during the day, by the joint action of two bar magnets, and by terrestrial magnetism, reduced to their true positions at the standard temperature 60° of the magnet.

Mr. Christie repeated his observations *in the open air* in June, and from these it appears that the minimum intensity happened nearly at the time the sun passed the magnetic meridian, and rather later than in May, which was also the case with the time of the sun's passage over the meridian. The intensity increased till about six o'clock in the afternoon, after which it appears to have decreased during the evening, and to have been decreasing from an early hour in the morning. The following table shews the results of the observations of Mr. Christie, compared with those of M. Hansteen :—

Intensity deduced from Hansteen's Observations in 1820.			Intensity deduced from Mr. Christie's Observations in 1823.		
TIME.	MAY.	JUNE.	TIME.	MAY.	JUNE.
8h 0′ A.M.	1·00034	1·00010	7h 30′	1·00114	1·00061
10 30	1·00000	1·00000	10 30	1·00000	1·00000
4 0 P.M.	1·00299	1·00251	4 30	1·00175	1·00223
7 0	1·00294	1·00304	7 30	1·00220	1·00339
10 30	1·00191	1·00267	9 30	1·00231	1·00209

* Philosophical Transactions, 1825, p. 49-51.

The principal difference between these results is, that in Mr. Christie's observations the intensity seems to diminish more rapidly in the morning, and increase more slowly in the afternoon than it does in those of Hansteen.

(562) In April 1836, Baron Alexander Von Humboldt addressed a letter to His Royal Highness the Duke of Sussex, president of the Royal Society of London, to solicit the co-operation of that body towards the advancement of the knowledge of terrestrial magnetism, by the establishment of magnetic stations, and corresponding observations, and soliciting it to extend, in the colonies of Great Britain, the line of simultaneous observations; these stations he proposed to be established, either in the tropical regions on each side of the magnetic equator, or in the high latitudes of the southern hemisphere, and Canada.

This distinguished traveller has been for many years much occupied with the phenomena of the intensity of the magnetic forces, and the inclination and declination of the magnetic needle. During the years 1806 and 1807, particularly at the periods of the equinoxes and solstices, he measured the angular alterations of the magnetic meridian, at intervals of an hour, often of half an hour, without interruption, during four, five, and six days, and as many nights, in a large garden at Berlin. The instrument he employed was a magnetic telescope *(lunette aimantée)* of Prony, capable of being reversed upon its axis, suspended according to the method of Coulomb, placed in a glass frame, and directed towards a very distant meridian mark, the divisions of which, illuminated during the night, indicated even six or seven seconds of *horary* variation. In verifying the habitual regularity of a *nocturnal period*, he was struck with the frequency of the perturbations, especially of oscillations, the amplitude of which extended beyond all the divisions of the scale, and which occurred repeatedly at the same hours before sun-rise, and the violent and accelerated movements of which could not be attributed to any accidental mechanical cause. These vagaries of the needle, the almost periodical return of which has been recently confirmed by M. Kupffer, appeared to Humboldt the effect of a re-action of the interior of the earth towards the surface, of *magnetic storms*, which indicate a rapid change of tension. From that time, he says, he has been anxious to establish on the east and west of the meridian of Berlin, apparatus similar to his own, in order to obtain corresponding observations, made at great distances, and at the same hours.

(563) In 1827, Baron Humboldt renewed these observations at Berlin, and endeavoured at the same time to generalize the means of simultaneous observations, the accidental employment of which had

produced such important results. One of Gambey's compasses was placed in the *magnetic pavilion*, in which no portion of iron was introduced, which had been erected in the middle of a garden. At his request, the Imperial Academy and the Curator of the University of Cazan, erected *magnetic houses* at St. Petersburgh and at Cazan, and the Imperial Department for Mines having concurred in the same object, *magnetic stations* have been successively established at Moscow, Barnaoul, and at Nertschinsk. The Academy of St. Petersburgh has done still more, and has sent a courageous and clever astronomer, M. George Fuss, the brother of its perpetual secretary, to Pekin, and has procured the erection there of a *magnetic pavilion* in the convent garden of the Monks of the Greek church. Since the return of M. Fuss, M. Kowanko continues the observations of horary declination, corresponding to those of Germany, St. Petersburgh, Cazan, and Nicolajeff in the Crimea, where Admiral Greigh has established one of Gambey's compasses, the care of which is confided to the director of the observatory, Mr. Knorre. A magnetic apparatus has also been established at the depth of thirty-five fathoms, in an adit in the mines of Freiberg, in Saxony, where M. Reich, to whom we are indebted for valuable experiments and observations upon the mean temperature of the earth at different depths, is assiduously engaged in making observations at regulated intervals. Observations of horary declination, made at Marmato, in the province of Antioquia, in South America, in north lat. 5 27', in a place where, as at Cazan and Barnaoul, in Asia, the declination is eastern, have been transmitted by M. Boussingault; while on the north-western coasts of the new continent, at Sitka, in the Russian settlements, Baron Von Wrangal, also provided with one of Gambey's compasses, has taken part in the simultaneous observations made at the time of the solstices and equinoxes. Magnetic apparatus have also been sent by Baron Humboldt to Havannah and Cuba; and M. Arago has erected a compass, at his own expence, in the interior of Mexico, where the soil is elevated six thousand feet above the level of the sea. Preparations are likewise making for the establishment of a magnetic station in Iceland.

(564) The suggestions of Baron Humboldt received from the Royal Society the attention which they merited; and a committee was appointed for carrying his recommendations into effect. Conformably with the report made by this committee, the ten following places were fixed on by the council, as being the most eligible for carrying on magnetic observations, according to the plan recommended by Baron Von Humboldt:—Gibraltar, Corfu, Ceylon, Hobart Town, Jamaica, Barbadoes, Newfoundland, Toronto, Bagdad, and the Cape of Good Hope; these places being permanent stations, where officers of engineers and

clerks are always to be found. The council also determined that for the present, the observations of magnetism may be limited to those of the direction of the magnetic needle ; and the meteorological observations restricted to those made on the four days, and in the manner recommended in Sir John Herschel's instructions.*

(565) A magnetical observatory has lately been erected at Dublin : the following notice of it was read by Professor Lloyd at a meeting of the British Association at Liverpool.

"The magnetical observatory, now in progress in Dublin, is situated in an open space in the gardens of Trinity College, and sufficiently remote from all disturbing influences. The building is forty feet in length, by thirty in depth. It is constructed of the dark-coloured argillaceous lime-stone, which abounds in the valley of Dublin, and which has been ascertained to be perfectly devoid of any influence on the needle. This is faced with Portland stones ; and within the walls are to be *studded*, to protect from cold and damp. No iron whatever is used throughout the building. With reference to the materials, Professor Lloyd mentioned, that in the course of the arrangements now making for the erection of a magnetical observatory at Greenwich, Mr. Airy had rejected bricks in the construction of the building, finding that they were in all cases magnetic, and sometimes even *polar*. Mr. Lloyd has since confirmed this observation, by the examination of specimens of bricks from various localities ; and though there appeared to be great diversity in the amount of their action on the needle, he met with none entirely free from such influence.

"The building consists of one principal room, and two smaller rooms, one of which serves as a vestibule. The principal room is thirty-six feet in length, by sixteen in breadth, and has projections in its longer side, which increase the breadth of the central part to twenty feet. This room will contain four principal instruments, suitably supported on stone pillars : viz., a transit instrument, a theodolite, a variation instrument, and a dipping circle. The transit instrument (four feet in

* See a full account of the " Instructions for the Scientific Expedition to the Antarctic Regions, prepared by the President and Council of the Royal Society," in the 15th volume of the L. & E. Phil. Mag. See also a Letter from M. Kreil, containing a succinct account of the results of his magnetic observations at Milan, Vol. xvi. Also " Deductions from the First Year's Observations at the Magnetic Observatory at Prague," in a Letter from Professor Kreil to Major Sabine, Vol. xvii. An account of magnetic-term observations, taken in conformity with the instructions given by the Council of the Royal Society at Kerguelen's Land, also of hourly magnetic observations from May 25th to June 27th, was communicated by the Lords Commissioners of the Admiralty to the Royal Society, March 18th, 1841. See also Sir John Herchel's report on " Magnetic and Meteorological Observations," in the Transactions of the Manchester Meeting of the British Association ; and the valuable report of Major Sabine, published in the Seventh Report, Brit. Assoc. 1838.

focal length), will be stationed close to the southern window of the room. In this position it will serve for the determination of the time ; and a small trap-door in the ceiling will enable the observer to adjust it to the meridian.   The theodolite will be situated towards the other end of the room, and its centre will be on the meridian line of the transit. The limit of the theodolite is twelve inches in diameter, and is read off by three verniers to ten seconds.   Its telescope has a focal length of twenty inches, and is furnished with a micrometer reading to a single second, for the purpose of observing the *diurnal variation*.

"The variation instrument will be placed in the magnetic meridian, with respect to the theodolite ; the distance between these instruments being about seven feet.   The needle is a rectangular bar, twelve inches long, suspended by parallel silk fibres, and enclosed in a box to protect it from the agitation of the air.   The magnetic bar is furnished with an achromatic lens at one end, and a cross of wires at the other, after the principle of the collimator.   This will be observed with the telescope of the theodolite in the usual manner ; and the deviation of the line of collimation of the collimator from the magnetic axis, will be ascertained by reversal.   The direction of the *magnetic* meridian being thus found, that of the true meridian will be given by the transit.   It is only necessary to turn over the transit telescope, and, using it also as a collimator, to make a similar reading of its central wire, by the telescope of the theodolite.   The angle read off on the limb of the theodolite, is obviously the supplement of the variation.   The use of the transit has been suggested by Dr. Robinson ; and it is anticipated that much advantage will result from the circumstance ; that the two extremities of the arc are observed by precisely the same instrumental means.   With this apparatus it is intended to make observations of the *absolute variation* twice each day, as is done in the observatory of Professor Gauss, at Göttingen, the course of the *diurnal variation*, and the hours of maxima and minima, having been ascertained by a series of preliminary observations with the same instrument.

"A dipping circle, constructed by Gambey, will be placed on a pillar at the remote end of the room ; and will be furnished with a needle, whose axis is formed into a knife edge, for the purpose of observing the diurnal variations of the dip.   Gauss's large apparatus will also be set up in the same room, and will be used occasionally, especially in observations of the *absolute intensity*, made according to the method proposed by that distinguished philosopher.

"The bars are too large to be employed in conjunction with other magnetical apparatus.

"It is intended to combine a regular series of meteorological observations with those on the direction and intensity of the terrestrial and magnetic force just spoken of ; and every care and precaution has been adopted in the construction of the instruments."

(566) The use of bricks in the construction of these observatories, should be particularly guarded against, as when built into large edifices, they are found, not only to influence the magnetic needle, but to acquire magnetic polarity : this has been observed in the chimnies of factories. The material from which they were made (observes Prof. Stevelley) must be largely impregnated with iron : the mud of rivers was the *detritus* from hills whose rocks are often highly magnetic. The engineers employed in the trigonometrical survey of Ireland, had erected a mound of stones, composed of basalt, to sustain the signal staff which they had erected on the highest hill near Belfast : the effect of that heap of stones on the magnetic needle was so great, that in walking round it, the needle would veer round to every point of the compass.

(567) *Theory of Terrestrial Magnetism.*—Since the earth then is evidently a source of magnetic action, it becomes important to investigate the cause or causes from which it derives its origin, and what the nature of its magnetic condition may be.

Dr. Gilbert first suggested that the earth may contain within itself, and in a position nearly coinciding with its axis of rotation, a powerful magnet. If this were the case, that pole of the magnet which acts in our northern hemisphere, must have a southern polarity ; and that pole in the southern a northern, the former attracting the north, and the latter the south pole of the needle. The ordinary phenomena of terrestrial magnetism, may, agreeably to this hypothesis, be represented by placing a bar magnet within a terrestrial globe, and holding a small needle suspended by a fibre at different parts of its surface. When held in the northern hemisphere, it will always point to the north end of the enclosed magnet, exhibiting all the phenomena of the variation of the needle as usually observed : at the equator it will have no dip, each pole being equally attracted by the corresponding pole of the enclosed magnet ; and at the poles, the dip will be 90°, as observed by Commander Ross, at the northern magnetic pole.

(568) The phenomena of the dip are, however, more complicated than this hypothesis will allow us to suppose. Some observations of Hansteen, in Siberia, have led him to imagine that there is another magnetic pole in that country, which regulates the magnetic phenomena. In order to reconcile this with the theory of Gilbert, another magnet must be supposed to pass through the globe in the direction of a diameter, whose pole coincides with the Siberian magnetic pole ; but even this addition would be inadequate. The theory which has been long gaining ground is, that the magnetism of the earth is as that of a

spherical shell of iron, on which magnetism is induced. In such a
mass, and, indeed, in every mass of iron, either regular or irregular,
solid or hollow, the centres of action are always coincident with the
centre of attraction of the surface of the mass ; whereas, the centres of
action in *regular* magnets are placed *at,* or close to, their extremities.
The magnetism of the earth, recent and numerous observations have
shown, cannot be explained by the action of two magnetic poles at a
distance from each other. On the contrary, Biot has observed, that
the nearer the poles were taken to each other, the greater was the
agreement betweeen the computed and observed results. This idea has
been adopted by almost every philosopher who has investigated the
subject ; and the only difficulty is to assign a cause for the induced
magnetism. The following are the speculations of Hansteen :—

" For these reasons it appears most natural to seek their origin in the
sun, the source of all living activity : and our conjecture gains probability from the preceding remarks on the daily oscillations of the needle.
Upon this principle the sun may be conceived as possessing one or more
magnetic axes, which, by distributing the force, occasion a magnetic
difference in the earth, in the moon, and in all those planets whose
internal structure admits of such a difference. Yet allowing all this,
the main difficulty seems not to be overcome, but merely removed from
the eyes to a greater distance ; for the question may still be asked with
equal justice, *whence did the sun acquire its magnetic force ?* And if
from the sun we have recourse to a central sun, and from that again to
a general magnetic direction throughout the universe, having the milky
way for its equator, we but lengthen an unrestricted chain, every link
of which hangs on the preceding link, no one of them on a point of support. All things considered, the following mode of considering the
subject seems to me most plausible. If a single globe were left alone
to move freely in the immensity of space, the opposite forces existing in
its material structure would soon arrive at an equilibrium conformable
to their nature, if they were not so at first, and all activity would soon
come to an end. But if we imagine another globe to be introduced,
a mutual relation will arise between the two ; and one of its results
will be a reciprocal tendency to unite, which is designated, and sometimes thought to be explained, by the merely descriptive word *attraction.*
Now, would this tendency be the only consequence of that relation ?
Is it not more likely that the fundamental forces, being driven from
their state of indifference or rest, would exhibit their energy in all
possible directions, giving rise to all kinds of contrary action ? The
electric force is excited not by friction alone, but also by contact, and
probably also, although in smaller degrees, by the mutual action of two
bodies at a distance ; for contact is nothing but the smallest possible

distance, and that moreover only for a few small particles. Is it not conceivable that magnetic force may likewise originate in a similar manner? When the natural philosopher and the mathematician pay regard to no other effect of the reciprocal relation between two bodies at a distance, except the tendency to unite, they proceed logically, if their investigations require nothing more than a moving power; but should it be maintained that no other energy *can* be developed between two such bodies, the assertion will need proof, and the proof will be hard to find.

" I reckon it possible, therefore, that by means of the mutual relation subsisting between the sun and all the planets, as well as between the latter and their satellites, a magnetic action may be excited in every one of those globes, whose material structure admits of it, in a direction depending on the position of the rotatory axis with regard to the plane of the orbit. Each of the planets may thus give rise to a particular magnetic axis in the sun; but as their orbits make only small angles with the sun's equator and each other, these magnetic axes would perhaps, on the whole, correspond with the several rotatory axes. Such planets as have no moons would, on this principle, have but one magnetic axis; the rest would in all cases have one axis more than they have moons, if those different axes, by reason of the small angles which the orbits of their several moons form with each other, did not combine into a single axis. The conical motions by which the rotatory axes of the planets are carried round the pole of the ecliptic (the precession in the earth), joined to the revolving motion of the orbits about the sun's equator (which occasions the present diminution in the obliquity of the ecliptic), might perhaps, in this case, account for the change of position in the magnetic axis. It would greatly strengthen this hypothesis, if the above great magnetic period, after the lapse of which both axes again assume the same position, should in fact be found to coincide with the period of the precession, which, however, seems a little doubtful."

(569) Such were the speculations of Hansteen. But Sir David Brewster has proved, from an immense number of meteorological observations, that there were in our northern hemisphere two poles of maximum cold; that these poles coincided with the magnetic poles; that the circle of maximum heat, like the magnetic equator, did not coincide with the equinoctial line; that the isothermal lines, and that the lines of equal magnetic intensity, had the same general form surrounding and enclosing the magnetic poles, and those of maximum cold; and that by the same formula *mutatis mutandis*, we could calculate the temperature and the magnetic intensity of any point of the globe; thus there can be no doubt that there is a close connection between the phenomena of temperature and magnetism; and since the

discovery of Dr. Seebeck, that the mere application of heat to a circuit
of two metals is capable of developing magnetic effects, we may consider
that we have arrived a step nearer to an explanation of the earth's magnetism, by referring to the sun as the great agent of all these phenomena;
but we have yet to discover the metallic thermo-magnetic apparatus by
which they are produced.

(570) The electro-magnetic hypothesis was first advanced by the
able writer of the article from which most of the preceding remarks
were taken, Sir David Brewster; and Mr. Barlow has, by a beautiful
experiment, shown its application. It occurred to him that if he could
distribute over the surface of an artificial globe a series of galvanic currents, in such a way that their tangential power should everywhere give
a corresponding direction to the needle, this globe would exhibit, while
under electrical induction, all the magnetic phenomena of the earth upon
a needle freely suspended above it. The following is an account of the
experiment :—

" I procured a wooden globe, sixteen inches in diameter, which was
made hollow for the purpose of reducing its weight, and while still in
the lathe, grooves were cut to represent an equator and parallels of
latitude, at every $4\frac{1}{2}°$ each way from the equator to the poles; these
grooves were about one-eighth of an inch deep and broad; and lastly,
a groove of the same breadth, but of double the depth, was cut like a
meridian from pole to pole half round. These grooves were for the
purpose of laying in the wire, which was effected thus :—the middle of
a copper wire nearly ninety feet long, and one-tenth of an inch in
diameter, was applied to the equatorial groove, so as to meet in the
transverse meridian; it was then made to pass round this parallel,
returned again along the meridian to the next parallel, and then passed
round this again, and so on, till the wire was thus led in continuation
from pole to pole.

"The length of wire still remaining at each pole was bound with
varnished silk, to prevent contact, and then returned from each pole
along the meridian groove to the equator. At this point, each wire
being fastened down with small staples, the wires for the remaining
five feet were bound together to near their common extremity, where
they opened to form two points for connecting the poles of a powerful
compound voltaic battery.

" When this connection was made the wire became, of course, an
electric conductor, and the whole surface of the globe was put into a
state of transient magnetic induction; and consequently, agreeably to
the laws of action above described, a neutralized needle freely suspended
above such a globe, would arrange itself in a plane passing from pole to
pole through the centre, and take different angles of inclination, according to its situation between the equator and either pole.

" In order to render the experiment more strongly representative of the actual state of the earth, the globe in the state above described, was covered by the gores of a common globe, which were laid on so as to bring the poles of this wire arrangement into the situation of the earth's magnetic poles, according to the best observations we have for this determination; I therefore placed them in latitude 72° north, and 72° south, and on the meridian corresponding with 76° west, by which means the magnetic and true equators cut one another at about 14° east and 166° west longitude.

" The globe being thus completed, a delicate needle must be suspended above it, neutralized from the effect of the earth's magnetism, according to the principle I employed in my observations on the daily variation, and described in the Philosophical Transactions for 1823; by which means it will become entirely under the superficial galvanic arrangement just described. Conceive the globe now to be placed so as to bring London into the zenith, then the two ends of the conducting wire being connected with the poles of a powerful battery, it will be seen immediately that the needle, which was before indifferent to any direction, will have its north end depressed about 70°, as nearly as the eye can judge, *which is the actual dip in London.* If now we turn the globe about on its support, so as to bring to the zenith places equally distant with England from the magnetic pole, we shall find the dip remains the same; but the variation will continually change, being first zero, and then gradually increasing eastward as happens on the earth. If again we turn the globe so as to make the pole approach the zenith, the dip will increase, till at the pole itself the needle will become perfectly vertical. Making now this pole recede, the dip will decrease till at the equator it vanishes, the needle becoming horizontal; continuing the motion, and approaching the south pole, the south end of the needle will be found to dip, increasing continually from the equator to the pole, where it becomes again vertical, but reversed as regards its verticality at the north pole."

(571) But although a sphere thus arranged may be made to exhibit the phenomena of terrestrial magnetism without the aid of any magnetic body, we have yet to learn how such a system of electrical currents can have existence in the earth, unless we refer them to the action of the sun on a metallic thermo-electric apparatus distributed over the earth. It would still, however, remain to be shown what this thermo-electric apparatus is, and where and how it is distributed.

Whether we seek for a cause of terrestrial magnetism in electrical currents, induced on the earth's surface, or whether we refer it to magnetism induced on the ferruginous matter it contains, or in its atmosphere, we are limited to the *Sun*, if not as a primary cause, at least as an agent, to which magnetic phenomena have a distinct reference; future investigation must decide whether it acts by its heat, or by its light, or by specific rays or influences of a magnetic nature. Barlocci and Zantedeschi found that both natural and artificial magnets had their magnetism greatly increased by exposure to common solar rays; a result which could not arise from their *heating* power, as an increase of temperature invariably diminishes the power of magnets.

(572) It is an admitted fact that the *aurora borealis* is a powerful source of magnetism, and that the *south* pole of the needle has a distinct connexion with it. Dr. Dalton, in a work published in 1793, has advanced several ingenious hypothetical views, respecting the cause of the aurora, and its magnetic influence. He says, " the region of the aurora is one hundred and fifty miles above the earth's surface. Immediately above the earth's surface is the region of the clouds, then the region of the meteors, called falling stars and fire-balls, and beyond this region is that of the aurora."*

" We are under the necessity of considering the *beams* of the *aurora borealis* of a *ferruginous* nature, because nothing else is known to be magnetic; and consequently that there exists in the higher region of the atmosphere, an electric fluid partaking of the properties of *iron*, or rather of *magnetic steel;* and that this fluid, doubtless from its magnetic property, assumes the form of cylindrical beams.

" With regard to the exciting cause of the aurora, I believe it will be found in change of temperature. Nothing is known to affect the magnetism of steel : heat weakens and destroys it; electricity does more; it sometimes changes the pole of one denomination to that of another, or inverts the magnetism. Hence we are obliged to have recourse to one of these two agents, in accounting for the mutations above mentioned. As for heat, we should find it difficult, I believe, to assign a reason for such sudden and irregular productions of it in the higher regions of the atmosphere, without introducing electricity as an agent in these productions; but rather than make such a supposition, it would be more philosophical to suppose electricity to produce the effect on magnetic matter IMMEDIATELY. The beams of the *aurora* being magnetic will have their magnetism weakened, destroyed, or inverted, *pro tempore*, by the several shocks they receive during the aurora." In another place he says, " I conceive that a beam may have its mag-netism inverted, and exist so for a time, &c." Again, " As the beams are swimming in a fluid of equal density with themselves, they are in

* Dr. Dalton adduces in proof of the great height of the aurora its extremely attenuated light, which, he says, may spread over one-half of the hemisphere, and not yield more light than the full moon : this, he says, arises from the extreme rarefaction of the air.

the same predicament as a magnetic bar or needle swimming in a fluid
of the same specific gravity with itself : but this last will only rest in
equilibrio, when in the direction of the *dipping needle*, owing to what
is called the *earth's magnetism ;* and as the former also rests in that
position only, the effects being similar, we must, by the rules of philo-
sophising, ascribe them to the same cause. Hence then it follows,
THAT THE AURORA BOREALIS IS A MAGNETIC PHENOMENON, AND ITS
BEAMS ARE GOVERNED BY THE EARTH'S MAGNETISM. I am aware
that an objection may be stated to this ; if the beams be swimming in
a fluid of equal density, it will be said they ought to be drawn down
by the action of the earth's magnetism. Upon this I may observe,
that it is not my business to show why this is not the case, because I
propose the magnetism of the beams as a thing demonstrable, and not
as a hypothesis. We are not to deny the cause of gravity, because we
cannot show how the effect is produced. May not the difficulty be
removed by supposing the beams of *less* density than the surrounding
fluid ?"

Lastly, although it has been clearly proved that a source of
magnetism does exist in the atmosphere, yet it may be asked if there is
any reason for believing that the magnetism in the atmosphere is strong
enough to be considered as the *only* source of terrestrial magnetism. It
has been shown by M. Arago, that the auroræ which exist only at St.
Petersburg, in Siberia, and even in North America, actually disturb the
magnetic needle at Paris ; and he considers it highly probable that the
auroræ even round the south pole of our globe extend their influence to
Paris. Sufficient, however, has, I think, been said, to show that many
more enquiries are wanting, before the question as to the true cause or
causes of terrestrial magnetism can be satisfactorily answered.*

* See two papers entitled " Contributions to Terrestrial Magnetism," communicated to the Royal Society by Major Sabine, and read March 19th, 1840, and Feb.
11th, 1841. See also a full account of a very extensive series of observations made
at the magnetic observatories of Toronto, Trevandrum, St. Helena, and the Cape of
Good Hope, during a remarkable magnetic disturbance on the 25th and 26th of
September, 1841, in the 20th vol. of the L. and E. Phil. Mag. Also a notice of a
remarkable magnetic disturbance which occurred on the 2nd and 4th of July 1842,
observed at the Dublin Magnetical Observatory, and communicated to the Phil.
Mag. Vol. xxi. by Dr. Lloyd.

LECTURE VIII.

ON ELECTRO-MAGNETISM.

Discovery by Oersted—Mutual actions and relative motions of magnets and wires
carrying currents of Electricity—Induction of permanent magnetism on steel
and temporary magnetism on soft iron—Ritchie's Rotating Magnet—Rotating
coil—Electro-magnets of Roberts, Radford and Joule—Ritchie's observations
on Electro-magnets—Electro-magnetic Engines: Davenport's, Taylor's, Davidson's, Jacobi's, Talbot's, Wheatstone's, Henley's and Bain's—Galvanometers:
Ironmongers' hydrostatic galvanometer—Dr. Locke's Thermoscopic galvanometer — Electro-magnetic telegraphs : Alexander's, Morse's, Davy's, Wheatstone
and Cooke's needle telegraph ; their improved Electro-magnet telegraph ; their
printing telegraph — Methods of insulating the wires — Bain's Experiments ;
his Electro-magnetic printing telegraph ; his other applications of Electro-
magnetism ; his Electro magnetic clocks — The Rev. Mr. Lockey's contact
formers—Wheatstone's Electro-magnetic clock.

(573) The disturbance produced in the magnetic needle by the aurora
borealis and lightning, had long suggested to philosophers that the
agencies of Electricity and magnetism must be connected by some
close and intimate relation. For nearly half a century the discovery
of this relation was a favourite subject of speculation ; and it is curious
to compare the various opinions which were maintained by different
experimentalists. Magnetic properties were easily communicated to
bars of steel, by passing strong electrical shocks through them, but no
general law could be traced as governing the polarity thereby imparted.
D'Abilard imagined, that he had proved that the electric discharge imparts a northern polarity, to that point of a steel bar at which it enters,
and a southern polarity to that at which it makes its exit : and this
quite independently of the position of the needle, with respect to the
magnetic poles of the earth. Wilke, on the other hand, was equally
satisfied that an invariable connexion exists between the negative Electricity, and the northern polarity.

(574) In one of the essays (which received a prize) on the question
proposed by the electoral academy of Bavaria, in 1774, " Is there a
real and physical analogy between electric and magnetic forces; and if

such analogy exists, in what manner do these forces act on the animal body ?" Professor Van Swinden, of Franeker, after a long and elaborate discussion of the subject, arrived at the conclusion, that the similarity between Electricity and Magnetism, amounts merely to an *apparent* resemblance, and does not constitute a true physical analogy ; whence he infers that these two powers are essentially different and distinct from each other. On the other hand, Professors Steiglehner and Hubner, maintained that both classes of phenomena are referable to the *same* agent, varying only in consequence of a diversity of circumstances. In this unsettled state, the subject remained till some years after the discovery of galvanism, by which a fresh field of enquiry was opened, and a means of maintaining a large and continuous current of Electricity obtained. The first approach to a solution of the question, was the publication of Ritter; he asserted, "that a needle, composed of silver and zinc, arranged itself in the magnetic meridian, and was slightly attracted and repelled by the poles of a magnet :" he also stated, "that by placing a gold coin in the voltaic circuit, he had succeeded in giving to it positive and negative electric poles ; and that the polarity so communicated, was retained by the gold, after it had been in contact with other metals, and appeared therefore to partake of the nature of magnetism. A gold needle, under similar circumstances, acquired still more decided magnetic properties ;" and, " that a metallic wire, after being exposed to the voltaic current, took a direction, N.E. and S.E."

(575) In consequence of the vague and loose manner in which Ritter advanced his speculations, but little notice was taken of them, and no satisfactory results were obtained, till the year 1819, when Professor Oersted, of Copenhagen, made his famous discovery, which forms the basis of the science of electro-magnetism. The fact observed by Oersted was, that when a magnetic needle was brought near the connecting medium, (whether a metallic wire, or charcoal, or even saline fluids, of a closed voltaic circle,) it was immediately deflected from its natural position, and took up a new one, depending on the relative positions of the needle and wire. If the connecting medium was placed horizontally *over* the needle, that pole of the latter which was nearest to the *negative* end of the battery, always moved *westward;* if it was placed *under*, the same pole moved to the *east*. If the connecting wire was placed parallel with the needle, that is, brought into the same horizontal plane in which the needle was moving, then no motion of the needle in that plane took place, but a tendency was exhibited in it, to move in a vertical circle, the pole nearest the *negative* side of the battery being depressed when the wire was to the *west* of it, and elevated when it was placed on the *eastern* side. Fig. 200

represents a convenient arrangement for exhibiting the action of a wire
conducting a current of
Electricity on the magnetic
needle. *a a*, two turned
wooden pillars screwed into
a base board B, and sur-
mounted by two mercury
cups *c c*. *D*, a copper
wire, the ends of which dip
into the mercury, as do also
the wires connected with
the opposite extremities of

FIG. 200.

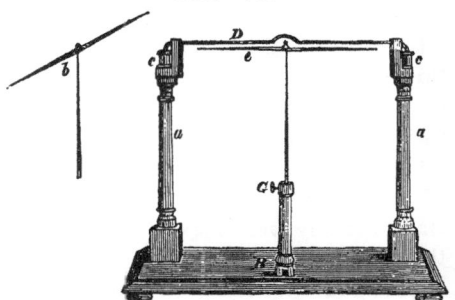

a simple voltaic battery. A current of Electricity can thus be made
to pass in either direction along the wire D : *e* is the magnetic needle
nicely poised on a wire, which by means of the screw G may be elevated or depressed, and the needle thus set either above or below the
wire D, or it may be removed and replaced by the dipping needle *b*.
As in all electro-magnetic researches, it is necessary to bear in mind
these affections of the needle and electrified wire ; several contrivances
have been made to assist the memory respecting the details. Fig. 201
represents the plan of Dr. Roget. *A B* is a slip of card, on each side of
which, a line *a b* is drawn along the
middle of its length, the end *a* being
marked $+$, the end *b* —, and the centre
c being crossed by an arrow, at right
angles to it, directed as in the figure.
Through the centre, and at right
angles to the plane of the slip of
card, there is made to pass, a slender
stem of wood, at the two ends of
which, are fixed in planes, parallel to
the slip of card *A B*, the circular
discs of card marked respectively
with the letters *N* and *S*, and with
arrows parallel to, but pointing in a

FIG. 201.

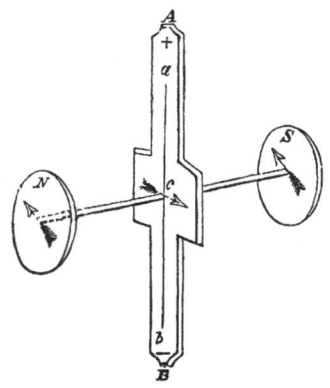

contrary direction to the one at *c*. The same marks must be put on
the reverse of each of the three pieces of card, so that when held in
different situations they may be seen without turning the instrument.

(576) If the line *a b*, be supposed to represent the connecting wire,
(the direction of the current of Electricity being denoted by the signs
$+$ and — at the ends of the line) the arrow at the centre will point
out the direction in which it tends to move, when under the influence
of the north pole of a magnet, situated at *N ;* or of a south pole

situated on the other side, as at *S* ; and vice versâ the arrows *N* and *S*, will indicate the directions in which the north and south pole respectively tends to revolve round the connecting wire in its vicinity, with relation to the direction of the current of Electricity, that is passing through it. It must be observed that the poles *N*, *S*, are not considered as in connexion with each other, or as forming parts of one magnet; their operations are exhibited singly and quite independently of each other. The advantage of the instrument consists in its being capable of being held in any situation, and thus easily adapted to the circumstances of any fact or experiment of which we may wish to examine the theory.

(577) A useful help to the memory has also been suggested by Ampère. Let the observer regard himself as the conductor or connecting wire, and imagine a positive electric current to pass from his head towards his feet in a direction parallel to the magnet; then its north pole in front of him will move to his right side, and its south pole to his left. The plane in which the magnet moves is always parallel to the plane in which the observer supposes himself to be placed. If the plane of his chest be horizontal, the plane of the magnet's motion will be horizontal; but if he lie on either side of the horizontally suspended magnet, his face being towards it, the plane of his chest will be vertical, and the magnet will tend to move in a vertical plane.

(578) The extent of the declination of the magnetic needle depends entirely on the *quantity of Electricity* passing along the connecting wire, and has nothing to do with the *tension* of that Electricity, nor is it increased by increasing the intensity of the current; hence the employment of galvanic batteries for the exhibition of the effects of electromagnetism; and hence, also, the reason why the first enquirers were foiled in their attempts to elicit these effects.

(579) That the conducting wire is capable of attracting and repelling the poles of a magnet, may be proved by the following experiment :— suspend horizontally a magnetic needle, and place it near the connecting wire of a closed circle, held *vertically* and at right angles to it; the positive current being supposed to flow upwards. When the wire is exactly midway between the poles of the needle, no effect will be produced, but on moving it about half-way to the north pole, it will be attracted; still continue to advance it, and when it reaches the extremity of the north pole, it will be repelled. Similar effects occur on advancing the wire towards the south pole. If the current of Electricity be made to *descend* the wire, or if the wire be placed on the *eastern* side of the needle, the current ascending, equal but opposite effects will be observed; then, repulsion will first take place, and afterwards attraction.

(580) From the manner in which the needle is affected, when placed

parallel to either side of the electric current, it was inferred that a *current of magnetism* is set in motion at right angles to the latter. This was termed by Dr. Wollaston *vertiginous* magnetism ; and by Mr. Barlow, the magnetic force was said to exert a *tangential action.*

(581) The figures 202 and 203, will represent the direction of the circulating current of magnetism. In Fig. 202, the connecting wire is

FIG. 202.   FIG. 203.

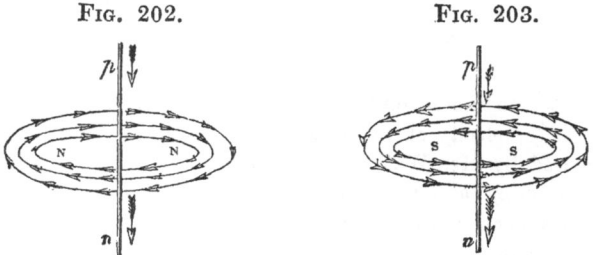

placed vertically, and the electric current is descending through it from *p*, to *n ;* the arrows denote the direction in which the north pole of a magnet will have a tendency to move round it, i. e., from left to right, or in the direction of the hands of a watch. Fig. 203 shows the motion impressed on the south pole by a similar current. When the direction of the Electricity is reversed, the wire still preserving its vertical position, the direction of the action is also reversed.

(582) Reasoning on these relative motions of the needle and electrified wire, Dr. Faraday conceived that the pole of a magnet ought to revolve about the conductor, and the conductor about the pole of a magnet, and by the following ingenious apparatus he succeeded in proving this to be the case.

Into the centre of the bottom of a cup, as in the vertical section, Fig. 204, a copper wire *c*, D, was inserted, a cy-   FIG. 204.
lindrical magnet *n*, *s*, was attached by a thread to
the copper wire, *c*, and the cup was nearly filled
with mercury, so that only the north pole of the
magnet projected. A conductor, *a*, *b*, was then
fixed in the mercury, perpendicularly over *c*. On
connecting the conducting wires with the opposite
ends of the battery, a current was transmitted from
one wire, through the mercury to the other. If
the positive current descended, the north pole of
this magnet immediately began to rotate round the
wire, *a*, *b*, passing from east through the south to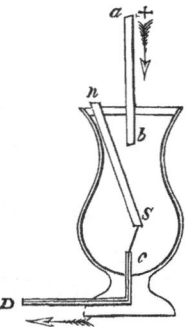
west, i. e., in the direction of the hands of a watch ; but if the current ascended, the line of rotation was reversed. Conversely, a magnet

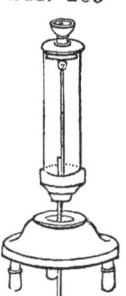

FIG. 205 was fixed in a vessel of mercury, and the conducting wire hung from a hook above it, the end just dipping into the fluid; the electric current being then transmitted through the moveable conductor, Faraday found that the free extremity instantly began to revolve round the pole of the magnet, in a direction similar to the last. A good contrivance for exhibiting this, is shown in Fig. 205.

(583) In order to obviate the necessity of employing so much quicksilver, which, by the resistance which it offers to the revolution of the magnet, greatly diminishes the velocity of the rotation, the apparatus in Fig. 206, has been devised by Mr. Watkins. It exhibits the contrary poles of two magnets rotating about two electrified wires. Two flat bar magnets, doubly bent in the middle, and having at the under part of the bend agate cups fixed, by which they are supported upon upright pointed wires, affixed in the basis of the apparatus, and upon which they turn round as upon an axis. Above the agate cups, on the upper part of the bend, small cisterns to hold mercury are also formed. Two circular troughs to contain mercury, are supported upon

FIG. 206.

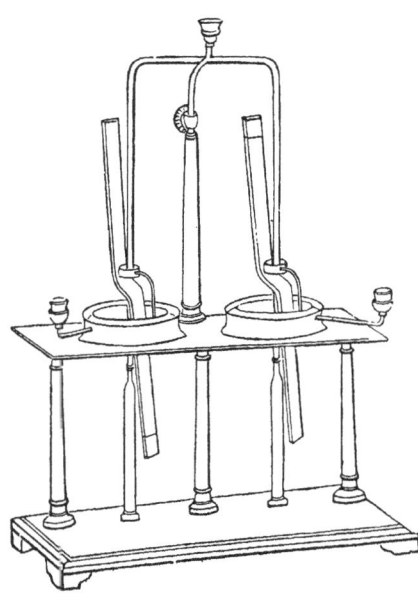

a stage, affixed to the basis, having holes in their centres, to allow the magnets to pass through them. A bent pointed wire is affixed into the cisterns of each magnet, the ends of which dip into the mercury contained in the troughs upon the stage; and through the sides of the trough, wires are passed, entering into the mercury contained in the troughs, and bearing at their ends other cups to hold mercury. To steady the motion of the magnets, wire loops are affixed to them, which embrace the upright pointed wires on which the magnets rest. A hollow pillar is firmly affixed to the stage, in which a bent wire supporting another cross wire is inserted, and is capable of

being raised or lowered, and secured at any required height by a binding screw.   The two ends of the cross wire are bent downwards and
pointed, and made to enter the two small cisterns affixed upon the magnets.
A third cup to contain mercury is also provided at the top of the cross
wire, and a communication being made with the battery by means of uniting wires dipping into the mercury in the cups, the wire from the positive
end of the battery being placed in the upper cup, and the wire from
the negative end in each of the lower cups, the magnets will begin to
rotate in opposite directions, and those directions may be reversed, by
changing the situations of the uniting wires.   Two batteries should
here be employed, in order to make both the magnets revolve with the
desired velocity ; and attention must be paid, when using two batteries,
that the currents of Electricity flow in the same direction ; otherwise,
the phenomena of the revolutions of the magnets in contrary directions
will not take place, but they will both revolve in the same direction.

(584) Thus it will be seen that the direction of the rotation imparted
by a fixed current to a moveable pole, will be the same as that which
the same pole imparts to the same current.   Sup-

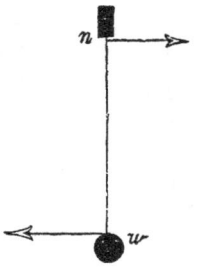

FIG. 207.

pose w, Fig. 207, to represent a section of a conducting wire, along which a positive current is
descending, and n the north pole of a magnet.
The influence of w on n will be to impel it in the
direction of the arrow ; but n will also re-act on
w, and tend to produce in it an opposite direction,
as exhibited by the arrow attached to w.   Each
is supposed to describe a circle round the other,
moving in the same direction as the hands of a
watch ; and if w and n were at liberty equally to
move, they would have a tendency to rotate round the line between
them.

(585) Ampère first succeeded in effecting the rotation of a magnet,
round its own axis.   In his original experiment, the magnet was
allowed to float, without a support, in a vessel of mercury, being kept
in a vertical position, by a weight of platinum attached to its lower end.
The object was, to make the electrical current pass through one half of
the magnet itself, and then to divert it from its course, and make it pass
away in such a direction, as that it should not affect the other half.   The
reason of this is evident : suppose a positive current be made to descend
a magnet placed vertical, its north pole being uppermost, it would tend
to urge that pole round from left to right ; but its influence on the south
pole, would be just the reverse, tending to urge it from right to left ; or
if two electrical currents be supposed, corresponding to the vitreous and

* Popular Sketch of Electro-Magnetism, by Francis Watkins.

resinous electricities, the tendencies would be the same ; and here it may
be as well to mention, that in describing the phenomena of electro-magnetism, I shall, to avoid tediousness, adopt the language of a single fluid,
and suppose, that in the connecting wire of a voltaic battery, the electrical
current is passing in one stream, from the positive to the negative end.

(586) In Ampère's experiment, the electric current, after traversing
the upper half of the magnet, passes into the mercury, and being
diffused through it, acts in no sensible degree on the lower half, and
does not interfere with the rotation produced by its influence on the
upper pole.

It is, however, better to carry off the current by a different channel,
and this is effected by adopting the form of apparatus, shown in Fig.
208. It is thus constructed by Mr. Watkins. A flat

Fɪɢ. 208.

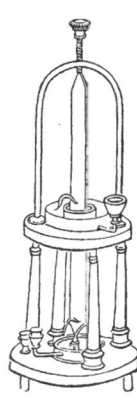

bar magnet is supported in a vertical position, by an
upright metal wire, affixed in the basis of the apparatus,
and having a hole in its centre, containing an agate
cup, to receive the lower pointed end of the magnet ;
its upper end turns in another hole, made in a vertical
screw, with the milled head to turn it by, which is
passed through a screw hole, made in an arched piece of
wire, affixed to the upper part of the basis. Around
the first mentioned vertical wire, a cistern to contain
mercury is provided ; and another, having a hole in its
centre, to allow the magnet to pass through, and revolve
within it, near the middle of the magnet. These cisterns have metal wires projecting into them, through
their sides, to support cups which contain mercury, to
effect the communication with the voltaic battery, by
means of uniting wires. Into the magnet, two small bent and pointed
wires are affixed, the ends of which dip into the mercury, contained in
the cisterns. When the voltaic circuit is complete, the magnet begins
to rotate within the Electricity, which it conducts itself, as it in fact
forms part of the circuit ; the rapidity of the revolutions of the magnet,
depending upon the delicacy of the sustaining point, the strength of the
magnet, and the power of the battery employed. If it be desired to
actuate a large magnet, it is necessary an addition to the apparatus
should be made, by providing a cup, affixed to the vertical screw, to
contain mercury, by which contrivance, and by employing an additional
battery, a current of Electricity can be passed from the top of the magnet to its equator ; and, as in the first mentioned case, an opposite current can be passed from its lower end to the equator, an additional
force is obtained. The current from the second battery must of course
be sent along the upper half of the magnet, in a direction contrary to

that which passes through the lower pole; but since the rotatory force
is proportional to the power of the voltaic battery employed, it is probable that the second battery would be equally efficacious, if it were
employed in increasing the strength of the first, by being joined to it.
The ends of the wires should be amalgamated, by rubbing them first
with nitrate of mercury, and then dipping them into the clean metal.

(587) Fig. 209 represents an apparatus to exhibit the rotation of a
conducting body round its own axis, and is exactly the converse of the
last experiment.   In the former case, the electric

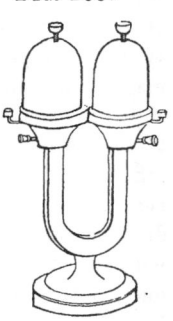

FIG. 209.

current was applied in the interior of the magnet,
but here means have been devised for procuring the
action of the magnet, from the interior of the conducting body.   In the place of the wire, therefore, a
hollow metallic cylinder is employed, in the axis of
which, the influencing magnet can be placed.   Mr.
Barlow devised this instrument, and the figure shows
the arrangement on a horse-shoe magnet, by Mr.
Watkins.   A horse-shoe magnet is supported vertically upon a stand, having holes formed in the centres of its ends.   Two wooden circular troughs are
secured by binding screws upon the arms of the magnet, to contain
mercury.   Into the holes in the centres of the ends of the magnet, two
conical pointed wires are inserted, which are affixed in the middle of
two hemispherical cups, united to cylinders, the rims of which are
formed into points, which are dipped into the mercury, contained in
the circular troughs.   Upon the top of each hemisphere, is placed a
small platinum cup, to contain mercury.   Other cups for holding mercury, are supported on the external ends of bent wires, which pass
through the sides of the circular troughs, into the mercury, contained
therein.   When a stream of voltaic Electricity is passed through this
apparatus, by means of connecting wires, placed in the mercury, contained in the upper and lower cup, the cylinders commence revolving
in opposite directions, that cylinder on the north pole, and down which
the current is descending, moving of course from left to right; but if
the two upper cups be united by a wire, and the lower cups connected
with the positive and negative extremities of the voltaic battery, the
same stream will traverse both sides of the apparatus, passing upwards
in one cylinder, and downwards in the other; and the rotations will
now, from the contrary influences of the two poles, be in the same
direction in both cylinders.

(588) Dr. Faraday has shown, that the results in this last experiment, are the same when the magnet and conductor are united together; for, on fixing a thin piece of wood on the upper end of a magnet,

loaded at its lower extremity with a platinum weight, and floating in a vessel of quicksilver, and attaching to the wood an arch of strong wire, the whole apparatus commenced revolving, on the transmission of the electric current through it ; on the other hand, when a hollow cylinder of metal was balanced on a vertical axis of wood, and acted on by the poles of a magnet placed *outside*, the rotatory force was very feeble. This affords us means of explaining the circumstances of the rotation of a magnet, about its own axis ; for the explanation of that experiment will very much depend on the course which the current of Electricity is supposed to take, in its passage through the magnet. If it be supposed to pass through the interior, along the axis of the magnet, it would then occasion rotation, by its influence on the parts of the magnet that are situated nearer the surface ; but if the course of the current be supposed to be along the surface, it will itself be influenced by the polarity of those portions of the magnet which lie near the axis, and the rotatory tendency impressed upon it will produce the rotation of the magnet, which will, of course, be carried along with it. This, it will be seen, corresponds with the rotation of a conducting body round its own axis, a magnet being in the centre ; and it has been shown above, that the circumstance of the magnet and conductor being immoveably joined, makes no difference in the results.

(589) Another fact is made apparent by this last experiment, which is, that the electro-magnetic influence of the conductor takes place equally, when the electrical current is diffused over a considerable surface, as when it is concentrated in a single wire ; in the cylinder, every filament of which it is composed may be supposed to conduct its share of the current, and thus contribute towards the general effect.

(590) A magnetic needle, is found to be influenced by the current of Electricity that is passing through the voltaic battery, from its *positive* to its *negative* pole, as well as by the wire that completes the circuit ; or, in other words, every part of the circuit exhibits the same electro-magnetic properties ; and, as action always implies an equal and corresponding re-action, the magnet may be supposed to have a tendency to move the battery, equal to that which the battery has to move it. This tendency was first actually exhibited, by a very ingenious contrivance of Ampère, and which Mr. Watkins has applied to each of the poles of a horse-shoe magnet, as shown in Fig. 210. It consists of a horse-shoe magnet, firmly fixed to a stand, at its bent part ; its two ends being made round, and having a small hole in the centre of each, at the bottom of which hole, an agate cup

Fig. 210.

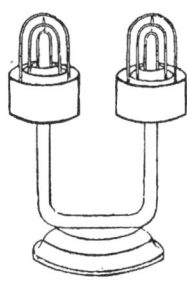

is placed, and in which pointed wires fixed to the parts presently to be described are made to revolve. A double cylindrical copper vessel, having a bent metal wire, fixed to the top of its innermost cylinder, with a vertical wire pointed at both ends, fixed in the middle of that bent wire, is hung upon the upper end of each pole of the magnet, the lower points of the vertical wires of each vessel, entering the holes, formed as above described, in the magnet for that purpose. Two hollow cylinders of zinc, each furnished with similar bent wires, having holes made in the under sides of each, are then placed within the double copper vessels ; the holes in the bent wires, being hung upon the uppermost pointed ends of the vertical wires, before mentioned. Diluted acid being then poured into the space between the copper cylinders, the voltaic action commences, and presents the phenomena of the whole of the four cylinders revolving upon their axes, the copper vessels revolving in opposite and contrary directions, and the zinc cylinders turning in opposite directions to them ; the rapidity of their revolutions depending upon the strength of the acid, and the delicacy of their suspension.*

(591) Numerous amusing experiments have been devised for exhibiting the vibratory tendencies of electrified wires, when under the influence of magnets. Fig. 211 represents an arrangement of Mr. Marsh. It consists of a slender wire, suspended from a loop, and capable of free motion ; its lower end is amalgamated, and dips into a small cistern of mercury; the cups a and b, are filled also with mercury, and through them the electrical current is passed down the loose wire ; no motion of this wire is perceptible, until a horse-shoe magnet is placed in a horizontal position, on the basis, with its poles enclosing the wire, when it is instantly urged either forwards towards c, or backwards towards d, according to the position of the poles, and the direction of the current. In either case it is thrown out of the mercury, and the circuit being thus broken, the effect ceases, until the wire falls back again, by its own weight, into the mercury ; when the current being re-established, the same influence is again exerted, the phenomenon is repeated, and the wire exhibits a quick succession of vibratory motions.

Fig. 211.

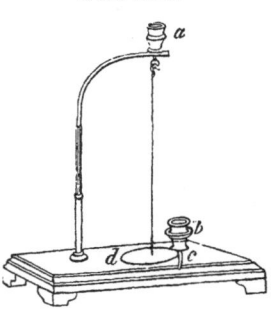

(592) This vibratory motion is easily converted into one of rotation,

* The zinc cylinders revolve with great rapidity, but from the superior weight of the copper cylinders, when filled with the exciting liquor, it is rarely that the rotatory tendency can be exhibited in them.

by employing a spur wheel, as in Fig. 212. The radii of the wheel

FIG. 212.

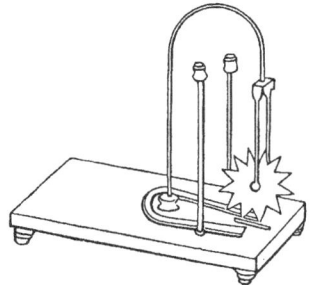

must be so arranged that each ray shall touch the surface of the mercury, before the preceding ray shall have quitted it. The direction of the motion depends of course on the same circumstances, as were before mentioned.

This forms a very brilliant experiment, when a powerful battery and a strong magnet are employed. The wheel revolves with immense velocity, and streams of sparks of a green colour, arising from the combustion of the copper points of the radii of the wheel, are thrown sometimes over the cups of the instrument.

Mr. Sturgeon found that the division of the wheel into rays was not necessary, and that if a circular metallic disc be substituted for the spur wheel, as shown in Fig. 213, it will revolve equally well. In all

FIG. 213.

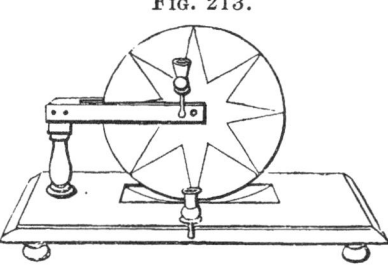

these experiments it is important that the ends of the wires and surface of the metals, which touch the mercury, should be well amalgamated, in order to ensure perfect contact.

By altering the direction of the electrical current all the vibrations and rotations that have been just described are *reversed :* this alteration may be effected very readily by means of the little apparatus shown in Fig. 214, which was the contrivance of Magnus.

FIG. 214.

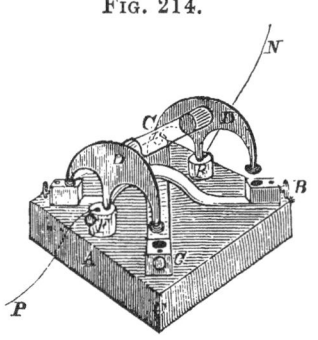

A A is a block of wood on which is fixed the two brass bands *B B, C C,* terminated at their extremities by square blocks of brass with binding screws and cup shaped holes. *D D,* two flat pieces of brass connected and insulated by the glass rod *E.* These are the break pieces, and move on joints at *F F,* where they are connected by the wires *P N* with the battery. The other binding screws at the termination of the brass bands serve to connect the arrangement through which the currents are to pass.

(593) We have now seen abundant instances of the mutual action of electrical currents and magnets on each other, and referring to the first experiment of Oersted, it will be remembered, that a positive current, passing horizontally over the needle, from north to south, has a tendency to cause the north pole to move to the east; consequently, if another current be made to pass under the wire, from south to north, the needle will be affected with twice the force that a single wire would have exerted; or, what is the same thing, a continuous wire may be bent back on itself, as in Fig. 215, and this constitutes the simplest form of the galvanometer.

Fig. 215.

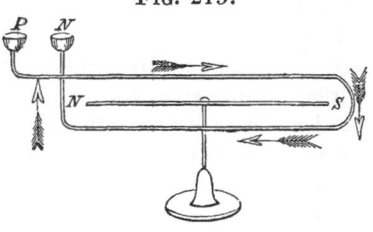

The current, passing in opposite directions, above and below the needle, will conspire, in both cases, to deflect it from its natural position, and in the same direction, and to bring it into a position nearer to right angles to the plane of the wires.

(594) If instead of a rectangle, the wire be supposed to be bent into a circular form, the pole of a magnetic needle, placed in the centre, or in a line passing through the centre, would be impelled in one uniform direction, by the electric current, transmitted through the wire, and could the current move in a perfect circle, from right to left, the north pole of a magnet, placed in the centre, would move to the right, and the south pole to the left. If the north pole of a magnet, therefore, were presented to the right hand side of this circular current, it would tend to move away from it, having the appearance of being repelled; just the contrary would happen, if the south pole of a magnet were presented on the same side, that is, there would be the appearance of a mutual attraction between them. But when either of these poles is presented on the other side of the plane of the circular current, effects of an opposite kind are produced; the north pole appears to be attracted, and the south pole to be repelled.

(595) Now, since a similar and reciprocal action takes place between the magnetic pole and the electric current, the latter, together with the wire that conveys it, will, if at liberty to move, recede from the north pole, or appear to be repelled by it, or advance towards the south pole, and appear to be attracted by it. Fig. 216 represents the very ingenious apparatus invented by M. De la Rive, for exhibiting this reciprocal action. It consists of a glass cylindrical vessel, having a cork float attached to its upper end. Into this vessel is inserted a small

Fig. 216.

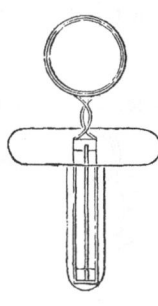

voltaic combination, formed on Dr. Wollaston's plan, and consisting of a plate of zinc surrounded by a copper plate, the zinc plate being insulated upon its edges. A copper wire affixed by soldering to both these plates, is made into the form of a ring, consisting of several coils of the wire, which is besides insulated, by being wrapped round with silk thread : upon pouring diluted acid into the glass vessel, and placing the plates in it, voltaic action commences, and is manifested by placing the apparatus afloat in water, when the coil will have a tendency to take a position in the plane of the magnetic meridian, and will exhibit all the effects of the attractive and repulsive tendencies which have been described above, when a strong bar magnet is brought near it on either side. If the magnet be sufficiently slender to pass through the ring, the following curious phenomenon will be observed :—If the pole be presented to it on the side where attraction takes place, the ring will move towards it till it arrives at the pole, and then proceed onwards in the same course, the magnet being held in the axis of the ring, till it reaches the middle of the magnet, but there it seems inclined to stop ; and then, after a few oscillations, it settles as in a position of equilibrium; for, if purposely displaced, by bringing it forwards towards the other pole, it returns with a force which shows that it is repelled from that other pole. Let the magnet now be withdrawn, and turning it half round, so that its poles are in directions the reverse of what they were at first, and holding the ring in one hand, let the magnet be again introduced into it with the other hand, until it is half way through. Under these circumstances, it is just possible that it may have been brought into such a situation as that the ring may again be in equilibrium, undetermined in what direction to move ; but the slightest change in this position, causes it to move with an accelerated velocity towards that pole which is nearest to it ; and getting entirely clear of the magnet, it is projected to a considerable distance from it. At length, however, it stops, and, gradually turning round, presents the opposite face to the magnet ; attraction now takes place, and the ring returns to the magnet with a force equal to that with which it had before fled from it ; and passing again over its pole, finally rests in its position of equilibrium, encircling the middle or what may be termed the equator of the magnet. In the former position it was equally attracted by the two poles of the magnet, in the latter it is equally repelled ; and accordingly, the

Fig. 217.   first was an unstable and the last a stable equilibrium.*   I

know of no experiment better calculated to exhibit to a class in a lecture-room the mutual affections of a magnet and an electrified wire than this : the motions of the floating coil are less impeded by setting it afloat in a small thin varnished wooden dish as shown in Fig. 217, where the

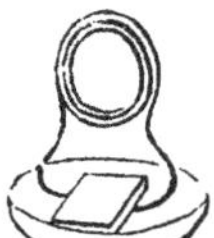

* Roget.

voltaic pair is represented as being placed in a horizontal position in a little bowl; the whole is then set afloat in a large basin or trough of water, and on pouring a little dilute sulphuric acid into the bowl the coil will be found to be surprisingly sensible to the influence of a magnet and will be attracted or repelled at the distance of several inches.

(596) The directive tendency of an electrified wire, FIG. 218. may also be strikingly exhibited, by bending it into the form of a spiral, and either connecting it with the floating galvanic arrangement, or suspending it delicately by a hook, as in Fig. 218, and passing the voltaic current through it, the plane of the spiral will be found to place itself east and west, the positive current ascending on the west side, and descending on the east; taking the same course as the hands of a watch, when it is held on edge, with the plane of the dial lying east and west, facing south. That side of the spiral which is towards the north, acts as the north pole; and the south side

has an opposite polarity. *Each side powerfully attracts iron filings.*

(597) But a still closer imitation of a magnet is obtained by
making a voltaic conductor into
the form of a helix, as in Fig. 219,
which was done by Arago, at the
suggestion of Ampère. A needle
placed in the axis of this helix,

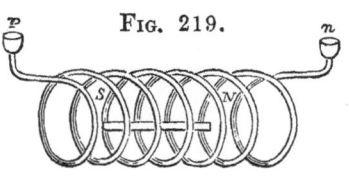

FIG. 219.

becomes fully magnetised in an instant; for in this arrangement, the current nearly in every part of its course, is at right angles to the needle, and as each coil adds its effect to that of the others, the united action of the helix is very powerful.

(598) The polarity given to the needle depends on two circumstances, first, the direction of the current with reference to the axis of the helix; and, secondly, the direction in which the helix is turned, which may be either in the form of a right handed screw, the turns proceeding downwards from right to left, or, of a left handed screw, the turns proceeding in a contrary direction. And if the current be descending on the side next to the spectator in the horizontal helix, (Fig. 219,) the north pole of a magnet in the axis will be determined to the right, and the south pole to the left, and this tendency will be given in the right handed helix, if the current be transmitted through it from left to right, but in the left handed helix from right to left. If a glass tube be thus surrounded by a coil of wire, and traversed by the electric current, its action on a small magnetised needle will be so powerful, that when

placed within it, it will actually start up and remain suspended in the air, in opposition to the force of gravity ; and this will sometimes take place even in a vertical position of the helix.*

(599) Thus then, a heliacal conductor may be regarded as a *magnet*, as long as the electric current is passing through it; and it was fully demonstrated by Ampère, soon after the discovery of Oersted, that exactly what would be anticipated from the magnetic influence of conductors, takes place, for two voltaic conductors, or two portions of the same conductor attract each other when the currents have the same direction, and are mutually repulsive, when traversed by opposite currents. Thus in the two parallel positive currents A B, and C D, Fig.

FIG. 220.

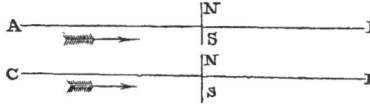

220, which flow in the same direction, the contiguous sides are affected with an opposite polarity, one being south, and the other north; whereas, in the two contrary currents, E F, and G H, Fig. 221, the adjacent sides have the same polarity, and therefore repel each other. Similarly, when two currents cross each other, as A B, C D, Fig. 222, it is obvious that at two of the four corners, A D, and C B, similar poles are contiguous ; while at the other corners different poles concur. Hence the wires tend to revolve round E, and place themselves parallel to the currents, so that both may flow in the same direction.

FIG. 221.

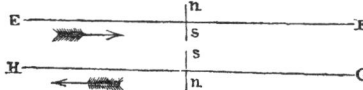

FIG. 222.

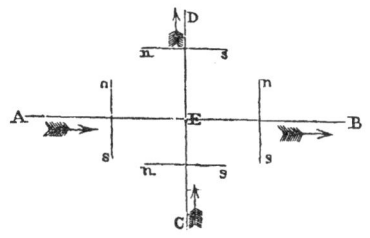

(600) If instead of a steel needle a bar of soft iron be placed in the axis of a helix it will become powerfully magnetic, but it will remain so only as long as the voltaic current continues to flow through the wire ; the moment that the current ceases, the magnetism of the bar disappears, and this the more completely in proportion as the iron is softer and freer from carbon. A very ingenious and beautiful piece of apparatus was contrived by the late Dr. Ritchie for illustrating this induction of magnetism on soft

* In order to exhibit this curious experiment, it is necessary to employ a helix of small internal diameter, and consisting of at least six strands of insulated copper wire.

iron. It is shown in Fig. 223, where a bar of iron is represented covered with a helix of insulated copper wire and mounted horizontally on a wire the extremity of which is finely pointed, so as to allow the bar to rotate freely. The two ends of the helix are bent downwards so as just to dip into a small channel of mercury divided into two parts by a diaphragm of wood, one end of the wire dips into each division of the trough

FIG. 223.

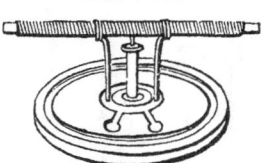

and a sufficient quantity of mercury is poured into the trough to fill it, without however allowing the two portions to become united; the mercury in each division will be found to rise a little above the level of the partition by capillary repulsion. It will thus be immediately seen that on connecting the two cells of mercury with the two plates of a battery the current must pass through the helix enclosing the iron bar before the circuit can be completed and that the iron will consequently become for the time magnetic. Now suppose each end of the iron bar to be opposed to the pole of a powerful steel bar magnet, an opposite pole on each side, and suppose the connexions with the battery to be made so that the north pole of the iron bar is formed opposite to the north pole of the steel bar, then as a necessary consequence, the opposite end of the iron bar will be a *south* pole and will be opposed to the south pole of the steel bar; *repulsion* will accordingly take place on each side and the iron bar will move through half a revolution. Here the wires of the helix surrounding it pass over the wooden partition and dipping into the opposite cells of mercury, the polarity of the bar becomes reversed and so on, the bar soon revolving with great rapidity in consequence of its polarity being reversed twice during each revolution. Sometimes the bar is arranged to rotate vertically as shown in Fig. 224.

FIG. 224.

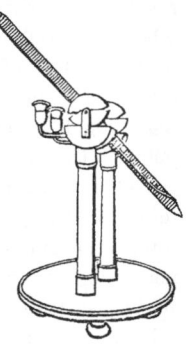

(601) In all these pieces of apparatus the employment of mercury is essential. Messrs. Knight have, however, devised a method of arranging the rotating magnet whereby the use of the fluid metal is dispensed with. A round plate of brass is divided by two small stripes of ivory and the wires of the helix are terminated by two small metallic rollers which thus pass easily over the brass surface, contact being broken at the proper place by the ivory strips. Fig. 225 exhibits this useful modification of Ritchie's rotating magnet.

FIG. 225.

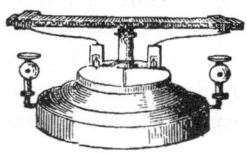

(602) In Fig. 226 a horse-shoe magnet is represented supported on a tripod stand with levelling screws, in which state it is well adapted for exhibiting the rotation of coils, wires, helices, &c. $A\ A$ is the magnet; B the tripod stand, $C\ C$ two circular wooden cisterns for holding mercury and capable of being adjusted at any required height by binding screws, $E\ E$ are two light wire frames, $F\ F$ two helices, H a Ritchie's rotating magnet; on the tops of the wire frames and helices are small cups to contain a drop of mercury, G is a piece of brass wire bent twice at right angles and terminated at each end by a fine point to dip into the globules of mercury : it can be raised or depressed without disturbing the general arrangement of the apparatus as a simple inspection of the figure will show.

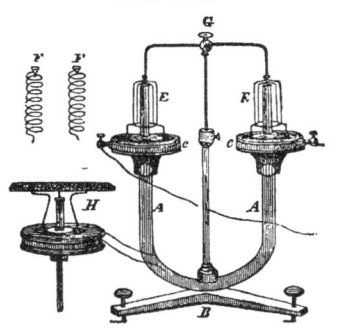

FIG. 226.

When the rotating magnet is set in action in this apparatus a loud humming noise and sometimes a loud musical sound is excited by the rapid vibratory motion assumed by the fixed magnet during the rapid revolution of the electro-magnet. This musical sound is best observed when the levelling screws of the tripod are placed on a mahogany table in the middle of a large room. For the electro-magnet H a simple coil of wire may be substituted, the rotation of which will be exceedingly rapid, its faces becoming alternately attracted and repelled by the poles of the magnet.

(603) In Sturgeon's " Annals of Electricity," vol. 3, p. 426, will be found a description and engraving of an ingenious apparatus for exhibiting the simultaneous rotation in opposite directions of the permanent magnet and the electro-magnet. The current from the voltaic battery instead of going directly to the mercury flood in connexion with the electro-magnet, is made to enter two concentric troughs containing mercury placed immediately under the former, communication between the upper and lower cups being established by means of wires ; the electro and permanent magnet may thus be placed on one spindle, and the former being put in motion it was found that the permanent magnet immediately commenced rotating in an opposite direction. This instrument was the contrivance of Mr. C. W. Collins.

(604) *Electro-Magnets.*—In order to produce the greatest effect of electro-magnetic induction on soft iron, the current from the battery must be made to envelope it by passing through a considerable length of insulated copper wire wound round the iron ; but since a great length of wire is found

to diminish the influence of the current, as has been proved by Cumming, Barlow, and Ritchie, it is better that the total length of wire intended to be used, should be cut into several portions, each of which, covered with silk or cotton thread, to prevent lateral communication, is to be coiled separately on the iron. The ends of all the wires must then be collected into two separate parcels, and made to communicate with the same voltaic battery, taking care that the current shall pass along each wire in the same direction. Fig. 227 exhibits a simple arrangement of the electro magnet mounted on a wooden stand with a small scale pan attached to the bitt or keeper.

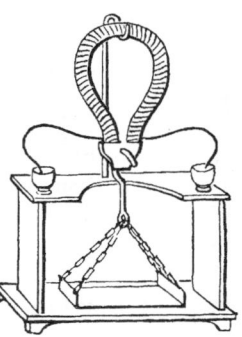

FIG. 227.

(605) The magnetic power thus induced on the iron is immense. The author has for some years been in the habit of exhibiting at lectures an electro-magnet, constructed on the above principle, of an horse shoe form, which when excited by an energetic compound battery, sustains from 10 to 14 cwt. A much greater lifting power has, however, been obtained with other varieties of the electro-magnet. There are three different arrangements described in the 6th volume of Sturgeon's "Annals," which particularly deserve to be mentioned. The first is the contrivance of Mr. Richard Roberts; its peculiarity consists in the great extent of the area of the face, on the surface of which, a series of grooves are formed, into which the conducting wire is coiled. The magnet is 2 and 7-16ths inches thick, and 6 and 5-8ths inches square, on its face into which are planed (at equal distances from each other, across its surface) four grooves, one and a quarter inch deep, and nearly three-eighths of an inch broad. Into these grooves was coiled, three-fold deep, a bundle of thirty-six copper wires, (No. 18,) wrapped with cotton tape, to prevent contact with the iron, the wires having no insulation from each other. The magnet, with the conducting wire, weighed 35 lbs. The armature was 1½ inch thick, and the same size as the magnet on the face; its weight was 23 lbs. The upper side of the iron, which constituted the magnet, was formed into an eye or bow, by which the whole was suspended; and a similar bow was formed on the back of the armature, to which the weight scale was attached. This electro-magnet, when excited by a battery of eight pair of Sturgeon's cast-iron jars, (286) is reported to have sustained the enormous weight of 2950 lbs.* which is nearly double the weight which the author's large magnet, the weight of which is about 112 lbs. will sustain with any battery that has been tried.

(606) The second electro-magnet mentioned in Sturgeon's "Annals,"†

* Sturgeon's "Annals," vol. vi. p. 168.　　　　† Vol. vi. page 231.

is that of Mr. Joseph Radford. Its peculiarities consist in the convoluted figure of its face, and in the unusual arrangement of its poles, both of which are on the same convoluted strip of iron, one pole occupying the whole length on one edge, and the other the whole length of the opposite edge. Its diameter is 9 inches, and it weighs, with its copper coil, eighteen pounds four ounces. The keeper, or armature, weighs 14 lbs. $4\frac{3}{4}$ oz. The depth of the convoluted groove or recess, is $\frac{3}{8}$ of an inch, and $\frac{1}{4}$ of an inch wide. The width or breadth of the metal between the grooves is $\frac{1}{2}$ an inch ; the thickness of the magnet is one inch, at the outside edge, and about $\frac{3}{8}$ in the centre. When excited by a battery of twelve of Sturgeon's jars, this electro-magnet is stated to have sustained 2500 lbs. avoirdupois ; it is, therefore, in proportion to its weight, much more powerful than Mr. Roberts's magnet.

(607) The third electro-magnet alluded to, is that of Mr. J. P. Joule, and is shown in Fig. 228. *B B*, are two rings of brass, each 12

FIG. 228.

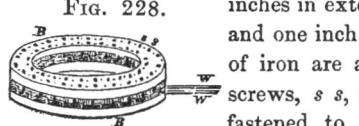

inches in exterior diameter, two inches in breadth and one inch in thickness ; to each of these, pieces of iron are affixed, by means of the bolt headed screws, *s s*, &c. : 24 of these are *grooved*, and fastened to the upper ring ; 24 are plain and affixed to the lower ring. A bundle *W W*, consisting of sixteen copper wires, (each of which was 16 feet long, and one-twentieth of an inch thick), covered with a double fold of thick cotton tape, was bent in a zig-zag direction about the grooved pieces. Fig. 229

FIG 229.

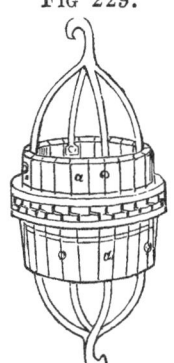

represents the method adopted for giving the electro-magnetic ring a firm and equable suspension : *a, a,* are hoops of wrought iron, to each. of which four bars of the same metal are riveted and welded together at the other end into a very strong hook. The hoops are bound down to the brass rings by means of copper wires. The weight of the pieces of grooved iron was 7·025 lbs., and that of the plain pieces 4·55 lbs ; and when excited by 16 pairs of the cast-iron battery, arranged into a series of four, a weight of 2,710 lbs. was suspended from the armature, without separating it from the electro-magnet ; and Mr. Joule thinks, that by the use of some precautions, which have occurred to him since making his first experiments, the actual power will be very considerably augmented.

(608) It has been mentioned that when very soft iron is employed in the construction of the electro-magnet, its magnetism nearly disappears, when the voltaic current ceases to flow through the helix surrounding

it. It was, however, discovered, by the late Dr. Ritchie, that there are other circumstances which modify the retaining power; the most remarkable of which is the *length of the magnetic circuit.* When the electro-magnet is very short, and the poles near each other, the retaining power is exceedingly small;—when the magnet is very long, the retaining power is very great, the reason of which appeared to Dr. Ritchie to be this :*—the molecules of the electric fluid acting on each other with the same force, will obviously return to their natural position most rapidly, when the length of the circuit through which the action takes place, is diminished. If it be diminished till the coercitive force of the iron be overbalanced by the tendency of the molecules to return to their natural state of equilibrium, from which they have been forced by the action of the conducting wire, the electro-magnet will lose all its retaining power.

(609) Another singular fact, discovered by Dr. Ritchie was—that a *short* electro-magnet, though its lifting power be very considerable, is incapable of inducing permanent magnetism on an unmagnetized horse-shoe of tempered steel ; while an electro-magnet of four feet in length, though of no greater lifting power than the small one, is capable of inducing a very considerable permanent effect. It was likewise found, by Dr. Ritchie, that a bar electro-magnet, four feet long, which scarcely retained any power when its connexion with the battery was broken, on being *re-connected* with it, in the *same* direction as before, was *rapidly* converted into a powerful magnet ; but after being removed, and its wires now connected with the opposite poles, it required a long time to convert it into a magnet of much inferior power, as if the atoms of Electricity, having been first put in motion in one direction, are afterwards more easily turned in that direction than in the contrary.

(610) *Electro-Magnetic Engines.*—One of the first enquiries that the discovery of this wonderful property of iron to become magnetic, by the action of voltaic Electricity gave rise to, was—Can it not be applied as a moving power ? A vast deal of time and money have been expended in attempting to solve this question, and a great number of exceedingly ingenious machines have been invented and described. Up to the present time, however, all attempts to produce a really powerful engine, or one that by its performances can be said to hold out any prospects of success, have failed, and the principle on which an electro-magnetic motive engine should be constructed, is still a desideratum. It would occupy far more space than we can afford, to give a description of all the machines for producing motion by electro-magnetism that have been successively brought before the public, most of them can be ranked only among philosophical toys. Two or three shall, however, be briefly described, and we shall select those that have by their inventors been thought worthy of being patented.

* L. and E. Phil. Mag., vol. iii. p. 123.

(611) In Silliman's Journal, for April, 1837, there is a notice of a ro-
tative engine, invented and patented by Mr. Thomas Davenport of Bran-
don, in the county of Rutland, and State of Vermont, United States. The
following is a general outline of its construction :—the moving part is
composed of two iron bars, placed horizontally, and crossing each other
at right angles ; they are covered with insulated copper wire, and sustained by a vertical axis : proper connection with the voltaic battery
being made in the usual manner. Two semicircles of strongly magnetized steel form an entire circle, interrupted only at the two opposite
poles ; and within this circle, which lies horizontally, the galvanized
iron cross moves in such a manner, that its iron segments revolve
parallel, and very near to the magnetic circle, and in the same plane.
Its axis, at its upper end, is fitted by a horizontal cog wheel to another
and larger vertical wheel, to whose horizontal axis the weight is attached,
and raised by the winding of a rope. By the galvanic connexion, these
crosses, and their connected segments are magnetized, acquiring north
and south polarity at their opposite ends ; and being thus subjected to
the attracting and repelling force of the circular fixed magnets a rapid
horizontal movement is produced, at the rate of six-hundred revolutions
in a minute, when a large calorimotor is employed. The movement is
instantly stopped by breaking the contact with the battery, and then
reversed by simply interchanging the connexion of the wires of the
battery with those of the machine, when it becomes equally rapid in the
opposite direction. Another machine, composed entirely of electro-
magnets, both in its fixed and revolving members, is also described.

(612) In a subsequent number of the same Journal, it is stated that
the proprietors had been engaged in experiments on magnets of different
modifications, as well as on the proper distance between the magnetic
poles of the circle, that they had entirely altered the form and arrangement of the magnets, greatly increasing thereby the energy of the
machine. The use of magnets in the form of segments of a circle, was
discontinued, and horse-shoe formed magnets substituted; the poles
being changed once in every $3\frac{1}{2}$ inches of the circle. On this arrangement, a machine with a wheel seven inches in diameter, elevated 90 lbs.
one foot per minute, and performed about twelve-hundred revolutions
in the same time. It is also stated that the proprietors were engaged
in constructing a machine with a motive wheel of about $2\frac{1}{2}$ feet in diameter, from which, it was expected, that sufficient power to propel a
Napier's printing press (requiring a two-horse power), would be obtained. The author is not aware whether the new machine realized the
anticipations that were indulged in, nor has he heard what success
crowned the subsequent efforts of the ingenious American.

(613) In 1838, Captain Taylor obtained a patent for an electro-

magnetic engine, in the United States; and on November, 2nd 1839, he patented the same engine in England.   In the London " Mechanics' Magazine," vol. xxxii. p. 694, there is a full description and drawing of this engine, a working model of which was for some time exhibiting in the Collosseum, in active operation, turning to the wonder and admiration of thousands, articles in wood, ivory, and iron.   Referring to the above periodical for a detailed description of this engine, we shall confine ourselves to the peculiar principle of its action, as explained by Mr. Taylor. " The generality of the plans which have been hitherto devised for obtaining a working power from electro-magnetism, have depended on taking advantage of the *change of polarity*, of which masses of iron fitted as electro-magnets, are susceptible so as to cause them alternately to attract and repel certain other electro-magnets, brought successively within the sphere of their influence, and thus to produce a continuous rotatory movement, and the failure of these attempts are owing to the difficulty, if not impossibility of accumulating power by such means. Instead of this, Mr. Taylor employs as his prime movers, a series of electro-magnets, which are alternately and almost instantaneously magnetized and de magnetized, without any change of polarity whatever taking place, and in bringing certain other masses of iron or electro-magnets successively under the influence of the said prime movers when in a magnetised state, and in de-magnetizing the said prime movers as soon, (or nearly so,) and as often as their attractive power ceases to operate with advantage; or in other, and perhaps plainer words, his invention consists in letting on or cutting off a stream of the electric fluid in such alternate, quick, and regular succession, to and from a series of electro-magnets, that they act always attractively or positively only, or with such a preponderance of positive attraction, as to exercise an uniform moving force upon any number of masses of iron or magnets placed so as to be conveniently acted upon."   The power of the machine constructed on this principle which was exhibiting at the Collosseum, was small, certainly much below that of a single man; and we are unable to say whether Mr. Taylor has since succeeded in increasing it.*

(614) It appears from a letter in the Phil. Mag.† from Professor P. Forbes of Aberdeen, and also from a communication from Mr. Robert Davidson to the Mechanics' Magazine,‡ that the latter individual had

---

* Since writing the above, I have been informed by Mr. Henley, that he constructed a very large electro-magnetic engine, on Captain Taylor's plan, at the time that that gentleman took out his patent.   The wheel was 7 feet in diameter, and weighed 4 cwt.   This machine did some work, but at an enormous expense, 6 cwt. of sulphate of copper having been consumed in one week, in experiments alone.   The battery employed contained 13 cwt. of metal.

† Vol. xv. p. 250.   ‡ Vol. xxxii. p. 63.

anticipated Mr. Taylor in the principle of his machine, having in 1837 employed the electro-magnetic power in producing motion by *simply suspending the magnetism without a change of the poles :* " so close," says Mr. Davidson, " is the resemblance between Mr. Taylor's machine and one of mine, that independently of the frame work, I believe the chief differences are, first, that the circumference of Mr. Taylor's revolving disk is composed of alternate parts of copper and ivory, while the circumference of mine is composed of alternate parts of copper and box-wood ; and second, that in Mr. Taylor's machine the armatures appear to be sunk to about half their depth in the periphery of the wheel to which they are attached, while in mine they are sunk their whole depth, so as to be *flush* with the cylindrical surface."

(615) In the Practical Mechanics' and Engineers' Magazine for November, 1842, there is a full account and drawing of a large Electromagnetic Locomotive constructed by Mr. Davidson and tried on the Edinburgh and Glasgow Railway. The carriage is sixteen feet long and six feet broad and weighs above five tons, including batteries, magnets, &c. The electro-magnets are not one solid piece of iron, nor are they rounded behind. Each of the side parts or arms is constructed of four plates of soft iron put together, so as to form as it were a box for the sake of lightness. The arms are twenty-five inches long and joined together behind by plates of iron. Their rectangular poles measure eight by five inches, and at their nearest points are only about four inches asunder. The coils with which they are surrounded do not consist of a single copper wire, but of bundles of wire wrapped round with cloth to insure insulation. According to Mr. Davidson's first arrangement these magnets were placed so that their poles were nearly in contact with the revolving masses of iron in their transit : but so prodigious was the mutual attraction, that the means taken to retain the magnets and iron in their assigned position were insufficient. They required to be more firmly secured, and their distances had to be somewhat increased, which perhaps contributed very materially to the failure of the machine which when put in motion on the rails travelled about four miles an hour only, thus exhibiting a power less than that of a single man who on a level railway could certainly move a carriage of this weight at as great a velocity.

(616) In 1838, Professor Jacobi, of St. Petersburgh, at the expence of an imperial commission, tried the grand experiment of propelling a boat by the agency of electro-magnetism. The vessel was a ten-oared shallop, equipped with paddle wheels, to which rotatory motion was communicated by an electro-magnetic engine. The boat was 28 feet long and $7\frac{1}{2}$ feet in width, and drew $2\frac{3}{4}$ feet of water. In general there were ten or twelve persons on board, and the voyage was con-

tinued (on the Neva) during entire days. The difficulty of then managing the batteries, and the imperfect construction of the engine were sources of frequent interruption and could not be well remedied on the spot. After these difficulties were in some degree removed, the professor gives as the result of his experiments, that a battery of twenty square feet of platinum will produce power equivalent to one horse : but he hoped to be able to obtain the same power with about half that amount of battery surface. The vessel went at the rate of *four miles* per hour, which is certainly more than was accomplished by the *first* little boat that was propelled by the power of steam. In 1839, Jacobi tried a second experiment in the same boat, the machine which was the same as that used on the previous occasion, and which occupied little space, was worked by a battery of sixty-four pairs of platinum plates, each having thirty-six square inches of surface* and charged according to the plan of Grove with nitric and sulphuric acid. The boat, with a party of twelve or fourteen persons on board, went against the stream at the rate of three miles an hour.

(617) These spirited experiments are certainly deserving of the greatest praise and encouragement, for it is quite clear that nothing conclusive as to the practicability of employing electro-magnetism as a moving power can be gathered from mere models, and that trials on the large scale are alone to be depended upon. We fully concur, (and we believe our readers will do so likewise) with the following remarks of the editor of the Engineers' Magazine. " The results of Professor Jacobi's experiments on the water, and of Mr. Davidson's on the railway, will no doubt disappoint the exaggerated expectations of some, who may view them as rather unfavourable to the claims of electro-magnetism : but it must be kept in mind that it is only a few years since, that it in any degree engaged attention as a means of obtaining mechanical power. Should it ever lead to the results anticipated from it as a prime mover, there are many advantages which it will possess over steam. The clash, din, and concussion occasioned by steam engine machinery—the dread of explosions—and the smoke, dust, and danger of fire would all be got rid of. The only noise in an electro-magnetic locomotive or boat would be that of the wheels, and the batteries could be charged in such a manner as to avoid all disagreeable smell. But even if the method of exciting them should be such as to produce hydrogen gas, this, instead of being permitted to escape and annoy passengers, could be collected and rendered available as a means of producing light and heat when required. So far, however, as light is concerned it could be obtained otherwise at no additional expence ; for a

* This battery, it will be observed, was not about one-fifth part the size of that employed in the experiment of 1838.

piece of charcoal, being interposed at a small breach in the wires connected with the batteries, would, by its ignition, afford the most intense and brilliant light imaginable, and furnish the means also of communicating signals to an immense distance. We are inclined, however, to think that the application of this new prime mover to navigation, particularly on the ocean, holds out better hopes of success than its application to locomotives on the land. Iron vessels have now been proved well adapted for duty at sea : and since that metal and salt water constitute two important elements of a voltaic battery, may not some means of introducing a *third* element be suggested, so that a great part, if not the whole of the surface of the ship may be called into action for the purpose of furthering her progress, thus making the ocean so far her propeller as well as support, while her own body also performed two important offices ? Much less weight would also require to be carried by an electro-magnetic boat than by a steamer, and she could therefore undertake much longer voyages."* †

* See " Practical Mechanics' and Engineers' Magazine." Part 14, p. 52.

† For the convenience of those who feel inclined to see and examine the descriptions of the numerous electro-magnetic machines that have been proposed by different experimentalists, a list of references to the periodicals containing some of the principal ones, is subjoined.

Sturgeon's Electro-magnetic Engine for turning Machinery. " Annals of Electricity." Vol. i. p. 75.

Jacobi's valuable paper on the application of Electro-magnetism to the moving of machines, with a description of an Electro-magnetic Engine. " Annals of Electricity." Vol. i. p. 408-419.

Mr. Joule's Electro-magnetic Engine. " Annals of Electricity." Vol. ii. p. 122.

Mr. Davenport's Electro-magnetic Engine. " Annals of Electricity." Vol. ii. p. 257.

The Rev. F. Lockey's Electro-magnetic Engine. " Annals of Electricity," Vol. iii. p. 14.

Dr. Page on Electro-magnetism as a moving power. " Annals of Electricity." Vol. iii. p. 554.

Mr. Joule's second Engine. " Annals of Electricity." Vol. iv. p. 203.

Mr. Uriah Clarke's Engine. " Annals of Electricity." Vol. v. p. 33.

Mr. Thomas Wright's Engine. " Annals of Electricity." Vol. v. p. 108.

Mr. U. Clarke's Electro-magnetic Locomotive Carriage. " Annals of Electricity." Vol. v. p. 304.

Jacobi on the " Principles of Electro-magnetical Machines." Report of the Meeting of the British Association in Glasgow in September, 1840. " Annals of Electricity." Vol. vi. p. 152. (This is a most valuable paper, and is well deserving of attentive study.)

Mr. Robert Davidson's Electro-magnetic Locomotive. " Engineers Magazine,' &c. Part 14. p. 48.

Mr. Taylor's Engine. " Mechanics' Magazine." Vol. xxxii. p. 694.

Mr. Watkins's Electro-motive Machine. " Phil. Mag." Vol. xii. p. 190.

An Inquiry into the possibility and advantage of the application of Electro-magnetism as a moving power, by the Rev. James William M'Gauley. " Report

Fig. 229 represents a small working model of an electro-magneto motive engine constructed by Mr. Bain, with some few improvements by the publishers of this work. On to a stout mahogany board are fixed the brass uprights EE; to these are attached the electro magnets A B, covered with stout

FIG. 230.

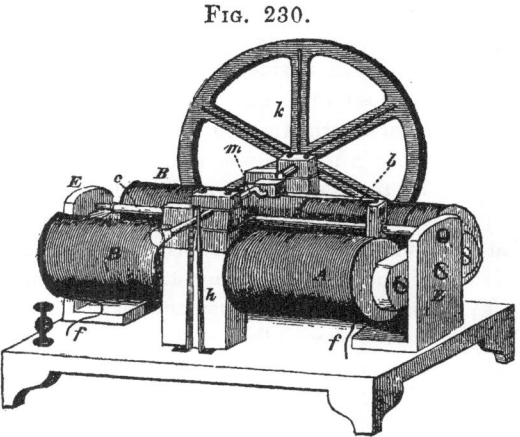

wire; through the upper part of these uprights, and above the magnets, the two ends of the steel spindle c work; this spindle carries about its centre an iron bit, which is alternately attracted by the two magnets A and B, but prevented from absolute contact by pieces of paper; another spindle m, at right angles with c, and supported by the uprights h h, carrying at one end the fly wheel k, and on the other a small pully, is cranked in the centre and connected with c by the spring and hook b. At h are seen two brass springs bearing lightly on the spindle, which is divided in the middle by a small piece of ivory, so that one only is in contact at the same time. The connections are formed thus :—one termination of the electro-magnet A is connected to one of the upright springs bearing on the spindle, and the other termination to the binding screws seen at the end of the board. The one termination of the electro-magnet B is connected with the other spring, and the other extremity to the same binding screw to which one end of A was attached, the remaining binding screw being in connection by means of a wire with the brass box in which m works. The working of this machine is greatly assisted by two spiral springs fixed underneath the board attached to the moving bit. The whole arrangement performs extremely well, and no doubt if made on a large scale would be very powerful.

(618) We are indebted to Mr. Henley for the following descriptions of two electro-magnetic engines which he has lately constructed for Mr. Talbot and Professor Wheatstone. The descriptions were accompanied by drawings, which we regret that time does not allow of our having engraved for the present occasion : from the clearness of the accounts,

of the Proceedings of the British Association for the Advancement of Science a the Dublin Meeting, August, 1835."

however, we have no doubt that our readers will be enabled to form a good general idea of the structure of the machines :—

"Mr. Talbot's engine consisted of 6 powerful horse-shoe electromagnets, placed in a line with their poles upwards; to each magnet was adapted an armature, to which was affixed a jointed arm : in the centre of the armature was a hole which was fitted to a six throw crank, the throws set at an angle of 60° with each other, the currents acting on the magnets in succession from 1 to 6; at the time of breaking at 6 it commences again at 1 : contact is made when the armature is at a small distance from the magnet, and is continued till it reaches it; at this moment contact is broken and made with the succeeding one ; the connecting rod playing through the hole in the armature allows the crank to pass down ; when it remains stationary on the poles of the magnets, a piece of paper prevents adhesion. There is a knob at the end of the connecting rod, by which it lifts the armature in one position, and is pulled in the other. In this machine the magnet acts when the crank is in the very best position, and were it not for the additional friction from the great number of rubbing parts, it would certainly be the best form of machine. On the shaft is mounted an A shaped frame, and at one end carries a heavy fly wheel, and at the other a contact breaking apparatus, which consists of a wheel and six levers, the points of which dip in mercury.*"

(619) The original of Professor Wheatstone's machine consisted of a brass ring, within which were placed eight magnets ; an eccentric wheel revolves within, the longest radius of which passes close to each magnet successively, following the current as it were, which acts on each magnet a little in advance of the wheel ; the break piece, which is stationary, is made of a piece of ivory, into which is let eight pieces of brass ; the shaft, which passes through this without touching, carries a spring which presses on the break piece. The shaft and frame work is in connection with one pole of the battery, one end of the coil on each magnet, and the other end of each with its corresponding piece of brass; the shaft also carries a fly-wheel and pulley to transfer the power. If there were a hundred magnets, of course the same battery would be sufficient, as they only act one at a time. The author had the pleasure

---

* This engine was 3 feet 6 inches long, and 2 feet 6 inches wide, but its power was not equal to the expectations that were formed of it ; when excited by a Grove's battery, consisting of 4 cells with double plates of zinc 9 inches by $6\frac{1}{2}$, platinum plates 9 inches by $5\frac{1}{2}$, excited by diluted sulphuric acid 1 to 4, and concentrated nitric acid, Mr. Henley drove with it a lathe in which he turned a gunmetal pulley 5 inches in diameter, but in three quarters of an hour the battery was quite exhausted. This machine, it will be observed, was something more than a *model*.

of seeing an improvement on this engine at work at Mr. Gassiot's house. *Two* magnets were here made to answer the same purpose as eight in the other, and there is only one portion of the eccentric made use of. It is divided into four parts, and fits on a wooden wheel ; two magnets are placed underneath. This amounts to the same thing, with the advantage of being balanced by having the same weight on each side of the axis. If eight magnets were used with this, four would act at a time ; and with proportional increase in the battery, this would be a powerful machine. The contact is broken by points dipping in mercury.

(620) Mr. Henley has also constructed a machine which works a lathe : it is made with three horse-shoe magnets, with their poles upwards, the bent parts crossing each other ; a bolt passes through them, and holds them firmly to a base ; within the poles revolves a soft iron cross : one magnet acts at a time, but the cross is attracted continually. The cross is suspended by four stout brass columns. At first there were but two ; it was found necessary, however, to add two more to resist the strain of the magnets, as the poles are curved, and the cross passes very close. There is a piece of apparatus to stop it immediately : it is a lever which makes contact with one of the magnets independent of the break-piece.

(621) At the meeting of the British Association for the Advancement of Science, held at Dublin in 1835, the following notice of an economic application of Electro-Magnetism to manufacturing purposes, was introduced by Mr. Robert Mallet.

The separation of iron from brass and copper filings, &c., in workshops, for the purpose of the re-fusion of them into brass, is commonly effected by tedious manual labour. Several bar or horse-shoe magnets are fixed in a wooden handle, and are thrust in various directions through a dish or other vessel containing the brass and iron turnings, &c. ; and when the magnets have become loaded with iron, it is swept off from them by frequent strokes of a brush. This is an exceedingly troublesome and inefficacious process. It appeared to Mr. Mallet that a temporary magnet of great power, formed by the circulation of an electric current round a bar of iron, might be substituted advantageously. The following is the arrangement which he has adopted. Several large round bars of iron are bent into the form of the capital letter U, each leg being about six inches long. They are all coated with coils of silk-covered wire in the usual way, and are then arranged vertically at the interval of five or six inches apart. All the wires from these coils are collected into one bundle at their respective poles, and there joined into one by soldering, a large wire being placed in the midst of them and amalga-

mated. A galvanic battery is provided, with the poles of which the
wires from the electro-magnets are connected.

The rest of the arrangement is purely mechanical. The required
motions are taken from any first mover, usually a steam-engine. The
previously described arrangement being complete, a chain of buckets is
so contrived as to carry up and discharge over the top of the magnets a
quantity of the mixed metallic particles : most of the iron adheres to
the magnets, while the so-far-purified brass falls into a dish or tray,
placed beneath to receive it. This latter is also one of a chain of dishes,
the horizontal motion of which is so regulated that the interval between
the two dishes is immediately under the magnets, in the interval of time
between two successive discharges of the mixed particles on the bars.
At this juncture the communication between the galvanic battery and
the magnets is interrupted by withdrawing the wires from the cups of
mercury forming the poles ; and the result is, that the greatest part of
the adhering iron drops off, and falls in the space between the two
dishes. The next dish now comes under the magnets, the communication is restored, and a fresh discharge from the buckets takes place, and
so the process is continued. Some iron constantly adheres to the magnets ; but this is found of no inconvenience, as it bears but a small proportion to the quantity separated.

(622) *Galvanometers.*—This instrument, the simplest form of which
is shewn in Fig. 215, was originally suggested by Professor Schweigger
of Halle, very soon after the discovery of electro-magnetism, and was
by him called an *electro-magnetic multiplier*. The effect of passing an
electric current through the wire, in the direction pointed out by the
arrows, is, as we have seen, to deflect the needle from its natural position,
and to bring it into a position nearer to a right angle to the plane of the
wire. To multiply this effect, and to render the instrument a more
susceptible indicator of feeble electrical currents, various forms have
been given to the instrument ; in all, the convolutions of the wire are
multipled, and the lateral transfer of Electricity prevented by coating
it with sealing-wax or silk. Fig. 231 is a vertical section of the torsion
galvanometer of the late Professor Ritchie. The following is his
description of its construction :—" Take a fine copper wire, and cover
it with a thin coating of sealing-wax, roll it about a heated cylinder, an
inch or two in diameter, ten, twenty, and any number of times, according to the delicacy of the instrument required. Press together the
opposite sides of the circular coil till they become parallel, and about
an inch, or an inch and a half long. Fix the coil in a proper sole, and
connect the ends of the wires with two small metallic cups, for holding
each a drop of mercury. Paste a circular slip of paper, divided into

equal parts horizontally, on the upper half of the coil, and having a black line drawn through its centre, and in the same direction with the middle of the coil. Fix a small magnet, made of a common sewing needle, or piece of steel wire, to the lower end of a fine glass thread, while the upper end is securely fixed with sealing-wax in the centre of a moveable index, as in the common torsion balance. The glass thread should be inclosed in a tube of glass, which fits into a disc of thick plate glass, covering the upper side of the wooden box containing the coil and magnetic needle."*

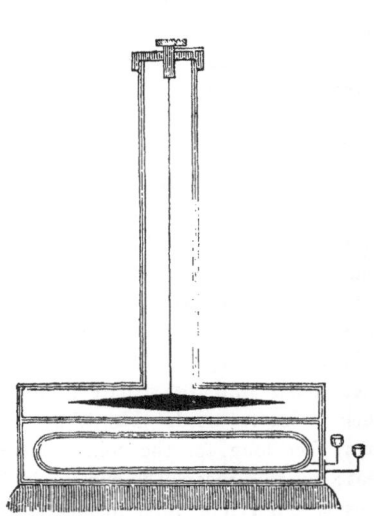

Fig. 231.

(623) The sensibility of this instrument is very much increased by neutralizing the magnetic influence of the earth, by employing two needles, which was first done by Professor Cumming of Cambridge, and afterwards on an improved principle by Nobili. The neutralizing needle in his instrument is attached to the principal one; placing them one above another and parallel to each other, but with their poles in opposite directions. They are fixed by being passed through a straw, suspended from a thread. The distance between the needles is such as to allow the upper coil of the wires to pass between them, an opening being purposely left, by the separation of the wires at the middle of that coil, to allow the middle of the straw to pass freely through it. A graduated circle, on which the deviation of the needle is measured, is placed over the wire, on the upper surface of the frame of the instrument, having an aperture in its centre for the free passage of the needle and straw. The whole of this arrangement will be easily understood, by imagining another needle to be suspended to the one above the coil in Fig. 231, moving *within* the wire, and having its poles turned the reverse of those of the upper needle.* In Nobili's instrument, the frame was twenty-two lines long, twelve wide, and six high. The wire was of copper, covered with silk, one-fifth of a line in diameter, and from twenty-nine to thirty feet in length, making seventy-two

* Phil. Trans. 1830, p. 218.

† The instrument as thus constructed is called the astatic needle galvanometer.

revolutions round the frame. The needles were twenty-two lines long, three lines wide, a quarter of a line thick, and they were placed on the straw five lines apart from each other.*

(624) The advantages of Nobili's instrument consist in the directive force, arising from the influence of the earth's magnetism being nearly balanced, and a double rotatory tendency being given to the needles. The lower needle is acted upon by the sum of the forces of the currents in every part of the coil, and the upper needle is acted upon by the excess of force in the upper current which is nearest to it, which force, of course, acts in a direction the reverse of that in which it acts upon the lower needle, being situated on the opposite side; but since the poles are also in a reversed position, the rotatory tendency becomes the same in both needles. M. Lebailiff has extended the principle of Nobili's galvanometer, by employing four needles, two within the coil, having their poles similarly situated, and one above, and one below, having their poles reversed. He likewise employs five parallel wires, each sixty feet long for the coil, instead of one length of three hundred feet; by this means the current is divided into five parts, and made to flow through five different channels, with the alleged advantage of increasing the quantity, and diminishing the intensity of the electricity; it is not decided whether this is the case, nor is the advantage of employing four needles sufficiently obvious.

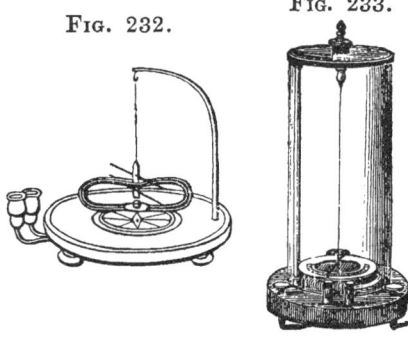

FIG. 232.   FIG. 233.

In Fig. 232 is shown a simple arrangement of the galvanometer, with astatic needle; and in Fig. 233 the same instrument is represented, with delicate and elegant adjustments, forming a galvanic multiplier, the sensibility of which may be illustrated by making with it the following experiment : —Twist a piece of iron or zinc-wire round one of the binding screws, connected with one extremity of the coil, place on the top of the other binding screw a drop of spring-water, into which dip the other end of the wire, the needles will immediately be moved by the weak current thus set in motion. So exquisite a test indeed of the presence of minute quantities of Electricity is a well made galvanometer, that by it Schoenbein was able to prove a change in the composition of chloride of cobalt, when that salt in solution was changed blue by the action of heat.†

* Roget's Treatise.   † See Pog. Annalen, xlv. p. 263.

(625) We have seen (37) that the lightness and flexibility of gold leaf have rendered that metal highly valuable to the electrician in the construction of instruments for appreciating minute quantities of statical Electricity. The same material, with the addition of a magnet, may be arranged so as to form probably one of the most delicate tests possible, of the existence and direction of a weak *galvanic* current. A slip of gold leaf is retained in the axis of a glass tube by a metallic forceps at each end, and a strong horse-shoe magnet is fixed with its poles on either side of the middle of the tube; on causing the electrical current to pass down the gold leaf it will be attracted or repelled, laterally by the poles of the magnet, according as the current is ascending or descending.*

(626) Mr. Sturgeon also describes† an instrument in which a single gold leaf is employed, but instead of a magnet a dry electric pile is used : "a glass phial has its neck cut off and is perforated on its two opposite sides, for the introduction of two horizontal wires. These wires are formed into screws and work in box-wood necks which are firmly cemented to the bottle; with their centres directly over the perforations. Through the centre of a wooden cap cemented to the top of the bottle, passes a brass wire tapped at its upper extremity for the reception of a metallic plate, and from its lower extremity hangs a very narrow slip of gold leaf pointed at its lower end, which reaches just as low as the inner balls of the horizontal wires. The bottle stands upon, and is cemented to a box-wood pedestal. Upon two glass pillars fixed to a wooden base, is placed, horizontally, a dry electric pile, consisting of about one hundred pairs, or rather single pieces of zinc with bright and dull surfaces. The poles of this pile are connected with the two horizontal wires by thin copper wires." The sensibility of this instrument Mr. Sturgeon states to be very great. A zinc plate about the size of a sixpence being attached to the upper end of the axial wire, on pressing upon it a similar sized copper plate, the pendant leaf leans towards the *negative* ball, and when the copper is suddenly lifted up, the leaf will strike; when the plates are reversed the leaf leans towards and strikes the *positive* ball.

(627) In October, 1841, Mr. Iremonger communicated to the London Electrical Society the following description of a novel and ingenious hydrostatic galvanometer.‡ "A small bar magnet is attached to the bottom of an areometer; this apparatus being so weighted that the ball may float just below the surface of pure water. Over the proof glass, containing the said areometer, is passed a De la Rive's ring placed

---

* Cumming's Manual of Electro-dynamics. † Lectures on Galvanism, p. 80.

‡ Proc. London Elec. Soc., p. 175.

rather below the level of the lower pole of the magnet. Now, on passing a voltaic current through the ring, the magnet and areometer are forced downwards : but at the same time I accompany this motion by a corresponding movement of the ring, by which means the descent of the floating apparatus is continued till the electro-magnetic forces are in equilibrium with the upward pressure of the liquid. Now, the pressure of liquids being simply as their height, the different degrees of any *equally* divided scale attached to this instrument will be of *equal* value —no slight advantage. The delicacy of the instrument will depend on several circumstances, such as the size of the stem of the areometer, the strength of the magnet and also on the length of wire and number of turns in the ring." Mr. Iremonger gives a detailed account of the method of constructing this instrument, for which, as it could not be well understood without a drawing, the reader is referred to his original paper in the "Proceedings of the Electrical Society."

(628) A large and very sensible thermoscopic galvanometer was invented by Dr. Locke, Professor of Chemistry in the Medical College of Ohio, and by him communicated to the Phil. Mag., in August, 1837. The object proposed by Dr. Locke in the invention of this instrument was to construct a thermoscope so large that its indications might be conspicuously seen on the lecture table by a numerous assembly, and at the same time so delicate as to show extremely small changes of temperature. How far he succeeded will appear from the following very popular experiment he was in the habit of making with it. By means of the warmth of the finger applied to a single pair of bismuth and copper discs, there was transmitted a sufficient quantity of Electricity to keep an eleven inch needle weighing an ounce and a half in a continued revolution, the connexions and reversals being properly made at every half turn.

The greater part of this effect was due to the *massiveness* of the coil which was made of a copper fillet about fifty feet long, one-fourth of an inch wide, and one-eighth of an inch thick, weighing between four and five pounds. This coil was not made in a pile at the diameter of the circle in which the needle revolved, but was spread out, the several turns lying side by side and covering almost the whole of that circle above and below. It was wound closely in parallel turns on a circular piece of board eleven and a half-inches in diameter, and half an inch in thickness, covering the whole of it except two small opposite segments of about ninety degrees each ; on extracting the board a cavity of its own shape was left in which the needle was placed.

The copper fillet was not covered by silk or otherwise coated for insulation, but the several turns of it were separated at their ends by veneers of wood just so far as to prevent contact throughout. In the massiveness of the coil this instrument is perhaps peculiar, and by this means it affords a free passage to currents of the most feeble intensity, enabling them to deflect a very heavy needle. The coil was supported on a wooden ring furnished with brass feet and levelling screws, and surrounded by a brass hoop with a flat glass top or cover, in the centre of which was inserted a brass tube for the suspension of the needle by a cocoon filament. The needle was the double astatic one of Nobili, each part being about eleven inches long, one-fourth wide, and one fortieth in thickness. The lower part played within the coil and the upper one above it and the thin white dial placed upon it, thus performing the office of a conspicuous index underneath the glass. For experiments in which large quantities of Electricity are concerned, this instrument is quite unfit : but it is well adapted to show to a class, experiments on radiant heat with Pictet's conjugate reflectors, in which the differential or air thermometer affords to spectators at a distance but an unsatisfactory indication. For this purpose the electrical element necessary is merely a disc of bismuth as large as a shilling soldered to a corresponding one of copper, blackened and erected in the focus of the reflector, while the conductors pass from each disc to the poles of the galvanometer. With this arrangement the heat of a non-luminous ball at the distance of twelve feet will impel the needle near 180°, and if the connexions and reversals are properly made will keep it in continued revolution.[*]

(629) Hitherto, as we have seen, our scientific mechanists have not succeeded in constructing on electro-magnetic principles, an engine which, as a prime mover, can be said to have any claim to public attention : other applications of this force have, however, been successfully made, and to them we will now devote a little attention.

(630) It is with regret however that we here find occasion to allude to the observations with which we prefaced our remarks on " Lightning Conductors ;" (178) it is with pain that we see so much acrimony and ill-will indulged in by parties of high scientific standing, and unquestionable talent, and much does it diminish the pleasure with which we examine the beautiful details of some of the recently contrived electro-magnetic telegraphs and clocks, to find that the respective inventors are bitterly contending for " priority of discovery," and leaving the calm and dignified pursuit of philosophy to follow the unworthy employment of mutual invective and detraction. Luckily we are not called upon to express any opinion relative to the respective claims of the rival parties, and we escape with pleasure from the invidious task of so doing, and though we have not by any means been inattentive to the merits of the case and may have formed our own

[*] L. and E. Phil. Mag. Vol. ii. p. 378.

private opinions thereon, we shall cautiously abstain from taking any part in the controversy, and endeavour, as far as our space will allow, to do full justice to the labours of all.

(631) *Electro-Magnetic Telegraphs.*—The idea of employing Electricity as an agent to effect communication between distant places, is of no recent date; for almost as soon as it became known that conducting wires had the power of transmitting Electricity instantaneously to the distances of several miles, the idea occurred to several Electricians, that correspondence between distant parties might be accomplished by electric action. In 1748,[*] Dr. Watson, Bishop of Llandaff, with several other philosophers, made experiments at Shooter's Hill, which showed that Electrical discharges from a Leyden jar could be propagated through a distance of upwards of four miles, without any appreciable loss of time, although a considerable portion of the circuit was formed of land and water. The success of these experiments appears to have given rise to the first ideas of forming electric telegraphs, by means of which, distant parties might hold correspondence. From the time that Dr. Watson made his experiments at Shooter's Hill, there have been many contrivances for applying electric agency to telegraphic communication. Before 1750, *Winkler*, at *Leipsic*, discharged Leyden jars through very long circuits, in some of which a river formed a part, *Le Monnier*, at *Paris*, produced shocks through 12,789 feet of wire:— and it is said that *Betancourt*, at *Madrid*, discharged electric jars through a distance of 26 miles.

(632) In 1816, Mr. Ronalds, of Hammersmith, invented and constructed an electric telegraph, which he worked by a single circuit, through eight miles of wire, in the presence of several scientific men; and in 1823 he published a work,[†] in which he very fully described his telegraph, in both letter press and plates, together with several other Electrical instruments of his invention. Mr. Ronalds employed *clocks* to work his telegraph; a revolving disc was fixed upon the seconds' *arbor*, the signals being engraved upon it in divisions, from its centre to the circumference, each division being in size and shape similar to an opening in a *fixed* plate, behind which it revolved, so that only one division or signal could be seen at one time. An ingenious application of the voltaic battery to telegraphic purposes was made by M. Sommering:—a series of gold pins were arranged for the decomposition of water, and by touching a key, any of them could be brought into play, and thus signals could be communicated.

(633) In the year 1819, the famous discovery of electro-magnetism was made by Oersted, and since that time, nearly all the telegraphs

* See Sturgeon's Annals of Electricity, vol. v. p. 299.

† See a reference to it in the last Edition of the Encyclopædia Britannica, p. 582.

that have been brought before the public are based on the deflection of the magnetic needle by the voltaic current. It was Ampère who first suggested this application, and Mr. Alexander, of Edinburgh, who first took advantage of the suggestion. His telegraph* consisted of thirty-one wires, for the purpose of showing the alphabet in full, with stops, &c., in all thirty signals, which were shown upon a distant dial. A voltaic battery was provided, and a series of troughs of mercury to which were attached keys, to be pressed down by the finger of the operator, by which the voltaic circuit was completed; thirty magnetic needles, each carrying a screen which concealed a letter, were fixed on the dial, and each needle had its corresponding key. When no Electricity was passing, these screens remained stationary over the several letters, and consequently concealed them from view; but when the current was made to flow, by the depression of a key, the corresponding needle in the distant instrument was deflected, carrying the screen with it, and uncovering the letter, which became exposed to view. For this telegraph, a caveat for Great Britain and Ireland was lodged by Mr. Alexander, in April, 1837.

(634) In the same year, a public exhibition of an electric recording register Telegraph, in which deflected needles and pencils recorded signals, was made in America, by Mr. Morse; it is described in Silliman's Journal of Science, for October, 1837, and also in Franklin's Journal. In November of the same year, Mr. Davy exhibited a telegraph at Exeter Hall, which attracted considerable attention. In this apparatus, the signals appeared as luminous characters within a dark aperture: and in July of the following year, the same individual took out a patent for improvements in apparatus for making telegraphic communications or signals, by means of electric currents.†

(635) The first patent of Messrs. Wheatstone and Cooke, "for improvements in giving signals and sounding alarms at distant places, by means of electric currents, transmitted through metallic circuits," was sealed on the 12th of June, 1837. The telegraph here patented they call their *needle telegraph;* it is thus described by Mr. Wheatstone, in his examination before the Parliamentary Committee on Railways.—
" Upon a dial are arranged five magnetic needles in a vertical position; twenty letters of the alphabet are marked upon the face of the dial, and the various letters are indicated by the mutual conveyance of two needles, when they are caused to move. These magnetic needles are acted upon by Electrical currents passing through coils of wire placed immediately behind them. Each of the coils forms a portion of a communicating wire, which may extend to any distance whatever; these

* See Mr. Finlaison's pamphlet on "the Applications of the Electric Fluid to the Useful Arts," &c., in which there is an engraving of Mr. Alexander's Telegraph.
† For a description and engraving of this Telegraph, see Finlaison's pamphlet.

wires at their termination are connected with an apparatus which may be called a communicator because by means of it, the signals are communicated. It consists of five longitudinal, and two transverse metal bars, fixed in a wooden frame; the latter are united to the poles of a voltaic battery, and, in the ordinary condition of the instrument, have no metallic communication with the longitudinal bars which are each immediately connected with a different wire of the line; on each of these longitudinal bars, two stops are placed, forming together two parallel rows. When a stop of the upper row is pressed down the bar upon which it is placed, forms metallic communication with the transverse bar below it, which is connected with one of the poles of the battery; and when one of the stops of the lower row is touched, another of the longitudinal bars forms a metallic communication with the other pole of the voltaic battery, and the current flows through the two wires connected with the longitudinal bars to whatever distance they may be extended, passing up one and down the other, provided they be connected together at their opposite extremities, and affecting magnetic needles placed before the coils, which are interposed in the circuit.*

A second patent for improvements on the needle t legraph was specified by Messrs. Wheatstone and Cooke, in October, 1838; and in July, 1840, a third patent was taken out by the same gentlemen for an electro-magnetic telegraph, in which *clock-work* acted upon by electro-magnets producing a step by step motion, similar to the seconds' hand, was used, the number of wires being two, or sometimes three.

(636) In his improved telegraph, Mr. Wheatstone has turned to account the property possessed by soft iron of immediately acquiring and losing magnetic properties, by the establishment or interruption of a current in a wire covered with silk with which it is surrounded. It is sufficient now to have two conducting wires between one station and another, to be able to transmit all the letters of the alphabet and the figures which may be required in a telegraphic communication. By means of a commutator which serves to interrupt or establish the circuit at one of the stations, soft iron is magnetized and de-magnetized an equal number of times at the other station. The commutator is a wheel turning on its axis, and the circumference of which presents forty-eight portions alternately conductors; so that for one complete revolution of the wheel the current is twenty-four times interrupted and re-established; a letter of the alphabet corresponds to each of these twenty-four alternations; the soft iron at the other station is in like manner magnetized and de-magnetized twenty-four times. This alternate state of the magnetization and non-mag-

---

* See the fifth Report of the Parliamentary Committee on Railways; also, the Mechanics' Magazine, for 1840, which contains an engraving of Wheatstone's Needle Telegraph.

netization of the soft iron permits of an oscillating motion being given to a small appendix, also of soft iron, which results from the alternate attraction and non-attraction exercised on it by the electro-magnet ; and this alternate movement which is communicated to a wheel successively brings before the observer each of the 24 letters of the alphabet which are engraved on this wheel. Care must be exercised that there is agreement between the letters corresponding to the alternations of the commutator, and those produced by the alternate movement produced by the magnetization and demagnetization of the electro-magnet. This telegraph works with great ease and facility. Suppose, for example, the commutator placed at A, and the letter A brought to the observer by the electro-magnet ; (at each station they agree to arrange the apparatus that the starting point is the same) ; we wish to transmit the letter D ; the commutator must be moved onward three alternations ; B, C, and D, have been successively introduced at the other station ; then we stop, and D remains fixed for an instant at the other station ; and so on for each of the other letters. A large bell which is struck by a piece of soft iron, that is attracted by the electro-magnet at the moment when the circuit is established, serves to give the signal. It is evident that in order that the transmission be reciprocal, there must be a double set of apparatus, so that each station may possess those necessary to transmit, and those necessary to receive the communication.*

* For this popular account of Wheatstone's improved telegraph we are indebted to a paper by Prof. de la Rive, translated in the second number of Mr. Walker's valuable " Electrical Magazine." No other account has, we believe, hitherto been published, and we can cordially join in the following remark of the Genevese philosopher : " It is a pity that so many beautiful researches, of which several are terminated, received so rare and so incomplete a publication." We shall take the liberty of borrowing from M. de la Rive's " Notes " a few more references to the labours of the ingenious Mr. Wheatstone : " Mr. Wheatstone has moreover applied the same principle to the transmission of motion to the hands of a clock, and to other hands arranged on a dial, in the same manner as those which are moved directly by the clock ; and they progress at precisely the same rate which enables us to employ only a single clock-movement to make many pairs of hands journey round their respective dials at places widely distant and apart. But the most ingenious of the applications of the principle which the talented English philosopher has made, is in respect to the use he has derived from it to register meteorological observations. The motion of the barometer and thermometer, &c., may thus be estimated every half hour. The apparatus for producing these results is complicated but perfect. A mechanical motion is given to many parts of the apparatus ; then the contact of the mercury with a fine platinum wire placed in the tube of the meteorological instruments closes the circuit, and determines certain mechanical effects, by means of the magnetization of soft iron. The desired result is obtained by the combination of these effects with those resulting from the movement constantly impressed on the different parts of the apparatus by a mechanical force such as clock motion."

(637) An addition to the electro-magnetic telegraph was afterwards made and patented by Mr. Wheatstone, by which the letters are *printed* instead of their being merely presented to the eye. The following are the means by which this was effected :*—For the paper disc of the telegraph, on the circumference of which the letters are printed, a thin disc of brass is substituted, cut from the circumference to the centre, so as to form four-and-twenty springs, on the extremities of which types or punches are fixed : this type-wheel is brought into any desired position just as the paper disc is. The additional part consists of a mechanism, which, acted upon by an electro-magnet, occasions a hammer to strike the punch, brought opposite to it, against a cylinder, round which are rolled alternately several sheets of thin white paper, and of the blackened paper used in the manifold writing apparatus ; by this means, without presenting any resistance to the type wheel, several distinct copies of the message transmitted are obtained.

(638) The method of laying the telegraph wires which was first adopted, consisted in covering them with cotton carefully varnished, and then depositing them in smooth iron tubes, with frequent arrangements for obtaining access to the wires, and for the facility of examination and repair. The tubes, after being carefully tarred, were either buried in the ground, or fixed on low posts and covered with a wooden rail. The total cost per mile was estimated at about £290. Messrs. Cooke and Wheatstone, have now, however, rejected this method, and taken out a patent† for a new plan, by which the cost is reduced to £150 per mile. The present method of proceeding in laying down the telegraph, is first to fix firmly in the ground, at every 500 or 600 yards, strong posts of timber, from 16 to 18 feet in height, by 8 inches square at bottom, and tapering off to six by seven inches at top, fixed into stout sills, and properly strutted. Attached to the heads of these posts, are a number of winding apparatus, corresponding to the number of conducting wires to be employed ; and between every two of such posts, upright wooden standards are fixed, about 60 or 70 yards apart. A ring of iron wire, (No. 7 or 8,) which has been formed by welding the short lengths in which it is made together, is then placed upon a reel carried on a hand-barrow, and one end being attached to the winder at one draw-post, the wire is extended to the adjoining draw-post, and fixed to its corresponding winder at that post. By turning the pin of the ratched wheel with a proper key, the wire is tightened to the necessary degree ; thus the greatest accuracy may be attained in drawing the wires up till they hang perfectly parallel with each other. To effect the perfect insulation of the wires at the draw posts, wooden boxes are employed to enclose that portion of the post to which the winders are

* See Literary Gazette, 18th June, 1842.   † Specified 11th March, 1843.

attached, and small openings are left for the free passage of the wires, without risking any contact with the outer box. The standards are furnished with covers, either parted off by an overhanging roof between each wire, and again between the lowest wire and the earth, or by a series of metal shields. An eye of metal, with a slit on the upper side, forms a hook to support the wire ; and to insulate the wire from the hook, a *split quill* is slipped over the wire, on which it rests; the whole is then carefully painted with several coats of anti-corrosive paint, or asphalt varnish may be employed for the wires. For long distances, earthenware or glass is employed for insulation. The *earth itself* is employed as half of the conducting circuit, by which one wire is saved in each circuit. This plan is being carried out on the Great Western Railway, and eleven miles of it are now completed.

(639) On the 7th of June, 1842, Mr. Bain patented a different method of insulating the conducting wires of a telegraph, viz.—by imbedding them in *asphaltum* two or three feet under ground; a trench is to be dug two feet in depth, the bottom of which is to be coated with a covering of boiling asphaltum, *half an inch thick*, which is to be allowed to harden. The wires having been then laid on this bed of asphalt, are to be covered in with an upper stratum of the same material. The cost of this method is calculated by Mr. Bain not to exceed £50 per mile. Mr. Cooke, however, (a gentleman whose long practical experience entitles his opinion to much consideration), declares his conviction that £150 per mile would not cover the expense of the asphalt bed at the lowest rate of charge that has been proposed by the Companies ;* and more than this, that when completed, the plan would be liable to considerable objection. " Asphalt," he observes, " except at a very low temperature, is a conductor of Electricity ; cracks are liable to form, through which water would find its way, and a very few of such cracks in a distance of ten miles, would allow the whole of the Electricity to escape, and thus entirely to cut off the communication. Hence would arise a constant necessity of taking up and replacing defective parts,—that is, when the locality of the defect should be ascertained by the aid of my detector. Dearly bought experience has convinced me that no compound of a pitchy or resinous nature is adapted for this purpose,—it is either too brittle in cold, or too soft in warm weather, admitting moisture through its fissures in the former case, or becoming itself a conductor in the latter. These evils increase rapidly as the distance increases, because the number of the battery plates (and consequently their energy to overcome the partial resistance of an inferior conducting medium,) must be increased to transmit the electric current through the lengthened conducting wire."

* Mechanics' Mag., vol. xxxix. p. 109.

(640) Mr. Alexander Bain (in whose behalf Mr. Finlaison's pamphlet was written,) has considerably distinguished himself by his ingenious applications of Electricity to practical purposes. Some of the facts which he has described relative to the earth as a conductor and permanent generator of voltaic Electricity are new, and not unlikely to be of importance in a practical point of view. Whilst prosecuting some experiments with an electro-magnetic sounding apparatus, in the year 1841, it was found that if the conducting wires were not perfectly insulated from the water in which they were immersed, the attractive power of the electro-magnet did not entirely cease when the circuit was broken. With a view of ascertaining the true cause of the phenomenon, Mr. Bain, in conjunction with Lieutenant Wright, made a series of experiments on the Serpentine river in Hyde Park, and after verifying their former observations relative to the remnant of power in the electro-magnet when contact with the battery was broken, the electro-magnet being on one side of the river, and the battery on the other, the wires passing *through* the river; and after making other experiments, in which the water and the moist earth formed part of the circuit, and wire the remainder, it occurred to Mr. Bain, that if a positive metal were attached to one end of the conducting wire, and a negative metal to the other, and if the two metals were then placed in water, or buried in the moist earth while the connecting wire was properly insulated, a current might be generated. This was found to be the case, for when a large surface of copper was placed within Kensington Gardens at the one end of the river, and within Hyde Park at the other end, a similar surface of zinc, and the metals connected by a wire, in the circuit of which was a galvanometer, a current of considerable intensity was found to be passing. The experiment was next tried on a more extended plan; a surface of zinc was buried in the moist earth of Hyde Park, and at rather more than a *mile distant*, a surface of copper was buried, and the metals were connected by a wire suspended on the railings; when the plates were large, Mr. Bain not only obtained the usual electro-magnetic effects in an enhanced degree, but also succeeded in the performance of electrotype operations; for in the course of a few minutes he coated a half-crown with copper. Subsequent experiments have shown him that if the metals are thus buried, and connecting wires are employed, electrotype depositions may be effected, and electro-magnetic apparatus worked for a great length of time.

(641) Mr. Bain has patented several applications of Electricity to useful purposes, amongst others an electro-magnetic *printing* telegraph, which, in July 1841, was exhibited and lectured on at the Polytechnic Institution, and an electro-magnetic clock, which was exhibited and lectured on in March of the same year: both of these instruments

evince a very superior degree of ingenuity on the part of the inventor. The printing telegraph (for a full description and engraving of which we must refer to Mr. Finlaison's pamphlet) consists of three principal parts. First, the rotatory motion given to the type wheel, step after step like the seconds hand of a clock until the required letter arrives opposite the paper. This motion, Mr. Bain has since superseded by a continuous uniform motion regulated by centrifugal force. Secondly, the means of inking the types, or otherwise making permanent the imprint of the type upon the paper. Thirdly, the motion communicated to the paper so as to bring a fresh surface under the types and receive the printed intelligence in a continuous spiral line, until the paper is filled ; thus producing in print, precisely as in the pages of a book, the letters composing the message. A peculiar feature in Mr. Bain's telegraph, and one in which it differs from all others, is the substitution of *wire coils* freely suspended on centres for electro-magnets ; these coils, within and in the vicinity of which, are fixed powerful permanent magnets are deflected as long as the electrical current is passing through them, but when the electric current is broken they are drawn upwards by the force of spiral springs ; levers are released, and the machinery of the telegraph worked by mainsprings, are left free to rotate. The only battery proposed to be employed by Mr. Bain is a pair of copper and zinc plates, one of which is to be buried in the earth at one station, and the other at the distant station, where there is to be a telegraph the exact counterpart of the first, and with this he expects to obtain an electric current of the required energy.* A continuous flow of Electricity through the wire coils, when the telegraph is *not* at work keeps them constantly deflected ; but when a message is

* We fear that Mr. Bain will be disappointed in this, should he erect a pair of his telegraphs at great distances apart. Mr. Finlaison, indeed says, (see page 34 of his pamphlet,) " if a copper wire one-sixth of an inch in thickness, be imbedded in a bar of boiling asphaltum, and sent along the railway (for its better protection) from London to Liverpool—if two tons weight of zinc plates be immersed in the Mersey at Liverpool and attached to that end of the wire — and if a ton weight of copper be sunk in the river Thames and attached to this end of the wire, no rational man can doubt that an electric current would be established of ten times the power necessary to work a telegraph." Every electrician, however, knows that the resistance which the Electricity would meet with in traversing such a length of wire would be very great, and that such resistance could not be overcome by increasing the *size* of the battery plates, but only by adding to their *number ;* should, therefore, the experiment be tried, is it exceedingly probable that however carefully the uniting wire were insulated, no trace of Electricity would be detected. Mr. Finlaison, does not, however, lay any claim to the title of an electrician ; had he been one he would not probably have written the following : " May not the mere contact of the respective plates with moisture excite an electrical activity of the metallic particles without any oxidation taking place ?—and

to be transmitted, the operator by drawing out a metal pin from a hole in the dial of his machine, interrupts the circuit and the machinery is put in motion and continues so, until by inserting the pin in the hole under the signal which he wishes to communicate, the operator closes the circuit and both machines stop instantly.

(642) Mr. Bain has also invented a pendulum which is moved by a metallic surface in the moist earth, of no more than four or five feet; he intends to apply it to telegraphic purposes, and expects by its agency, to be able to discard wheels of any and every description, as well as electro-magnets. Another of his inventions is an instrument which he calls a voltaic governor, from its power of controlling the electric force as the governor of an engine controls the force of the steam. Hitherto, the only method of adjusting the action of the constant voltaic battery to the work to be done, has been by taking advantage of the modifications of a chemical character, of which the various elements are susceptible; by means of Mr. Bain's instrument another power is brought to bear, viz. a *mechanical,* and the two forces are made to counterbalance each other and produce an equilibrium or given constant action. As our space does not admit of our giving descriptions of these two ingenious instruments, we refer our readers to the second number of the Electrical Magazine, and proceed to give a short account of another of Mr. Bain's inventions, viz. his electro-magnetic clock.

(643) *B*, Fig. 234, is a back view of an ordinary clock, with a

FIG. 234.

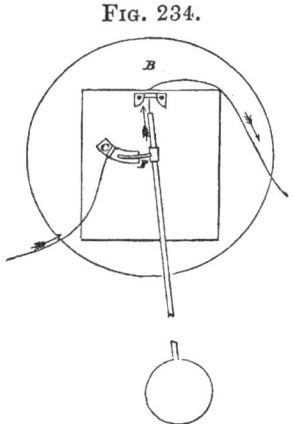

pendulum vibrating seconds; *C*, a plate of ivory affixed to the frame of the clock, in the middle of which is inserted a slip of brass in connection with the positive pole of the battery. To the pendulum is attached a very light brass spring *F*, in such a manner, that every vibration of the pendulum brings the free end of the spring into contact with the strip of brass, thus completing the electric circuit, which is broken as soon as the spring touches the ivory. A series of *electric clocks* may be connected, by means of the wires, with this clock, and if a voltaic battery be included in the circuit they will all go together.

may not oxidation itself be the effect, not the cause of such electrical action, so originating in the mere contact of the metals with moisture, the action being greatly increased in energy when the liquid is acidulated? The writer has seen the current produced through the earth so instantaneously that there was *not a moment's time for oxidation.*" (See page 35 of his pamphlet.)

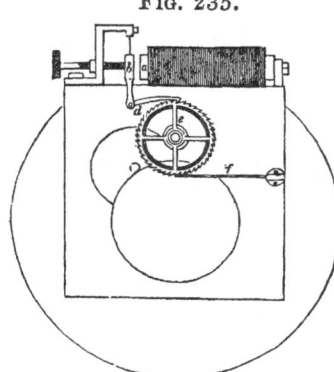

Fig. 235.

Fig. 235, is a back view of one of the electric clocks. *a*, is an electro-magnet, and *b* its feeder suspended by a spring, pendulum fashion; *c*, is a small screw to regulate the distance of the feeder from the electro-magnet. At the lower end of the feeder is jointed a light click lever *d*, falling into the teeth of a ratchet wheel *e*; *f* is a spring to keep the ratchet wheel steady. When the pendulum of the clock B, Fig. 234, sends an electric current through the conducting wire, the feeder is attracted by the magnet, and the click lever *d*, takes over one tooth of the ratchet wheel; upon the current being arrested (by the spring *F* of the pendulum, leaving the slip of brass in the primary clock,) the feeder falls back into its former position, and causes the click lever to draw the ratchet wheel one tooth forward. The arbor of the ratchet wheel carries the *seconds' hand* which is thus taken forward one degree every second, corresponding to the vibration of the clock *B*. A pinion on the ratchet-arbor gives motion to other simple wheel-work which carries the minute and hour hands. When a large number of clocks are to be worked the ratchet wheel is placed on the arbor of the *minute hand* and is moved every minute instead of every second. An ivory circle with slips or studs of metal, inserted flush with its face, corresponding to the number of clocks or group of clocks intended to be worked, is fixed on the face of the regulating or primary clock; in the centre of this circle is placed the arbor of the seconds hand of the clock, upon which is fixed a slight metal spring with its free end in contact with the ivory circle. The conducting wire from the positive pole of the battery is in connexion with the framework of the clock; every time, therefore, that the seconds hand passes over a metal stud in the ivory circle an electric circuit is completed and a current transmitted to the clock or group of clocks in connexion with that particular stud. As the seconds hand passes over every portion of the circle once in each minute, the whole number of clocks thus connected with the regulating clock will be moved forward one degree every minute. By this means a large proportion of electric power is saved, for the battery has only a single clock or a small group of clocks to work at the same instant of time.

(644) Mr. Bain has also invented an apparatus for making ordinary clocks keep correct time; also a method of working the electric clock

by the deflection of the wire coil.   In conjunction with Mr. Barwise,
he took out a patent for these inventions, which was sealed 8th January,
1841.   On the 28th of March, his clock was exhibited at the Polytechnic Institution.*

Fig. 236, (5, plate 3, of Finlaison's pamphlet) shows the method
adopted by Mr. Bain for working the electric clock by the deflection of
the *wire coil*, instead of the attractive power of the electro-magnet.
A, is a coil of insulated copper wire, freely suspended on centres.   *B*,
is a compound permanent steel magnet, immoveably fixed within the
coil.   *C C*, are two spiral springs, one on each side, for the pupose of
conveying the electric current from the stationary conducting wire *D*,

FIG. 236.

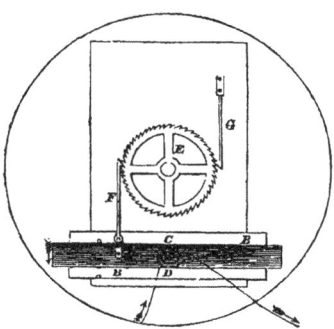

to the moveable coil.   *F*, is a click
lever attached to the coil.   *E*, is a
ratchet-wheel fixed upon the minute
hand arbor of the clock, and *G*, a
wheel to keep the spring steady.
The regulating clock transmits the
electric current to the wire coil, upon
which the left hand end is instantly
depressed, and the click lever *F*,
draws the wheel *E* forward one tooth.
When the flow of Electricity from
the regulating clock is discontinued,
the wire-coil resumes its original
horizontal position by the action of the spring *C*.   If the clock receives
an electric current once in every second, the wheel *E* is placed on the
arbor of the seconds' hand ; but if the Electricity is only transmitted
once in each minute, then the wheel *E* must be placed on the spindle
of the minute hand.

(645) The Rev. F. Lockey, to whose kindness in directing his attention to several novelties in electrical science the author is under great
obligations, has furnished this work with the following drawings of a
very ingenious contact former which he has contrived for the electro-
magnetic clock ; the author has frequently seen it in action, and has great
pleasure in saying that it works most satisfactorily. In Fig. 237, Fig. 1
shows the contact former entire ; in each figure similar letters refer to similar parts.   On the base board *A* (about three by two inches) is fixed the
circular box or trough *B* : Fig. 2 exhibits a section of this box-wood
trough wherein is turned the channel *R R*, and the central part *C C*,
is left as a solid cylinder on which to place rather firmly the glass
tube *G G*.   This tube, of which *G G* in Fig. 2 shows the section,

---

* The reader will find a full description with engravings of Mr. Bain's electro-
magnetic clock in Mr. Finlaison's pamphlet, so frequently alluded to.

rests on a rim or shoulder just below the rim at *B*, which shows the
level to which mercury is poured into the channel *R R*, for the purpose of closing the bottom of the tube and preventing all access
of dust to its interior. The glass tube is surmounted by an ivory
cap *H*, Fig. 1, cemented thereon ; through the ivory cap pass the
two wires *I I*, furnished with screw connecting pieces for the purpose
of uniting them with the *p* and *n* wires of the galvanic battery. The
lower end of these wires terminate in two very thin and flexible copper
springs, of which the lower portions are seen in Fig. 2, *E F*; they
are tied together at *K*, Fig. 1, a piece of ivory being interposed
to prevent metallic contact, as well as to place them parallel to
each other in the tube ; they are tipped with platinum foil at
E and *F*, and one of them is a little longer than the other. The
spring *F* is so set as to have a slight tendency to advance towards
E, but it is prevented from
doing so by the ivory stud *D*,
Fig. 2. Part of the central
cylinder *C C* is supposed to be
broken away in Fig. 2 to show
the lever (formed of iron wire
No. 18) bent somewhat in the
form *E C P L*, the fulcrum *P*
being beneath the end of the
tube, and the part *E C P*
working freely in a slit in the
cylinder *C C*. If this contact
breaker is intended to work
an electro-magnetic clock, it is
so placed at the side of the
central or regulating clock, as
that the pendulum *M N O*, just
before it swings into a perpendicular position, shall begin
to act on the lever at *L*.
During the remaining half of
its vibration towards the right
hand (in the figure) as well as
during the first half of its returning vibration towards the
left hand (see Fig. 3) it will
maintain contact between the

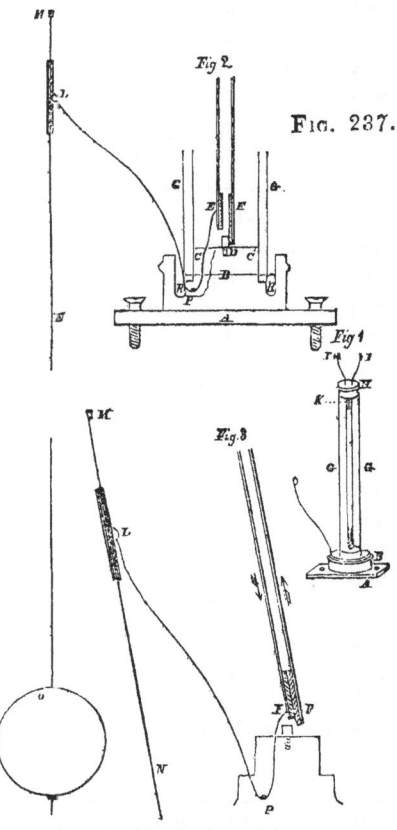

Fig. 237.

free or platinum ends of the springs *E* and *F*; during the other two
halves of its vibrations, the lever is unacted upon, and the springs return-

ing to their parallelism, contact is broken and the battery current ceases. Such an instrument does not impede the action of the clock to which it is applied; Mr. Lockey's had been upwards of six months in continuous use and acting with unfailing accuracy, when the author first saw it.*

(646) Mr. Wheatstone's electro-magnetic clock which was exhibited and explained at the Royal Society, 25th November, 1840, is thus constructed : all the parts employed in a clock for maintaining and regulating the power are entirely dispensed with. It consists simply of a face with its second, minute, and hour hands, and of a train of wheels which communicate motion from the arbor of the second's hand to that of the hour hand, in the same manner as in an ordinary clock train ; a small electro-magnet is caused to act upon a peculiarly constructed wheel, placed on the second's arbor, in such a manner that whenever the temporary magnetism is either produced or destroyed, the wheel, and consequently the second's hand, advance a sixtieth part of its revolution. On the axis which carries the scape wheel of the primary clock, a small disc of brass is fixed, which is divided on its circumference into sixty equal parts ; each alternate division is then cut out and filled with a piece of wood, so that the circumference consists of thirty regular alternations, of wood and metal. An extremely light brass spring, which is screwed to a block of ivory or hard wood, and which has no connexion with the metallic parts of the clock, rests by its free end on the circumference of the disc. A copper wire is fastened to the end of the spring, and proceeds to one end of the wire of the electromagnet ; while another wire attached to the clock frame is continued until it joins the other end of that of the same electro-magnet. A constant voltaic battery, consisting of a few elements of very small dimensions, is interposed in any part of the circuit. By this arrangement the circuit is periodically made and broken, in consequence of the spring resting for one second on a metal division and the next second on a wooden division. The circuit may be extended to any length, and any number of electro-magnetic instruments may be thus brought into sympathetic action with the standard clock. It is necessary to observe, that the force of the battery and the proportion between the resistances of the electro-magnetic coils and those of the other parts of the circuit, must, in order to produce the maximum effect with the least expenditure of power, be varied to suit each particular case.

We must here terminate this brief sketch of the applications which have been made of the electric fluid to useful purposes.

* The cost of working an Electro-magnetic clock, according to Mr. Tylee's observations, is under a penny per week, the battery employed being one on Smee's construction, platinized silver $3\frac{1}{2}$ inches square, and the exciting fluid water, with $\frac{1}{20}$ part of sulphuric acid. Such a battery Mr. Tylee finds will work his clock for fourteen days without being interfered with.

LECTURE IX.

MAGNETO-ELECTRICITY.

Electro-dynamic and magneto-electric induction — Terrestrial magneto-electric induction—Electric spark from a magnet—Dr. Ritchie's method of detonating oxygen and hydrogen gases by the magnetic spark—The magneto-electrical machine—Saxton's improved instrument—Clarke's large machine —Woolrich's application of the magneto-electrical machine—Theory of the magneto-electrical machine—Secondary currents—Experiments with a ribbon coil—Professor Henry's experiments on induced electrical currents—Induced currents from the discharge of a Leyden phial—Electro-magnetic coil machines—Dr. Bird's contact breaker—Different forms of the electro magnetic coil machine—The scintillating circle — Lockey's water regulator — Henley's electro-magnetic coil machine—Thermo-electricity—Thermo-battery—Dr. Andrews's experiments on thermo-currents.

(647) *Electro-dynamic and magneto-electric induction.*—When a current of Electricity from a single voltaic pair (241) is sent through a metallic wire, it induces a current of Electricity in a second wire placed near the first, at the moment that contact with the battery is made, and when it is broken; but while the Electricity *continues* to flow through the first wire, no inductive effect on the second wire can be perceived. The force of the induced current is stronger on making, than on breaking battery contact, and its direction in one case the reverse of the other. By arranging a length of about two hundred feet of copper wire in a coil round a block of wood, and a second similar coil as a spiral between the coils of the first, metallic contact being every where prevented by twine; by connecting the ends of the second coil with a small helix formed round a glass tube, in which was placed a common sewing needle, and then causing a current of Electricity from a voltaic battery to pass through the first coil, Dr. Faraday found that the needle became a magnet, provided it was removed from the helix *before* battery-contact was broken; if, however, it was allowed to remain, its magnetism was entirely or very nearly destroyed. If the needle was introduced into the helix *after* battery-contact had been made with the first coil, it acquired no magnetic properties unless

c c 2

it was allowed to remain till battery-contact was broken, it then
became a magnet, though with its poles in a contrary direction from the
first, thus proving that it is only at the moment of making and breaking battery-contact that a current of Electricity is induced in the
second coil, and that the direction of the induced current is opposite in
the two cases.

(648) If, instead of coiling the wires round a block of wood, they
are arranged round a ring of iron as shown in Fig. 238, where A and
B represent the compound helices, being
separated by about half an inch of uncovered iron; a current of Electricity from
the battery sent through one helix A, induces a current in the second helix B,
much more powerful than when the arrangement is made round wood, but only

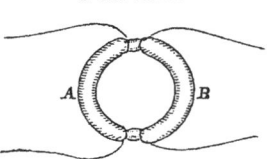

Fig. 238.

on making and breaking contact as before; if the battery be large, a
minute *spark* may be perceived between charcoal points fastened to the
ends of B, at the moment of making, and sometimes, though not often,
on breaking contact with the battery, but never while a continuous
current is passing through A.

(649) To prove that the increased inductive power is occasioned by
the iron and is not a common effect of metals, an arrangement of
helices of copper wire may be wound round a hollow cylinder of thin
wood or pasteboard, and the power of the induced current tested first
with the helices alone; then, after inserting a bar of copper, lead, tin, or
any other metal except iron and perhaps nickel, in the axis of the
cylinder, no effect beyond that of the helices alone will be found to be
produced; but when a bar of soft iron is inserted, the power of the
induced current will be found to be surprisingly increased.

(650) By the following beautiful experiment of Faraday,* the property of *ordinary magnets* to induce electrical currents without the
intervention of any galvanic arrangement, is clearly demonstrated; a
long compound helix wound round a cylinder of pasteboard was connected with a galvanometer (622) by two copper wires, each five feet in
length, a soft iron bar was introduced into its axis; a couple of bar-
magnets, each twenty-four inches long, were arranged with their opposite poles at one end in contact, so as to resemble a horse-shoe magnet,
and then contact made between the other poles and the ends of the
iron cylinder so as to convert it into a magnet, as shown in Fig. 239.
By breaking the magnetic contacts, or reversing them, the magnetism
of the iron cylinder could be destroyed or reversed at pleasure. Upon
making magnetic contact, the needle was deflected; continuing the

* Experimental Researches, 36, 37.

contact, the needle became indifferent and resumed its first position; on
breaking contact, it was again deflected, but in the opposite direction;

FIG. 239,

and then it again became indifferent; when the magnetic contacts were
reversed, the deflections were reversed. In order to prove that the
induced electrical current was not occasioned by any peculiar effect
taking place during the formation of the magnet, Dr. Faraday made
another experiment in which soft iron was rejected, and nothing but a
permanent steel magnet employed. The ends of the compound helices
being connected with the galvanometer, either pole of a cylindrical
magnet was thrust into the axis, as shown in Fig. 240, the needle of
the galvanometer was immediately deflected,   FIG. 240.
but soon resumed its first position; on withdrawing the magnet a second disturbance of
the needle took place, but in an opposite direction.

(651) When a powerful magnet is employed, induced electrical currents are evinced by the galvanometer, when the helix with its iron
cylinder is brought near, but *without touching* the magnetic poles : and
by experimenting with the large compound magnet belonging to the
Royal Society,* Faraday was able to throw the needle of the
galvanometer 80° or 90° from its natural position by placing the copper
helix *without the iron cylinder* between the poles; and by using an
armed loadstone capable of lifting about thirty pounds, he succeeded in
powerfully convulsing the limbs of a frog by the induced electrical
current.

(652) The discovery of M. Arago of the influence of various substances on the oscillations of the magnetic needle, and the experiments of
Mr. Babbage and Sir John Herschel on the supposed development of
magnetism by rotation, have been described (518 et seq.) The experiments of the French philosopher led him to the conclusion that all substances whatever exert an influence on the needle when made to rotate
in a plane parallel to it, or vice versâ; but Mr. Babbage, and Sir John
Herschel could only obtain the effects with bodies that were good con-

* This magnet is composed of about 450 bar magnets, each fifteen inches long,
one inch wide, and half an inch thick, arranged in a box so as to present at one
of its extremities two external poles. It requires a force of nearly one hundred
pounds to break the contact of an iron cylinder three-quarters of an inch in diameter and twelve inches long, put across the poles. It formerly belonged to Dr.
Gowin Knight.

ductors of Electricity : they refer the phenomenon to magnetism *induced* in the plate by the magnet,—the essential circumstance being that the revolving substance shall acquire and lose its magnetism in sensible time,—to an *attractive* force. Arago and Ampère, on the other hand, conceive the action to be always repulsive.

(653) Faraday was, however, the first to demonstrate that a permanent current of Electricity may be produced by ordinary magnets. Fig. 241 represents the form of apparatus employed. A copper plate

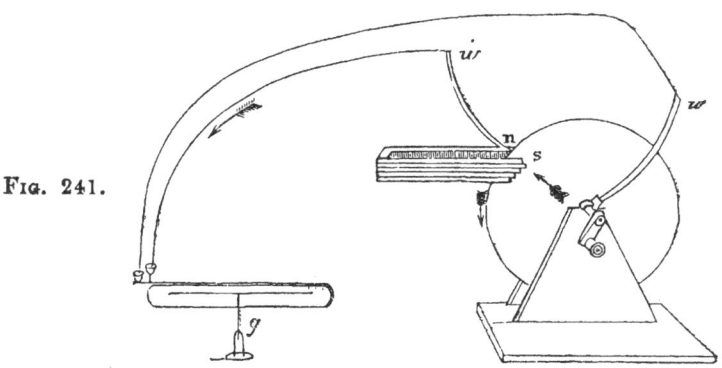

FIG. 241.

mounted on an axis, is furnished with a handle for giving it motion ; *w w*, are conducting wires,—the one retained in perfect metallic contact with the axis, and the other with the circumference of the disc. A powerful horse-shoe magnet is then placed so as to allow of the revolution of the disc between its poles, and the wires *w w*, are connected with the galvanometer, *g* ; the wire *w*, is retained on the circumference of the disc, at the point between the poles of the magnet. When this machine is made to revolve from right to left, a current of Electricity from the centre to the circumference is determined in the direction of the arrows, and the galvanometer is deflected accordingly. If the revolution of the disc, or the poles of the magnet be reversed, the electric current moves in an opposite direction :—while the plate is at rest, there is no disturbance of the needle of the galvanometer. The same effects are produced when electro-magnets, or coils of wire are substituted for the permanent magnetic poles ; and when instead of employing a circular disc of metal, a strip of copper plate is placed between the magnetic poles, while two conductors from the galvanometer are held in contact with its edges, a current of Electricity is shown to be produced by simply drawing the slip of metal between the poles of the magnet.

(654) The *law* which governs the evolutions of Electricity by magneto-electric induction, is thus illustrated by Faraday.* If in Fig. 242,

* Experimental Researches, 114.

P N, represent a horizontal wire passing by a
marked magnetic pole, so that the direction of
its motion shall coincide with the curved line
proceeding from below upwards; or if its motion parallel to itself be in a line tangential to
the curved line, but in the general direction of
the arrows, or if it pass the pole in other direc-

FIG. 242.

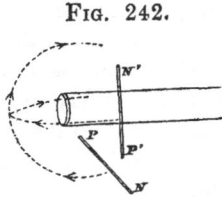

tions, but so as to cut the magnetic curves in the same general direction,
or on the same side as they would be cut by the wire if moving along
the dotted curved line; then the current of electricity in the wire is
from P to N. If it be carried in the reverse directions, the electric current will be from N to P; or if the wire be in the vertical position
P′ N′ and it be carried in similar directions, coinciding with the
dotted horizontal curve so far as to cut the magnetic curves on the
same side with it, the current will be from P′ to N′. If the wire be
considered as a tangent to the curved surface of the cylindrical magnet,
and it be carried round that surface into any other position; or if the
magnet itself be revolved on its axis so as to bring any part opposite to
the tangential wire; still if afterwards the wire be moved in the directions indicated, the current of Electricity will be from P to N; or if it
be moved in the opposite direction, from N to P; so that, as regards the
motions of the wire past the pole, they may be reduced to two,
directly opposite to each other, one of which produces a current from
P to N, and the other from N to P.

The same holds true of the unmarked pole of the magnet, except that
if it be substituted for the one in the figure, then, as the wires are moved
in the direction of the arrows, the current of Electricity would be from
N to P, and when they move in the reverse direction, from P to N.

(655) The direction of the current of Electricity which is excited in
a metal when moving in the neighbourhood of a magnet is thus shown
to depend upon its relation to the magnetic curves. Faraday, with his
usual happy method of illustration, has given us this popular expression
of it. Let A B, Fig. 243, re-
present a cylinder magnet, A
being the marked, and B the unmarked pole; let P N, be a
silver knife-blade, resting across
the magnet with its edge upward, and with its marked or
notched side towards the pole,
A; then in whatever direction

FIG. 243.

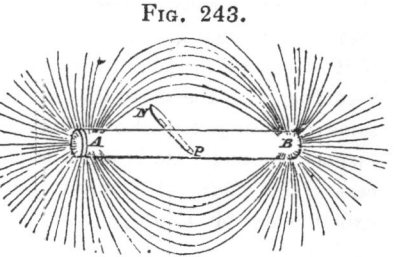

or position this knife be moved, edge foremost, either about the marked
or unmarked pole, the current of Electricity produced will be from P to

N, provided the intersecting curves proceeding from A, abut upon the notched surface of the knife, and those from B upon the un-notched side; or if the knife be moved with its back foremost, the current will be from N to P, in every possible position and direction, provided the intersected curves abut on the same surfaces as before. A little model is easily constructed, by using a cylinder of wood for a magnet, a flat piece for the blade, and a piece of thread connecting one end of the cylinder with the other, and passing through a hole in the blade for the magnetic curves; this readily gives the result of any possible direction.

(656) From this discovery of Faraday, then, viz. that when a piece of metal is passed before a single pole or between the opposite poles of a magnet, electrical currents transverse to the direction of motion are produced across it, a very satisfactory explanation of the phenomenon first observed by Arago, and afterwards examined in detail by Babbage and Herschel can be given without having recourse to the supposition of the formation in the revolving copper of a pole of the opposite kind to that approximated, surrounded by a diffuse polarity of the same kind. It is evident that as the plate revolves, in the neighbourhood of the magnet or *vice versâ*, electrical currents are produced from the centre to the circumference, or from the circumference to the centre, in the direction of the radii; and the effect is precisely the same as in electro-magnetic rotations, which as we have seen (581), are governed by the following law;—if a wire $P N$, Fig. 244, be connected with the positive and negative ends of a voltaic battery so that the positive electricity shall pass from P to N, and a marked magnetic pole N be placed near the wire, between it and the spectator, the pole will move in a direction tangential to the wire that is towards the right, and the wire will move tangentially towards the left, according to the direction of the arrows. So also when a plate of metal is made to rotate beneath a magnetic pole (suppose an N pole), a series of currents of Electricity will pass from the centre to the circumference of the plate, if it is rotating in the direction of the hands of a watch, or from the circumference to the centre if it is rotating in the contrary direction; and it is at once evident that, according with the above law, both magnet and plate must move in the same direction; it is also evident why the phenomena cease when the magnet and metal are brought to rest, for then the electrical currents cease; the effects of a solution of the continuity of the disc in the experiments of Babbage and Herschel (520) are likewise readily explained.

FIG. 244.

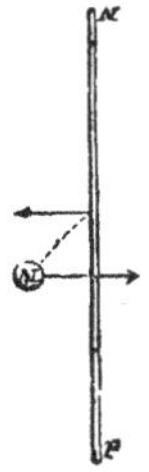

(657) *Terrestrial magneto-electric induction.*—When a soft iron bar is held in the direction of the magnetic meridian, and inclined in the

position of the *dip* of the needle, it becomes a temporary magnet, the lower end acquiring the properties of the north pole (543); if the bar be inverted its polarity is at the same time changed. Faraday took a soft iron cylinder, and having carefully deprived it of all traces of magnetism by heat, he placed it in the axis of a coil of wire, the ends of which were connected with a galvanometer by wires eight feet long. The coil was then held in the line of the dip, and then suddenly inverted ; the needle of the galvanometer was immediately deflected, proving that a current of Electricity was evolved by means of the magnetism of the globe ; he afterwards succeeded in obtaining indications of Electricity *without the iron cylinder ;* and by causing a circular plate of copper to rotate in a horizontal plane, electric phenomena were produced without any other magnet than the earth : when the plate was revolved in the same direction as the hands of a watch move, the current of Electricity was from the centre to the circumference ; when in the contrary direction, the current was from the circumference to the centre.

(658) *A new electrical machine* was thus formed, differing remarkably from the common machine in the circumstance of the plate being a most perfect conductor, and in the absolute necessity of a good conducting communication with the earth. When the plate was revolved in the magnetic meridian no electrical effects were developed, and they became most powerful when the angle formed by the plane of the plate with the dip was 90°. It was likewise shown by Faraday that a current of Electricity is produced in a wire by merely moving it from right to left, or from left to right, over a galvanometer, and he states it to be a remarkable consequence of the universality of the magnetic influence of the earth, that scarcely any piece of metal can be moved in contact with others, either at rest or in motion with different velocities or in varying directions, without an electric current existing within them ; further researches likewise proved that the currents produced by the magneto-electric induction in bodies, is exactly proportional to and altogether dependent upon their conducting power.

(659) By the aid of the diagram, Fig. 245, the relation between volta-electric and magneto-electric induction may easily be understood. Suppose an electrical current to be passing through the middle wire from *P* to *N*, this wire is surrounded at every part by *magnetic curves*, diminishing in intensity according to their distance from the wire, and which in idea may be likened to rings situated in planes perpendicular to the wire, or rather to the electric current within it. The dotted rings may represent the magnetic curves round the wire *N P*, and if small magnetic needles be placed as tangents to it, they will become arranged as in the figure. But if instead of causing the needles to be influenced

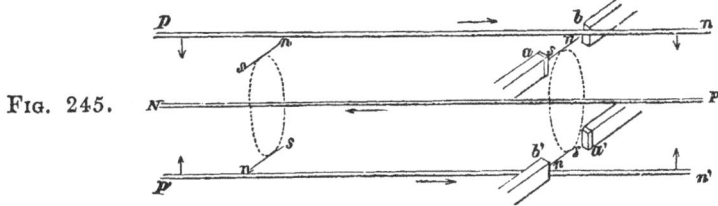

FIG. 245.

by an electric current, they are acted on by magnets, then in order that
they shall take up the same position as before, the magnets must be
placed as shown in the figure, the marked and unmarked poles *a b*
above the wire being in opposite directions to those *a' b'* below it ; in
such a position, therefore, the magnetic curves between the poles *a b*
and *a' b'* have the same general direction with the corresponding parts
of the ring magnetic curve surrounding the wire $N P$ carrying the
electric current.   Now if a second wire *p n* be brought near the wire
carrying the electric current, it will cut an infinity of magnetic curves
in the same manner as it would the magnet curves if passed from above
downwards between the poles, and the electric current induced in the
wire will obviously be the same in both cases ; if the wire *p' n'* be carried
up from below, it will pass in the opposite direction between the magnetic poles, but then the magnetic poles themselves are reversed, consequently the induced current is in the same direction as before ; it is
also for equally evident reasons in the same direction, if produced by the
influence of the curves dependent on the wire.

(660) *Electric spark from a magnet.*—Faraday first obtained a spark
from a temporary or electro-magnet in November, 1831 ; but the first
person who obtained the spark from a natural or permanent magnet in
this country, was Professor Forbes of Edinburgh ; the experiment was
made on the 13th of April, 1832, with a powerful natural magnet
capable of supporting 170 lbs., presented to the University of Edinburgh, by Dr. Hope.*   The arrangement of the apparatus is shown
in Fig. 246.   *A* is the magnet ; *a b* a cylindrical collector of soft iron
passing through the axis of the helix *c*, and connecting the poles of the
magnet ; accuracy of contact was found to be of considerable importance in the success of the experiment, and one side of the cylinder was
carefully formed to a curve of about two inches radius, for this purpose.   Great advantage was found from a mechanical guide not
represented in the figure to enable an assistant to bring up the connector
rapidly and accurately to the magnet in the dark.   The helix *c*, con-

* The *first* spark was obtained by Professor Forbes on the 30th of March.   It
appears that the first document, giving an account of the excitation of a spark
from a permanent magnet, is by Signor Nobili, and another dated from the Museum at Florence, 31st Jan. 1832.

sisted of about 150 feet of copper wire, nearly one-twentieth of an inch in diameter, seven and a half inches long, and containing four layers in thickness, which were carefully separated by insulating partitions of cloth and

FIG. 246.

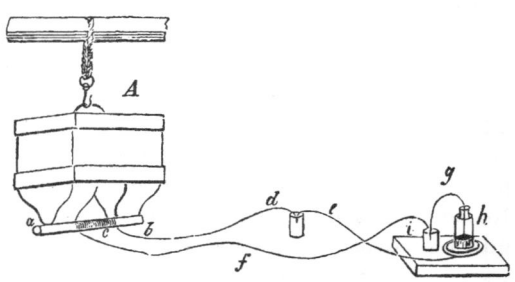

sealing-wax. The one termination *d e* of the wire passed into the bottom of a glass tube *h*, half filled with mercury, in which the wire terminated, and the purity of the mercurial surface was found to be of great consequence. The other extremity *f*, of the helical wire communicated by means of the cup of mercury *i*, with the iron wire *g*, the fine point of which may be brought by the hand into contact with the surface of the mercury in *h*, and separated from it at the instant when the contact of the connector *a b* with the poles of the magnet is effected. The spark is produced in the tube *h*.

The success of the experiment obviously depended on the synchronism of the production of the momentary current by connecting the magnetic poles, and the interruption of the galvanic circuit at the surface of the mercury; with a little practice, Mr. Forbes was able to produce for many times in succession, at least two sparks of a fine green colour, from every three successive contacts.

(661) The magnetic spark may be produced with great ease and certainty, and with a magnet of moderate strength by employing the little arrangement shown in Fig. 247. It consists merely of a cylinder of soft iron, round the centre of which is wound a few feet of small

FIG. 247.

insulated copper wire; to one end of this wire is soldered a small disc of copper which is well amalgamated, the other end is bent up, the point cleaned and amalgamated, and brought into contact with the disc. On laying this cylinder across the poles of the magnet, and then suddenly breaking contact, the point and the disc become separated at the same time, and the spark appears. Another excellent method of showing the spark from a single pole, and which is well adapted for the lecture table, is to mount a strong bar magnet (about two feet long) horizontally on a stand, to wind eighteen or twenty feet of wire, the ends of which are prepared as in the last arrangement round a piece of wood, through which a hole is cut large enough to allow the end of the magnetic bar to work freely, rapid horizontal motion is then given to the coil by means of a multiplying wheel, and contact between the point and disc is broken by the end of the bar striking a small piece of wood loosely placed at one end of the aperture in the wood, through which one end of the copper coil passes. A series of sparks which, if the magnet is powerful, are very brilliant, appear with such rapidity as to keep up a constant light.

(662) In the fourth volume of the London and Edinburgh Philosophical Magazine, page 104, a very simple method of detonating a mixture of oxygen and hydrogen gases by the magneto-electric spark was described by the late Dr. Ritchie. It forms an excellent class experiment. Round the soft iron lifter of a horse-shoe magnet capable of carrying fifteen or twenty pounds, ten or twelve feet of insulated copper wire are wound. To the ends of the coil two thick copper wires are to be soldered in order to form a complete metallic circuit when the lifter is in contact with the poles of the magnet. The magnet is mounted, poles upwards, on a wood stand, having a pillar with an arm or lever passing through a mortice in the top of it, for the purpose of removing, by a sudden jerk, the lifter from the poles of the magnet. In front of the magnet a glass tube is fixed, having its top closed by a cap of box-wood, through which the copper wires soldered to the extremities of the coil pass, as near air tight as possible, into the glass tube; the

end of one wire being flattened, is bent at right angles and well amalgamated. The other, which is straight, can be brought down or removed from it by means of the lever. The whole arrangement will be readily understood by a simple inspection of Fig. 248. The mixed gases are introduced into the tube G by means of a bent or flexible tube. On giving the lever E

FIG. 248.

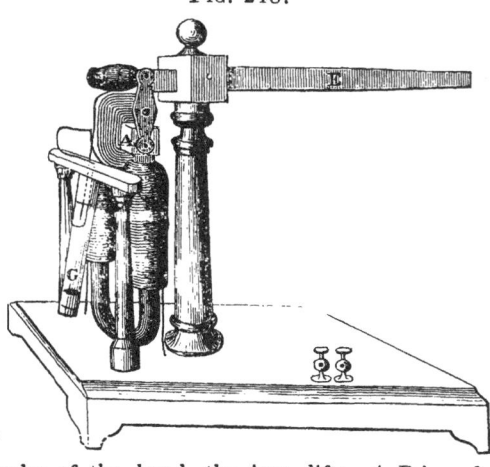

a smart blow with the palm of the hand, the iron lifter $A\ B$ is suddenly removed from the poles of the magnet, a current of Electricity is induced in the coil, contact between the wires in the tube G is broken, a spark appears, and the gases are immediately exploded.

(663) *The Magneto-Electrical Machine.*—The first magneto-electric machine,—that is, an instrument by which a continuous and rapid suc-

cession of sparks could be obtained from a permanent magnet, was invented by M. Hipolyte Pixii, of Paris, and was first made public at the meeting of the *Académie des Sciences*, on Sept. 3rd 1832. A description of this invention will be found in the *Annales des Chimie*, for July, 1832, and a representation of it in Becquerel's Traité de l'Electricité, vol. iii. With this machine, furnished with a coil about 3000 feet in length, sparks and strong shocks were obtained; a gold leaf electrometer was made to diverge, a Leyden jar was weakly charged, and water was decomposed.

(664) At the meeting of the British Association at Cambridge, in June, 1833, Mr. Saxton exhibited his improvement on Pixii's machine, and in the August of the same year, a large instrument on the new construction was placed in the Gallery of Practical Science in Adelaide Street. With this machine were exhibited, the ignition and fusion of platinum wire, and the excitation of an electro-magnet of soft iron; and in December, 1835, there was added to the instrument the double armature producing at pleasure either the most brilliant sparks, and strongest heating power, or the most violent shocks, and effecting chemical decompositions. Saxton's machine differs principally from Pixii's in two respects: first—in M. Pixii's instrument, the magnet itself revolves, and not the armature; and secondly, the interruptions, instead of being produced by the revolution of points, were made by bringing one of the ends of the wire over a cup of mercury, and depending on the jerks given to the instrument by its rotation for making and breaking the contact with the mercury.*

(665) In the Philosophical Magazine for October, 1836, Mr. E. M. Clarke describes his ingenious arrangement of the magneto-electrical machine, in which the battery of magnets is placed in a vertical, instead of a horizontal position, whereby vibration (known to be so injurious to magnets,) is materially lessened. Two soft iron armatures are employed; one is covered with 40 yards of thick copper bell wire, and is used for *quantity* effects, such as igniting platinum wire, magnetizing iron, producing the spark, deflagrating metals, &c.; and the other, the iron of which is only half the weight of the former, has 1,500 yards of fine insulated copper wire on it, and is used for the exhibition of those effects usually ascribed to intensity, viz., giving the shock, and effecting chemical decompositions.

(666) In the last Edition of this work, Mr. Clarke's magneto-electric-machine was fully described with engravings; and we avail ourselves of the present opportunity to give publicity to the instrument of Mr. Saxton as now constructed. Fig. 249 represents the complete machine, and

* A description and engraving of the machine deposited by Mr. Saxton in the Adelaide Gallery, in Aug. 1833, (and where it may still be seen), will be found in L. and E. Phil. Mag. vol. ix. p. 360.

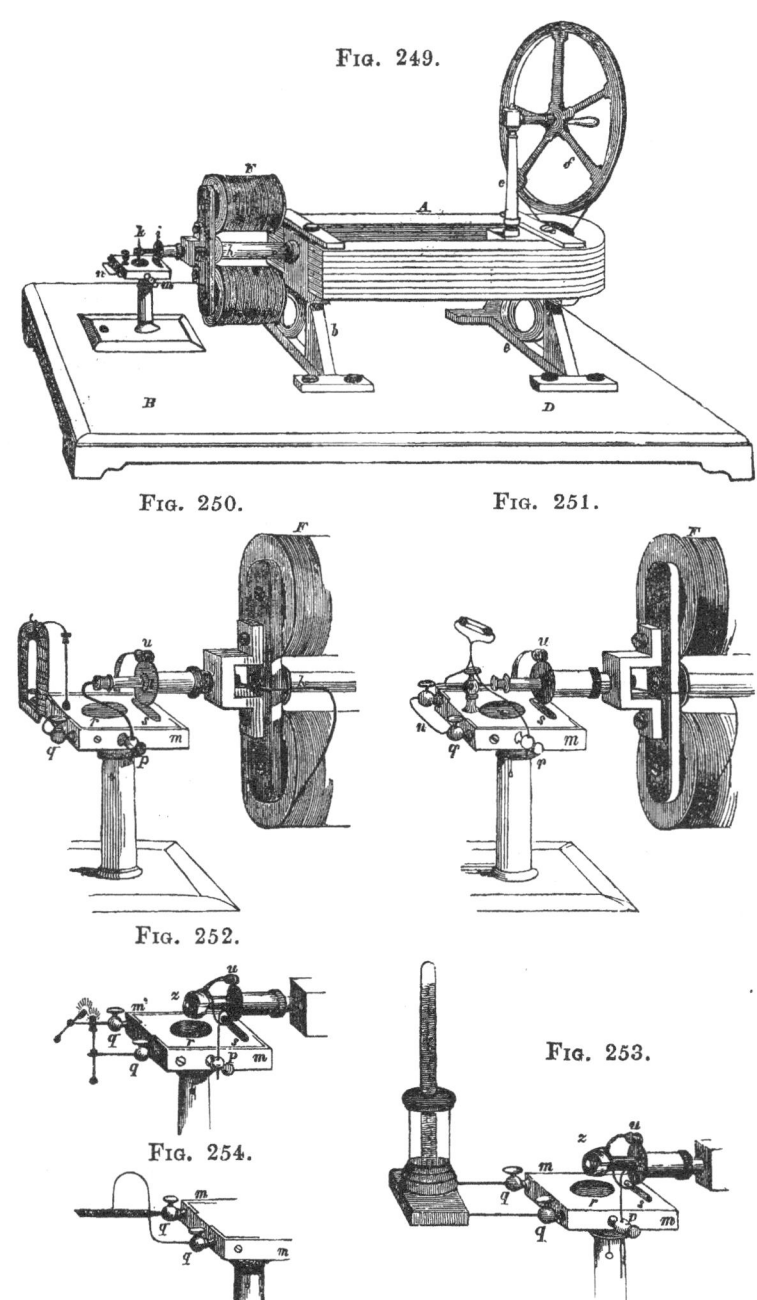

Fig. 249.

Fig. 250.   Fig. 251.

Fig. 252.

Fig. 253.

Fig. 254.

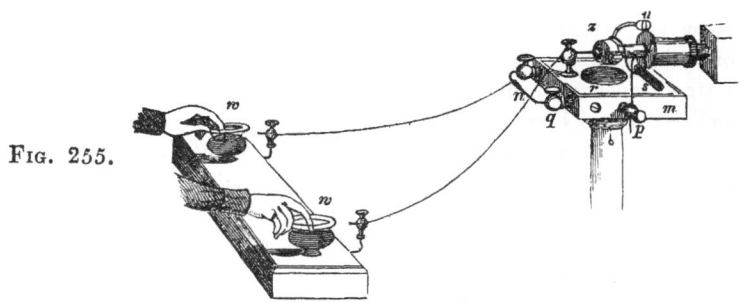

FIG. 255.

Figs. 250, 251, 252, 253, 254, 255, show the different arrangements and their application to illustrate various phenomena, The letters in Fig. 249, answer to the same in the other figures. A, is a compound horse-shoe magnet, composed of six or more bars, and supported on the rests, *b*, *e*, which are screwed firmly on the board, B D ; into the rest, *e*, is screwed the brass pillar, *c*, carrying the large wheel, *f*, having a groove in its circumference, and a handle by which it can readily be revolved on its axis ; a spindle passes from one end of the magnet to the other between the poles, and projects beyond them about three inches, where it terminates in a screw at *h*, to which the armatures to be described immediately are attached ; at the further extremity is a small pulley, over which a gut band passes, by means of which, and the multiplying wheel, *f*, the armatures can be revolved with great velocity.

(667) The armatures or *inductors*, as seen at F, are nothing more than electro-magnets ; two pieces of round iron are attached to a cross piece, into the centre of which the spindle, *h*, screws ; round each of these bars is wound in a continuous circuit a quantity of insulated copper wire, one end being soldered to the round disc, *i*, the other connected with the copper wire passing through, but insulated from it by an ivory ring. By means of the wheel and spindle, each pole of the armature is brought in rapid succession opposite each pole of the magnet, and that as near as possible without absolutely touching. The two armatures differ from one another. The one termed the *quantity* armature is constructed of stout iron, and covered with thick insulated wire. The other, the *intensity* armature, is constructed of slighter iron, and covered with from 1000 to 2000 yards, according to the size of the instrument, of fine insulated wire.

The quantity armature is adapted for exhibiting the magnetic spark, inducing magnetism in soft iron, heating platinum wire, &c. The intensity is best adapted for administering the magnetic shock, (which from only a moderate sized machine, is so powerful, that few will venture to take it a second time,) and for effecting chemical decompositions.

(668) *The Flood Cup*, is that part of the instrument to which the different arrangements and apparatus used to illustrate various pheno-

mena are attached.   The one here represented can be used either with
or without mercury : it consists of a square block of wood supported on
a stand capable of being raised or lowered to the height required.   Two
hollows, *r* and *s*, are made on the top, into which mercury is put when
that medium is required ; the round metal disc, *i*, Fig. 249, revolves in *s*,
and the point, *h*, just dips into *r* ; the wire fork, *n*, connects the two
floods of mercury together.   On revolving, the armature contact is continually broken and renewed at the point, *k*, and a most brilliant succession of sparks, forming almost a continuous light, is produced.   Two
pieces of stout brass, *m*, bent at two right angles, are fixed to the sides
of the wood block, but insulated from each other ; to these are attached
binding screws, which answer in every respect the same purpose as the
mercury.

   (669) Fig. 250 exhibits a small electro-magnet, connected with the
apparatus.   A small roller, *u*, attached to a spring, presses on the circumference of the metal disc, *i*, at the same time a bent piece of wire
is fixed by the binding screw, *p*, into the centre of the copper pin, the
terminating wires of the electro-magnet being connected with the
binding screws, *q q*.   Fig. 251 exhibits the arrangement for heating platinum wire, which is enclosed in a small glass tube.   Fig. 252,—the
arrangement for igniting charcoal points, the break piece, *z*, being screwed
to the copper pin, and a wire spring inserted into the binding screw, *p* ;
by this means contact is broken and renewed, as with the mercury and
points shown in Fig. 249.   Fig. 253 illustrates the method of decomposing
water.   Fig. 254, the combustion of steel by inserting a piece of wire in
one of the binding screws, *q q*, and a small file into the other ; on
turning the wheel of the machine, and drawing the wire over the
surface of the file, brilliant scintillations are produced.   Fig. 255 exhibits
the arrangement for administering the magneto-electric shock.   *w w*,
represent two glass cups containing a little acidulated water, into these
the fingers or hands are to be dipped ; a small metal bottom connects
them with the binding screws, one of which is connected by a wire with
the fork, *n* ; the other is inserted into the centre of the copper pin : on
turning the wheel the magneto-electric shock is communicated.

   (670) The magneto-electric machine is certainly one of the most beautiful and instructive instruments of modern science ; by it we see exemplified
the close connection between, if not the identity of, the electric and magnetic forces ; by it the same heating, magnetizing, and decomposing power,
—the same velocity of motion, the same physiological and the same chemical effects are shown to be common to both ; and let us not forget that
it is to the indefatigable labours, and splendid talent of an *English*
philosopher that we are indebted for the development of the leading
principles on which this beautiful instrument is constructed.   Let us
remember that the unfolding of the laws of electro-dynamic, and mag-

neto-electric induction was effected by a countryman of our own, and that these brilliant discoveries were not, (as many of the first importance are known to have been) the offspring of accidental or fortuitous circumstances, but purely the result of, and affording fine illustrations of that method of physical research introduced by the great reformer of philosophy, (Lord Bacon) viz:—*well-founded and well-verified inductions and deductions*.

(671) The largest magneto-electrical machine that has yet been constructed, is probably the one which was exhibited at the Electrical Society, in September, 1838, by Mr. E. M. Clarke. The following account of some experiments performed with it on that occasion, appeared in the *Morning Chronicle*. " The experiments submitted were performed with a machine greatly exceeding in dimensions any other that has yet been constructed, the magnetic batteries being separated into two parts, connected together by the inductors rotating at their sides; the quantity arrangement being on one side, and the intensity on the other. The results with the machine in this form were so opposite to what was anticipated, that Mr. Clarke at first suspected his arrangement to be defective. As usual, his quantity-inductor was furnished with a short coil of thick insulated copper wire, and the intensity inductor with 15,375 yards of fine copper wire. On trying the voltameter with the intensity arrangement, to his astonishment, no decomposition took place, though the shock obtained by it was most excruciating, nay, even dangerous. Mr. Clarke next tried the decomposing power of the quantity inductor, and obtained one cubic inch of gas in four minutes. This being a novel fact, he was induced to imagine that the cause might be traced to a compound action produced by the rotation of the two inductors ; and he therefore determined to arrange the magnets similarly to those of the machines he had been in the habit of constructing, the only difference consisting in the size of the instrument, and in the means of communicating motion to the inductors, namely, by a crank and treadles, similar to a lathe. The battery consisted of 106 cast steel bars, each four feet long, the whole weighing 156 lbs. Ivory was made use of to retain the wires on the conductors in lieu of brass plates, which gave uncertain results, owing to their conducting property. The novel results of the experiments were—*first*, the great amount of gas obtained from the quantity-inductor ; in one instance, one cubic inch in one minute and a-half, which result confirmed the correctness of the original arrangement. Second, the trifling decomposing effect from the intensity-inductor, as appeared from the table accompanying the paper read by Mr. Clarke on the occasion. The voltameter employed in the experiments was furnished with two slips of platinum, one inch in length, and three-eighths of an inch in breadth ; the decomposing power

was, however, increased at least fifteen times by the substitution of
fine pointed wires of platinum. The next experiments referred to the
different appearance of the spark with different modifications of the
inductors. With the intensity inductor, a long, noiseless, straggling
spark was obtained, having much resemblance to the spark which
passes from the prime conductor of an electrical machine to a body
placed at what is called the striking distance. The quantity-inductor
gave a spark which not only had the usual stellar form, but was accompanied with a loud snapping resembling the discharge of a Leyden jar.
Although these distinctions existed between the sparks, they both
appeared equally luminous. The experiments were most brilliant."

(672) An application of the magneto-electrical machine to the art
of *electro-plating* has been made and patented by Mr. J. P. Woolrich,
of Birmingham.* In the 38th volume of the Mechanics' Magazine,
page 146, will be found a full and illustrated account of the machine
which he employs, and of the method of conducting the process. Mr.
Woolrich makes use of sulphate of potash as the solvent for the gold,
silver, and copper, with which the articles are plated; and he gives in
his specification a detailed account of the method of preparing the
different " liquors." The thickness of the metallic coating to be deposited, depends on the time during which the article is submitted to the
operation of the magnetic apparatus and solutions; a thin coating will
be deposited in a few seconds, whilst to obtain a thick coating the
article must be submitted to the constant operation of the magnet and
solution for several hours. The distance at which the poles of the
magnet should be placed from the ends of the armature will depend on
the superficies of the article to be coated; the larger the superficies,
the nearer must the magnet be placed to the armature; and the smaller
the article, the greater must the distance be increased, the distance being
inversely as the superficies of the article to be coated. If the surface
of the article under the operation of coating becomes, while in connexion with the magnetic apparatus, of a brownish or darkish appearance, or if gas be evolved from its surface, the distance between the
poles of the magnet and the ends of the armature must be increased
until the metal contained in the solution is properly deposited. Mr.
Woolrich states† that he has introduced such improvements into the
magneto-electric machine, that at the cost of about 15*l*. he can construct an apparatus which is capable of depositing 60 ounces of silver
per week, and he hopes still to reduce its cost considerably.‡

---

* Patent, dated Aug. 1st, 1842. Specification enrolled, Feb. 1st., 1843.

† Mech. Mag. Vol. xxxvii. p. 15.

‡ It is proper to mention that in the early part of 1842, Mr. Sturgeon published
a work in which the following passage occurs. " It is now more than seven years

(673) A very few words will suffice to explain the theory of the mag-neto-electrical machine, as at present understood; as often as the bent ends of the armatures or inductors F, F, F, Figs. 249, 250, 251, are brought by the rotation of the wheel opposite the poles of the magnet, they become by induction, magnetic; but they cease to be so when they are in the position shown in Fig. 249, viz. at right angles to it. Now, we have seen (650) that at the moment of the *induction* as well as of the *destruction* of the magnetism in an iron bar surrounded by copper wire, opposite currents of Electricity are induced in the wire if the circuit be complete; the points k, Fig. 249, are therefore so arranged that they shall leave the mercury and thus break the circuit in the wire surround-ing the armature F, at the moment that its ends become opposed to the poles of the magnet; for which purpose they must be placed nearly at right angles to it : the circuit is thus broken at the precise moment that a rush or wave of Electricity is determined in the wire, and hence the electrical effects that are obtained. As, however, the currents alternate in opposite directions, we cannot obtain the oxygen and hydrogen gases from decomposed water separate when the full power of the machine is employed; but by causing the wire p, Fig. 254, to rub on the break-piece z, instead of on the cylindrical part, one-half of the power of the machine is destroyed, but as the induced electrical current will be in one uniform direction, the usual results of polar decomposition may be obtained.

(674) *Secondary currents.*—If a small pair of voltaic plates be moderately excited, and a small short wire used to connect its mercury cups, no spark, or only a very minute one, will be perceived, either on making or breaking contact. As, however, the length of the connect-ing wire increases, the spark becomes proportionally brighter, "*until, from extreme length, the resistance offered by the metal as a conductor, begins to interfere with the principal result.*"[*] If two equal lengths of wire be taken, and one made up into a helix, and the other laid out on the floor, and each used to connect the mercury cups of a small battery, very great difference will be observed in the size of the spark afforded by each, on breaking contact. Supposing the length of each to be sixty feet, the wire laid on the floor will give a small bright spark, while the wire wound into a helix, will produce a large brilliant spark

ago that I contrived a magnetic-electric machine, by means of which I coated metals with tin, copper, &c.; and I have employed the same machine to advan-tage in gilding, silvering, and platinizing various kinds of metal of inferior value; and I have no doubt that in this capacity the magnetic-electric machine may be-come generally useful. I have produced good electrotypes on a small scale by its employment."

* Faraday's Experimental Researches, 1067.

accompanied by a snap.   Again, to render the fact still more decisive, take one hundred feet of covered wire, and bend it in the middle, so as to form a double termination, which can be communicated with the electrometer ; wind one half into a helix, and let the other remain in its extended condition ; use these alternately as the connecting wire, and the helix will be found to give by far the strongest spark.

(675) The spark and snap are much increased when a bar of soft iron, or what is better, a bundle of iron wires, are introduced into the axis of the helix ; but it is only on breaking contact with the battery that the effect is produced ; the reason is, that the iron, magnetised by the power of the continuing current, loses its magnetism at the moment the current ceases to pass, and in so doing tends to produce an electric current in the wire round it.

(676) These effects are evidently dependent on some affections of the current in the conducting wire, and the spark produced when the cups of the electrometer are connected by a *short* wire, is the only one that can be considered as produced by the direct power of the battery ; that the increase of the spark when the wire is lengthened does not depend on any thing analogous to *momentum* in the Electricity circulating through it, and consequently producing effects at the instant the current is stopped, is proved by the fact that the same length of wire produces the effects in very different degrees, according as it is simply extended or made into a helix, or forms the circuit of an electro-magnet.   How then is it to be accounted for?   The ingenuity of Dr. Faraday has provided an answer.

(677) In his ninth series of Experimental Researches, he has shown, that if a current be established in a wire, and another wire forming a complete circuit be placed parallel to the first, at the moment the current in the first is stopped, it induces a current in the same direction in the second, the first exhibiting then but a feeble spark ; but if the second wire be away, disjunction of the first wire induces a current in itself in the same direction, producing a strong spark.

(678) The strong spark in the single long wire, or helix, at the moment of disjunction, is, therefore, the equivalent of the current which would be produced in a neighbouring wire, if such second current were permitted.   Viewing the phenomena, therefore, as the results of the induction of electrical currents, the different effects of short wires, long wires, helices, and electro-magnets, may be comprehended. If the inductive action of a wire a foot long, upon a collateral wire also a foot in length, be observed, it will be found very small ; but if the same current be sent through a wire fifty feet long, it will induce in a neighbouring wire of fifty feet a far more powerful current, at the moment of making or breaking contact, each successive foot of wire

adding to the sum of action ; and by a parity of reasoning a similar effect should take place when the conducting wire is also that in which the induced current is formed ; hence the reason why a long wire gives a brighter spark on breaking contact than a short one, although it carries much less Electricity.

(679) If the long wire be made into a helix, it will then be still more effective in producing sparks and shocks, on breaking contact ; for by the mutual inductive action of the convolutions, each aids its neighbour, and will be aided in turn, and the sum of effect will be greatly increased.

(680) On repeating, soon after their publication, the beautiful experiments of Faraday, which led him to the above conclusion, the author was induced to try the effects of strong and weak electrical currents, on a long flat coil, of considerable breadth of surface. For this purpose, two sheets of thin copper, each four feet and a-half long, and twenty-six inches wide, were cut into ribbons one inch wide ; all the lengths were soldered together, and formed into a single coil, with list intervening. A continuous coil of copper ribbon, two hundred and thirty-four feet long, was thus provided. At the commencement of the coil, and at intervals of twenty-five feet through its whole length, wires were soldered, which projected about two inches, and supported small cups to contain mercury. By this arrangement, the current could be sent through any length of ribbon, from two hundred and thirty-four to twenty-five feet, and by the aid of the mercury cups, the effect produced on one part of the wire, by the action of an electrical current sent through any other part, could be examined. In describing some of the experiments made with this coil, which seem to bear on the present subject, we shall distinguish the cups by figures, indicating their position on the coil, thus :—

1. 2. 3. 4. 5. 6. 7. 8. 9. 10.

corresponding to   25. 50. 75. 100. 125. 150. 175. 200. 234. feet.

Exp. 1. When communication was made with the positive end of a pair of plates, contained in a pint cup, and excited by dilute sulphuric acid, and cup 1, and the wire from the negative end, dipped in succession into 2, 3, &c., to the end of the arrangement, the spark and snap increased in intensity to 4 ; at 5, it appeared the same, and afterwards went on decreasing to 10, where the spark was not nearly so large, nor was the snap nearly so loud : the maximum effect, therefore, seems to be produced with the battery, when the current has traversed between 75 and 100 feet.

Exp. 2. When a half-pint battery was employed, no difference in the size of the spark could be perceived in 4, 5, 6, 7, 8 ; but it was larger and brighter in these cups than in 2, 3, and 9, and 10, in which it seemed about the same : with a pair of plates, containing about one-fourth the surface of metal, no difference in the size or appearance of the spark, could be perceived throughout the whole arrangement, after cup 2, being as bright at 10, as at 3.

Exp. 3. When a pair of plates, each two inches square, was employed, the spark seemed brighter at 9 and 10 than either 4, 5, or 6 ; with a pair one inch square, the difference was more marked ; with a pair half an inch square, it was feeble in 2, 3, 4, after which it went on increasing, and at 10 it was much larger ; with a pair a quarter of an inch, a slight snap was several times heard accompanying the spark in the last three cups ; but the sparks produced in the first six cups were decidedly smaller, and less bright.

Exp. 4. A pair one-eighth of an inch square was then tried : sparks were produced in all the cups ; feeble in 2, and 3, but in 7, 8, 9, 10, bright ; largest and brightest in 10.

Exp. 5. Strips of copper and zinc, half an inch long, and one-tenth of an inch wide, were immersed in the acidulated liquor, and connected with the coil : sparks were obtained in all the cups, bright in 8, 9, 10, ; the strips were then cut in half, and by rapidly breaking contact, sparks were obtained in 5, 6, 7, 8, 9, 10 ; they could only occasionally be got from 4, and not at all from 2 and 3. The strips were then reduced to about one-sixth of an inch long, and one-twentieth wide ; in 9 and 10 several sparks were obtained, but none could be got from any of the other cups.

Exp. 6. A large calorimotor, highly excited with nitrous acid, was then tried ; the brightest and loudest spark was at 2 ; the snap was very loud, and could be distinctly heard in a room, at the bottom of a flight of stairs, the door being shut, and also the door of the room in which the experiments were made ; it was as loud, every time contact was broken, as the explosion from a quart Leyden phial ; the sparks were very vivid, and evolved copious fumes of mercury : these effects rapidly diminished as the connecting wire approached the termination of the coil, and at 8, 9, 10, the sparks were not more brilliant, nor louder, than those produced by the half-pint battery.

Exp. 7. The shock with this apparatus did not increase with the spark : when a brass conductor was grasped by the hand, and kept permanently in 1, and another held in the left hand, and dipped successively into each of the other cups, in which contact with the battery was rapidly broken by an assistant, the large calorimotor being employed, the following were the results :—

Exp. 8. The left hand conductor being in cup 2, and contact with the battery broken in it, the shock was very slight ; in 3 it was stronger, and went on increasing to 10, where it was strong enough to be felt at the elbows; now the spark and snap were most intense in 2, and least in 10; hence the shock appears to be inversely as the spark.

Exp. 9. The left hand conductor was then dipped into the cup *next* to that in which contact was broken, being out of the circuit of the current, as for instance in 3, while contact was broken in 2; the shock went on increasing, as before, to the end of the arrangement; it was then kept permanently in 6, while contact was broken in all the other cups : in 2, 3, 4, the shocks were distinctly felt, and went on increasing to 9.

Exp. 10. The current was then passed from 2 to 9, while the conductors were held in 1 and 10, then from 3 to 8, then from 4 to 7, then from 5 to 6, shocks were felt in all cases; strongest when the current was from 2 to 9, and weakest when from 5 to 6.

Exp. 11. The wires from the battery were then connected with 1 and 2, and the conductors held in 3 and 4, then in 1 and 3, while the conductors were in 4 and 5, 1 and 4, 5 and 6, &c.; but no shocks were felt, even when the wires from the battery were in 1 and 8, and the conductors held in 9 and 10.

Exp. 12. Although, however, no shock could be obtained *directly* from this arrangement, yet the existence of a secondary current was easily proved by the galvanometer, when the positive wire from the battery was in 1 and contact broken by the negative wire in 2, one wire of the galvanometer being in 3, and the other in 10, the needle was strongly deflected in a direction indicating the passage of a current in the same direction as the inducing current, that is, from 3 to 10. When the needle had taken up its position, it was retained in it by a pin; contact was then broken in 2, and the needle was immediately deflected in a contrary direction, showing now the passage of a current from 10 to 3.

Exp. 13. By employing Callan's coil, (an arrangement to be described presently), powerful shocks were obtained from the secondary current; the wires from the large calorimotor being in 1 and 2, and the terminal wires of the primary coil in 3 and 10, shocks felt at the elbows occurred every time contact was broken, and when this was done rapidly by a revolving wheel, or by Ritchie's rotating magnet, the succession of shocks was almost intolerable; and when one of the wires of the primary coil was dipped in the same cup in which contact with the battery was broken, the shocks were very violent, even with a half-pint battery. When a wheel was employed to break contact, the scintillations were very brilliant, when it was connected with the first four cups of the coil. The well known optical illusion, of a body in rapid motion appearing stationary, was beautifully shown when the room was darkened, and the large battery used. Without the intervention of Callan's coil, the shocks obtained by breaking contact, by means of the wheel in the arrangement (8.), were not so strong as was expected. The most efficient method in this case was to draw the end of the connecting wire rapidly over the edges of the ribbon, from the centre to the circumference.

Exp. 14. Secondary shocks were obtained from the coil *immediately ;* i. e., without the intervention of Callan's apparatus, by dipping the conductors grasped by the hand, in 10, and the negative cup of the battery, while contact was broken in 2, 3, 4, 5, 6, 7, 8, and 9 ; in the last three, the shocks were as strong as could be given with any other arrangement of the apparatus, and when Callan's coil was interposed, very severe, even when contact was broken by 2, and a half-pint battery employed.

Exp. 15. The wire from the zinc end of the battery was then kept permanently in 10 , and contact with the copper end broken in 9, 8, 7, 6, 5, 4, 3, and 2, the left hand conductor being in 1, and the right in the positive cup. Shocks were felt in all ; slight in 9 and 8, and strong in 4, 3, 2.*

(681) These results agree very closely with those obtained by Professor Henry, of New Jersey, Princeton, to whom we are indebted for a very elaborate investigation of the phenomena of the induction of galvanic currents, and for the discovery of analogous results in the discharge of ordinary Electricity.

(682) It was found by this experimentalist that when the length of the coil is increased, the battery continuing the same, the deflagrating power decreases, while the intensity of the shock continually increases, but that there is a limit to the increase of the intensity of the shock ; and this takes place when the increased resistance or diminished conduction of the lengthened coil begins to counteract the influence of the increasing length of the current. When the intensity of the battery is increased, the action of the short ribbon coil decreases, but it is surprisingly increased when the length of the coil is increased in proportion. Thus Dr. Henry found that the current from a battery of 10 pairs of Cruickshank's trough, which, when sent through the ribbon,

---

* After these experiments were made, the ribbon was cut down the middle, and soldered together so as to form a length of about four hundred and sixty feet, half an inch wide, which was insulated and wound as before, into a spiral : the maximum spark, with this arrangement, a pint battery being used, was at about one hundred and fifty feet ; and with a very small voltaic pair, the largest and brightest spark was still at the extremity of the spiral ; on the whole, the effect of dividing the copper ribbon was to diminish the *size* of the spark, but greatly to increase the intensity of the shock, which, when obtained from the last cup, and the negative end of the battery, was very severe. Water was readily decomposed by the secondary current, developed by this arrangement.

a, *b*, or *c*, Fig. 256, produced scarcely any effect ; when passed through a spool of copper wire $\frac{1}{16}$ of an inch in diameter, and *five miles* long, gave shocks too strong to be taken through the body ; and that a battery composed of six pieces of copper bell-wire $1\frac{1}{2}$ inch long, and an equal number of pieces of zinc of the san.e size, was capable, through the medium of this long spool, to give shocks at once to *twenty-six persons* joining hands ; though when a simple battery, exposing a zinc surface of one square foot and three-quarters was employed, no shock, or at most a very feeble one only could be obtained.

(683) Fig. 256 repre-sents the method by which battery-contact was broken and renewed in Professor Henry's experiments : *a* is the ribbon coil about 100 feet long ; *d*, a rasp, the end of which communicates with the zinc cylinder of the battery through the medium of a cup of mer-cury ; one end of the ribbon is placed perma-

FIG. 256.

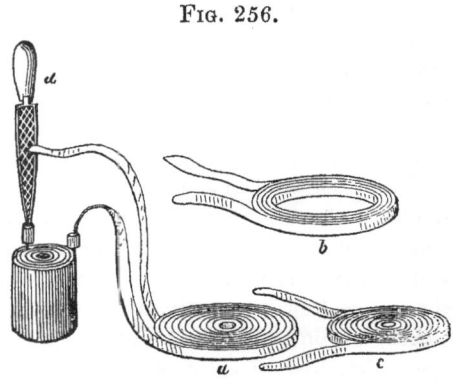

nently in the cup connected with the copper element, and by drawing the other end smartly over the surface of the rasp, a series of brilliant sparks are produced, and the electrical current through the coil is rapidly broken and renewed.

(684) Now on placing coil *c* containing about sixty feet of insulated copper ribbon on coil *a*, a plate of glass being interposed, and sending the electrical current from the battery through *a*, it was found that as often as the circuit was interrupted, a powerful secondary current was induced in *c*, and that when the ends of this coil were rubbed together, sparks were produced ; when a small coil of wire enclosing a needle was interposed, the needle became magnetic ; when a small horse-shoe of soft iron, surrounded by a coil of wire, was interposed, magnetism was developed : when the ends of the coil were attached to a small decomposing apparatus, gas was given off at each pole ; and when the body was interposed, a shock, though a feeble one, was expe-rienced.

(685) When, instead of a ribbon coil, a helix containing two thousand six hundred and fifty yards of fine insulated wire was placed on coil *a*, the magnetising effects disappeared, the sparks were smaller, and the decomposition less ; but the shock even from a single rupture of the battery current was sufficiently powerful to be sent through fifty-six

persons, and was too strong to be received with impunity by a single individual. When a helix containing 1500 yards of wire $\frac{1}{125}$ of an inch in diameter was placed on coil a, scarcely any effects could be obtained ; nevertheless when the ribbon coil was made into a ring as shown in b, Fig. 256, and the long spool of wire placed in the centre, shocks sufficiently intense to be felt at the shoulder, when passed only through the forefinger and thumb, were obtained, sparks and decomposition were produced, and needles were rendered magnetic.

(686) By these experiments it is proved that the induced current is diminished by increasing the *length* of the wire, and that the more so in proportion as the *size* of the wire is diminished ; it is also shown that when a ribbon coil is employed, the induced or secondary current has the properties of one of considerable *quantity* but of moderate *intensity :* and that when a helix of small copper wire is used, the secondary current is one of small *quantity*, but of great *intensity :* the difference between the currents induced in the two cases is precisely similar to the difference which we notice between the Electricity from a voltaic battery consisting of a single pair of plates, and that from a voltaic arrangement consisting of a number of pairs in series weakly charged, as from the water battery (248). The fact that the induced current is diminished by a further increase of the wire after a certain length has been attained is, as Dr. Henry observes, important in the construction of the magneto-electrical machine, since the same effect is produced in the induction of magnetism. The wire on the inductor of the machine may therefore be of such a length relative to its diameter as to produce shocks but no decomposition ; as we have seen was the case in the large magneto-electrical machine constructed by Mr. Clarke (671) ; and if the length of the coil be still further increased, the power of giving shocks may also become neutralized, from the increased resistance offered to the induced current by the great length of wire : a good idea of the difference between *quantity* and *intensity* may be formed by supposing that in the first there is a large current moving slowly without any or but small resistance, and in the second a small current moving with great velocity, but still having a resistance to overcome.

(687) The experiments of Professor Henry, relative to the induction of secondary currents at a distance, are exceedingly striking. By sending an intermitting current of Electricity through the spiral a, Fig. 257, and placing the helix of thin wire b over it, a plate of glass being interposed, shocks may be obtained on grasping the handles attached to the

Fig. 257.

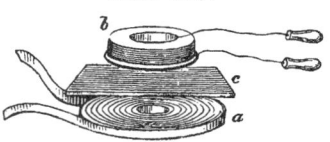

coil, when *a*, consisted of about 300 feet of copper ribbon, $1\frac{1}{2}$ inch in width, and *b*, a helix of copper wire five miles long: Dr. Henry found that shocks might be obtained when the coils were *four feet apart;* and at a distance of twelve inches, they were too strong to be taken through the body : the Professor also mentions a very instructive method of exhibiting these astonishing experiments, which the author has frequently adopted in the lecture room, viz. : to cause the induction to take place through the partition-walls of two rooms, for which purpose a coil about 100 feet of ribbon is suspended against the wall in one room, while a person in the adjoining room receives the shock by grasping the handles of a helix, of about 300 yards of thin wire, and approaching it to the spot opposite to which the coil is suspended. The effect is as if by magic, without a visible cause. It is best produced through a door or thin wooden partition.

(688) When an intermitting electrical current is sent through the ribbon coil *a*, Fig. 258, a secondary current is, as we have seen, induced

FIG. 258.

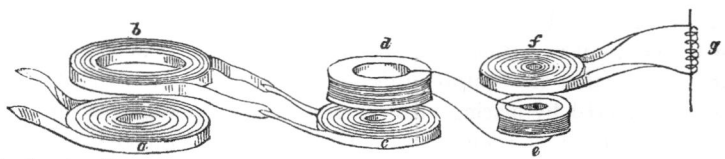

in *b*, placed at a distance above it : Dr. Henry further ascertained that this induced secondary current, by passing through the coil *c*, was capable of inducing a current of the *third order* in the helix *d ;* and that this coil again, by passing through the helix *e*, induced a current of the *fourth order* in the coil *f*, as was proved by the power possessed by this coil of magnetizing the needle *g ;* he further determined that there existed an *alternation* in the direction of the currents of the several orders, commencing with the secondary,—it was as follows :

Primary current .... +
Secondary current .... +
Current of the third order .... —
Current of the fourth order .... +
Current of the fifth order................... .... —

From our previous knowledge of this subject, the induction of currents of different orders, of sufficient intensity to give shocks, &c., could scarcely have been anticipated. The secondary current consists, as it were, of a single wave of the natural Electricity of the wire, disturbed but for an instant by the induction of the primary : yet this has the power of inducing another current, but little inferior in energy to itself ; and thus produces effects apparently much greater, in proportion to the quantity of Electricity in motion, than the primary current.

(689) In Fig. 258, the current induced in f by the helix e, is one of *quantity* ; the effects, however, of the induced tertiary current in d would be those of *intensity* ; and by grasping metallic handles, attached to the ends of that helix, shocks may be received : thus a quantity current can be induced from one of intensity and the converse.   Dr. Henry found that on interposing a screen of any conducting substance, between a and b, no secondary currents could be obtained ; a circular plate of lead, for instance, caused the induction in b almost entirely to disappear; but when a slip of the metal was cut out in the direction of a radius of the circle, the induction was not in the least interfered with : again, the coil b being placed upon a with the two ends separated, and on the coil the helix d, shocks could be obtained from the latter as if the coil were not present ; but when the ends of b were joined, so as to form a perfect metallic circuit, no shocks could be obtained.   The explanation of this apparent mystery was at first obscure ; it was, however, subsequently referred by Dr. Henry to the changes in the direction of the induced currents : the secondary current, which is induced in the screening plate, or closed ribbon coil, is in the same direction as the current from the battery ;—it nevertheless tends to induce a current in the adjacent conducting matter of a contrary direction. A similar re-action, as it were, may be observed by placing on a flat ribbon coil, containing about 100 feet of metal, another similar coil, and then taking the shock from the first when the ends of the second are joined, the intensity will be found to be very materially diminished ; although, if the ends of the second coil be not joined, no difference in the intensity of the shocks will be perceived.

FIG. 259.

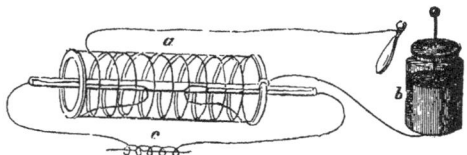

(690) By employing the arrangement shown in Fig. 259, Professor Henry succeeded in demonstrating that the discharge from the Leyden jar possesses the property of inducing a secondary current precisely the same as the galvanic apparatus. A hollow glass cylinder a, of about six inches in diameter, was prepared with a narrow riband of tinfoil, about thirty feet long, pasted spirally around the outside, and a similar riband of the same length pasted on the inside ; so that the corresponding spires of each were directly opposite each other. The ends of the inner spiral passed out of the cylinder through a glass tube, to prevent all direct communication between the two. When the ends of the inner riband were joined by the magnetizing spiral c, containing a needle, and a discharge from a half-gallon jar sent through the outer riband, the

needle was strongly magnetized in such a manner as to indicate an induced current through the inner riband, in the same direction as that of
the current of the jar. When a second cylinder, similarly prepared,
was added, a *tertiary* current was induced in the inner riband of the
second ; and by the addition of a third cylinder, a current of the *fourth*
order was developed.

(691) In all the experiments that were tried, the results with ordinary and galvanic Electricity proved to be similar. A most interesting
fact, however, came out in the course of the investigation : when the
Leyden experiments were made with the glass cylinders, the currents,
instead of *alternating*, as was the case in the galvanic experiments,
were all in the same direction as the discharge from the jar ; but when
the arrangement of coils and helices, Fig. 258, was used, the coils being
furnished with a double coating of silk, and the contiguous conductors
separated by a large plate of glass, the discrepancy vanished, and the
alternations were found the same as in the case of galvanism : thus the
cylinders gave currents all in one direction ; the coils in alternate directions. Dr. Henry made a great number of experiments, in order to get
some explanation of these apparently anomalous results ; and he at last
succeeded satisfactorily in tracing them to the *distances* of *the conductors*.
Thus two narrow strips of tinfoil, about twelve feet long, were stretched
parallel to each other, and separated by thin plates of mica to the distance of about $\frac{1}{50}$th of an inch.   When a discharge from the half gallon
jar was passed through one of these, an induced current in the same
direction was obtained from the other. The ribands were then separated by plates of glass, to the distance of $\frac{1}{20}$th of an inch,—the current
was still in the same direction, or *plus*. When the distance was increased to about one-eighth of an inch no induced current could be obtained ; and when they were still further separated the current again
appeared, *but was now found to have a different direction, or to be*
minus : no other change was observed in the direction of the current,
and the intensity of the induction decreased as the ribands were separated. Thus, when the conductors are gradually separated, there is, it
appears, a distance at which the current begins to change its direction,
and this distance depends on the amount of the discharge, and on the
length and thickness of the conductors. With a battery of eight half-
gallon jars, and parallel wires of about ten feet long, Dr. Henry found
that the charge in the direction did not take place at a less distance
than from twelve to fifteen inches ; and with a still larger battery, and
longer conductors, no change was found, although the induction was
produced at a distance of several feet.

(692) The reader, after the perusal of these beautiful experiments
of Dr. Henry, will find no difficulty in understanding the rationale

of the electro-magnetic coil machines, which we now have to present to his notice. It was Mr. Callan, of Maynooth College, who first contrived a convenient apparatus for the illustration of secondary currents. A coil of thick insulated copper bell-wire is wound on a small bobbin, and on a larger rod with a hollow axis, in which the bobbin may be introduced at pleasure, a length of about 1500 feet of thin wire is wound; the two coils are thus perfectly distinct from each other, and by sending the current from the battery through the interior coil, the Electricity present in the exterior coil is set in motion by its inductive influence; and from it both physiological and electrolytical effects may be obtained. The spark obtained on breaking battery-contact with the primary wire is generally fine and bright, in consequence of the *reflex wave* of Electricity which is generated in it; but if the ends of the exterior coil be brought into metallic contact, then little or no spark is seen when the primary circuit is broken, in consequence of the re-action of the induced current. If 100 yards of fine insulated copper wire be wound on a reel, and contact with an electrometer rapidly broken, shocks may be obtained by grasping metallic cylinders in connexion with the ends of the coil, from the reflex wave of Electricity which is generated; and if a bundle of iron wires be placed in the axis of the helix, the brilliancy of the sparks and the intensity of the shocks will be much increased in consequence of the *second wave* of Electricity which we have seen (673) is produced at the moment of the de-magnetization of the iron.

(693) Fig. 260 represents a very convenient little apparatus, devised by Dr. Golding Bird,* for rapidly breaking and renewing battery-contact with the primary coil. It is thus constructed : — A base board, about eight inches in length, and three in breadth, is furnished at both ends with a piece of hard wood, A and B, each having two holes excavated, for the purpose of holding mercury; each of these holes communicates by means

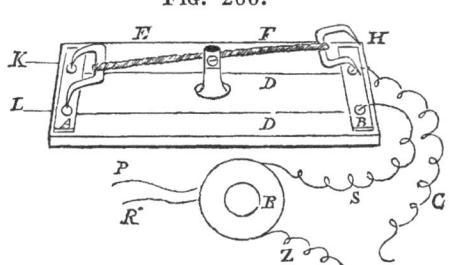

FIG. 260.

of thick copper wires, D D, with that opposed to it, in the other piece of wood at the opposite end of the board. Midway, between the receptacles for mercury, is a wooden support, so contrived, that a piece of soft iron wire, one-eighth of an inch in diameter, and five inches in length, may oscillate between the cheeks cut in its upper part, with as

* L. and E. Phil. Mag. vol. xii. p. 18.

little friction as possible. Around this iron wire, E F, are wound two helices of thin insulated copper wire, in the same direction, (from right to left), in such a manner, that the two ends of one helix may terminate in the copper points, G H, and the ends of the other helix in the points K L. Two small horse-shoe magnets, (not shown in Fig. 260, but shown in their proper places, in Fig. 261), are then fixed on proper supports, so

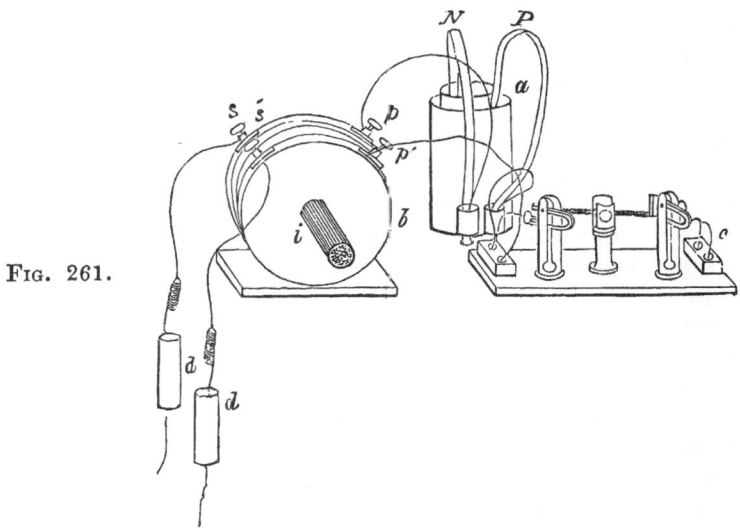

FIG. 261.

that they may each be placed near an end of the iron bar, E F, in a vertical plane just posterior to it, so that on depressing the end F of the bar, it may be opposite one pole (say the south) of one magnet, and consequently the end E will be opposite the other pole of the second magnet. On elevating the end E, the contrary will of course take place, and for this purpose, it is hardly necessary to say that the similar poles of the magnets should be in the same direction.

(694) From this description, it is evident that on connecting the cup of mercury in A or B, with the two plates of a single voltaic battery, the bar E F will become a temporary magnet, if the ends of either helix are allowed to dip in the mercury : and if connection with the battery be properly made, the ends or poles of the temporary magnet will be repelled by the poles of the permanent magnet, to which they are opposed ; the bar will consequently move, and so cause the immersion of the ends of the second helix in the other cups of mercury; repulsion will again occur, and so on ; about three hundred oscillations of the iron bar can thus be obtained in a minute.

The coil is thus arranged :—on a reel, with a hollow axis, three inches in length, are wound about sixty feet of copper wire, one-twelfth of an inch in diameter, covered with cotton thread, for the purpose of insulation, the two ends being connected with *p p*, Fig. 261, so that by means of the binding screws, communication may be made, by wires, with the contact-breaker and battery ; this is the primary coil ; over it, a second insulated copper wire, one-eightieth of an inch in diameter, and about one thousand five hundred feet in length, is wound, and the two ends are connected with *s s*, furnished with binding screws, for forming connection with wires, for communicating the shock, &c.

(695) The connection with the battery is best made in the way shown in Fig. 260, in which R represents a section of the reel, S an end of the short helix, connected with the mercury cup in B, Z the other end of the short helix, connected with one plate of the battery, whilst the wire C connects the other cup of mercury, in H, with the other plate of the voltaic couple. When this is properly arranged, a series of induced currents may be obtained from *s s*, the extremities of the long helix, capable not only of communicating a series of intense shocks, but of exerting powerful electrolytic action ; and when a bundle of soft iron wires is inserted in the hollow axis of the reel *i*, Fig. 261, the dynamic power of the coil is considerably increased. In this case, indeed, the sparks produced, when the ends of the helices round the iron bar, E F, leave the mercury, are very brilliant, accompanied by a loud snapping noise and a vivid combustion of mercury, clouds of the oxide of that metal being copiously evolved.

(696) If the ends P R', of the long and thin coil are furnished with platinum points, and immersed in water, acidulated with sulphuric acid, rapid electrolytic action ensues, torrents of minute bubbles of oxygen and hydrogen gases being evolved. If, instead of water, the points are pressed on paper moistened with iodide of potassium, electrolytic action ensues, iodine and oxide of potassium being separated. Solutions of neutral salts, such as sulphate of potash and soda, chloride of potassium, sodium, antimony, and copper, are also rapidly decomposed. In these experiments, as Dr. Bird remarks, it will be found that the great majority of the electro-positive elements (for example) appear constantly at one termination of the coil, *cæteris paribus*, but not all, for it must not be forgotten, that on making, as well as breaking contact with the electrometer, an induced current takes place in the long coil, though of far weaker intensity than the latter, to which it is opposed in direction, and consequently in electrolytic effects.

(697) If the brass conducting tubes *d d'*, Fig. 261, are grasped, even with the unmoistened hands, the intensity of the rapid succession of shocks will be found intolerable, even when the battery used consists but of two plates, presenting four or five square inches of surface ; and

with pairs half an inch, the shocks obtained, when contact is rapidly broken, (which in this case, is best done by a rotating magnet, motion being communicated by the fingers), are very disagreeable.

When the wires at the end of the conducting tubes, d d', are made to touch, small bright sparks are produced; and if while the oscillating bar of the contact-breaker is vibrating rapidly, a large calorimeter being employed, a piece of well-burnt charcoal is fixed on one of the terminations, and the other drawn lightly over it, a rapid succession of brilliant sparks is obtained. These sparks depend entirely upon the induced currents, as the fine coil has no connection with the calorimeter. For the exhibition of this, as well as of the electric light of an energetic arrangement, Mr. Bird finds pencils of that kind of artificial graphite, found lining the interior of the iron cylinders, used for the distillation of coal in gas manufactories, very far superior to box-wood, or indeed any other form of charcoal.

By connecting the ends of the primary coil of this arrangement, with the quantity inductor of the magneto-electrical machine, powerful shocks, and strong electrolytic effects are obtained; the spring must rub on the double break, which in this experiment performs the same office as Bird's contact-breaker; the coil, which, as we have seen, is, when revolving before the magnets, a powerful source of Electricity, supplying the place of the voltaic couple.

(698) There are several other forms of the electro-magnetic coil machine, and many other modes of breaking battery-contact with the primary coil. Fig. 262 is a very elegant arrangement; the primary coil consists of about thirty-five feet of insulated copper wire, No. 12, and the secondary of 1400 feet of copper wire, No. 20, battery-contact is broken and renewed by the rotation of the soft iron bar, h, which, mounted between two brass pillars, is situated immediately over the axis of the coil, in which is placed a bundle of iron wires; the electrical current from the battery passes through the pillar d, and the axis carrying the iron bar; and contact is broken and renewed by the point i dipping as h revolves into and out of the mercury contained in the brass cup, g, mounted on the brass pillar, a, through which the circuit is completed; communication with the voltaic battery is made through one pair of the binding screws on the base of the instrument; and the shocks, electrolytical effects, &c., are obtained from wires attached to the other pair.

Fig. (262.)

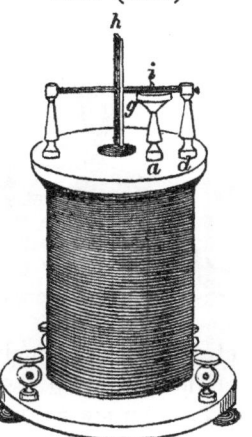

(694) Fig. 263 is another form of the instrument which possesses this advantage over the last, viz.,—that it does not require mercury. The current from the battery passes from the binding screw *p*, up the wire, *a*, which terminates in a small disc of iron, arranged immediately over

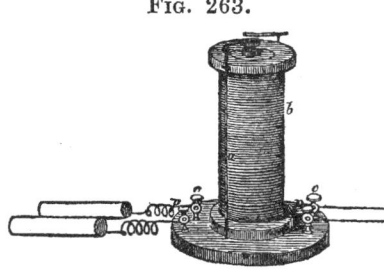

FIG. 263.

the bundle of iron wires in the axis of the coil, from which, however, it is prevented from coming immediately into contact when the machine is not in action by the horizontal spring by which it is connected with the wire *a*. The binding screw *c*, is connected with a wire, the top of which is seen in the figure rising above the coil. On the top of this wire is a horizontal strip of metal tipped with platinum, and with this by the action of the spring the disc of iron is kept in contact ; now when connection is made with the battery through the wires *p* and *c*, the central core of iron wires becomes magnetized, and consequently attracts the disc of iron, thus breaking battery contact ; the current being shut off, the disc of iron, is again raised by the spring, and thus contact is broken and renewed with amazing rapidity. The secondary effects are obtained from the handles attached to *p c*. For medical purposes this is decidedly the best arrangement of the coil machine, as it is more compact than any other, and dispenses with the use of mercury.

(695) In Fig. 264 we have a representation of the Rev. F. Lockey's

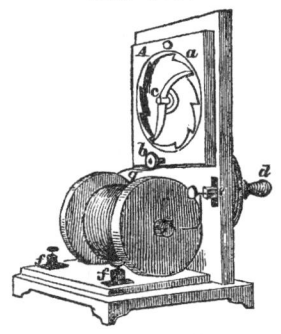

FIG. 264.

electro-magnetic coil machine, to which is attached an apparatus for producing luminous galvanic rings. The contact-breaker is the curved spring C, which is carried rapidly round by the multiplying wheel and handle *d*, striking in its course against the notches in the interior of the metallic circle *b*. This circle must have an odd number of teeth or notches in order that the ends of the *S* shaped spring may produce the spark at opposite parts of the ring ; when there are twenty-five or thirty-five breaks the resulting ring of sparks is exceedingly beautiful.

The diameter of the ring *a b* may be about five inches ; rings a foot in diameter produce very brilliant effects ; they may be made of different metals, and if corresponding springs be used, there will be a different light for each. The rings are secured in the circular rabbet of the square piece

of wood *A*, by small turn buttons; one end of the primary coil is in communication with the ring, the other is in connexion with the binding screw
e, where one of the battery wires is to be fixed. The spring *c* has metallic
communication with the other pole of the battery, by means of its metallic
socket, to which a wire is soldered and brought down to another connecting piece symmetrical with *e*, but not visible in the figure; a small
portion of this wire is seen at *g*; *f f* are the usual connecting pieces for
administering the shock.

(696) The *scintillating circle*, Fig. 265, is another of

Fig. 265.

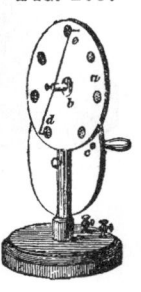

Mr. Lockey's instruments. Seven or any other uneven
number of fine files are procured, such as are used in
certain engravers' work; they are of a flat and circular form, about $\frac{3}{8}$ of an inch in diameter; they are
arranged and fixed upon an immoveable disc *a*; each is
in metallic contact with the other, and the whole with
the terminal wire of the battery, a powerful coil being
interposed. In the centre of the disc is the brass axis
b, carrying a small pulley, to which rapid rotation can
be communicated by the multiplying wheel *c*. This
pulley carries the steel wire *d*, *e*, pointed at its extremities, and bent at
such angles as lightly to drag over the faces of the files in rotation.
The wire is removable at pleasure to admit of its adjustment or replacement. The central axis has a wire leading to the other terminal of the
battery. On putting the apparatus in action, contact is made and
broken alternately at nearly opposite diameters of the disc. The effect
to the eye is a continuous circle of radiant and splendid scintillations.
This is perhaps one of the most delicately beautiful of electrical experiments. The effect is much improved if instead of separate pieces of
steel inserted in the brass ring, the *whole circle* is formed of steel either
cleaned off with a fine file and left purposely unpolished, or better still,
formed into a continuous and fine file over its whole surface.

(697) The *Water regulator*, Fig. 266, also contrived by Mr. Lockey,
is a useful piece of apparatus for the purpose of modifying the physiological effects of the galvanic shock obtained by the medium of the
self-acting *coil* or other source of power. The most powerful shock
can by this regulator be readily reduced to one in the mildest form.
In the medical administration of galvanic Electricity, this is a point of
some importance, as obviating the necessity of any adjustment of the
size of the battery, the depth to which its plates are immersed, &c. *h* is
a glass tube about five inches in length, capped at each extremity. This
tube rests upon, but is not fastened to, the base *g*, as it may sometimes be
desirable to attach the regulator *directly* to the connector of the coil, by the

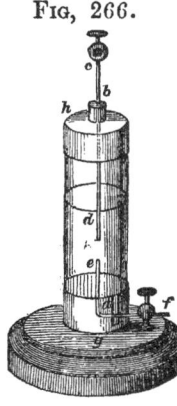

FIG, 266.

wire *f.* The fixed copper wire *e a f* bent to a right angle at *a*, passes out close to the bottom of the cap, and is fastened by the screw-connector *f.* This part is not quite correctly represented in the figure. The base *g* should be so shaped that the hole of the connector be on a level with the bottom of the cap *a g.* The copper connecting wire *c b d* passes through a stuffing-box in the centre of the upper cap *h* (which unscrews in order to introduce any convenient quantity of water). The apparatus is interposed in the circuit between one of the extremities of the coil and the person about to receive the shock ; and, according as the points of the wire *d* and *e* are approximated to, or separated from one another, will be the strength or the gentleness of the shock.

FIG. 267.

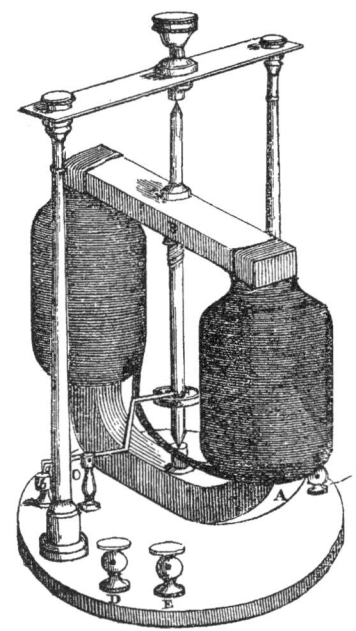

(703) In Fig. 267, we have an engraving of a very powerful arrangement of the electro-magnetic coil machine, made and presented to the author by Mr. W. T. Henley. A, a series of U-shaped bars of soft iron, bolted down to a base board, and wound with four coils of No. 14 covered copper wire to within an inch from either extremity ; over this is wound a thousand yards of No. 34 covered wire in one continuous length. B, the revolving armature which rotates between the poles of the magnet fixed on an axis, the lower end resting on a hard steel cap, the upper kept in its position by a screw passing through a flat piece of metal mounted on two brass columns. O, the apparatus for breaking contact, consisting of a small lever *a*, suspended on a pillar, one end dipping into a mercury cup, *b*, and the other end provided with a friction roller, running on an undulating wheel *c*, the prominent part of which, raising the end of the lever, dips the other end into the mercury, a spring *d* raising it out when the roller falls on the lower parts of the wheel. A break-piece formed of ivory and brass may be substituted for this, but the oxide of the metal formed by the spark is such an imperfect con-

ductor, that three cells have no more effect than one with the mercury. D E are the binding screws for forming connexion with the battery; the opposite ones are the ends of the secondary coil. On the same side of the base with the last (not seen in the wood-cut), is an ivory knob, which being turned, connects the ends of the secondary coil, either to diminish the primary spark, as the armature will then rotate for hours without burning the mercury, or to prevent the operator from receiving an unpleasant shock while adjusting the instrument. The ends of the thick wire are passed through the base, those from one pole soldered to E, the other to the mercury cup; the pillar C, and binding screw D, are connected together. This machine works very well with one or two cells of Smee's battery (273); and with an intensity series of eight, the secondary current is exceedingly powerful, the spark passing $\frac{1}{8}$ of an inch through air; with a battery of ten of Sturgeon's cast iron pots (286), the spark from the secondary coil passes $\frac{1}{5}$ of an inch through air, and brilliantly deflagrates gold and silver leaf; the shock would be far too powerful to be taken through the body, for when only two fingers are included in the circuit, it is sufficiently intense to be felt at the shoulders. With such a battery power the sparks from the primary coil are brilliant in the extreme; and from the ease with which the ends of the secondary coil are united and disunited, viz., by merely turning the ivory knob, the instrument is admirably adapted for demonstrating at the lecture table the induction and reaction of electrical currents; when the ends of the secondary are disunited, the sparks from the primary are large and brilliant; when united, they are small and faint.*

(704) *Thermo-Electricity.*—Hitherto, we have been treating of and endeavouring to show the connexion between Electricity as derived from four different sources. First, from friction under the head of Statical Electricity, and Electricity from effluent steam. Second, from chemical action under the head of Galvanism or Voltaic Electricity. Third, from some peculiar power possessed by certain animals, as

---

* A writer in the London, Edinburgh, and Dublin Phil. Mag., (Mr. J. E. Ashby,) recommends fine *iron* wire covered with cotton, as a substitute for *copper* in the secondary coils, and states that an increase rather than a diminution of effect is occasioned thereby, at less than one-sixth of the price, and with a great saving of space. Half a pound of this wire costs 1s. 3d., and measures nearly 1400 feet. With secondary coils so constructed, he has been able, he states, to make the magnetic spark pass through nearly one-hundredth of an inch between two wires, as in Mr. Crosse's experiment (249); and by means of a battery of about four square inches negative plate, and a length of only 1000 feet in the secondary, to excite a current in the primary coil. Mr. Gassiot, Mr. Ashby observes, used for the same purpose 2100 feet of copper wire, and twenty large cells of Daniell's battery. See L. and E. Phil. Mag., vol. xxi. p. 411.

the torpedo, and gymnotus, under the head of Animal Electricity. Fourth, from certain arrangements with magnets, under the head of Magneto Electricity. But there is a fifth remarkable source of Electricity remaining to be noticed, viz., that developed by the disturbance of temperature, which was first observed and made known by M. Seebeck, in communications to the Academy of Berlin, in the years 1821 and 1822, and to which the labours of intelligent experimentalists have drawn of late years an increasing attention in this country.

(700) The original observation of M. Seebeck, was, that electric currents can be excited in all metallic bodies by disturbing the equilibrium of temperature, the essential conditions being, that the extremities should be in opposite states as regards temperature. His apparatus was remarkably simple; it consisted of two different metals (antimony and bismuth were found the most efficient) soldered together at their extremities and formed into frames of either a circular or a rectangular figure. Electricity was excited by the application of heat to the places at which the metals were united and evinced by the disturbance of the magnetic needle, balanced on a point between the extremities. Fig. 268 shows the disposition of the apparatus; the needle is astatic. The best effect is produced by heating one of the compound corners by the flame of a spirit-lamp, and cooling the opposite corner by wrapping a few folds of filtering paper round it, and moistening it with ether. In Fig. 269, two frames composed of platinum and silver wires are represented delicately poised on the poles of a horse-shoe magnet, a spirit-lamp being placed between them, the flame of which causes the circulation of thermo-electric currents in the wires as evinced by their rotation round the poles of the magnet.

FIG. 268.

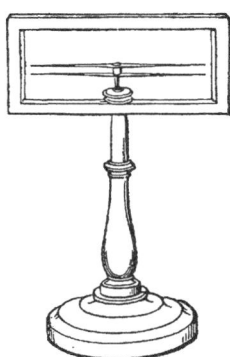

FIG. 269.

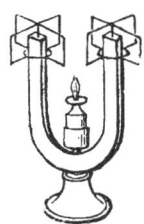

(701) Experiments have shown that the thermo-electric properties of metals have no connexion with their galvanic relations, or their capacity of conducting heat or Electricity, neither do they accord either with their specific gravities or atomic weights. In forming a thermo-electric series, it is desirable to combine an extreme positive with an extreme negative metal. The subjoined table, by Professor Cumming, exhibits the series, every substance being plus to that which precedes it, and minus to that which follows it.

Thermo electric series.	Volta series by acids.	Series of conductors, of Electricity.	of heat.
Galena	Potassium	Silver	Silver
Bismuth	Barium	Copper	Gold
Mercury ⎱	Zinc	Lead	Tin
Nickel ⎰	Cadmium	Gold ⎱	Copper
Platinum	Tin	Brass ⎬	Platinum
Palladium	Iron	Zinc ⎰	Iron
Cobalt ⎱	Bismuth	Tin	Lead
Manganese ⎰	Antimony	Platinum	
Tin	Lead	Palladium	
Lead	Copper	Iron	
Brass	Silver		
Rhodium	Palladium		
Gold	Tellurium		
Copper	Gold		
Silver	Charcoal		
Zinc	Platinum		
Cadmium	Iridium		
Charcoal ⎱	Rhodium		
Plumbago ⎰			
Iron			
Arsenic			
Antimony			

(702) Many trials have been made to construct thermo-electric piles, that would operate similar to the admirable instrument for which we are indebted to the genius of Volta. It appears that the labours of MM. Nobili and Melloni were first crowned with the greatest success. These two philosophers constructed conjointly, a thermo-electric pile, with which they have made some very interesting experiments on radiant heat. The pile was composed of fifty small bars of bismuth and antimony, placed parallel side by side, forming one prismatic bundle thirty millimetres* long, and something less in diameter. The two terminal faces were blackened. The bars of bismuth, which succeeded alternately to those of antimony, were soldered at their extremities to the latter metal, and separated at every other part of their surfaces, by some insulating substance, such as silk or paper. The first and last bars had each a copper wire which terminated in a peg of the same metal passing through a piece of ivory, fixed in a ring. The space between this ring and the elements of the pile was filled with some insulating substance. The loose extremities of the two wires were connected with the ends of the wire of a multiplier, which indicated

* A metre is 39·37 inches ; a diametre 3·9 inches ; a centimetre 0·39 inches ; and a millimetre 0·039 inches.

by the motion of the needle when the temperature of the farthest face of the pile was above or below that of the other.

In Fig. 270, we have a representation of this thermo-electric pile as arranged by Melloni for his experiments on radiant heat. *t*, a brass cylinder containing the compound bars, having the wires from the poles

FIG. 270.

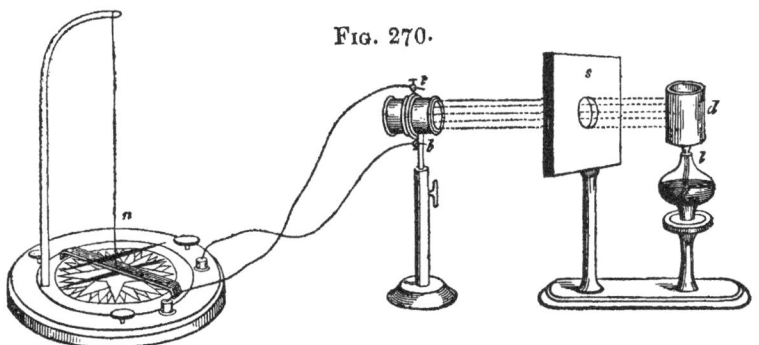

connected with the galvanometer or multiplier *n*. The extremities of the bars at *b* being exposed to any source of radiant heat, such as the copper cylinder *d* heated by the lamp *l*, while the temperature of the other extremity of the bars remains unchanged, a current of Electricity passes through the wires from the poles of the pile and causes the needle of the galvanometer to be deflected. The quantity of Electricity circulating, increases in proportion to the difference of the temperature of the two ends ; that is, in proportion to the quantity of heat falling on *b*, and the effect of this current of Electricity on the needle, or the deviation produced, is proportional to the quantity of Electricity circulating, and consequently to the heat itself—at least Melloni finds this correspondence to be exact through the whole arc from zero to twenty degrees when the needle is truly astatic.*

FIG. 271.

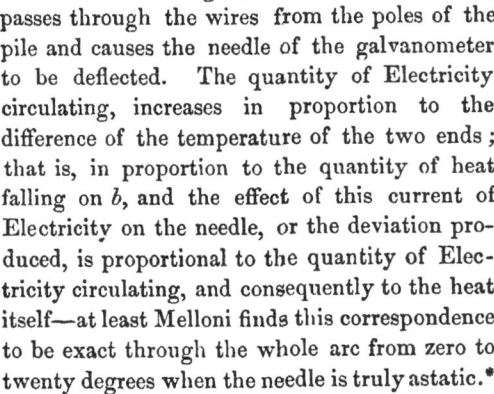

(708) Thermo-piles are now constructed by soldering together at their alternate edges, bars of antimony and bismuth, with squares of cardboard or thick paper intervening to prevent contact, the terminal metals being furnished with wires for the convenience of connexion.

Fig. 271 is a representation of a convenient form of the thermo-electric battery. It is composed of from 30 to 100 series of bars of antimony and bis-

* Graham's Chemistry.

muth soldered together at their extremities and placed in a metallic cylinder which is then filled with plaster of Paris, leaving merely the extremities of the bars exposed. The first bar of bismuth is connected with one mercury cup, and the last antimony bar with the other cup. The instrument is put in action by placing it in a vessel of ice and then laying the hot iron plate on the top.

(709) The first account we have of the production of a spark from a thermo-electric apparatus, appears in a communication from Professor Wheatstone to the London and Edinburgh Philosophical Magazine, vol. x. page 414. The following is the simple statement.

(710) The Cav. Antinori, Director of the Museum at Florence, having heard that Professor Linari, cf the University of Siena, had succeeded in obtaining the electric spark from the torpedo, by means of an electro-dynamic helix, and a temporary magnet, conceived that a spark might be obtained by applying the same means to the thermo-electric pile. Appealing to experiment, his anticipations were fully realized. No account of the original investigations of Antinori has reached, we believe, this country, but Professor Linari, to whom he early communicated the results he had obtained, immediately repeated them, and published the following additional observations of his own, in L'Indicatore Sanese, No. 50, Dec. 13, 1836.

1°. " With an apparatus consisting of temporary magnets, and electro-dynamic spirals, the wire of which was five hundred and five feet in length, he obtained a brilliant spark from a thermo-electric pile, of Nobili's construction ; consisting only of twenty-five elements, which was also observed in open day-light.

2°. " With a wire eight feet long, coiled into a simple helix, the spark constantly appeared in the dark, on breaking contact, at every interruption of the current ; with a wire fifteen inches long, he saw it seldom, but distinctly; and with a double pile, even when the wire was only eight inches long. In all the above mentioned cases, the spark was observed only on breaking contact, however much the length of the wire was diminished.

3°. " The pile, consisting merely of these few elements, readily decomposed water, within such restricted limits of temperature, as those of ice and boiling water. Short wires were employed, having oxidable extremities ; the hydrogen was sensibly evolved at one of the poles.

4°. " A mixture of marine salt moistened with water, and of nitrate of silver, being placed between two horizontal plates of gold, communicating respectively with the wires of the pile, the latter, after having acted on the mixture, gave evident signs of the appearance of revivified silver on the plate which was next the antimony.

5°. " An unmagnetic needle, placed within a close helix, formed by the wire of the circuit, being well magnetised by the current.

6°. " Under the action of the same current, the phenomenon of the palpitation of mercury was distinctly observed."

(706) The principal results here stated, were verified by Professor Wheatstone; he employed a thermo-electric pile, consisting of thirty-three elements of bismuth and antimony, formed into a cylindrical bundle, three-fourths of an inch in diameter, and one inch and one-fifth in length ; the poles of this pile were connected by means of two thick wires, with a *spiral of copper ribbon*, fifty feet in length and one inch and a-half broad, the coils being well insulated by brown paper and silk. One face of the pile was heated by means of a red-hot iron, brought within a short distance of it; and the other face was kept cool by contact with ice. Two stout wires formed the communication between the poles of the pile and the spiral, and the contact was broken when required in a mercury cup, between one of the extremities of the spiral and one of these wires. Whenever contact was thus broken, a *small but distinct spark was seen ;* it was visible even in day-light. Professors Daniell, Henry, and Bache, assisted in the experiments, and were all equally satisfied with the reality of the appearance.

At another trial, Professor Wheatstone obtained the spark from the same spiral, connected with a small pile of fifty elements, on which occasion Dr. Faraday and Professor Johnston were present. On connecting two such piles together, so that the similar poles of each were connected with the same wires, the same was seen brighter.

(707) Some experiments on the chemical action of the thermo-electric pile, were made *anterior* to those above described, by Professor G. D. Botto, of the university of Turin, with a different arrangement of metals; his experiments are published in the Bibliotheque Universelle for September, 1832. His thermo-electric apparatus was a metallic wire or chain, consisting of twenty pieces of platinum wire, each one inch in length, and one-hundredth of an inch in diameter, alternating with the same number of pieces of soft iron wire, of the same dimensions. This wire was coiled as a helix round a wooden rule, eighteen inches long, in such a manner that the joints were placed alternately at each side of the rule, being removed from the wood at one side, to the distance of four lines. Employing a spirit-lamp of the same length as the helix, and one of Nobili's galvanometers, a very energetic current was shown to exist ; acidulated water was decomposed, and the decomposition was much more abundant, when copper instead of platinum poles were used; in this case hydrogen only was liberated. The current and decomposition were augmented when the joints were heated more highly. Better effects were obtained with a pile of bismuth and antimony, consisting of one hundred and forty elements, bound together into a parallelopiped, having for its base a square of two inches, three lines, and an inch in height.

(708) For developing Electricity of feeble intensity, it is always best to employ a flat copper ribbon coil. Mr. Watkins found that he could always show a larger spark with it, than with an elongated wire coil and large temporary magnet; and that the snapping noise accompanying the thermo-electric spark was more discernible.

Mr. Watkins arranges one of the extremities of his pile of strong sheet copper, cut like a comb, and covered with soft solder; and when the moveable extremity of the flat coil is passed over the comb, and the thermo-electric pile in action, splendid sparks are seen every time the moving part of the coil breaks the circuit, by leaving a tooth of the comb. With a pile consisting of thirty pairs of bismuth and antimony, one inch and a-half square, and one-eighth thick, with the radiation of red-hot iron at one extremity, and ice at the other, a soft iron electro-magnet under the inductive influence of the Electricity thus generated, supported ninety-eight pounds weight. The same experimentalist states that he has thermo-electric piles in his possession, varying from fifteen to thirty pairs of metallic elements, which give brilliant sparks by simply pouring hot water on one end, while the other end is at the temperature of the atmosphere; and that sparks are exhibited by the same piles, when the temperature is reduced at one end by the aid of ice, while the other end is at the temperature of the surrounding air. In order to effect the decomposition of water, Mr. Watkins employs a massive thermo-battery, with pairs of bismuth and antimony, a small apparatus for the decomposition of water, of the ordinary description, and an electro-dynamic heliacal apparatus. The primary coil of wire is ninety feet long, and when the thermo-electric current simply pervades this coil, he does not notice any disengagement of the gases; but as soon as the contrivance for making and breaking battery-contact is put in action, then an evolution of the gases takes place, while at the same time powerful shocks are received from the secondary coil of wire one thousand five hundred feet long.

(709) From the interesting discovery made by Faraday, of the high conducting power of certain fused salts, for voltaic electricity, Dr. Andrews was led to imagine, that thermo-currents may be excited, by bringing them into contact with metals, and he succeeded in verifying this conjecture in the following manner:—

(710) Having taken two similar wires of platina, (such as are used in experiments with the blowpipe), and connected them with the extremities of the copper wire of a delicate galvanometer, he fused a small globule of borax, in the flame of a spirit-lamp, on the free extremity of one of the platinum wires, and introducing the free extremity of the other wire into the flame, he brought the latter, raised to a higher temperature than the former, into contact with the fused globule; the

needle of the instrument was instantly driven with great violence to the limit of the scale. The direction of the current was from the hotter platinum wire through the fused salt, to the colder wire. A permanent electrical current in the same direction, was obtained, by simply fusing the globule between the two wires, and applying the flame of the lamp in such a manner, that, at the points of contact with the fused salt, the wires were at different temperatures.

(716) Dr. Andrews also succeeded in obtaining chemical decompositions, by this peculiar thermo-current. A piece of bibulous paper, exposing on each side a surface of one-fourth of a square inch, was moistened with a solution of the iodide of potassium, and laid on a *platinum plate*, which was in metallic connection with one of the platinum wires used in the previous experiments. The extremity of the other platinum wire in contact with the globule, was applied to the surface of the bibulous paper, and the flame of the lamp was so directed, that the latter was the colder of the wires, between which the globule of borax, or carbonate of soda, was fused. The platinum plate in this arrrangement, therefore, constituted the *negative pole*, and the extremity of the wire applied to the bibulous paper, the *positive pole.** Accordingly when the circuit was completed, an abundant deposition of iodine occurred beneath the platinum wire. When a similar wire of platinum was substituted for the plate on the negative side, the effect was either *none* or scarcely perceptible.

(717) Dr. Andrews next formed a compound arrangement, by placing a series of platinum wires on supports, in the same horizontal line, and fusing between their adjacent extremities small globules of borax. The globules and wires were exactly similar to those that are used in blow-pipe experiments. A spirit-lamp was applied to each

---

* Dr. Andrews found that by using a *platinum wire*, exposing an extensive surface, as one pole of a voltaic pair, and a fine wire of the same metal as the other, he could effect the decomposition of water ; when, by employing a pair of similar platinum plates, or similar fine wires as poles, he could obtain no such result. After the evolution of gas has ceased, he finds that an additional quantity is procurable, either by increasing the surface of the broad pole, or by removing it, and heating it to redness, or by reversing the direction of the current. Dr. Andrews accounts for this, by supposing that when the poles exposed on both sides equal surfaces, the *gases were dissolved in the nascent state, by the surrounding liquid ;* but when the polar surfaces were unequal, the solution of the gas being greatly facilitated by the broader pole, the element of water separated there was dissolved, while the other element was disengaged in the gaseous state at the wire, which served as the opposite pole. In order, therefore, to discover, in case of difficulty, whether an electrical current is capable of decomposing water, or other substances, it is necessary to employ poles, having very unequal surfaces ; and this will be effected in the most perfect manner, by opposing a thick wire, or plate of platinum, to one of Wollaston's guarded points.—(See page 76, *et seq.*)

globule, so as to heat unequally the wires in contact with it; and the corresponding extremity of each wire being preserved at the higher temperature, the current was transmitted in the same direction through the whole series. By connecting the extremities of four cells of this arrangement with an apparatus for decomposing water, in which the opposite poles consisted of a thick platinum wire, and a guarded platinum point, (both being immersed in dilute sulphuric acid), very minute bubbles of gas soon appeared at the guarded point, and slowly separating from it, ascended through the liquid. They were obtained in whichever direction the current was passed, but rather more abundantly when the point was negative and the wire positive. With only two cells, similar bubbles formed in a visible manner on the guarded point, but in such exceedingly small quantity, that they did not separate from it. With an arrangement, containing twenty cells, a doubtful sensation was communicated to the tongue, when the poles were applied to it : but no spark was visible, although the current was passed through a helix of copper wire, surrounding a bar of iron, and the contact was broken with great rapidity, by means of a revolving apparatus. It is necessary to observe, however, that the lamps were unprotected, and that it was impossible to render the flames of such a number of spirit-lamps, burning near each other, so steady, as to heat at the same moment, in the required manner, all the globules and wires. With an enlarged and more perfect apparatus, Dr. Andrews thinks a spark might be obtained.

(718) Hence it appears, that an electrical current is always produced, when a fused salt, capable of conducting Electricity, is brought into contact with two metals, at different temperatures, and that powerful chemical affinities can be overcome by this current, *quite independently of chemical action*. The direction of the current is not influenced by the nature of the salt or metal, being always, from the hotter metal, through the fused salt to the colder ; its intensity is inferior to that of the hydro-electric current developed, by platinum and zinc plates, but greatly superior to that of the common thermo-electric currents, and is capable of decomposing, with great facility, water and other electrolytes. Dr. Andrews found also, that currents were produced *before* the salt becomes actually fused, but that their direction no longer follows the simple law before enunciated, but varies in the most perplexing manner, being first from the hot metal to the cold, then with an addition of heat, from the cold to the hot; and again, with a second addition of heat, from the hot to the cold.—(See Dr. Andrew's paper, in tenth vol., and 433 page, of the L. and E. Phil. Mag.)

(719) Since the phenomena of Thermo-Electricity seem to account in a very satisfactory manner for the general distribution of Electricity

and magnetism over the earth, the interest attached to this peculiar development of the subtil agent we have been engaged with, is exceedingly great. That the earth may be considered as a great magnet the
phenomena of the dip of the needle (544) sufficiently show : and the
facts connected with electro-magnetism, as set forth in the preceding
pages, lead to the conclusion that, when a magnetic needle is in its
natural position of north and south, there exist electrical currents in
planes at right angles to the needle descending on its east side, and ascending on its west side ; we must hence suppose that currents of
Electricity are constantly circulating within the earth, especially near its
surface, from east to west, in planes parallel to the magnetic equator.

(715) The cause of these electrical currents has been thus explained.* The earth, during its diurnal motion on its axis from west
to east, has its surface successively exposed to the solar rays, in an opposite direction, or from east to west. The surface of the earth, therefore, particularly between the tropics, will be heated and cooled in
succession, from east to west, and currents of Electricity on thermo-
electric principles will, at the same time, be established in the same
direction : now, these currents once established, from east to west, will,
of course, give occasion to the magnetism of the earth from north to
south. Hence the magnetic directive power of the earth, in a direction
nearly parallel with its axis, is derived from the thermo-electric currents
induced in its equatorial regions by the unequal distribution of heat
there present, and depending principally on its diurnal motion (see pars.
570, 571.) It does not belong to the question to consider whether the
phenomena of thermo-electricity actually depend on the decomposition
of heat, latent or sensible, as some suppose, the *facts* are well established, and the simple manner in which they account for terrestrial
magnetism, gives them a high degree of importance and interest.

(716) The whole subject cannot perhaps be better concluded than
with the following quotation, from the work of a highly distinguished
philosopher.† " These recent and beautiful discoveries show in the
most striking manner, that the operations of nature are more extraordinary, and indicate more of simplicity and wisdom of design in
proportion as they are better understood. By what simple expedients,
when known, are those wonderful phenomena of the earth's Electricity
and Magnetism produced, which formerly appeared so anomalous and
perplexing ? And what encouragement do these discoveries hold out
to us with respect to future discoveries, that may throw still further
light upon the operations of the great Architect of the universe ?"

---

* Prout's Bridgewater Treatise, p. 233.   † Ibid.

APPENDIX.

(*See Par.* 41, *et seq.*)

SINCE this portion of our work passed through the press, Dr. Faraday has published some further experimental demonstrations of his views relative to static electrical inductive action.* A pewter ice-pail, 10½ inches high and 7 inches diameter, was insulated and connected by a wire with a delicate gold-leaf electrometer; a round brass ball insulated by a dry thread of white silk was charged at a distance by a machine, or Leyden jar, and introduced into the ice-pail; the leaves of the electrometer immediately diverged with the same kind of Electricity as that with which the brass ball was charged : when the ball was removed the leaves collapsed. On introducing the electrified ball into the pail, the divergence of the electroscope increased, until the ball was about three inches below the edge of the vessel, after which the leaves remained steady and unchanged for any lower distance ; this shows that at that distance the inductive action of the ball is entirely exerted upon the interior of the ice-pail, and not in any degree directly upon external objects. If the ball be made to touch the bottom of the pail, *all* its charge is communicated to it ; there is no longer any inductive action between the ball and the pail, and the ball, upon being withdrawn and examined, is found perfectly discharged. Now when the ball is merely suspended in the pail, it acts upon it by *induction,* evolving Electricity of its own kind on the outside ; but when the ball touches the pail it communicates Electricity to it, and the Electricity that is afterwards on the outside may be considered as that which was originally on the ball. As this charge, however, produces no effect upon the leaves of the electrometer, it proves that the Electricity *induced* by the ball, and the Electricity *in* the ball, are accurately equal in amount and power.

Dr. Faraday then arranged four ice-pails, one within another, insulated by plates of shell-lac, on which they respectively stood ; with this system the charged ball acted precisely as with a single vessel, neither did it make any difference whether the three interior pails were insulated, or whether they were in metallic connexion, nor when for the pails a thick vessel of shell-lac or sulphur was substituted.

When in the place of one carrier, many carrier-balls in different positions were introduced within the inner vessel, no interference of one with the other was found : they acted with the same amount of force outwardly as if the Elec-

* L. E. & D. Phil. Mag. vol. xxii. p. 200.

tricity were spread uniformly over one carrier-ball, however much the distribution on each carrier may have been disturbed by its neighbours. Thus a
certain amount of Electricity acting within the centre of the ice-pail, exerts
exactly the same power *externally*, whether it act by induction through the
space between it and the pail, or whether it be transferred by conduction to the
pail so as absolutely to destroy the previous induction within.

A curious consideration arises from this perfection of inductive action.
" Suppose a thin uncharged metallic globe, two or three feet in diameter, insulated in the middle of a chamber, and then suppose the space within this globe
occupied by myriads of little vesicles or particles charged alike with Electricity
(or differently), but each insulated from its neighbour and the globe; their
inductive power would be such that the outside of the globe would be charged
with a force equal to the sum of *all* their forces, and any part of this globe (not
charged of itself) would give as long and powerful a spark to a body brought
near it, as if the Electricity of all the particles near and distant were on the
surface of the globe itself. If we pass from this consideration to the case of a
cloud, then, though we cannot altogether compare the external surface of the
cloud to the metallic surface of the globe, yet the previous inductive effects
upon the *earth* and its buildings, are the same; and when a charged cloud is
over the earth, although its Electricity may be diffused over every one of its
particles, and no important part of the *inductric* charge be accumulated upon
its under surface, yet the induction upon the earth will be as strong as if all
that portion of force which is directed towards the earth *were* upon that surface;
and the state of the earth, and its tendency to discharge to the cloud, will be
also as strong in the former as in the latter case. As to whether lightning-
discharge (concludes the learned professor) begins first at the cloud or at the
earth, that is a matter far more difficult to decide than is usually supposed ;
theoretical notions would lead me to expect that, in most cases, perhaps in all,
it begins at the earth."

(*See* 169—174.)

Since these paragraphs were printed, a valuable essay " on the Nature of
Thunderstorms," &c., has appeared from the pen of William Snow Harris, Esq.
The following remarks on the phenomena of lightning are so important and
interesting, that the author makes no excuse for transplanting them into this
Appendix.

" Arago divides the phenomena of lightning into three classes. In the first
he places those luminous discharges characterized by a long streak of light, very
thin, and well defined at the edges; they are not always white, but are sometimes of a violet or purple hue; they do not move in a straight line, but have a
deviating track of a zigzag form. They frequently divide in striking terrestrial
objects, into two or more distinct streams, but invariably proceed from a single
point. . . . Under the second class Arago has placed those luminous effects not
having any apparent depth, but expanding over a vast surface; they are frequently coloured red, blue and violet; they have not the activity of the former
class, and are generally confined to the edges of the cloud from which they
appear to proceed   The third class comprises those more concentrated masses
of light which he has termed globular lightning. The long zigzag and expanded

flashes exist but for a moment, but these seem to endure for many seconds; they appear to occupy time, and to have a progressive motion.

" It is more than probable that many of these phenomena are at last reducible to the common progress of the disruptive discharge, modified by the quantity of passing Electricity, the density and condition of the air, and the brilliancy of the attendant light. When the state of the atmosphere is such that a moderately intense discharge can proceed in an occasionally deviating zigzag line, the great nucleus, or head of the discharge, becomes drawn out as it were into a line of light visible through the whole track; and if the discharge divides on approaching a terrestrial object, we have what sailors call *forked lightning ;* if it does not divide, but exhibits a long rippling line, with but little deviation, they call it *chain lightning.* What sailors term *sheet lightning,* is the light of a vivid discharge reflected from the surfaces of distant clouds, the spark itself being concealed by a dense intermediate mass of cloud, behind which the discharge has taken place. In this way an extensive range of cloud may appear in a blaze of light, producing a truly sublime effect. The apearance termed *globular lightning* may be the result of similar discharges; it is no doubt always attended by a diffusely luminous track; this may, however, be completely eclipsed in the mind of the observer by the great concentration and density of the discharge, in the points immediately through which it continues to force its way, and where the condensation of the air immediately before it, is often extremely great. It is this intensely illuminated point which gives the notion of globular discharge : and it is clear, from the circumference of air which may become illuminated, the apparent diameter will often be great. Mr. Hearder, of Plymouth, once witnessed a discharge of lightning of this kind on the Dartmoor hills, very near him. Several vivid flashes had occurred before the mass of clouds approached the hill on which he was standing ; before he had time to retreat from his dangerous position, a tremendous crash and explosion burst close to him. To use his own words, " the spark had the appearance of a nucleus of intensely ignited matter, followed by a flood of light; it struck the path near me, and dashed with fearful brilliancy down its whole length, to a rivulet at the foot of the hill, where it terminated.' "

Appearances termed Fire Balls.—" A great deal has been said relative to these appearances, and some doubts have been entertained of their real existence as mere balls of electrical light. Nevertheless, the evidence of the existence of a form of disruptive discharge, faithfully conveying to the observer such an impression, is beyond question. A curious instance is given by Mr. Chalmers whilst on board the *Montague,* of seventy-four guns, bearing the flag of Admiral Chambers. In the account read at the Royal Society, (Phil. Trans. vol. xlvi. p. 366) he states that 'November 4th, 1749, whilst taking an observation on the quarter-deck, one of the quarter-masters requested him to look to windward : upon which he observed a large ball of blue fire rolling along on the surface of the water, as big as a mill-stone, at about three miles' distance. Before they could raise the main-tack, the ball had reached within forty yards of the main-chains, when it rose perpendicularly with a fearful explosion, and shattered the main-topmast to pieces.' In an account of the fatal effects of lightning in June, 1826, on the Malvern Hills, when two young ladies were struck dead, it is stated (Lloyd's Evening Post) that the electric discharge

F F

' appeared as a mass of fire rolling along the hill towards the building in which the party had taken shelter.'...... "It is by no means easy to explain these appearances on the principles applicable to the ordinary electric spark : the amazing rapidity with which this proceeds, and the momentary duration of the light, renders it almost a matter of impossibility that the discharge should appear under the form of a ball of fire; it would be a transient line of light:— we must look, therefore, to some other source for an explanation of these appearances.

"Now it is not improbable that in many cases, in which distinct balls of fire of sensible duration have been perceived, the appearance has resulted from the species of brush or glow discharge (described 67—75) which may often precede the main shock. In short, it is not difficult to conceive, that before a discharge of the whole system takes place, that is to say, before the constrained condition of the dielectric particles of air intermediate between the clouds and the earth becomes as it were overturned, the particles nearest one of the terminating planes or other bodies situate on them may begin to discharge upon the succeeding particles, and make an effort to restore the natural condition of the system by a gradual process.

" If therefore we conceive the discharging particles to have a progressive motion from any cause, then we shall immediately obtain such a result as that observed by Mr. Chalmers, on board the *Montague,* in which a large ball of blue fire was observed rolling on the surface of the water towards the ship from *to-windward.* This was evidently a sort of *glow discharge,* or *St. Helmo's fire,* produced by some of the polarized atmospheric particles yielding up their Electricity to the surface of the water. The clouds were in rapid motion : the discharging particles had motion towards the ship, the rate of which appears from the account, to correspond with the velocity of the breeze. On nearing the ship, the point of discharge became transferred to the head of the mast : and the striking distance being thus diminished, the whole system returned to its normal state, that is to say, a disruptive discharge ensued between the sea and the clouds, producing the usual phenomena of thunder and lightning, termed by the observers, the 'rising of the ball through the mast of the ship.' The fatal occurrence on the Malvern Hills, is another instance of the same kind. It is therefore highly probable that these appearances so decidedly marked as concentrated balls of fire, are produced by the glow or brush discharge, producing a St. Helmo's fire in a given point or points of a charged system previously to the more general and rapid union of the electrical forces : whilst the greater number of discharges described as globular lightning, are, as already observed, most probably nothing more than a vivid and dense electrical spark in the act of breaking through the air, which, coming suddenly on the eye, and again vanishing in an extremely small portion of time, has been designated a ball of light."

One more extract from this valuable work: speaking of the construction of lightning-rods, Mr. Harris says :—"Conductors for the protection of buildings are not unfrequently insulated in their course by means of glass, pitch, or some bad conductor, or otherwise are applied at short distances from the walls so as to interpose a stratum of air between them and the building. This practice is not only useless but disadvantageous, and is manifestly inconsistent with the principle on which conductors are applied, viz., that of providing a line of conduction to the earth, which, by a law of nature, the electrical discharge will follow in preference to any other. To imagine the possibility of its leaving such a line to move in a more difficult path through the building, is virtually to admit that we distrust the facts upon which our reasoning is founded. But if an electrical discharge is never found to have an easy line of transit, in order to pass upon matter out of such a line (186), the insulation of conducting rods is evidently unnecessary. If the conductor be placed at a short distance from the building, it is decidedly badly placed, since experience has shown that discharges of lightning are often uncertain, and determined through the air in other directions than that of the conductor. Hence we cannot apply a conductor too closely to the walls of a building which we are desirous to defend. Besides, it may be observed, that if by any law of electrical action the discharge could pass out of the line of the conductor, it is very unlikely to become arrested in its course by a few inches of any solid insulating substance, or by a few feet of air. A flash of lightning which can break through a distance of several hundred yards, in an atmosphere of the ordinary density, and shiver the most compact bodies into fragments, would scarcely be arrested by such insignificant means in any course it may be determined in. The principle of a lightning-rod is either absolutely true and in accordance with certain laws of nature; or it is a mere assumption unsupported by such laws: but it cannot be both.

" Whilst on this point it may not be out of place to notice an opinion prevalent with many, viz., that by placing a ball of glass on the projecting points of buildings and ships, lightning would not fall on them, glass being considered as a *repeller* of Electricity. Under this impression thick balls of glass have been in some instances placed on lighthouses and on ships' masts. Its application is, however, manifestly absurd on these grounds : the glass being merely a substance of low conducting power, and not dissimilar to the air in its electrical properties, it may in this sense be considered as a mass of air of unusual density. The employment of glass as a repeller of Electricity being founded altogether upon an assumption, and totally at variance with the first principles of electrical science, it is not without regret we find it employed in some of our public buildings of great national importance, as, for example, in some of our lighthouses."

(*See Par.* 175—209.)

Subsequent to the printing of this portion of our work, an excellent paper " On the difference between Leyden discharges and Lightning Flashes," &c. has been published in the " Proceedings of the Electrical Society," by the Editor of the Electrical Magazine. It would be impossible to do justice to this memoir, extending as it does over forty closely printed pages, in so brief an analysis of it as we could afford space to give in this Appendix. There are, however, one or two passages in immediate connexion with what has before been said, which, in justice to its author, we must not omit to extract for our readers. In a note in this work (see page 119), the following sentence occurs : " It is worthy of remark that Mr. Walker uses the expression, ' division of charge,' in preference to ' lateral discharge,' a very different affair, the existence of which was the principal topic of discussion between Mr. Harris

and Mr. Sturgeon." On this subject Mr. Walker remarks, " As I have now
had occasion to use the term 'lateral discharge,' it will not be out of place to
halt here, and at the expense of some digression, to develope the ideas represented by that term: I am especially called upon to do this, as Mr. Harris, in
allusion to the contents of my paper ' On the action of Lightning Conductors,'
says, ' he is quite led to modify the views originally entertained on the question, and so in his second page mixes up lateral discharge, about which there
was between myself and Mr. Sturgeon, a difference of opinion with ' division
of charge,' about which there really has never been any difference either with
myself or others." I have already expressed my opinion " that sparks will
pass from a lightning rod to vicinal conducting bodies," but whether this idea
be to be designated by one name or the other, I do not care to determine ; its
nature, which I hope to shew, would indicate a preference for the latter name
.... (Mr. Walker then quotes from Daniell and Faraday, to show that these
electricians consider the terms somewhat synonymous) he then proceeds—
" now Mr. Harris admits that if a passing discharge of lightning on its way
down the rod A B (see Fig. 82), meets with less resistance at e or c, than in
the conductor A B, it would divide amongst them ; is it not then equally clear
that it would still pass by them, if half the said resistance were at e and c, and
the other half at d and f. And would not this be a lightning rod (even according to Mr. Harris's showing) sending off a destructive explosion to semi-
insulated bodies ? against which he protests, saying there is not ' any such
instance in common Electricity, or in the operations of nature.' For be it
remembered that the whole tenour of my papers has been to show that the
lightning discharge pursues the path or paths offering in toto the least resistance,—to show the practical results of ' the well known property which an
electrical charge has, of passing that way where it meets with the least resistance, and consequently of dividing itself, and of preferring a very short passage
through indifferent conductors (or even through the air) to a long passage
through metals.' " Viscount Mahon, *vide* Principles of Electricity, p. 216, § 537.
The reader interested in this intricate question, should now carefully read what
Mr. Harris says on " lateral discharge," and " division of charge," in his Essay,
from page 194—208.

Opportunity may here be taken to correct a little mistake in Fig. 81, in
which the wire O is represented as passing through a glass tube, and thus as
insulated from the disc; this is Mr. Walker's variation of the experiment. In
the original one of Mr. Harris, the wire O was in metallic connexion with the
lower disc, and then no sparks passed between C C and O, when the upper
disc was electrified ; but when the wire O was insulated by passing it through
a glass tube, as shown in the figure, Mr. Walker found that a spark was
visible at O every time one was thrown on C ; but when metallic communication was made by touching the edge of the disc with the wire D, the sparks
ceased. " Now," says Mr. Walker, " Mr. Harris's experiment with the discs
is exactly the condition of a *copper-bottomed* vessel, fitted up with Mr. Harris's
lightning conductors ; and the same reasons which operate in preventing the
spark in the one case, also operate in the other; the metallic sheathing is the
means of communication with the ocean, and, therefore, to the sheathing will
the charge tend. Now the whole area of this sheathing is small—small, that is,

compared with the disc of the earth elsewhere considered; and many perfect metallic paths unite the conductors with this sheathing. If then no division of the flash would pass in our illustrative experiment to O, in *actual metallic connexion* with the lower disc, *à fortiori*, no division would pass to a metallic body in the ship not in *metallic connexion* with the sheathing. 'Tis true that if the path of resistance by means of such a body, were less than that by the conductor, it would pass; but such is physically impossible under the circumstances, for the increasing mass of conductors, as they reach the hull and the many perfect metallic connections which unite them with the sheathing, offer far less resistance, and, therefore, *offer a much more ready transit than any lateral path whatever.* In other cases, the conductors abut upon a disc of water, and are subject to the same laws of comparative existence as the discs on shore." Mr. Walker concludes his Essay with the following handsome eulogium on Mr. Harris's conductors.

" But for *copper-bottomed* vessels, Mr. Harris has devised as good a system of conductors as ingenuity could suggest; at least, such they appear to me. *Humanum est errare*—and I am not exempt from the common lot; yet if a patient and very laborious investigation into the state of our knowlege of the action of lightning, has given me any insight into the character of lightning-rods —if such study has not been in vain, and I have acquired aught of confidence in forming an opinion, and aught of influence in urging that opinion on others, I give my decided testimony in favour of ' The Plan of William Snow Harris, Esq., F.R.S., relating to the Protection of Ships from Lightning.' And should the day come which finds me on board a vessel thus protected, I could offer no better evidence of faith in my opinions, as expressed throughout this memoir, and of conviction of the efficacy of Mr. Harris's conductors, than of free will to make my couch even within the powder-magazine—so assured should I feel of SAFETY."

(*See Par.* 280, 281.)

Through the kindness of Professor Grove, I have been furnished with a copy of his paper on the " Gas Voltaic Battery," read before the Royal Society, May 11th, 1843, and I gladly take this opportunity of presenting to my readers a brief extract of its contents.

In the Professor's first experiments (280) it was found that 26 pairs were the smallest number that would decompose water; the arrangements have, however, subsequently been so much improved, that the same results can now be obtained with 4 cells, and iodide of potassium decomposed with a single one. After detailing some experiments, from the result of which the occasional unequal rise of liquid in the hydrogen tube was referred to local action, Mr. Grove says : " As a general result the *equivalent* action of the battery was beautiful; with fifty cells in action there was but a trifling difference in the rise of liquid in all the cells, and the rise of gas in the voltameter appeared so directly proportional, that an observer unacquainted with the rationale of a voltaic battery, would have said the gases from the exterior cells of the battery were conveyed through the solid wires and evolved in the voltameter, and had this been the first voltaic battery ever invented, this probably would have been the theory of its action."

Further experiments have also shown that the principal points of action in the gas battery are those which Mr. Grove originally believed (468), for when the platinum plates were wholly immersed in the liquid in the tubes, the action was extremely feeble. When a single pair was charged with oxygen and hydrogen, and a second with hydrogen in one tube, the other being filled with dilute sulphuric acid, the hydrogen of the second being connected metallically with the oxygen of the first, and the liquid of the second with the hydrogen of the first, water was decomposed; and when the battery was charged with hydrogen and *nitric acid*, in alternate cells, three cells were found sufficient to decompose water, the gaseous hydrogen deoxidating the nitric acid in the same way as *nascent* hydrogen does in the nitric acid battery (275).

It has been stated (281) that it is the opinion of Mr. Schœnbein that " the oxygen does not immediately contribute to the production of the current in the gas voltaic battery." Mr. Grove has, however, shown that a battery charged with hydrogen and dilute sulphuric acid in the alternate cells, will not act in an atmosphere of nitrogen ; " and," says Mr. Grove, " even if we assume the action of the oxygen to be a *depolarizing* one (281), this comes to the same thing, as this depolarization can only be accounted for as being effected by the combination of the oxygen with the hydrogen ; and we might conversely assume this combination to be the efficient cause of the current, and the depolarization to take place in the hydrogen tubes. It seems to me that the effects at both anode and cathode are reciprocally dependent."

Mr. Grove describes a series of experiments with other gases ; the following is a general account of his results with ten cells charged in series.

Oxygen and protoxide of nitrogen..	No effect on iodide of potassium.
Oxygen and deutoxide of ditto	Very slight, soon ceasing.
Oxygen and olefiant gas..........	Very feeble, but continuous.
Oxygen and carbonic oxide	Notable effects. Slight symptoms of decomposing water.
Oxygen and chlorine	Considerable action at first, scarcely perceptible in 24 hours.
Chlorine and dilute sulphuric acid ..	About the same.
Chlorine and hydrogen	Powerful effects. Two cells decomposing water.
Chlorine and carbonic oxide	Good. Ten cells decomposing water.
Chlorine and olefiant gas	Feeble.

The most interesting practical result of Mr. Grove's experiments on the gas battery, will probably be its application to Eudiometric purposes. " Two narrow cubic inch tubes of seven inches long, were carefully graduated into 100 parts. These were immersed in separate vessels of dilute sulphuric acid, and filled with atmospheric air exactly to the extreme graduation; the water-mark within the tube was examined when exactly at the same level as the exterior surface of the liquid: folds of paper were used to protect them from the warmth of the hands, and thus prevent expansion ; the barometer and thermometer were examined, and every precaution taken for accurate admeasurement. One of these tubes was left empty, in order to ascertain and eliminate from the result, the effect of solubility. Into the other was placed a slip of platinized platinum foil, one quarter of an inch wide. This strip of foil was connected by a platinum wire with another strip placed in a tube of hydrogen, and inserted in the same vessel. After the circuit had been closed for two days, the liquid was found to have risen in the graduated tube 22 parts out of the 100 in the tube placed by its side, it had risen one division. The tubes were allowed to remain several days longer, but no further alteration took place. This analysis gives, therefore, 21 parts in 100 as the amount of oxygen in a given portion of air." In these experiments, it must be observed, that only a single pair of the gas battery can be used, as, if more be employed, the electrolyte is likely to be decomposed, and gas added to the compound.

Another useful application of this interesting battery is the means which it affords of obtaining perfectly pure *nitrogen*. All the oxygen in a given quantity of air may be abstracted, as well as the free oxygen contained in the l quid which confines it, and by subsequently introducing into the tube a little lime water, the trifling quantity of carbonic acid may be removed.

With respect to the theory of the gas battery, Mr. Grove says : " Applying the theory of Grotthus to the gas battery, we may suppose that when the circuit is completed at each point of contact of oxygen, water and platinum in the oxygen tube, a molecule of hydrogen leaves its associated molecule of oxygen to unite with one of the free gas ; the oxygen thus thrown off unites with the hydrogen of the adjoining molecule of water, and so on until the last molecule of oxygen unites with a molecule of the free hydrogen ; or we may conversely assume that the action commences in the hydrogen tube.".... "There are one or two other theoretical points as to which the gas battery offers ground of interesting speculation ; the contact theory is one (468). If my notion of that theory be correct, I am at a loss to know how the action of this battery will be found consistent with it. If, indeed, the contact theory assumes contact as the efficient cause of voltaic action ; but admit that this can only be circulated by chemical action, I see little difference, save in the mere hypothetical expression, between the contact and chemical theories : any conclusion which would flow from the one, would likewise be deducible from the other. There is no observed sequence of time in the phenomena, the contact or completion of the circuit, and the electrolytical action are synchronous. If this be the view of contact theorists, the rival theories are mere disputes about terms : if, however, the contact theory connects with the term contact an idea of force which does or may produce a voltaic current, independently of chemical action, a force without consumption, I cannot but regard it as inconsistent with the whole tenor of voltaic facts and general experience."

In a postscript appended to this paper, bearing date July 7th, Mr. Grove details some further experiments, the theory of which seems at present by no means clear. On repeating the Eudiometrical experiment already described with an apparatus in which the external air was shut out, it was found, after the expiration of three days, that the volume of gas in the air tube which had previously contracted had now *increased* and continued to do so. Mr. Grove at first believed that *nitrogen was decomposed ;* he subsequently, however, found that the increase was due to the addition of *hydrogen,* and that in order to obtain the effect with certainty, two points were essential ; first, the exclusion of any notable quantity of atmospheric air from solution, and secondly, great purity in the hydrogen; it hence becomes necessary, in order to ensure accuracy in Eudiometric experiments, either purposely to use common hydrogen or to employ closed vessels, the tubes of which are long and narrow ; and having first charged the tubes with hydrogen and atmospheric air, to allow these to remain in closed circuit until all the oxygen is abstracted, and a little hydrogen added by the electrolytic effect to the residual nitrogen ; then to substitute oxygen for the original hydrogen, which will in its turn abstract hydrogen from the nitrogen, and leave only *pure* nitrogen. This, Mr. Grove says, he has frequently done with perfect success.

The Professor thinks that the only way at present of accounting for this fact disclosed in these last experiments is, contrary to the views of Dalton, to regard mixed gas as in a state of feeble chemical union, the effect being produced by the affinity of the nitrogen or carbonic acid for the hydrogen ; the affinity of the oxygen of the water, being balanced between the hydrogen in the liquid and that in the tube, would enable the resultant feeble affinity of the nitrogen for hydrogen to prevail. Mr. Grove does not, however, venture a positive opinion ; the fact, as he says, " that gaseous hydrogen should abstract *oxygen from hydrogen* without the latter forming any combination, being so novel, that attempted explanation is likely to prove premature."

(See Par. 348, et. seq.)

Having been obliged to put an end to the experiments with sulphate of barytes and sulphate of strontia, I avail myself of the opportunity afforded by this Appendix to describe briefly the results of these and some other similar experiments ; some of the crystals that have been transferred to the gallipot and glass are exceedingly beautiful and perfect rhomboidal prisms ; they are about the size of the head of a small pin, and adhere firmly to the vessels on which they have been deposited. The experiments have been ten months in progress ; and could they have been continued another year, would doubtless have furnished most satisfactory results. It is a fact especially worthy of remark, that the sulphates of barytes and strontia, particularly the former, are distinguished by their almost total insolubility in water ; water, nevertheless, was the only electrolyte employed ; it consequently appears that a feeble voltaic current possesses the wonderful power of giving solubility to the most insoluble salts in nature, and by a series of decompositions and recompositions of yielding them up again unaltered and in their characteristic crystalline forms. The experiment in which a piece of white marble is under the influence of the voltaic current in a basin of spring water, continues still in action and progresses satisfactorily, though slowly ; the action on the positive marble has been very great : it is now, after ten months' action, cut nearly half through, and the negative marble, and the sides of the glass are covered with myriads of crystals; the carbonate of strontia has likewise transferred a prodigious number of crystals to the negative slate, and to the sides of the basin. I was rather disappointed at perceiving, two or three months since, that many of the crystals that had formed very beautifully on the edges of the slate *nearest* the positive crystal had gradually disappeared ; but I have now much satisfaction in observing that they are re-appearing on the further side of the slate to which they seem to adhere very firmly. It was many months before any crystals were transferred from the carbonate of barytes; they are now, however, making their appearance in numbers scattered over the negative side of the basin, the positive crystal is also beginning to be acted upon, and numerous crystals are forming on its surface.

The experiments relating to the development of acari (352) are still going on; the voltaic current has now been passing through the solutions uninterruptedly for sixteen months, during the whole of which time gas has not ceased to rise plentifully from both electrodes: nearly the whole of the silica appears to be separated from the alcali, and is collected partly round the positive electrodes, but principally in a mass at the bottom of the vessels: the liquid is now quite clear. The sides of the bell glasses are covered with white spots or splashes, as in Mr. Weekes' experiments, these spots consist of assemblages of minute crystals of silica. No acari have as yet been detected within the apparatus, though I have lately found several of these curious little insects *on* and *about* the cells of the battery. On communicating this fact to Mr. Weekes, he informed me that he has had several instances in which the acari were found outside the close vessel sometime before they appeared within, and that he has always traced them to the voltaic apparatus: that an electric current is necessary to their production, seems proved by the fact that within arrangements exactly similar but without a voltaic current, they do *not* appear, as Mr. Weekes has proved in numerous instances. Mr. Crosse also informs me that he has ascertained from long experience that two opposite electric currents will extend over the whole of a large table on which a charged voltaic battery is placed, and more especially when the extremities are connected by only imperfect secondary conductors, such as silicate of potash, &c. In this case a drop of water overset on the table, or any splash, or even moisture would partake of the electrical action.

While on this subject I avail myself of the permission kindly granted by my friend, Mr. Weekes, to notice some of the experiments on which he is at present engaged, and which cannot fail to be read with interest. "In some of my recent experiments connected with this curious inquiry, after swarms of the usual acari had appeared and continued for three or four months, a host of other insects followed, and very shortly after, *all the acari disappeared.* I am not quite certain, but I think they were *eaten* by the new comers, which are altogether of another genus; but as I am not much of an entomologist, I cannot pretend to assign them a proper place in such arrangements. This I know, they are very strange creatures; so swift too, on foot, that I could only secure one specimen, which I have preserved for microscopic examination in a drop of Canada balsam.......The experiment was an *open* one, that is to say, the solution from which the acari were developed, was exposed to the atmosphere under a close screen, and the whole apparatus in perfect darkness. It is possible that the appearance of the creature last mentioned, may be unconnected with electric action in this experiment, but there are various circumstances which lead me to an opposite conclusion."......" My experiments on the decomposition of sugar in solution by means of a voltaic current are progressing beautifully; one experiment has now been going on fourteen months, and the carbon of the sugar is attached in abundance to the positive platinum wire, in a most beautiful foliated form. The solution is placed in a cylindrical glass vessel having a porous bottom of baked earth, through which filtration incessantly goes on. The current proceeds from a water battery of twenty-five pairs, and as it continues to work admirably, I shall let the experiment go on some time longer. The electrodes at the extremities of the platina wires are of hard gas carbon, but they are so much obscured in a dark brownish cloud, that unless I were to disturb the proceeding, I could not describe their present condition."

In a recent communication, Mr. Weekes has favoured me with the details of a remarkable experiment in frictional Electricity which, as it is new (to me at least), I shall here describe.

"A friend at Washington has given me the particulars of a remarkable experiment in frictional Electricity translated by him from a paper by Petrina of Linz. Last night (10th November, 1843), about eleven o'clock, the wind being east, a beautiful and uniform current set in from my atmospheric apparatus, and lasted nearly two hours. This enabled me (probably for the first time under such circumstances) to try the experiment with signal success. Petrina's experiment consists in charging a jar of a foot square external coating from a plate machine of about thirty inches in diameter by means of the inductive power of the flame of a candle, the flame being brought in contact with the knob of the jar, the coating of which was in communication with the earth, but placed at the distance of six or eight feet from the machine, instead of being in contact or nearly so with the prime conductor. It is advisable to remove the flame from the knob of the jar before ceasing to turn the machine, otherwise the flame subsequently acts the part of a discharger. Petrina says, he obtained distinct indications of a charge at the distance of twelve feet from the machine. In my experiment last night by means of an atmospheric current and a two-gallon jar, the knob of which was placed in contact with the flame of a tallow candle (six to the pound) fourteen feet removed from the terminus of my apparatus, *seven spontaneous discharges took place within four minutes.*"

(See Par. 638.)

I am informed by Mr. Tylee, Chemist of Bath, that the terrestrial battery is admirably adapted for electro-plating purposes. The articles covered in this way, (a weak solution of cyanide of silver being employed,) do not, he says, peel off like those plated in a strong solution, and by a Smee's battery, the time required is of course much longer. Mr. Tylee has buried a series of thirty pairs of zinc and copper plates in his garden, and by bringing the wires attached to the terminal plates into his laboratory, he has his electro-plating apparatus constantly ready for action. He informs me that he has never found the slightest variation in the power of his battery (as indicated by the deflection of the magnetic needle) since he first arranged it, nor does it appear to make any difference whether the earth is perfectly dry or saturated with moisture. Mr. Tylee has also communicated to me a very simple and excellent method of depositing a surface of copper on tin medals, which as is well known, become instantly discoloured when immersed in an acid solution. He dissolves 2 oz. of cyanide of potassium in a pint of water, and then adds as much oxide of copper as the solution will take up: the medal to be coated is then made the cathode of a voltaic pair, and immersed in the cupreous solution, a piece of copper forming the anode being likewise immersed; the medal speedily receives a brilliant copper covering, which does not subsequently tarnish, and it can then be used to receive a thick deposit of copper in the ordinary electrotype apparatus.

(See Par. 697.)

The Rev. Mr. Lockey having kindly communicated to me a drawing and description of an ingenious electro-magnetic *signal* or *bell-striking* apparatus of his invention, I feel much pleasure in here giving it publicity. On the base board S, (Fig. 272.) is fixed the electro-magnet c, d, its terminal wires a, b, pass through the base board and lead off to the poles of the exciting battery. The armature e, is prevented from actual contact with the magnetic poles by an interposed piece of leather. The armature is capable of slight angular motion, (*i. e.*, like the lid of a box,) the bar to which it is affixed being centered at f and g. At the end towards f a toothed segment of a circle h is fixed, which, when the armature is attracted by the electro-magnet, traverses in the direction of the arrow near h. The toothed segment acts on the *pinion wheel, i.* On the axis of this pinion, and at right angles to it, is fixed the arm k, carrying the hammer k. The stroke of the hammer is given on the central stem (and interiorly) of the French bell-spring n, n, n, a piece of leather being dove-tailed in the edge of the hammer to give mellowness to the tone, which, if the spring be about five inches diameter, will have the effect of an abbey clock striking at a distance. This spring is firmly fixed by a square piece of brass to a block of wood at o, this block being also firmly attached to the base board.

On the contacts with the battery being completed the armature descends and the hammer is withdrawn. Contact being broken, the hammer stroke is given under the influence of a light spiral watch-spring contained in the box m. By means of a ratchet wheel and catch at the bottom part of the box l, this spring can be slackened and tightened, and thus regulated exactly to the power of the battery used. The ingenious mechanism of this part of the apparatus was contrived by a talented young watch-maker of Bath, Mr. George Wadham. By means of the contact breaker (645) applied to the bell-hammer of an ordinary eight-day house clock, this apparatus will synchronically strike the hours in any apartment of the house in which it can be interposed in a galvanic circuit.

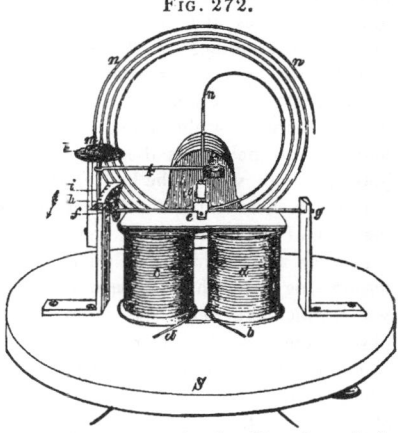

FIG. 272.

(*See Par.* 419.)

Shortly after these experiments (415-420) the directors of the Polytechnic
Institution determined on constructing a machine on a large scale for the
purpose of producing Electricity by the escape of steam, and under the
superintendence of Mr. Armstrong, assisted by Captain Ibbetson, the " Hydro-
Electric Machine" was finished and placed in the theatre of the Institution,
where by its extraordinary power it now daily excites the astonishment
of all who behold it.   The representation in our frontispiece was taken
from the machine itself by the kind permission of the directors—by the aid of
this and the following description, we trust to make it tolerably clear to such
of our readers as have not yet had the opportunity of seeing it and hearing it
more ably described by Professor Bachoffner.   The machine consists of a
cylindrical-shaped boiler, similar in form to a steam-engine boiler, constructed
of iron plate ⅜ in. thick ; its extreme length is 7 feet 6 inches, one foot of which
being occupied by the smoke chamber, makes the actual length of the boiler
only 6 feet 6 inches, its diameter is 3 feet 6 inches.   The furnace and ash-
hole are both within the boiler, as seen in the figure ; when it is required
entirely to exclude the light, a metal screen is readily placed over these ; by
the side of the door is seen the water-gauge and feed-valve.   On the top of
the boiler and running nearly its entire length are forty-six bent iron tubes,
terminating in jets having peculiar shaped apertures and formed of Partridge
wood, which experience has shown Mr. Armstrong to be the best for the purpose ; from these the steam issues—the tubes spring from one common pipe,
which is divided in the middle and communicates with the boiler by two
elbows : by this contrivance the steam is admitted either to the whole or part
of the tubes ; the steam being shut off or admitted by raising or lowering the
two lever handles seen in the front of the boiler.   Between the two elbows is
placed the safety-valve for regulating the pressure, and outside them on one
side is a cap covering a jet employed for illustrating a certain mechanical
action of a jet of steam, and on the other a loaded valve for liberating the
steam when approaching its maximum degree of pressure.   At the further
extremity of the boiler is seen the funnel-pipe or chimney, so contrived that, by
the aid of pullies and a balance weight, the upper part can be raised and made
to slide into itself (similar to a telescope) so as to leave the boiler entirely
insulated.   To prevent as much as possible the radiation of heat, the boiler is
cased in wood, and the whole is supported on six stout glass legs 3½ inches
diameter and three feet long.   In front of the jets, and covering the flue for
conveying away the steam, is placed a long zinc box, in which are fixed four
rows of metallic points for the purpose of collecting the Electricity from the
ejected vapour, and thus prevent its returning to restore the equilibrium of the
boiler—the box is so contrived that it can be drawn out or in, so as to bring
the points nearer or further from the jets of steam ; the mouth or opening can
also be rendered wider or narrower, as seen in our cut : by these contrivances
the power and intensity of the spark is greatly modified.   A ball and socket-
joint, furnished with a long conducting rod, has lately been added to the
machine, so that by its aid the Electricity can be readily conveyed to the
different pieces of apparatus used to exhibit various phenomena.   The pres-

sure at which the machine is usually worked, is 60 lbs. on the square inch. As it is now fully established that the Electricity of the Hydro-Electric Machine is occasioned by the friction of the particles of water (417), the latter may be regarded as the glass plate of the common electrical machine; the partridge wood as the rubber, and the steam as the rubbing power. The Electricity produced by this engine is not so remarkable for its high intensity, as for its enormous quantity. The maximum spark obtained by Mr. Armstrong in the open air was 22 inches; the extreme length under present circumstances has been 12 or 14 inches; but the large battery belonging to the Polytechnic Institution, exposing nearly 80 feet of coated glass, which under favourable circumstances, was charged by the large plate machine 7 feet in diameter, in about 50 seconds, is commonly charged by the Hydro-Electric Engine in 6 or 8 seconds. The sparks which pass between the boiler and a conductor are exceedingly dense in appearance; and especially when short, more resemble the discharge from a coated surface than from a prime conductor. They not only ignite gunpowder, but even inflame paper and wood-shavings when placed in their course between two points. In the 151st number of the Philosophical Magazine, a series of electrolytic experiments made with this machine are described by Mr. Armstrong: true polar decomposition of water was effected in the clearest and most decisive manner, not only in one tube, but in ten different vessels arranged in series and filled respectively with distilled water, water acidified with sulphuric acid, solution of sulphate of soda, tinged blue and red, solution of sulphate of magnesia, &c. &c., and the gases were obtained in sufficient quantities for examination.

The following curious experiments are likewise described :—Two glass vessels containing pure water were connected together by means of wet cotton; on causing the electric current to pass through the glasses, the water rose above its original level in the vessel containing the negative pole, and subsided below it in that which contained the positive pole, indicating the transmission of water in the direction of a current flowing from the positive to the negative wire.

Two wine glasses were then filled nearly to the edge with distilled water, and placed about 4-10ths of an inch from each other, being connected together by a wet silk thread of sufficient length to allow a portion of it to be coiled up in each glass. The negative wire, or that which communicated with the boiler, was inserted in one glass, and the positive wire, or that which communicated with the ground, was placed in the other. The machine being then put in action, the following singular effects presented themselves :—

1st,—A slender column of water, inclosing the silk thread in its centre, was instantly formed between the two glasses, and the silk thread began to move from the negative towards the positive pole, and was quickly all drawn over and deposited in the positive glass.

2nd,—The column of water after this continued for a few seconds suspended between the glasses as before, but without the support of the thread; and when it broke, the Electricity passed in sparks.

3rd,—When one end of the silk thread was made fast in the negative glass, the water diminished in the positive glass and increased in the negative one; showing apparently that the motion of the thread, when free to move, was in the reverse direction of the current of water.

4th,—By scattering some particles of dust upon the surface of the water, it was soon perceived by their motions that there were two opposite currents passing between the glasses, which, judging from the action upon the silk thread in the centre of the column, as well as from other less striking indications, were concluded to be *concentric*, the inner one flowing from negative to positive, and the outer one from positive to negative. Sometimes that which was assumed to be the outer current was not carried over into the negative glass, but trickled down outside of the positive one; and then the water, instead of accumulating as before in the negative glass, diminished both in it and in the positive glass.

5th,—After many unsuccessful attempts, Mr. Armstrong succeeded in causing the water to pass between the glasses, without the intervention of a thread for a period of several minutes, at the end of which time he could not perceive that any material variation had taken place in the quantity of water contained in either glass. It appeared therefore that the two currents were *nearly*, if not *exactly* equal, while the inner one was not retarded by the friction of the thread. Mr. Armstrong likewise succeeded in coating a small silver coin with copper; in deflecting the needle of a galvanometer, between 20° and 30°; and in making an electro-magnet by means of the Electricity from this novel machine.

INDEX.

A

ERRATA.

The Author greatly regrets that the following Errata escaped him whilst correcting the proof sheets.

Page   3,   2 lines from bottom, *for* Réaumer *read* Réaumur.
      10,   18 lines from bottom, *insert* the, *after* with.
      12,   11 lines from top, *read thus* " that the mutual attraction or repulsion of two electrified bodies is directly proportional to the quantity of electricity on the one multiplied by the quantity of electricity on the other, and inversely proportional to the square of the distance between them.
      23,   13 lines from top, *for* acquired *read* acquire.
      23,   17 lines from top, *for* body *read* bodies.
      29,   7 lines from bottom, *for* Electricities *read* Electricity.
      46,   2 lines from top, *for* phenomenon *read* phenomena.
      49,   5 lines from bottom, *for* take *read* takes.
      67,   18 lines from bottom, *for* changed *read* charged.
      95,   3 lines from top, *for* interrupted *read* uninterrupted.
     132,   5 lines from top, *for* Woollaston *read* Wollaston.
     132,   17 lines from top, *for* Boulogue *read* Bologna.
     133,   20 lines from top, *for* gums *read* gum.
     142,   14 lines from top, *for* with ball *read* the pith ball.
     144,   8 lines from top, the statement that no physiological or chemical effects have been obtained from the column is incorrect ; Mr. Gassiot having succeeded in charging a Leyden battery, and in decomposing iodide of potassium, *see* (246).
     190,   17 lines from top, *for* hole *read* pole.
     192,   4 lines from bottom, *erase* the sentence " The wires covered with ropes, were spread upon the platinum, which became red-hot when the electrical current traversed it."
     199,   17 lines from top, *for* than even *read* that over.
     199,   14 lines from bottom, *for* particles *read* particle.
     231,   13 lines from top, *for* Cyanogon *read* Cyanogen.
     236,   4 lines from bottom, *for* soldering *read* silvering.
     254,   10 lines from top, *for* is *read* are.
     280,   13 lines from top, *for* quality *read* equality.
     290,   5 lines from bottom, *for* 9·36 *read* 1·36.
     302,   15 lines from bottom, *for* were *read* was.
     339,   7 lines from top, *for* Ironmonger's *read* Iremonger's.
     381,   20 lines from top, *for* are *read* is.
     393,   13 lines from bottom, *for* currents *read* current.
     402,   17 lines from top, *for* sulphate *read* sulphite.

In Fig. 9, Page 16, the condensing plate is shown as fixed to the cap of the instrument, instead of which it ought to be attached to the upper disk, and the other plate raised to a corresponding height.

London : STEWART and MURRAY, Green Arbour Court, Old Bailey.